OXFORD SERIES IN OPTICAL AND IMAGING SCIENCES

OXFORD SERIES IN OPTICAL AND IMAGING SCIENCES

EDITORS
Akira Hasegawa
Marshall Lapp
Benjamin B. Snavely

Henry Stark
Andrew C. Tam
Tony Wilson

1. D. M. Lubman (ed.). *Lasers and Mass Spectrometry*
2. D. Sarid. *Scanning Force Microscopy With Applications to Electric, Magnetic, and Atomic Forces*
3. A. B. Schvartsburg. *Non-linear Pulses in Integrated and Waveguide Optics*
4. C. J. Chen. *Introduction to Scanning Tunneling Microscopy*
5. D. Sarid. *Scanning Force Microscopy With Applications to Electric, Magnetic, and Atomic Forces, revised edition*
6. S. Mukamel. *Principles of Nonlinear Optical Spectroscopy*
7. A. Hasegawa and Y. Kodama. *Solitons in Optical Communications*
8. T. Tani. *Photographic Sensitivity: Theory and Mechanisms*
9. D. Joy. *Monte Carlo Modeling for Electron Microscopy and Microanalysis*
10. A. B. Shvartsburg. *Time-Domain Optics of Ultrashort Waveforms*
11. L. Solymar, D. J. Webb, and A. Grunnet-Jepsen. *The Physics and Applications of Photorefractive Materials*

The Physics and Applications of Photorefractive Materials

L. SOLYMAR
Department of Engineering Science
University of Oxford

D. J. WEBB
Physics Laboratory, University of Kent

A. GRUNNET-JEPSEN
Department of Chemistry and Biochemistry
University of California, San Diego

CLARENDON PRESS · OXFORD
1996

Oxford University Press, Walton Street, Oxford OX2 6DP

Oxford New York
Athens Auckland Bangkok Bombay
Calcutta Cape Town Dar es Salaam Delhi
Florence Hong Kong Istanbul Karachi
Kuala Lumpur Madras Madrid Melbourne
Mexico City Nairobi Paris Singapore
Taipei Tokyo Toronto

and associated companies in
Berlin Ibadan

Oxford is a trade mark of Oxford University Press

Published in the United States
by Oxford University Press Inc., New York

© L. Solymar, D. J. Webb, and A. Grunnet-Jepsen, 1996

All rights reserved. No part of this publication may be
reproduced, stored in a retrieval system, or transmitted, in any
form or by any means, without the prior permission in writing of Oxford
University Press. Within the UK, exceptions are allowed in respect of any
fair dealing for the purpose of research or private study, or criticism or
review, as permitted under the Copyright, Designs and Patents Act, 1988, or
in the case of reprographic reproduction in accordance with the terms of
licences issued by the Copyright Licensing Agency. Enquiries concerning
reproduction outside those terms and in other countries should be sent to
the Rights Department, Oxford University Press, at the address above.

This book is sold subject to the condition that it shall not,
by way of trade or otherwise, be lent, re-sold, hired out, or otherwise
circulated without the publisher's prior consent in any form of binding
or cover other than that in which it is published and without a similar
condition including this condition being imposed
on the subsequent purchaser.

A catalogue record for this book is available from the British Library

Library of Congress Cataloging in Publication Data
Solymar, L. (Laszlo)
The physics and applications of photorefractive materials/
L. Solymar, D. J. Webb, A. Grunnet-Jepsen.
(Oxford series in optical and imaging sciences; 11)
Includes bibliographical references and index.
1. Electrooptical devices–Materials.   2. Photorefractive
materials.   3. Image processing.   I. Webb, D. J.   II. Grunnet
-Jepsen, A.   III. Title.   IV. Series.
TA1750.S67   1996   537.5'4–dc20   95-48216
ISBN 0 19 856501 1

Typeset by Keyword Typesetting Services, Wallington, Surrey
Printed in Great Britain by
Bookcraft (Bath) Ltd
Midsomer Norton, Avon

PREFACE

The properties of photorefractive materials have been studied with great interest, and one might say with great enthusiasm, for two decades at least. The number of papers written on the subject is over two thousand, there are a few books available which contain selections of papers, a few books in which each chapter is written by another author, there is one introductory book, and there are a large number of review papers looking at various aspects of the phenomena. To our knowledge no book has been written with the aim of describing and evaluating all the major effects and giving an account of all the major applications. Perhaps such an aim is over-ambitious: one may argue that the range of phenomena is so wide that any attempt at comprehensive coverage is bound to fail. Have we aimed at comprehensive coverage? Nearly. There was only one topic, namely phase conjugation, which we did not think we could adequately cover in the space available, so we disposed of it in two brief sections, one for forward phase conjugation and one for the reverse variety. Otherwise we did make the effort. It is for the reader to judge how well we achieved our aim.

The next question is how to organize the material. Can we be comprehensive and comprehensible at the same time? Can we *include* everything and can we *explain* everything? Well, one way of doing it is to separate the two things. Part I tries to be comprehensible whereas Part II tries to be comprehensive. Part I is designed as an introduction to the photorefractive field. There is emphasis on the physics accompanied by simple mathematical models. It is a bird's eye view of the whole landscape which shows the major features but does not dwell on the details. Part II covers a wider field, so necessarily the explanations are more concise. There is more detail there but at the same time more references to the literature. It is intended to be comprehensive.

Part III is concerned with applications. The aim is always to explain why photorefractive materials are suitable for that particular application, give the results achieved, and possibly finish with the state of the art. Obviously, more space is devoted to applications with higher potential but attempts have been made to include those with more modest immediate prospects, based on the principle that one never knows which application will turn out to be really important.

Our list of references is long but far from being exhaustive. As mentioned before, the total number of papers written about photorefractives is over 2000. At no stage in writing this book did we ever believe that we would be able to refer to all of them. We had to be selective. We ended up with a mere 900. It is quite possible that we did not include some important papers. To those who find their favourite paper not included we can only offer our apologies. It is not a value judgement. It very often happens when one follows a particular line of argument that some papers just do not fit in. They would represent too much of a digression for which there is no adequate space. If we create the impression that we omitted too many papers we would like to note that we had the courage to omit about 20 of our own papers as well.

It is customary to say at this stage what audience the authors had in mind. In many a preface the claim is made that the book is written for the widest spectrum possible, from beginners to experts, from research students who have just entered the field to researchers who grew old during the decades devoted to those particular studies. We would like to make the same claim. We believe that any research student entering the field will find a gentle introduction with no more than a modicum of references in Part I. For the more advanced student Part I might serve as a reminder that there are other topics too besides his/her favourite one. For the expert Part I should be suitable for bedtime reading. Part II is certainly not for the beginner and it is unlikely that even advanced students might want to read through it section by section. We hope though that our attempt at ordering the material will help the advanced student and our list of references should reduce the time spent on searching the literature. What will the expert make of Part II? He/she will be interested to see the impact of his/her own work: how it is introduced, how it is presented, how comprehensive the coverage is, and how representative the references are. Part III, we hope, will appeal to most of those in the field. The beginner will find information, the advanced student may find inspiration (Isn't this device similar to the one I invented yesterday? If I modify it just a little . . .) and the expert might still want to skip through it just to see whether he/she has missed anything.

Acknowledgements should first go to our wives, Marianne, Jane, and Helene, who were willing to put up with the long hours we spent on this book. Our thanks must also go to colleagues who made some crucial experiments, who pointed out inconsistencies in theories, who appeared in the morning waving a new set of calculations, in fact to all those who helped us to form the ideas jotted down in the next few hundred pages. It was our privilege to work with Bun Au, Ivo Aubrecht, Hannah Ellin, Steve Elston, Don Erbschloe, Grahame Faulkner, John Heaton, Per Johansen, David Jones, Malgosia Kaczmarek, Victor Kalinin, Chong-Hoon Kwak, Serguey Lyuksyutov, Toby McClelland, Ted Paige, Chris Raymond, Ivan Richter, Klaus Ringhofer, Mike Schaub, Misha Shamonin, Boris Sturman, Jeno Takacs, Sarah Tao, and Tony Wilson.

Oxford	L. S.
Canterbury	D. J. W.
San Diego	A. G.-J.

September 1995

CONTENTS

PART I A BIRD'S EYE VIEW OF PHOTOREFRACTIVE PHENOMENA ... 1

General introduction ... 3

1 Diffraction phenomena by static and dynamic volume gratings ... 7
1.1 Introduction ... 7
1.2 Static volume gratings ... 7
1.3 The basic phenomena ... 16
    1.3.1 The creation of the grating ... 16
    1.3.2 Grating diffraction ... 20

2 How light influences material properties ... 22
2.1 The basic differential equations ... 22
2.2 Linearization ... 24
2.3 Solution of the linearized equations ... 28
2.4 The photovoltaic effect ... 38
2.5 Characteristic lengths and characteristic times ... 41
2.6 Figures of merit ... 43
2.7 The very short time limit ... 47
2.8 Enhancement mechanisms ... 48
2.9 Large modulation effects ... 55
2.10 External circuit current ... 60
2.11 Electrons and holes ... 62

3 How material properties influence light ... 66
3.1 The change in the dielectric constant ... 66
3.2 The wave equation ... 67
3.3 Beam coupling: derivation of the coupled wave equations ... 68
3.4 Solution of the coupled wave equations ... 72
3.5 Direction of power transfer ... 74
3.6 Reflection gratings ... 76
3.7 Diffraction efficiency ... 80
3.8 Higher diffraction orders ... 84
3.9 Transients for optical beams ... 86
3.10 Temporal modulation ... 89
3.11 Scattering and resonators ... 92
3.12 Phase conjugation ... 93

PART II DISCUSSION OF THE PHYSICS 99

General Introduction 101

4. A history of the photorefractive effect 103
- 4.1 Introduction 105
- 4.2 Early history 106
- 4.3 The emergence of a new set of equations, and some related thoughts 106
- 4.4 Modern history 109
- 4.5 A review of the reviews 112
- 4.6 Some statistics 113

5 The materials equations 117
- 5.1 Introduction 117
- 5.2 Models and equations 121
  - 5.2.1 Standard model 124
  - 5.2.2 Bipolar transport 127
  - 5.2.3 Two or more impurity species 130
  - 5.2.4 Shallow trap models and the influence of temperature 135
  - 5.2.5 Electron–ion 'fixing' model 144
  - 5.2.6 Two-photon writing model 145
  - 5.2.7 Quantum well models 146
  - 5.2.8 Trapless photorefractive model 148
  - 5.2.9 Charge exchange between species (trap intercommunication) 149
  - 5.2.10 Valley's second-order differential equation 151
- 5.3 Space charge waves 151
  - 5.3.1 Introduction 151
  - 5.3.2 Derivation of the dispersion equation 153
  - 5.3.3 Effect of holes 157
- 5.4 Time-varying interference patterns and applied voltages 157
  - 5.4.1 Introduction 157
  - 5.4.2 DC applied field plus detuning: low frequency resonance 157
  - 5.4.3 DC field: variation of space charge field with intensity 159
  - 5.4.4 DC field plus detuning: bipolar low frequency resonance 159
  - 5.4.5 DC field plus detuning: high frequency resonance 160
  - 5.4.6 AC applied field 161
  - 5.4.7 Detuning plus AC applied field 168
  - 5.4.8 External circuit current due to phase modulation 169
- 5.5 Optical pulses and high powers 170
  - 5.5.1 Introduction 170
  - 5.5.2 Predictions of the standard model 170
  - 5.5.3 More complex models 173
  - 5.5.4 Non-photorefractive effects 179
- 5.6 Other topics 182

		5.6.1 Large modulation solutions	182
		5.6.2 Equivalent circuits	184

6 The field equations for two-wave mixing — 187
    6.1 Introduction — 187
    6.2 Derivation of the coupled wave equations — 188
    6.3 Further simplifications — 192
    6.4 Crystal orientation — 193
    6.5 Jones vector description — 194
    6.6 Degenerate two-wave mixing in the diffusion regime — 197
    6.7 Space charge field enhancement — 201
    6.8 Limitations of the simple theory — 207
    6.9 Diffraction efficiency — 215
    6.10 The piezoelectric effect — 226
    6.11 The interaction between the space charge field and the optical field — 230
        6.11.1 An analytical solution — 230
        6.11.2 Numerical solution — 232

7 Multi-wave mixing — 236
    7.1 Introduction — 236
    7.2 Forward three-wave mixing (I) — 236
    7.3 Forward three-wave mixing (II) — 245
    7.4 Forward four-wave mixing — 247
    7.5 Forward phase conjugation — 251
    7.6 Phase conjugation — 253
        7.6.1 Four-wave mixing — 256
        7.6.2 Self-pumped phase conjugate resonators — 260
        7.6.3 Mutually pumped phase conjugators — 264

8 Spurious beams — 268
    8.1 Introduction — 268
    8.2 Scattering — 269
        8.2.1 Introduction — 269
        8.2.2 Beam distortion due to non-uniform transverse intensity distribution — 269
        8.2.3 Scattering due to two-wave amplification — 270
        8.2.4 Bragg diffraction from noise gratings — 270
        8.2.5 Combination of Bragg diffraction and two-wave mixing — 272
        8.2.6 Backward four-wave mixing — 273
        8.2.7 Anisotropic scattering — 274
        8.2.8 Further notes on scattering — 278
        8.2.9 Noise reduction techniques — 279
    8.3 Subharmonic instabilities — 281
        8.3.1 Introduction — 281
        8.3.2 Theory and experiment — 282

x   *Contents*

       8.3.3   Analogies   286
       8.3.4   Subharmonic space charge waves   287
       8.3.5   Other mechanisms for subharmonic generation   288
       8.3.6   Subharmonic domains   290
       8.3.7   Competition effects   291
   8.4   Spatio-temporal instabilities   293

9   Six topics in search of a chapter   297
   9.1   Introduction   297
   9.2   The photovoltaic effect   298
   9.3   Photorefractive polymers   302
   9.4   Band-edge photorefractivity   305
   9.5   Quantum well structures   306
   9.6   Stratified holographic optical elements   309
   9.7   Solitons   309

PART III   APPLICATIONS   313

General Introduction   315

10   Image amplification and image processing   317
   10.1   Introduction   317
   10.2   Image amplification   319
       10.2.1   Scattered light and beam fanning   320
       10.2.2   Pump beam depletion   322
       10.2.3   Modulation transfer function   325
   10.3   Image processing   328
       10.3.1   Image thresholding   328
       10.3.2   Edge enhancement   329
       10.3.3   Novelty filter   332
       10.3.4   Spatial light modulation   336
       10.3.5   Holographic interferometry   339

11   Correlation and associative memories   342
   11.1   Introduction   342
   11.2   Correlation and convolution   342
   11.3   Material response   344
   11.4   Historical background   349
   11.5   Departures from ideality in the correlation process   353
   11.6   Shift, scale, and rotation invariance   359
   11.7   Efficiency   362
   11.8   Associative memories   365

12 Storage — 372
- 12.1 Introduction — 372
- 12.2 Early work — 372
- 12.3 Hologram fixing — 374
  - 12.3.1 Thermal fixing — 374
  - 12.3.2 Electrical fixing — 378
  - 12.3.3 Two-photon storage — 381
  - 12.3.4 Two-wavelength storage — 382
  - 12.3.5 Refreshed memories — 384
  - 12.3.6 Storage–amplification scheme — 385
- 12.4 Hologram multiplexing — 385
  - 12.4.1 Spatial multiplexing — 386
  - 12.4.2 Angular multiplexing — 386
  - 12.4.3 Wavelength multiplexing — 390
  - 12.4.4 Phase encoding — 391
  - 12.4.5 Selective erasure — 394
  - 12.4.6 Recording schedules — 394
  - 12.4.7 Storage capacity — 399

13 Other applications — 400
- 13.1 Introduction — 400
- 13.2 Narrow band interference filter — 400
- 13.3 Reconfigurable array interconnection — 400
- 13.4 Adaptive interferometry — 403
- 13.5 Self-organizing optical circuits — 405
- 13.6 Some further applications — 406

Appendix A   Wave propagation in anisotropic materials — 409

Appendix B   The electro-optic effect, including the piezoelectric contribution — 416

Appendix C   Determination of photorefractive parameters — 427
- C.1 Introduction — 427
- C.2 Mobility, lifetime, electron–hole competition factor, and effective trap density — 428
- C.3 Sign of carriers — 435
- C.4 Electro-optic coefficient — 436
- C.5 The elastic, elasto-optic, dielectric, and piezoelectric properties — 440

Bibliography — 444

Index — 489

PART I
A bird's eye view of photorefractive phenomena

GENERAL INTRODUCTION

As mentioned in the Preface, the aim of Part I is to explain a wide range of photorefractive phenomena in the simplest possible terms. We shall always try to emphasize the underlying physical principles, use relatively simple mathematical models, plot lots of curves, and give the more important references.

The first problem is where to start. What can we assume to be known? How can we match the book to the existing knowledge of the expected audience? To be concrete, should we assume in the present case that the audience is familiar with the basic tenets of optics? With wave propagation in anisotropic materials? With basic semiconductor theory on donors and acceptors, with the generation and recombination of mobile carriers? With the theory of static gratings? To choose the starting point is not a trivial matter, and we do not wish to pretend that we have got it right for everybody. Some of the topics will be assumed to be known like basic optics or semiconductor theory; some other topics, e.g. wave propagation in anisotropic materials, will be relegated to appendices. Some other topics, in particular the theory of static volume gratings, will be treated in Chapter 1. Why that particular topic? Because if we wish to understand the properties of photorefractive gratings, which belong to the dynamic variety, we need to be familiar with static gratings first.

In addition, Chapter 1 will introduce the physics of photorefractive materials. We shall try to make it plausible why gratings are generated when a photorefractive material is illuminated by two optical waves. We shall also try to say something, on purely physical grounds, about the various stages of the appearance of the grating, how the various quantities, like electron concentration, ionized donor concentration, space charge electric field, develop with time.

One of the interesting properties of photorefractive materials, perhaps the most interesting one, is that a phase shift is needed between the imposed interference pattern and the spatial variation of the dielectric constant in order to have gain where by gain we mean the amplification of one of the input beams at the expense of the other one. We shall show that this property, in some sense, is shared with static gratings.

Chapter 2 is concerned with the response of the material to optical excitation in the form of an input interference pattern. In Section 2.1 we shall discuss the so-called band transport model. The physical arguments put in mathematical form yield then the band transport equations, which consist of a set of partial differential equations. Under some quite reasonable-looking assumptions (the most important among them being that the modulation of the incident intensity pattern is small) the differential equations are linearized in Section 2.2. The solution for the most important physical quantity, the space charge electric field, is given in Section 2.3. The next Section is concerned with an interesting effect manifested by the appearance of a current in response to an input optical beam, known as the photovoltaic effect. We shall give there the relevant equations and the linearized solution for the space charge electric field.

The results can be presented in terms of different sets of parameters. In general, our parameters will be the characteristic electric fields which are dependent on the grating spacing. It is of course possible to describe the same phenomena in terms of some other set of parameters. In addition to the parameters known for a long time in physics, like Debye length and diffusion length, we introduce in Section 2.5 the drift length and photovoltaic length favoured by some of the workers in the field. Section 2.6 is devoted to some figures of merit which may be used to compare the different photorefractive materials.

The photorefractive effect depends on the optical excitation of charge carriers and that takes time. It is obvious therefore that photorefractive phenomena occur faster at higher input intensities. This intensity-dependent behaviour is discussed briefly in Section 2.7.

There are two methods for enhancing the space charge electric field. It can be done by detuning slightly one of the input beams or by applying a time-varying electric field to the photorefractive crystal. This will be the subject of Section 2.8.

The small modulation assumption, and the subsequent linearization, discussed in Section 2.2 have quite general validity but they do not always apply and large modulation effects must be taken into account. This is shown in Section 2.9.

When the interference pattern is moving, a current appears in the external circuit as will be discussed in Section 2.10.

Finally we shall acknowledge the fact that mobile charge carriers may be present with both negative and positive polarities, i.e. holes can just as well contribute to the photorefractive effect as electrons. Their contributions are usually in the opposite directions leading to a diminished photorefractive effect. A brief discussion of this so-called electron–hole competition will be done in Section 2.11.

In Chapter 2 we discussed the effect of the input beams on the material. Chapter 3 is concerned with the inverse effect, namely the influence of the changing material properties on the optical beams. It is essentially about the interaction between the optical fields. The analysis starts in Section 3.1 with the variation of the dielectric constant, which is put in the wave equation in Section 3.2, yielding the coupled wave equations in Section 3.3. The differential equations are solved in Section 3.4 leading to transfer of power in the beams. The direction of this power transfer is discussed in Section 3.5.

Most of the devices proposed, most of the experiments performed, and most of the calculations done are for the transmission case when both beams are incident from the same side of the photorefractive material. When the beams are incident from the opposite side they generate reflection gratings, the subject of Section 3.6.

The gratings generated can of course be used for diffracting another beam with a certain diffraction efficiency. The relevant equations are derived and evaluated in Section 3.7. If the interbeam angle between the beams is small, then grating diffraction may give rise to multiple beams, the so-called higher orders. This is discussed in Section 3.8.

The time variation of the various physical quantities like ionized donor density or space charge electric field can of course be calculated on the basis of the materials equations alone, and were indeed calculated in Section 2.3. However, the spatial variation of the optical beams will also have an effect introducing some corrections. Mathematically this means that the materials and field equations need to be combined. Some brief observations on these combined effects will be made in Section 3.9. The input beam can of course be modulated in time and it may be shown that the modulation will also be amplified. Some examples will be shown in Section 3.10.

Imperfections in photorefractive materials lead to scattered radiation which may be selectively amplified in a certain direction. The effect can be utilized in resonators to be discussed in Section 3.11. The last section of Chapter 3 (Section 3.12) is concerned with one of the most interesting topics, phase conjugation, for which photorefractive materials are particularly suitable.

1
DIFFRACTION PHENOMENA BY STATIC AND DYNAMIC VOLUME GRATINGS

1.1 Introduction

Diffraction gratings have a long and distinguished history. They have been used for a century and a half on account of their abilities to resolve the various spectral components of light. What are they? Any periodic structure will act as a diffraction grating. Up to quite recently they were produced mechanically by ruling grooves on a surface. Nowadays the holographic variety, produced by the interference of two waves, is more in vogue. Holographic gratings can also be divided into two big families: thin gratings and thick or volume gratings. A thin grating is produced for example by exposing and etching a photoresist. The aim there is to achieve a surface modulation. Our interest is in the other variety. Volume gratings, as the name implies, represent a periodic variation in some property of a chunk of material. Most often it is the dielectric constant which undergoes periodic variation. The technique of producing such a grating is illumination by an interference pattern followed by some processing. It needs to be mentioned however that not all volume gratings are man-made. Nature also provides them in the form of single-crystal materials in which the atoms follow a regular pattern. They diffract electromagnetic waves of rather short wavelength which are in the X-ray region. The discipline is known as X-ray crystallography.

Our aim is to study photorefractive gratings. They can be classified as dynamic gratings, meaning that the characteristics of the grating may vary as a function of time. The diffracted light will actually affect the properties of the grating. Static gratings, in contrast, are independent of the input light. They are linear, and their properties are not a function of time. They will be studied in the next section.

1.2 Static volume gratings

In this section we shall assume that we are dealing with a static grating, a grating which is there for ever. It consists of a slab of material in which the relative dielectric constant, ε_r, varies as a function of z in the form

$$\varepsilon_r = \varepsilon_{ro} + \varepsilon_{r1} \cos Kz \tag{1.1}$$

where ε_{ro} is the background dielectric constant, ε_{r1} is the amplitude of the dielectric grating, and K is the spatial frequency of the variation, related to the spatial period (called the grating spacing) as $\Lambda = 2\pi/K$.

There are of course many ways of producing diffraction gratings. A periodic variation of the dielectric constant is one of them, the subject of the present section. What matters is that the grating should be able to diffract an incident wave, and dielectric gratings in a volume of material (hence the name of volume gratings) are far the most efficient representatives of the class. They are the most efficient in the sense that under certain conditions the incident beam can be completely extinguished, and its intensity converted into that of the diffracted wave.

We are now going to investigate the case when a wave is incident upon the material at an angle θ as shown in Fig. 1.1(a)[1]. One can solve the problem without major difficulties for any angle of incidence. We shall however restrict generality here and assume that the incident wave satisfies the condition for cumulative interactions, known as the Bragg condition. We shall present this condition in the form of a vector triangle, shown in Fig. 1.1(b) where $\mathbf{k}_1$ and $\mathbf{k}_2$ are the input wave vectors and $\mathbf{K} = K\mathbf{i}_z$ is the so-called grating vector ($\mathbf{i}_z$ is the unit vector in the z direction). The two wave vectors may be written as

$$\mathbf{k}_1 = k(\mathbf{i}_x \cos\theta + \mathbf{i}_z \sin\theta), \quad \mathbf{k}_2 = k(\mathbf{i}_x \cos\theta - \mathbf{i}_z \sin\theta) \tag{1.2}$$

and $k = 2\pi/\lambda$ where λ is the wavelength in the medium. The Bragg condition may then be written in the form

$$\mathbf{k}_1 - \mathbf{k}_2 = \mathbf{K} \tag{1.3}$$

which may easily be seen to be equivalent with the more often quoted form

$$\Lambda = \frac{\lambda}{2\sin\theta} \tag{1.4}$$

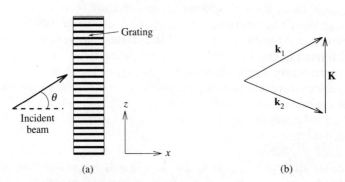

Fig. 1.1 (a) An optical beam incident on a material with a periodic variation of the dielectric constant. (b) The dielectric grating of wave vector $\mathbf{K}$ will diffract the incident beam of wave vector $\mathbf{k}_1$ in the direction $\mathbf{k}_2$ such that the Bragg condition, eqn (1.3), is satisfied.

---

[1] For simplicity we shall disregard refraction here so that θ may be regarded as the incident angle in the material *after* the refraction has taken place. Alternatively, one may think of it (e.g. Kogelnik 1969) as if the wave were incident from a material of the *same* dielectric constant.

1.2 Static volume gratings

Next, we shall embark upon the derivation of the coupled wave differential equations. They have a distinguished history. Although there have been quite similar formalisms in X-ray diffraction theory it is usual to date coupled wave equations from the work of Raman and Nath (1935). Notable continuations were given by Phariseau (1956) and W. R. Klein and Cook (1967) but the study known most widely is that of Kogelnik (1969), which is thorough, methodical, and very well presented. We shall follow it here in a very much simplified form. Any reader interested in more details should consult the original article. For wider problems of volume holography see Solymar and Cooke (1981) and Syms (1990).

The equation to solve is the wave equation, which we shall present in the well known scalar form for a temporal variation $\exp(j\omega t)$ as

$$\nabla^2 \mathcal{E} + \omega^2 \mu_o \varepsilon_o \varepsilon_r \mathcal{E} = 0 \tag{1.5}$$

where ∇ is now a two-dimensional vector operator, $\nabla = \dfrac{\partial}{\partial x} \mathbf{i}_x + \dfrac{\partial}{\partial z} \mathbf{i}_z$ (we shall assume that there is no change in the direction of the y coordinate so we do not need the third dimension), $\mathcal{E}$ is the optical electric field, μ_o and ε_o are the free space permeability (the material is assumed to be non-magnetic) and permittivity respectively, and ϵ_r is given by eqn. (1.1).

The crucial assumption is that the electric field may be written in the form of two plane waves,

$$\mathcal{E} = R(x) \exp(-jkp_1) + S(x) \exp(-jkp_2) \tag{1.6}$$

where

$$p_1 = x\cos\theta + z\sin\theta \quad \text{and} \quad p_2 = x\cos\theta - z\sin\theta \tag{1.7}$$

R and S are (using Kogelnik's notation) the reference wave and subject wave respectively. When the grating is illuminated by the reference wave it will diffract it into the subject wave. The diffraction will take place gradually, i.e. the amplitude of the subject wave will slowly increase from zero, and simultaneously, the amplitude of the reference wave will decline from its initial value. The changes in amplitudes are assumed to take place in one dimension only (i.e. x direction), which is in accordance with the plane wave assumption.

Next, we need to observe that with our new notations Kz in eqn (1.1) may be written as

$$Kz = k(p_1 - p_2) \tag{1.8}$$

The wave equation takes then the form

$$\frac{\partial^2 \mathcal{E}}{\partial x^2} + \frac{\partial^2 \mathcal{E}}{\partial z^2} + k^2 \left[1 + \frac{\varepsilon_{r1}}{\varepsilon_{ro}} \cos k(p_1 - p_2)\right] \mathcal{E} = 0 \tag{1.9}$$

Substituting eqn (1.6) into (1.9) yields

$$\exp(-jkp_1)\left[\frac{d^2R}{dx^2} - 2jk\cos\theta\frac{dR}{dx} - k^2R\right] + \exp(-jkp_2)\left[\frac{d^2S}{dx^2} - 2jk\cos\theta\frac{dS}{dx} - k^2S\right]$$
$$+ k^2\left\{1 + \frac{\varepsilon_{r1}}{2\varepsilon_{ro}}[\exp jk(p_1 - p_2) + \exp jk(p_2 - p_1)]\right\}\{R\exp(-jkp_1)$$
$$+ S\exp(-jkp_2)\} = 0 \qquad (1.10)$$

where terms in

$$k^2 R\exp(-jkp_1) + k^2 S\exp(-jkp_2) \qquad (1.11)$$

may be seen to cancel. Our next approximation is to neglect the second derivatives. This is equivalent to the assumption that the energy exchange between the two waves occurs on a large enough scale relative to the wavelength. The approximation is also known as the slowly varying envelope approximation, useful in other branches of physics too.

Let us concentrate now on the product

$$\{\exp jk(p_1 - p_2) + \exp jk(p_2 - p_1)\}\{R\exp(-jkp_1) + S\exp(-jkp_2)\} \qquad (1.12)$$

There will altogether be four terms with exponents

$$-jkp_2, \quad jk(p_2 - 2p_1), \quad jk(p_1 - 2p_2) \quad \text{and} \quad -jkp_1 \qquad (1.13)$$

Out of these, $jk(p_1 - 2p_2)$ and $jk(p_2 - 2p_1)$ represent waves travelling in directions other than θ. Further scrutiny would show them to represent higher diffraction orders. They matter in 'thin' holograms but not in volume holograms. We are going to neglect them. Having done so, we may satisfy eqn (1.10) by requiring that the coefficients of $\exp(-jkp_1)$ and $\exp(-jkp_2)$ both vanish. The resulting two differential equations are

$$-2jk\cos\theta\frac{dR}{dx} + k^2\frac{\varepsilon_{r1}}{2\varepsilon_{ro}}S = 0 \qquad (1.14)$$

and

$$-2jk\cos\theta\frac{dS}{dx} + k^2\frac{\varepsilon_{r1}}{2\varepsilon_{ro}}R = 0 \qquad (1.15)$$

Introducing the coupling constant

$$\kappa = \frac{\varepsilon_{r1}k}{4\varepsilon_{ro}\cos\theta} \qquad (1.16)$$

we have finally the coupled wave differential equations

$$\frac{dR}{dx} + j\kappa S = 0 \quad \text{and} \quad \frac{dS}{dx} + j\kappa R = 0 \qquad (1.17)$$

The boundary conditions are

$$R(0) = 1 \quad \text{and} \quad S(0) = 0 \qquad (1.18)$$

1.2 Static volume gratings

corresponding to the fact that the incident (reference) wave has an amplitude of unity and the diffracted (subject) wave is zero at the input to the grating. The solution of eqn (1.17) subject to the boundary conditions (1.18) is

$$R = \cos \kappa x, \quad S = -j \sin \kappa x \tag{1.19}$$

Intensity is proportional to the modulus square of the amplitude. They are plotted for both waves in Fig. 1.2 against κx, normalized distance in the grating. It may be seen that the complete incident power can be converted into that of the diffracted wave. It is also interesting to see that the grating amplitude can be 'too high' in the sense that for $\kappa x > \pi/2$ the intensity starts to be reconverted from the diffracted wave into the original reference wave. Such 'overmodulation' may indeed occur in a photorefractive material, as will be discussed in Section 6.10.

The usual aim is to convert as much intensity as possible into the diffracted wave. The measure is the diffraction efficiency, which we shall denote by η_d. It is defined as the diffracted intensity relative to the incident intensity. From eqn (1.19) we obtain

$$\eta_d = \frac{|S|^2}{|R|^2 + |S|^2} = |S|^2 = \sin^2 \kappa x \tag{1.20}$$

In fact, with our choice of unit input amplitude the intensity of the diffracted wave is equal to the diffraction efficiency.

For many applications a quite small diffraction efficiency is adequate, in which case the sinusoidal in eqn (1.20) may be replaced by its argument, yielding the approximate formula

$$\eta_d = (\kappa x)^2 \tag{1.21}$$

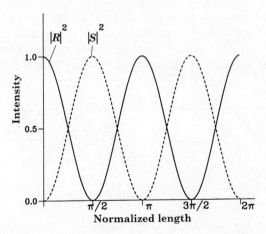

Fig. 1.2 Intensity of incident (Reference) and diffracted (Subject) beams as a function of normalized thickness of a volume grating.

So far we have considered the angle of incidence to be such as to satisfy the Bragg conditions. This is of course not true in general. Volume holograms are very often used (say, in multiple storage or in a channel dropping filter) at wavelengths and incident angles away from the Bragg conditions. In the usual jargon one refers to this as the off-Bragg case. The result is reduced efficiency which quickly approaches zero. The corresponding mathematics is quite straightforward, a generalization of the approach given above. We shall not do it here as it would offer no new physical insight. Instead, we shall give here an approximate derivation based on simple geometrical arguments, and we shall add to it later a discussion on the definition of the so-called off-Bragg parameter.

Let us take one period out of the grating as shown in Fig. 1.3(a). A ray is incident at point A at an angle θ. Part of it is diffracted at the same angle towards point C, the other part continues undisturbed to B where further diffraction takes place. The two diffracted rays (originating from A and B respectively) are in phase provided the path difference $\overline{AB} - \overline{AC}$ (where $\overline{BC}$ is perpendicular to $\overline{AC}$) is equal to one wavelength in the medium. From the geometry the condition may be given as

$$\overline{AB} - \overline{AC} = 2\Lambda \sin \theta = \lambda \tag{1.22}$$

which may again be recognized as the Bragg condition given by eqn (1.4). In terms of phase differences the above equation may be rewritten as

$$\frac{2\pi}{\lambda} 2\Lambda \sin \theta = 2\pi \tag{1.23}$$

If the Bragg condition is *not* satisfied, the diffracted rays are not in phase and the diffraction efficiency is bound to decline. Under what conditions will the diffraction efficiency be zero? We can make a simple construction under the assumption that the intensity associated with each diffracted ray is the same, and the total diffraction may be obtained by summing the contribution of each diffracted ray in the correct phase. If the incident ray traverses N periods as shown in Fig.1.3(b) then there will be N diffracted rays (to be exact, there would be $N + 1$ diffracted rays if we counted both the first one and the last one, but if N is large we can just as well take N diffractions). If each diffracted ray has a phase

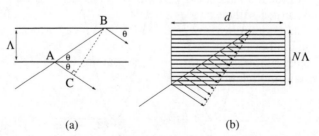

Fig. 1.3 (a) Diffraction from a single period of a dielectric grating. (b) The total diffracted light equals the sum of the N diffracted rays. These contributions will only be in-phase when the Bragg condition is satisfied.

1.2 Static volume gratings

difference of p relative to the previous one, and we take the phase of the first one equal to p, then the ith diffracted ray has a phase ip. The sum for all diffracted rays may be written, using an elementary amount of complex algebra, as

$$S = \left| \sum_{i=1}^{N} e^{iip} \right| = \left| \frac{\sin(Np/2)}{\sin(p/2)} \right| \quad (1.24)$$

whence the first zero occurs for $p = 2\pi/N$. Thus, the diffraction efficiency will be zero if we shall have a progressive phase difference of $2\pi/N$ in addition to the phase difference of 2π stipulated by eqn (1.23).

Considering first that the Bragg condition is violated by a wavelength deviation of $\Delta\lambda$, the equation to be satisfied is

$$\frac{2\pi}{\lambda + \Delta\lambda} 2\Lambda \sin\theta = 2\pi - \frac{2\pi}{N} \quad (1.25)$$

If $\Delta\lambda \ll \lambda$ (i.e. we have a really thick grating as is usually the case in a photorefractive material) we obtain from the above equation

$$\frac{\Delta\lambda}{\lambda} = \frac{1}{N} = \frac{\Lambda}{d\tan\theta} \quad (1.26)$$

If the violation of the Bragg condition occurs due to a change in the incident angle, then the corresponding equation is

$$\frac{2\pi}{\lambda} 2\Lambda \sin(\theta + \Delta\theta) = 2\pi + \frac{2\pi}{N} \quad (1.27)$$

Assuming again that $\Delta\theta \ll \theta$ we obtain from eqn (1.27)

$$\Delta\theta = \frac{\tan\theta}{N} = \frac{\Lambda}{d} \quad (1.28)$$

What would be the value for this angular deviation in a practical case? Assuming $d = 5$ mm and $\Lambda = 2\,\mu$m we find $\Delta\theta = 1.4$ milliradian. Hence two gratings whose orientations differ from each other by the above value can be independently reconstructed. There will be no 'cross-talk' between the two gratings. If we superimpose 100 gratings we have still not used more than 40 milliradian $= 2.3°$ of the total angular spectrum. We shall return to these problems in Chapter 10 concerned with storage applications.

If we record M gratings, what will be their diffraction efficiencies? Obviously, it depends on the total available change in the dielectric constant, $\varepsilon_{r1(max)}$. A rough picture envisaging simple superposition of the gratings would yield for the actual dielectric constant modulation[2]

$$\varepsilon_{r1} = \frac{\varepsilon_{r1\,max}}{M} \quad (1.29)$$

---

[2] Such superposition of gratings is possible in a so-called latent image material (e.g. dichromated gelatin) in which subsequent recordings are independent of each other. However in a photorefractive material, in which the recording of a new grating will result in some erasure of the existing gratings, the conclusions are only qualitatively true.

The corresponding diffraction efficiency may be expected to be small so we may replace the sinusoidal with its argument as was done in eqn (1.21), leading to the expression

$$\eta = (\kappa d)^2 = \left(\frac{\pi}{2} \frac{\varepsilon_{r1(max)} n_r}{\varepsilon_{ro}} \frac{1}{M} \frac{d}{\lambda} \right)^2 \quad (1.30)$$

where λ is now the free-space wavelength and n_r is the refractive index of the material, related to the dielectric constant as

$$n_r^2 = \varepsilon_{ro} \quad (1.31)$$

Taking, $d = 5$ mm, $\lambda = 0.5\,\mu$m, $M = 100$ and, perhaps a little optimistically, $n_r \varepsilon_{r1(max)}/\varepsilon_{ro} = 10^{-3}$ we obtain a very respectable 2.5% for diffraction efficiency.

Next, we shall discuss the deviation from the Bragg condition with the aid of the Ewald diagram which is well known both in X-ray diffraction theory and in the theory of optical gratings. It is defined as the geometrical locus of the end-points of all possible wave vectors, which is a circle for an isotropic material whenever a two-dimensional analysis suffices. If we need to take into account all three directions then we must talk about an Ewald sphere.

The diffraction efficiency depends on the off-Bragg parameter, for which two different constructions, leading to somewhat different results, are used in the literature. Let us first show (Fig. 1.4a) the construction of the diffracted wave vector based on eqn (1.3) when the Bragg conditions are satisfied. Then both the incident wave vector and the wave vector of the diffracted wave are on the circle of radius k. When the incident angle is different, the construction according to method 1 is shown in Fig. 1.4(b). Equation (1.3) is still assumed to be valid (Kogelnik 1969) but $\mathbf{k_2}$ is no longer on the Ewald circle. The off-Bragg parameter is measured by the difference between k and $\mathbf{k_2}$:

$$\psi = \frac{1}{2k}(k^2 - k_2^2) \quad (1.32)$$

According to method 2 the off-Bragg parameter is equal to

$$\psi = \overline{AB} \quad (1.33)$$

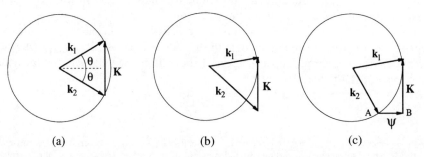

Fig. 1.4 Ewald diagrams for (a) on-Bragg diffraction, (b) off-Bragg diffraction according to method 1, and (c) off-Bragg diffraction according to method 2.

1.2 Static volume gratings

where points A and B are constructed as shown in Fig. 1.4(c). B is the end point of $\mathbf{k}_1 - \mathbf{K}$. A denotes the intersection of the Ewald circle with the line dropped from B perpendicularly to the surface of the grating (the x direction). The vectorial relationship is

$$\mathbf{k}_2 = \mathbf{k}_1 - \mathbf{K} - \psi \mathbf{i}_x \qquad (1.34)$$

The proponents of method 2 claim that the diffracted wave when it gets out of the grating will have to satisfy anyway the condition $k_2' = k$, so it makes good sense to assume that condition immediately. The proponents of method 1 claim that $\mathbf{k}_2$ obtained their way is the actual wave vector with which the diffracted wave propagates in the grating (the eigensolution of the problem) so it must give superior accuracy. When the diffracted wave leaves the grating it is always possible to satisfy the boundary conditions, as discussed by Sheppard (1976). For photorefractive materials no careful experiments have been done to compare the predictions of the two methods (for small incident angles and for small enough deviation from the Bragg condition they are equivalent). A study on volume holograms in photographic emulsions (Syms and Solymar 1983), however, came down clearly in favour of method 2. In the present book, in accordance with most of the treatments, we shall use method 2 when the problem arises.

We have restricted the analysis and the discussion to transmission gratings when the transmitted wave and the diffracted wave exit from the same side of the material. The principles and the technique of solution are very similar for reflection gratings (Kogelnik 1969). In the simplest case the fringes are then parallel to the z axis (see Fig. 1.5). A wave incident in the positive x direction will give rise to a reflected wave, i.e. a wave travelling in the negative x direction. The mechanism is of course still the same: cumulative interaction by Bragg reflection.

There is not much difference in principle between transmission and reflection gratings but there is a great difference in the actual magnitude of the grating spacing. As may be seen in Fig.1.5, the resulting grating vector is now twice as large as the individual wave vectors, yielding $\Lambda = \lambda/2$. For a free-space wavelength of 0.5 μm and an index of refraction equal to 2.5 the grating spacing is as small as 100 nm. It is still true that the wavelength selectivity is inversely proportional with the number of periods traversed. A material of 5 mm thickness then contains 50 000 grating periods. This piece of material may then serve as a frequency filter. At the above mentioned wavelength of 0.5 μm the wavelength

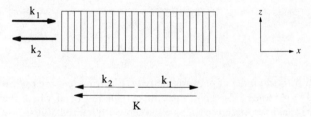

Fig. 1.5 Diffraction from a reflection grating.

sensitivity is 0.1 Å = 10 pm, a remarkably small figure. Such wavelength selectivity can be used in practical filters which will be discussed in Section 13.2.

1.3 The basic phenomena

1.3.1 The creation of the grating

We discussed diffraction by gratings in Section 1.2. The concepts discussed will become useful when we talk about the diffraction properties of gratings in photorefractive materials. The aim now is to give a simple explanation of the underlying physical phenomena, indicating how gratings in photorefractive materials are recorded.

First, what is a photorefractive material? A working definition would concentrate on two properties: it is photoconductive and electro-optic. What else do we need? In the simplest model the crystal will be regarded as a slab of material that contains one type of mobile carriers, electrons, and two types of impurities, donors and acceptors, which reside somewhere deep in the forbidden gap, as shown in Fig.1.6. Some of the donor atoms and all the acceptor atoms are assumed to be ionized. Hence our variables are n, the density of electrons, N_D the density of donors, N_D^+ the density of ionized donors, and N_A^- the density of ionized acceptors.

In the basic configuration corresponding to the simplest experiment two beams are incident symmetrically upon the crystal at angles $\pm\theta$ as shown schematically in Fig. 1.7(a). We shall regard them as two plane waves

$$\mathcal{E}_1 = \mathcal{E}_{10} e^{j\varphi_1} e^{-j\mathbf{k}_1 \cdot \mathbf{r}} \tag{1.35}$$

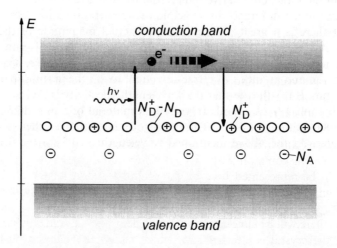

Fig. 1.6 Band transport model of photorefractivity. Electrons, e^-, are optically excited out of un-ionized donor atoms, $N_D^+ - N_D$, into the conduction band where they diffuse and drift in an electric field before recombining elsewhere with ionized donors, N_D^+. The ionized acceptors, N_A^-, are inactive.

1.3 The basic phenomena

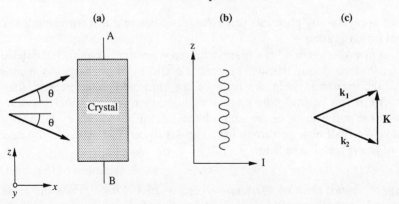

Fig. 1.7 (a) Basic experimental arrangement for two-wave mixing (TWM) in photorefractive materials. (b) The two indicent beams set up a sinusoidal intensity pattern inside the crystal. (c) An interference pattern of grating **K** is created by the two beams of wave vectors $\mathbf{k}_1$ and $\mathbf{k}_2$.

$$\mathcal{E}_2 = \mathcal{E}_{20} e^{\varphi_2} e^{-j\mathbf{k}_2 \cdot \mathbf{r}} \tag{1.36}$$

where $\mathcal{E}_1$ and $\mathcal{E}_2$ are the electric fields of beams 1 and 2, $\mathcal{E}_{10}$ and $\mathcal{E}_{20}$ are the respective amplitudes, φ_1 and φ_2 are the respective phases, **r** is a radius vector and the beams propagate in the directions given by the wave vectors $\mathbf{k}_1$ and $\mathbf{k}_2$.

The physical quantity that drives the whole process is the interference pattern, to which we must accord particular attention. Its spatial variation across the crystal, shown in Fig. 1.7(b), may be described by the modulus square of the electric field

$$I = |\mathcal{E}_1 + \mathcal{E}_2|^2 = \mathcal{E}_{10}^2 + \mathcal{E}_{20}^2 + 2\mathcal{E}_{10}\mathcal{E}_{20} \cos(\mathbf{K} \cdot \mathbf{r} + \varphi_2 - \varphi_1) \tag{1.37}$$

where we have already met the grating vector, $\mathbf{K} = \mathbf{k}_1 - \mathbf{k}_2$. This relationship between the wave vectors and the grating vector implies, as already mentioned in Section 1.2, that the Bragg conditions are satisfied.

How will the crystal respond to the incident optical fields? As mentioned before, the crystal is photoconductive, hence the input light will result in the excitation of pairs of electrons and ionized donors, their density being proportional to the intensity (which, in common with the overwhelming majority of the papers in this topic, will be represented here by the modulus square of the electric field, ignoring the impedance of the medium). However, the electrons are mobile, hence they will diffuse from higher to lower density. The ionized donors cannot move, consequently there will be a violation of charge neutrality and an electric field will appear. But the crystal is electro-optic, which means that an electric field will cause a small change in the dielectric constant. Thus the impressed periodic interference pattern will lead to a dielectric constant which varies with the same period. And a periodic variation in the dielectric constant, as agreed in Section 1.2, is what we call a dielectric grating. Thus, we have been able to show that without applying any chemical processing it is possible to turn the interference pattern into a grating.

Let us now discuss in a little more detail how we expect our variables (electron density, ionized donor density, current, electric field) to vary. As mentioned before, the excitation is in the form of the sinusoidal interference pattern of Fig.1.7(b). If this perturbation is small enough, then we may safely assume that the response will also be of the same shape, i.e. all our variables will vary sinusoidally. We shall now go through the various discernible stages in the development of the space charge field.

Stage 1. Initial creation of electrons and ionized donors. Since the crystal is photoconductive, there will be an equal number of electrons and ionized donors created, in phase with the interference pattern (one half of a period shown in Fig.1.8a). On a time scale short enough, there will be no time for either current flow or recombination. Hence space charge neutrality still prevails and there is no electric field.

Stage 2. Electrons and ionized donors increase together but ionized donors outpace electrons. The electrons are mobile and they will therefore diffuse from high density to low density (from region H to region L in Fig.1.8b). This diffusion current (Fig.1.8c) will obviously be 90 degrees out of phase with the electron distribution since it is proportional to the gradient of the electron density. In region H (where the electron density is above average) electrons will disappear by recombination and, in addition, will flow to region L (where the electron density is below average) leaving an excess of ionized donors behind them. In region L a certain proportion of the electrons (both those excited locally and the immigrant electrons) will recombine and will thereby reduce the ionized donor density. By these means, net space charge (Fig. 1.8d) and consequently an electric field will appear.

What will be the phase relationship between the emerging electric field and the net space charge distribution? Since it is the gradient of the electric field that is proportional to space charge, the electric field (Fig. 1.8e) will have the same phase as the diffusion current, 90 degrees out of phase with the net space charge distribution.

Stage 3. Electron density is saturated, ionized donor density keeps on growing. The electron density will reach saturation in a time known as the lifetime of the electron (about 10^{-9} s in $LiNbO_3$ and 10^{-5} s in $Bi_{12}SiO_{20}$, BSO). That means that all the newly created electrons will have to disappear either by flowing away or by recombination. Ionized donor density however will keep on growing by the same mechanism as in stage 2. The electric field will also grow. As the electric field grows, a new component of current, a drift current opposing the diffusion current, will also grow. Since the diffusion current remains constant and the drift current grows, the net electron current decreases.

1.3 The basic phenomena

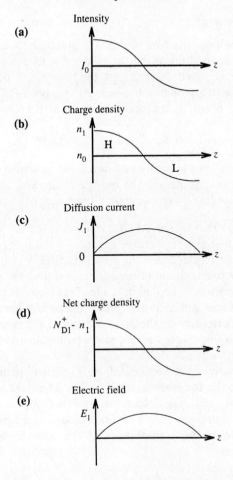

Fig. 1.8 The response of a material to light. (a) When the material is exposed to an interference pattern, with an average intensity of I_o, electrons (or holes) are generated by photoexcitation. (b) These mobile charges diffuse from high (H) to low (L) density regions, and (c) give rise to a diffusion current, J_1. (d) The subsequent recombination of electrons, n_1, and ionized donors, N_{D1}^+, leads to the appearance of a net space charge, and (e) an accompanying space charge field, E_1.

Stage 4. Stationary state. Electrons stop moving from high density to low density because the drift current exactly balances the diffusion current. Ionized donor density stops growing because the recombination of ionized donors and electrons occurs at exactly the same rate as their generation. The electric field also reaches a steady-state value.

An essential point to note from Fig. 1.8 is that the electric field has a phase difference of $\Phi = \pi/2$ relative to the interference pattern.

1.3.2 Grating diffraction

We have seen now how the two beams will give rise to a grating. But will the grating affect the beams? Since the Bragg conditions are automatically satisfied (because the beams created the grating), the beams will be diffracted into each other. Does that mean interaction between the two beams? Does that mean power transfer from one beam to the other one? What do we mean by interaction, anyway?

To arrive at a provisional definition let us consider the following experiment:

(i) measure the output intensity of beam 1 in the absence of beam 2;
(ii) measure the output intensity of beam 2 in the absence of beam 1;
(iii) measure the output intensities when both beams 1 and 2 are incident.

We shall now say that beams 1 and 2 interact if the intensity measured in one beam is dependent on the presence or absence of the other beam.

We shall show in Chapter 2 that the power transfer depends on sin Φ where Φ is the phase angle between the interference pattern and the space charge field. Is there a way that such a result could be made plausible by recourse to some simpler relationships? It turns out that the phase difference between the interference pattern and the distribution of the dielectric constant has a similar role in ordinary volume holography, where the recorded grating does not depend on the waves present.

Instead of one wave, as in Section 1.2, we shall now assume two waves R_1 and R_2 to be incident upon the grating (see Fig. 1.9). The wave vectors and the angles are again such that the Bragg conditions are satisfied. We shall, however, assign now phase angles ϕ_1 and ϕ_2 to the two input waves. Analogously to the derivation presented in Section 1.2 the amplitudes of the output waves are given by the expressions

$$R_1(d) = A_1 \exp(j\phi_1) \cos \nu, \quad S_2(d) = -jA_1 \exp(j\phi_1) \sin \nu \quad (1.38)$$

$$R_2(d) = A_2 \exp(j\phi_2) \cos \nu, \quad S_2(d) = -jA_2 \exp(j\phi_2) \sin \nu \quad (1.39)$$

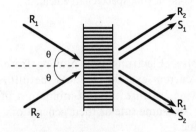

Fig. 1.9 Diffraction of light from a grating. R_1 and R_2 are reference waves, and S_1 and S_2 are the respective waves diffracted by the grating.

1.3 The basic phenomena

where

$$\nu = \kappa d \qquad (1.40)$$

A_1 and A_2 are the input amplitudes, S_1 and S_2 are the diffracted output beams, d is the thickness of the hologram, and the coupling constant κ is defined in eqn (1.16).

When only beam 1 is incident, the output intensities in the direction of beams 1 and 2 are

$$I_{11} = A_1^2 \cos^2 \nu, \quad I_{12} = A_1^2 \sin^2 \nu \qquad (1.41)$$

as discussed in Section 1.2. Analogously, when only beam 2 is incident, the output intensities in the direction of beams 1 and 2 are

$$I_{21} = A_2^2 \sin^2 \nu, \quad I_{22} = A_2^2 \cos^2 \nu \qquad (1.42)$$

What happens when both beams are incident? The output intensity in the direction of beam 1 is then

$$\begin{aligned} I_1 &= |A_1 \exp(j\phi_1) \cos \nu - jA_2 \exp(j\phi_2) \sin \nu|^2 \\ &= A_1^2 \cos^2 \nu + A_2^2 \sin^2 \nu - A_1 A_2 \sin^2 \nu \sin(\phi_2 - \phi_1) \end{aligned} \qquad (1.43)$$

and the output intensity in the direction of beam 2 is

$$\begin{aligned} I_2 &= |-jA_1 \exp(j\phi_1) \sin \nu + A_2 \exp(j\phi_2) \cos \nu|^2 \\ &= A_1^2 \sin^2 \nu + A_2^2 \cos^2 \nu + A_1 A_2 \sin^2 \nu \sin(\phi_2 - \phi_1) \end{aligned} \qquad (1.44)$$

When $\phi_2 - \phi_1 = 0$ the output intensities may be obtained by simply adding the intensities when only one or the other beam is present, i.e.

$$I_1 = I_{11} + I_{21} \quad \text{and} \quad I_2 = I_{12} + I_{22} \qquad (1.45)$$

When $\phi_2 - \phi_1 \neq 0$ then the output intensities cannot be obtained by simply superimposing I_{11} and I_{21} in the direction of beam 1, or superimposing I_{12} and I_{22} in the direction of beam 2. It is important to emphasize that when $\phi_2 - \phi_1 \neq 0$ we have the 'interaction' term $A_1 A_2 \sin^2 \nu \sin(\phi_2 - \phi_1)$ which adds to the intensity of one output beam and subtracts from the intensity of the other beam. The interaction term depends on $\sin(\phi_2 - \phi_1)$.

What is $\phi_2 - \phi_1$? Remember that the grating is given as $\cos Kz$. It is clear then that $\phi_2 - \phi_1$ is the phase difference between the interference pattern, generated by R_1 and R_2, and the grating.

2
HOW LIGHT INFLUENCES MATERIAL PROPERTIES

2.1 The basic differential equations

Our aim, as mentioned before, is to give a general overview of the whole field in this chapter, an overview which contains not only the basic physics but also the basic equations. We shall, of course, consider only those effects and those terms which are necessary for developing a 'feel' for photorefractive phenomena.

In this section we shall start the quantitative analysis by deriving a partial differential equation for the space charge field, the most important physical quantity in our analysis. All the phenomena displayed depend on this induced electric field. We shall follow the treatment as formulated by Kukhtarev *et al.* (1979a,b).

How will the ionized donor density change as a function of time in the presence of incident optical intensity? Obviously, the rate of change of the ionized donors will be equal to the rate of generation minus the rate of recombination. Let us now look at each of them separately. The obvious assumption is that the rate of generation is directly proportional both to the number of un-ionized donors (which are available for ionization) and to the incident intensity. There is, of course, thermal generation as well but it does not usually play a significant role (one exception will be commented on a little later) and will be disregarded here.

According to the argument above the generation rate per unit volume can be written as

$$g = sI(N_D - N_D^+) \qquad (2.1)$$

where s is the photoionization coefficient. Alternatively, we can argue that if α is the absorption constant then the intensity absorbed per unit time per unit volume is αI. The number of photons absorbed is then $\alpha I/\hbar\omega$ where $\hbar$ is Planck's constant divided by 2π. Assuming a quantum efficiency η_q (i.e. one photon knocks out on the average η_q electrons), the generation rate may be written as

$$g = \eta_q \alpha I/(\hbar\omega) \qquad (2.2)$$

For the rate of recombination two different expressions have been used in the literature. If the ionized donor density can be regarded as approximately constant then the rate of recombination may be written in the form of n/τ_e where τ_e represents a time constant usually referred to as recombination time. This was the implicit assumption in earlier photorefractive literature (see e.g. Alphonse *et al.* 1975). A more accurate expression is provided by quadratic recombination

2.1 The basic differential equations

involving both species, i.e. the rate of recombination is proportional both to the density of electrons and to the density of ionized donors. Hence the rate equation is

$$\frac{\partial N_D^+}{\partial t} = sI(N_D - N_D^+) - \gamma_R n N_D^+ \tag{2.3}$$

where γ_R is the recombination constant.

A similar equation can be written for the electrons which takes into account that electrons are mobile and can flow away. This equation is known as the continuity equation. It was used in the formulation of Kukhtarev et al. (1979a, b) and it is one of the basic equations in most treatments. We shall instead use the so-called equation for the total current, which is simpler both mathematically and conceptually. Mathematically, it comes from one of Maxwell's equations

$$\nabla \times \mathbf{H} = \mathbf{J} + \varepsilon_s \frac{\partial \mathbf{E}}{\partial t} \tag{2.4}$$

where $\mathbf{H}$ is the magnetic field, $\mathbf{J}$ is the current density, ε_s is the static dielectric constant[1], $\mathbf{E}$ is the electric field, and t is time. If we take now the divergence of both sides we obtain the relationship

$$\nabla \cdot \left(\mathbf{J} + \varepsilon_s \frac{\partial \mathbf{E}}{\partial t} \right) = 0 \tag{2.5}$$

From now on we shall abandon the vector operators and assume that there is a variation only in the transverse direction. Equation (2.5) may therefore be integrated to give

$$\mathbf{J} + \varepsilon_s \frac{\partial \mathbf{E}}{\partial t} = C \tag{2.6}$$

where C is independent of the spatial coordinate.

Physically, C can be regarded as the total current which has two constituents: J, the current due to carrier flow, and the displacement current $\varepsilon_s \partial \mathbf{E}/\partial t$. The total current may be zero as in Fig. 1.7(a) where points A and B are unconnected. If A and B are joined or a voltage source is inserted between them, then a current could flow, and that would be J_o the total current in the circuit. Hence, we may take

$$C = J_o \tag{2.7}$$

We shall consider two contributions to the current, the conduction current and the diffusion current in the form

$$J = e\mu n E + k_B T \mu \frac{\partial n}{\partial z} \tag{2.8}$$

---

[1] The reason for taking the static dielectric constant (in contrast to the optical dielectric constant) is that the space charge field varies only slowly in time.

where e is the charge of the electron, m is the mobility, E now includes the field set up by the applied voltage, k_B is Boltzmann's constant, T is the absolute temperature, and z is the transverse coordinate as shown in Fig. 1.7(b).

We still need one equation which relates the space charge electric field to the charge imbalance. It is given by Poisson's equation (or we may call it one of Maxwell's equations) in the form[2]

$$\frac{\partial}{\partial z}(\varepsilon_s E) = e(N_D^+ - N_A^-) \tag{2.9}$$

We have now a set of non-linear differential equations which will tell us how the material quantities will change in the presence of an input optical intensity. Recall that the optical intensity was given by eqn (1.37) in the form[2]

$$I = I_o + I_1 \cos(Kz + \phi_2 - \phi_1) \tag{2.10}$$

containing a spatially constant term and a periodically varying term which we may regard as a perturbation, denoted by

$$I_p = I_1 \cos(Kz + \phi_2 - \phi_1) \tag{2.11}$$

We have already argued in Section 1.3.1 that if the excitation, i.e. this periodic perturbation, is small enough then we may expect that the system will follow suit, the response will vary in the same manner.

Is that true for a non-linear system? Well, non-linear differential equations may lead to all kinds of solutions including higher harmonics, subharmonics and even chaos, so can we expect them to give the simple response mentioned? It seems unlikely. We shall of course be concerned with the non-linear differential equations at several places in the book. In this introductory chapter our best way of continuing is to tamper with the equations, turn them into a linear set. How can we do that? It is, in fact, quite easy and it uses the time-honoured method of neglecting the products of small quantities. We shall do that in the next section.

2.2 Linearization

We shall now make the assumption that the solution of our differential equations may be written in the form

$$X(z,t) = X_o(t) + X_p(z,t) \tag{2.12}$$

where X_o is independent of space and X_p is the periodic perturbation. X stands, of course, here for any of our variables N_D^+, n, J, or E (remember that the ionized acceptor density is constant).

Where are our non-linear terms? They may be found in eqns (2.3) and (2.8). IN_D^+ is the product of the excitation and one of our variables, whereas nN_D^+ and nE are products of two of our variables. In each case we shall linearize the

---

[2] Strictly speaking, we should have included here the electron density n. It turns out however that for the usual input intensities n is small and can be neglected. For the full treatment see Section 5.2.

2.2 Linearization

expressions as follows:

$$\begin{aligned}IN_D &= I_o N_{Do}^+ + I_o N_{Dp}^+ + I_p N_{Do}^+ + I_p N_{Dp}^+ \\ &\approx I_o N_{Do}^+ + I_o N_{Dp}^+ + I_p N_{Do}^+\end{aligned} \quad (2.13)$$

and similarly

$$nN_D^+ \approx n_o N_{Do}^+ + n_o N_{Dp}^+ + n_p N_{Do}^+ \quad (2.14)$$

and

$$nE \approx n_o E_o + n_o E_p + n_p E_o \quad (2.15)$$

With eqns (2.13) to (2.15) we may now rewrite our differential equations in the form

$$\begin{aligned}\frac{\partial N_{Do}^+}{\partial t} + \frac{\partial N_{Dp}^+}{\partial t} =& s(I_o + I_p)N_D - sI_o N_{Do}^+ - sI_o N_{Dp}^+ - sI_p N_{Do}^+ \\ &- \gamma_R (n_o N_{Do}^+ + n_o N_{Dp}^+ + n_p N_{Do}^+)\end{aligned} \quad (2.16)$$

$$e\mu n_o E_o + e\mu n_o E_p + e\mu n_p E_o + k_B T\mu \frac{\partial n_p}{\partial z} + \varepsilon_s \frac{\partial E_o}{\partial t} + \varepsilon_s \frac{\partial E_p}{\partial t} = J_o \quad (2.17)$$

and

$$\varepsilon_s \frac{\partial E_p}{\partial z} = e(N_{Do}^+ + N_{Dp}^+ - N_A^-) \quad (2.18)$$

The next step is to separate the space-independent and space-dependent terms in eqns (2.16 to 2.18). Those which do not depend on space are as follows

$$\frac{\partial N_{Do}^+}{\partial t} = sI_o(N_D - N_A^-) - \gamma_R n_o N_{Do}^+ \quad (2.19)$$

$$J_o = e\mu n_o E_o + \varepsilon_s \frac{\partial E_o}{\partial t} \quad (2.20)$$

$$0 = N_{Do}^+ - N_A^- \quad (2.21)$$

In eqn (2.19) we may substitute $\partial N_{Do}^+/\partial t = \partial n_o/\partial t$ since electrons and ionized donors vary at the same rate when the current is independent of space. Hence the equation to solve for the average electron density is

$$\frac{\partial n_o}{\partial t} + \gamma_R n_o N_A^- = sI_o(N_D - N_A^-) \quad (2.22)$$

which has the solution

$$n_o = sI_o(N_D - N_A^-)\tau_e[1 - \exp(-t/\tau_e)] \quad (2.23)$$

where

$$\tau_e = \frac{1}{\gamma_R N_A^-} \tag{2.24}$$

may now be regarded as the lifetime of the electron. This is simple enough. For times much shorter than the lifetime the average electron density increases linearly with time, and for long enough time it reaches the saturation density $sI_o(N_D - N_A^-)\tau_e$. In the following we shall assume that saturation has been reached. This is a perfectly good assumption for the materials used and for most intensities employed, because the grating formation time is usually much longer than the electron lifetime.

Equation (2.20) determines the current in the external circuit, which is the same thing as the average current in the crystal. If we know how the applied voltage varies with time (we would actually need some further information, namely the internal impedance of the voltage source) then we can work out the current in the external circuit with the aid of eqn (2.20).

Similarly, we obtain for the space-varying quantities:

$$\frac{\partial N_{Dp}^+}{\partial t} = sI_p(N_D - N_A^-) - sI_o N_{Dp}^+ - \gamma_R(n_o N_{Dp}^+ + n_p N_{Do}^+) \tag{2.25}$$

$$0 = e\mu(n_o E_p + n_p E_o) + k_B T \mu \frac{\partial n_p}{\partial z} + \varepsilon_s \frac{\partial E_p}{\partial t} \tag{2.26}$$

$$\varepsilon_s \frac{\partial E_p}{\partial z} = e N_{Dp}^+ \tag{2.27}$$

The above set of differential equations is now linear. We can therefore opt for the simple method of working in terms of exponential quantities, a standard method in many branches of physics and engineering. The actual excitation is given by eqn (2.11). In exponential form it modifies to

$$I_p = I_1 \exp(-jKz) \tag{2.28}$$

where I_1 is now complex and incorporates the phases ϕ_1 and ϕ_2. Mathematically, this is a great simplification. We should however always remember that the physically realizable values are always given by the real parts.

The response may now be given in the form

$$X_p = X_1(t) \exp(-jKz) \tag{2.29}$$

which means that spatial differentiation, $\partial/\partial z$, may be replaced by $-jK$. Equations (2.25) to (2.27) then take the form

$$\frac{\partial N_{D1}^+}{\partial t} = sI_1(N_D - N_A^-) - (sI_o + \gamma_R n_o)N_{D1}^+ - \gamma_R n_1 N_{Do}^+ \tag{2.30}$$

2.2 Linearization

$$0 = e\mu(n_o E_1 + n_1 E_o) - jKk_B T\mu n_1 + \varepsilon_s \frac{\partial E_1}{\partial t} \quad (2.31)$$

and

$$-jK\varepsilon_s E_1 = eN_{D1}^+ \quad (2.32)$$

We have three equations from which the three unknowns can be determined. Our main interest is the space charge electric field. Straightforward elimination of the variables will lead to the following differential equation in E_1:

$$\frac{\partial E_1}{\partial t_n} + pE_1 = mq \quad (2.33)$$

where $m = I_1/I_o$ is the modulation of the interference pattern,

$$p = \frac{1}{D}\left(1 + \frac{E_D + jE_o}{E_q}\right), \quad q = j\frac{E_D + jE_o}{D}, \quad D = 1 + \frac{E_D + jE_o}{E_M} \quad (2.34)$$

and our variable is now t_n, the normalized time, defined as

$$t_n = t/\tau_d \quad (2.35)$$

where

$$\tau_d = \frac{\varepsilon_s}{e\mu n_o} \quad (2.36)$$

is known as the dielectric relaxation time. The characteristic electric fields E_D, E_M, and E_q have been introduced by Kukhtarev et al. (1977). They may be written in terms of our earlier defined parameters as

$$E_D = \frac{k_B TK}{e}, \quad E_M = \frac{\gamma_R N_A^-}{\mu K}, \quad E_q = E_{qo}(1-a), \quad E_{qo} = \frac{eN_A^-}{\varepsilon_s K}, \quad a = \frac{N_A^-}{N_D} \quad (2.37)$$

It is quite easy to give physical meaning to these fields. E_D is the so-called diffusion field. It may be seen to be inversely proportional to grating spacing. Obviously, the smaller is the grating spacing, the greater is the gradient of the electron distribution and the stronger are the forces of diffusion. Under certain circumstances, as will be shown in the next section, the space charge field is equal to the diffusion field.

The physical meaning of E_M may be appreciated by writing it in the alternative form

$$E_M \mu \tau_e = \frac{\Lambda}{2\pi} \quad (2.38)$$

Thus E_M is the electric field which will move an electron a distance $\Lambda/2\pi$ during its lifetime.

A simple meaning can also be attached to E_{qo}, called the saturation field. It is the maximum space charge field that can exist in the material for a sinusoidal charge distribution when $a \ll 1$. How can we show that that is the maximum? Well, if mobile electrons have negligible contribution then all space charge must come from trapped electrons. In the absence of optical excitation the density of ionized donors is equal to the density of ionized acceptors. At a certain point in the crystal the maximum possible negative charge can occur when none of the donors are ionized (i.e. electrons have filled up the ionized donors) and we see the full negative charge of the acceptors. The maximum positive space charge occurs when the density of ionized donors is twice the density of the ionized acceptors. Hence the maximum sinusoidal distribution can have an amplitude of N_A^-. According to Poisson's equation

$$\varepsilon_s \frac{\partial E_1}{\partial z} = e N_A^- \cos Kz \tag{2.39}$$

yielding

$$E_1 = \frac{e N_A^-}{\varepsilon_s K} \sin Kz = E_{qo} \sin Kz \tag{2.40}$$

2.3 Solution of the linearized equations

We shall now discuss how the space charge electric field rises and decays on the basis of eqn (2.33). It looks very simple and very familiar. Indeed, all students of science are familiar with first-order linear differential equations which have solutions in the form of exponential functions. The solution is either decaying or growing. However, the present differential equation does not fit into that pattern, for the reason that the coefficients are complex. This makes the equation equivalent to a second-order differential equation with solutions which may be combinations of exponential and oscillatory functions. Formally, with an initial condition

$$E_1 = 0 \quad \text{at} \quad t_n = 0 \tag{2.41}$$

the solution is

$$E_1 = m \frac{q}{p} [1 - \exp(-p t_n)] \tag{2.42}$$

When $E_o = 0$ all the coefficients are real and the function will rise monotonically, but for $E_o \neq 0$ the imaginary part of p will be different from zero and an oscillatory behaviour occurs. The rise time may be seen to be given by the real part of p.

A similar solution exists for the decline of the electric field. Assuming that it has a value of $E_1(0)$ at $t = 0$, and the excitation is suddenly stopped (i.e. $m = 0$), then the solution of eqn (2.33) is

$$E_1 = E_1(0) \exp(-p t_n) \tag{2.43}$$

2.3 Solution of the linearized equations

The same is true for the decaying solution as for the rising solution: the decay time is given by Re(p) (i.e. the real part) and the imaginary part of p is responsible again for the oscillations. We must however exercise a little care here: the symmetry in the rising and decaying solution exists only if we use the normalized temporal variable t_n. In a practical case it is far from obvious that the dielectric relaxation time is the same both for rise and decay. It depends on the way the excitation comes to a stop. We may make the interference pattern vanish in a number of different ways, of which we shall consider four:

(i)   beam 2 is blocked, i.e. $|\mathcal{E}_{20}|^2$ is made zero but $|\mathcal{E}_{10}|^2$ remains the same;
(ii) beam 1 is blocked, i.e. $|\mathcal{E}_{10}|^2$ is made zero but $|\mathcal{E}_{20}|^2$ remains the same;
(iii) both beams are blocked but an incoherent beam of the same average intensity is incident;
(iv) both beams are blocked.

In all the above cases we shall also assume that the small modulation assumption is valid, $|\mathcal{E}_{20}|^2 \ll |\mathcal{E}_{10}|^2$. In case (i) when beam 2 is blocked, the average intensity has hardly changed and therefore the rise and decay times are the same. In case (ii) the average intensity suddenly drops. Since the electron lifetime is short we may assume that the electron density instantaneously adjusts to the lower average intensity, the dielectric relaxation time suddenly increases, and consequently the grating decays much more slowly. In case (iii) the rise and decay times are again identical. The photoexcited electron density is the same whether the intensity comes from a coherent or incoherent source.

Case (iv) needs to be considered in more detail. If all optical excitation stops, then according to our equations, the electron density drops to zero and the dielectric relaxation time becomes infinitely long. This is obviously wrong, and is a consequence of neglecting thermal excitations. The reason for neglecting thermal excitations in Section 2.1 was a desire to keep the equations simple. In this particular case we should restore thermal excitations to their rightful place. The generation of ionized donors must still be proportional to the density of unionized donors but we need to introduce a thermal generation constant β_t. Equation (2.22) should therefore take the form

$$\frac{\partial n_o}{\partial t} + \gamma_R n_o N_A^- = (\beta_t + sI_o)(N_D - N_A^-) \qquad (2.44)$$

In the absence of optical excitation the equilibrium density is then

$$n_o = \beta_t (N_D - N_A^-)\tau_e \qquad (2.45)$$

This is the electron density in the dark. This will determine the dark conductivity, through that the dielectric relaxation time, and consequently the decay time when the exciting beams are switched off.

The length of time for which a grating in a photorefractive material remains available is an important one for storage applications. Some materials, notably $LiNbO_3$, have very low dark conductivity, leading to storage times that can be measured in months. We shall return to these problems in Chapter 10. For the time being we shall concentrate on two aspects of the rising solution: the short time limit and the stationary case.

Short time limit. By this we mean the solution in the vicinity of $t = 0$. It may be obtained from eqn (2.42) by expanding the exponential or directly from the differential equation (2.33) by disregarding p. In either case we obtain

$$E_1 = mqt_n = jm \frac{E_M(E_D + jE_o)}{E_M + E_D + jE_o} t_n \qquad (2.46)$$

The limiting value of dE_1/dt_n is $jm(E_D + jE_o)$ when E_M is large, and jmE_M when E_M is small. Hence E_M represents the maximum rate of change of E_1.

Can we arrive at the same relation by simple physical arguments? One way of doing this was proposed by Yeh (1987a). Essentially, his argument was that the rise of the space charge field is limited by the time needed to generate photocarriers[3]. The space charge field is due to the ionized donor density N_{D1}^+, which may be obtained from eqn (2.32). But in order to ionize a donor at least one photon must be incident. Thus the fastest rate at which N_{D1}^+ can increase is $(\alpha\eta_q mI_o/\hbar\omega)t$. Consequently, the fastest rate at which E_1 can increase is

$$E_1 = j \frac{e}{K\varepsilon_s} m \frac{\eta_q \alpha I_o}{\hbar\omega} t \qquad (2.47)$$

which, after a few simple algebraic operations, reduces to

$$E_1 = jmE_M \frac{t}{\tau_d} \qquad (2.48)$$

in agreement with our previous argument.

Let us next look at the pure diffusion case when $E_o = 0$. Then

$$E_1 = jm \frac{E_M E_D}{E_M + E_D} t_n \qquad (2.49)$$

It may be seen from the above equation that the space charge electric field is purely imaginary. It also follows from eqn (2.49) that there must be an optimum value of K since E_D is proportional to K and E_M is inversely proportional to K. The optimum occurs when $E_D = E_M$. Remembering the definitions in eqn (2.37), this occurs when

$$K^2 = \frac{e}{\mu\tau_e k_B T} \qquad (2.50)$$

Using the Einstein relationship

$$eD = \mu k_B T \qquad (2.51)$$

(where D is the diffusion constant) and the definition of diffusion length

$$L_D = (D\tau_e)^{1/2} \qquad (2.52)$$

---

[3] See Glass et al. (1987), pointing out that this criterion had already been incorporated in the definition of some of the earlier figures of merit in general use.

2.3 Solution of the linearized equations

both well known from semiconductor theory (see e.g. Sze 1985) we obtain

$$K = \frac{1}{L_D} \qquad (2.53)$$

i.e. the optimum occurs when the grating spacing is 2π times the diffusion length.

For $E_o \neq 0$ the space charge field may be seen to be complex. As mentioned before, and will be derived in Section 3.4, the energy exchange between the two optical waves depends on the imaginary part of the space charge field. If the grating recorded is for the purpose of diffracting another optical beam, then what counts is the modulus of the space charge field. We shall plot them both. At this stage however we need to put in eqn (2.46) the actual crystal parameters. Unfortunately, neither the parameters themselves nor the range over which they might vary are well known. We shall choose here

$$\mu = 3 \times 10^{-6}\, m^2\, V^{-1}\, s^{-1}, \quad \gamma_R = 1.65 \times 10^{-17}\, m^3\, s^{-1},$$
$$N_A^- = 10^{22}\, m^{-3}, \qquad (2.54)$$
$$N_D = 10^{25}$$

for BSO, and

$$\mu = 5 \times 10^{-5}\, m^2\, V^{-1}\, s^{-1}, \quad \gamma_R = 5 \times 10^{-14}\, m^3\, s^{-1},$$
$$N_A^- = 2 \times 10^{22}\, m^{-3}, \qquad (2.55)$$
$$N_D = 10^{25}\, m^{-3}$$

for BaTiO$_3$. The most significant difference between them is in the value of the mobility–lifetime product. It is $1.8 \times 10^{-10}\, m^2\, V^{-1}$ for BSO and $5 \times 10^{-14}\, m^2\, V^{-1}$ for BaTiO$_3$. Accordingly, the value of E_M is higher for BaTiO$_3$ by about three orders of magnitude, which would affect significantly the rate at which the electric field rises at small grating spacings.

The values of dE_w/dt_n and $d(\text{Im}E_w)/dt_n$ are calculated from eqn (2.46) and plotted in Figs. 2.1(a,b) and 2.2(a,b) as a function of grating spacing for BSO and BaTiO$_3$ respectively. For convenience we have introduced here the notation $E_w = E_1/m$. It may be seen that the curves displayed in Figs 2.1(a) and 2.1(b) are nearly identical. The reason is that E_M is small for BSO (large $\mu\tau_e$ product) and therefore E_o dominates in the denominator of eqn (2.46), with the result that the whole expression becomes imaginary. Hence the modulus agrees with the imaginary part. In the absence of an applied electric field the maximum is at 2π times the diffusion length ($L_D = 0.68\,\mu m$) and the maximum may be seen to be shifting towards larger grating spacings as E_o increases.

The corresponding plots for BaTiO$_3$, Figs 2.2(a) and (b), look very different. There is now considerable difference between the slopes of the modulus and of the imaginary part. The reason is that for BaTiO$_3$ E_M is several orders of magnitude larger than that for BSO, thus E_o in the denominator of eqn (2.46) has little

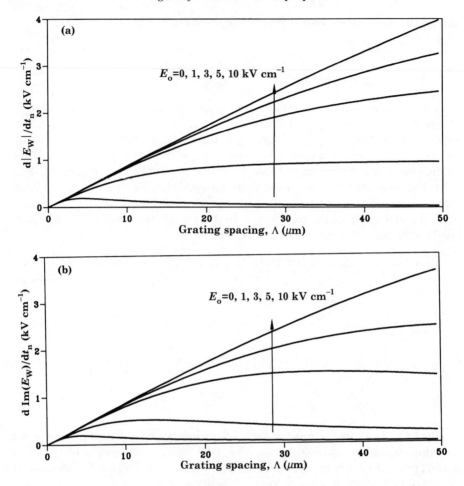

Fig. 2.1 The rate of change of (a) the modulus and (b) the imaginary part of the space charge field as a function of grating spacing for a crystal of BSO and different applied fields.

significance. Consequently, the phase angle will decline with increasing E_o, which will check the rise of the imaginary part.

It is still true of course that in the absence of an applied electric field the maximum is at 2π times the diffusion length. But the diffusion length is now much smaller ($L_D = 0.036\,\mu\text{m}$), yielding $0.226\,\mu\text{m}$ for the optimum grating spacing. If the aim is to obtain high diffraction efficiency fast (i.e. to obtain a large modulus) then the application of an electric field can help considerably. For the imaginary part though, particularly at larger grating spacings, an applied electric field is of little help.

2.3 Solution of the linearized equations

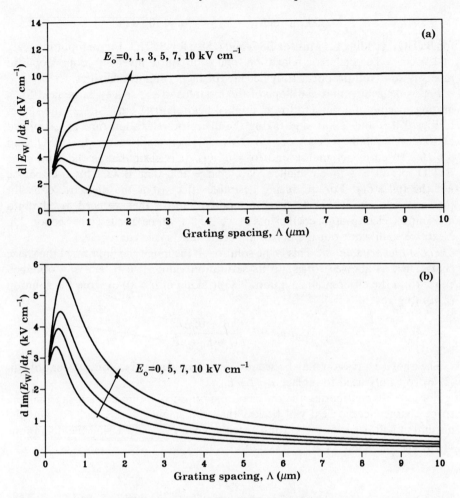

Fig. 2.2 Same as Fig. 2.1 but for BaTiO$_3$.

A general conclusion that can be drawn from the four sets of curves (Figs 2.1a,b and 2.2a,b) is that the application of an electric field will always speed up the appearance of the space charge field whether it is the modulus or the imaginary part. It is also interesting to look at the relative values shown. BaTiO$_3$ appears to be faster material if we think in terms of normalized time. For any application, however, we have to think in real time. BaTiO$_3$ will be much slower because our basis of normalization, the dielectric relaxation time, is much longer for BaTiO$_3$ than for BSO. In order to find the actual figures we need some further parameters. We shall choose them as

$$\varepsilon_s = 56\varepsilon_o, \quad s(N_D - N_A^-) = 10^{20}\, \text{J}\,\text{m}^{-1} \tag{2.56}$$

for BSO and

$$\varepsilon_s \approx 600\varepsilon_o, \quad s(N_D - N_A^-) = 7.6 \times 10^{19} \, \text{J m}^{-1} \tag{2.57}$$

for BaTiO$_3$, yielding 0.17 ms for BSO and 0.88 s for BaTiO$_3$ for an input intensity of 1 W cm^{-2}. The dielectric relaxation time is of course inversely proportional to intensity so it will proportionally decline for higher input intensities.

Let us determine now the slope of the modulus of the space charge field for a grating spacing of 5 μm and for an applied field of 10 kV cm^{-1}. Using the curves in Figs 2.1(a) and 2.2(a), and taking the dielectric relaxation time as calculated above, we obtain a rise of 2.6 kV cm^{-1} per millisecond for BSO and 8×10^{-3} kV cm^{-1} per millisecond for BaTiO$_3$. It is clear that in the same time BaTiO$_3$ produces a much smaller space charge field than BSO. This is of course not the full story. For producing desirable effects it is the dielectric constant (index of refraction) that we need to modify, and for that we need to take into account the electro-optic coefficients as well. This will be discussed in Section 2.6. Next, we shall work out the maximum achievable space charge field.

The stationary solution. This is the solution in the long time limit when the space charge field is allowed to rise to its saturation value. It can be easily obtained either from the differential equation (2.33) (taking d/d$t = 0$) or from the solution (2.42) to give

$$E_1 = jm \frac{(E_D + jE_o)E_q}{E_D + E_q + jE_o} \tag{2.58}$$

The above expression may be seen to be nearly identical in form to that of eqn (2.46). We only need to replace E_M by E_q.

For $E_o = 0$, considering the diffusion region only, we may again say that the space charge electric field will be less than the smaller of E_D and E_q, and the maximum will occur at the grating spacing when $E_D = E_q$. Using the definitions of E_D and E_q in eqn (2.37) the condition comes to

$$\Lambda_{\text{opt}} = 2\pi \left[\frac{k_B T \varepsilon_s}{e^2 N_A^-(1-a)} \right]^{1/2} = 2\pi L_{\text{Deb}} \tag{2.59}$$

where L_{Deb} is called the Debye screening length, corresponding to an effective charge density of $N_A^-(1-a)$. It is an important parameter in plasma physics. For plasma effects to be displayed, the size of the container has to be large relative to the Debye length. In our context it means that the space charge field must be small when the grating spacing is much less than the Debye length.

The limiting values are now E_q for small grating spacing and E_D for large grating spacing.

The modulus and imaginary part of the stationary space charge field are plotted against grating spacing in Figs 2.3(a,b) and 2.4(a,b) for BSO and BaTiO$_3$ respectively, using eqn (2.58). A quick look at the four sets of curves will show that there is a striking resemblance to the curves in Figs 2.1 and 2.2 but the roles of BSO and BaTiO$_3$ have been reversed. It is now BaTiO$_3$ for which the modulus is roughly equal to the imaginary part and BSO where the

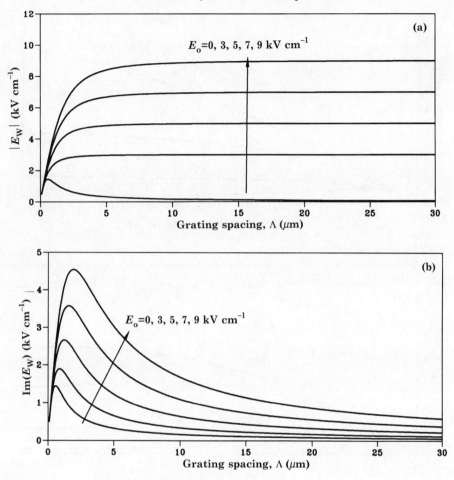

Fig. 2.3 (a) The modulus and (b) imaginary part of the steady-state space-charge field as a function of grating spacing for a crystal of BSO and different applied fields.

imaginary part is considerably less than the modulus. The reversal occurs since E_q has a much smaller value for $BaTiO_3$ than for BSO on account of its high static dielectric constant. There is a clear conclusion for BSO. Apart from the small grating spacing region the application of an electric field is of no great help for increasing the imaginary part of the space charge field but it will help with the modulus. For $BaTiO_3$ it may be seen equally clearly that an applied electric field always helps in reaching higher stationary space charge fields. The actual values of the achievable space charge field are also of interest. For $BaTiO_3$ for an applied field of 9 kV cm^{-1} at a grating spacing of 10 μm the space charge field is as high as 6.5 kV cm^{-1}.

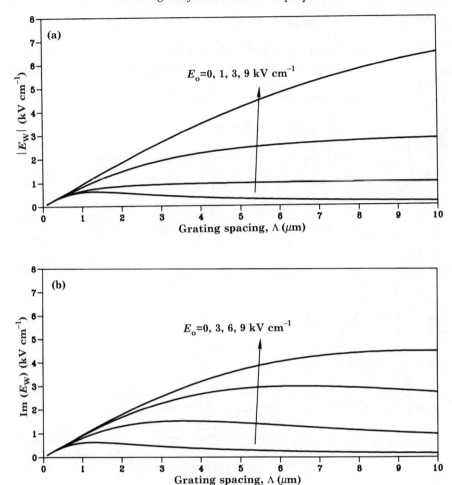

Fig. 2.4 Same as Fig. 2.3 but for BaTiO$_3$.

It is worth noting another limit, although it may be difficult to realize it in practice. It follows from eqn (2.58) that for sufficiently large E_o the space charge field may be made equal to E_q at any grating spacing.

Rise time. The time it takes for the space charge field to rise to 63% of its final value (the usual definition of rise time) may be obtained from eqn (2.42) in the form

$$\tau_{nr} = \frac{1}{\text{Re}(p)} = \frac{E_q}{E_M} \frac{(E_M + E_D)^2 + E_o^2}{(E_q + E_D)(E_M + E_D) + E_o^2} \quad (2.60)$$

This is of course the normalized rise time. The real rise time is proportional to the dielectric relaxation time, $\tau_r = \tau_{nr}\tau_d$, hence BaTiO$_3$ is bound to be a slower material than BSO in this aspect as well. When it comes to rise time, all the material parameters may be seen to play a role: both E_M and E_q have to be taken into account.

In the absence of an applied electric field and when $E_D \ll E_M, E_q$ (i.e. for large Λ) we find that eqn (2.60) yields $\tau_{nr} = 1$, but otherwise the rise time can vary in quite a wide range. Using our previously given parameters τ_{nr} is plotted against grating spacing for a range of E_o in Figs 2.5(a) and (b) for BSO and BaTiO$_3$ respectively. It may be seen that the trend for BSO is quite the opposite to that of BaTiO$_3$. For BSO the normalized rise time increases with increasing E_o and

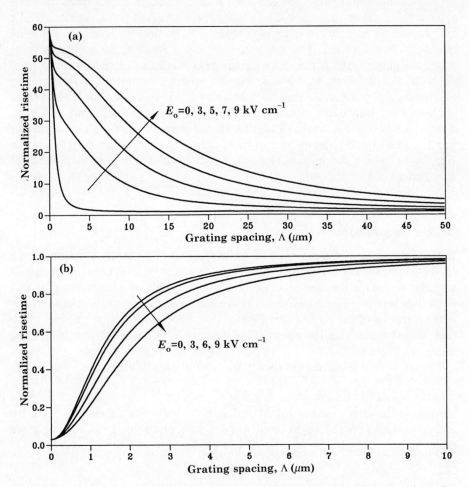

Fig. 2.5 The variation of the normalized rise time of the space charge field with grating spacing for (a) BSO and (b) BaTiO$_3$.

decreases with increasing grating spacing. For $BaTiO_3$ the normalized rise time decreases with increasing E_o and increases with increasing grating spacing. Numerically the normalized rise time for BSO is much larger than that for $BaTiO_3$. Considering however that the dielectric relaxation time is three orders of magnitude smaller in BSO than in $BaTiO_3$ we shall find that BSO is the faster material for the rise time as well.

Transients. We have so far investigated the slope at which the space charge field rises in the vicinity of $t = 0$, the stationary field, and the rise time. Now we shall discuss how it varies in the full range. In the absence of an electric field the solution is straightforward. Then $\text{Im}(p) = 0$ and the rise of the space charge field to its saturation value (see eqn 2.42) is monotonic. What happens for finite applied fields? Then the rise will be oscillatory. There will be an overshoot. The overshoot may actually be quite significant. Let us take for example a grating spacing of 3 μm for BSO. It may be seen from Fig. 2.3(b) that the stationary value of the imaginary part of the space charge field is relatively small. But the transients can be quite high, as shown in Fig. 2.6, where the imaginary part is plotted as a function of normalized time for a range of E_o. This is an example when there is a large effect only under transient conditions, a phenomenon that can be used in a novelty filter which we shall discuss in Chapter 10. $BaTiO_3$ shows smaller transients but the oscillatory behaviour is there nevertheless, as may be seen in Fig. 2.6(b), where the imaginary part of the space charge field is plotted for $\Lambda = 3\,\mu m$.

2.4 The photovoltaic effect

It was discovered by Glass *et al.* (1974) that an optical intensity incident upon a $LiNbO_3$ crystal gave rise to a current. They explained it as a consequence of an asymmetric charge transfer process. The carriers reside in asymmetric potential wells and therefore when excited they move preferentially in one direction, usually in the direction of the polar axis. In general the photocurrent depends on a third-rank tensor (known as the photovoltaic tensor). It is described in a little more detail in Chapter 9.

The photovoltaic current (in its scalar form) is usually written as

$$J_{ph} = \kappa_p \alpha I \tag{2.61}$$

where α is the attenuation constant and κ_p is a photovoltaic constant. Since the attenuation is proportional to the density of un-ionized donors it suits us much better to write the photovoltaic current in the form

$$J_{ph} = p_n I (N_D - N_D^+) \tag{2.62}$$

with p_n as the photovoltaic constant.

All our previous equations remain unchanged except the one for the current. We need to add J_{ph} to eqn (2.8) and repeat the analysis. Obviously, both the

2.4 The photovoltaic effect

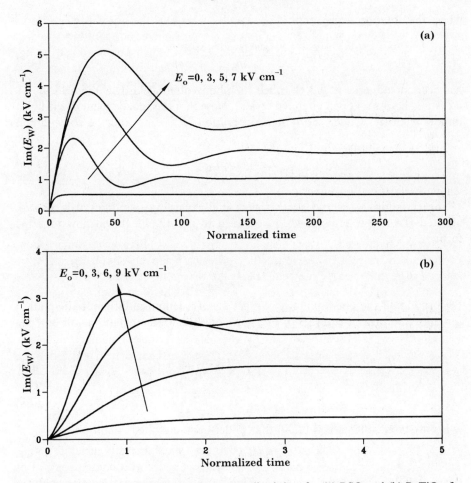

Fig. 2.6 The variation of $\text{Im}(E_w)$ against normalized time for (a) BSO and (b) BaTiO$_3$, for $\Lambda = 3\mu$m.

constant and the spatially periodic equations will be affected. Equation (2.20) will now take the form

$$J_o = e\mu n_o E_o + \varepsilon_s \frac{\partial E_o}{\partial t} + p_n I_o (N_D - N_A^-) \qquad (2.63)$$

When E_o is time-independent, the relationship is

$$E_o = \frac{J_o}{e\mu n_o} - E_{ph} \qquad (2.64)$$

where the notation

$$E_{\text{ph}} = \frac{p_n I_o (N_D - N_A^-)}{e\mu n_o} = \frac{p_n}{e\mu T_e s} \quad (2.65)$$

has been introduced. It will be called the photovoltaic field. It is the field that is measured across an open-circuited crystal. Note that it is independent of the input optical intensity. Under short-circuit condition E_o is zero and the short-circuit current is

$$J_{\text{sc}} = e\mu n_o E_{\text{ph}} \quad (2.66)$$

which does depend on intensity.

The equations concerned with the spatially varying quantities will also be affected. Two new terms have to be added to eqn (2.31). It will now take the form

$$0 = e\mu(n_o E_1 + n_1 E_o) + p_n[(N_D - N_{Do}^+)I_1 - N_{D1}^+ I_o] - jKk_B T\mu n_1 + \varepsilon_s \frac{\partial E_1}{\partial t} \quad (2.67)$$

Performing all the operations we can get again a temporal differential equation for E_1 in the form of eqn (2.33). The new constants in the equation will be

$$p = \frac{1}{D}\left(1 + \frac{E_D + j(E_o + aE_{\text{ph}})}{E_q}\right), \quad q = j\frac{E_D + j(E_o + E_{\text{ph}})}{D}, \quad D = 1 + \frac{E_D + jE_o}{E_M} \quad (2.68)$$

It may be seen from eqn (2.64) that under open-circuited conditions

$$E_o + E_{\text{ph}} = 0 \quad (2.69)$$

Consequently, q contains no photovoltaic excitation term. It is clear that the short-circuited condition, when $E_o = 0$, is much preferable. The space charge electric field is then

$$E_1 = jmE_M \frac{E_D + jE_{\text{ph}}}{E_M + E_D} t_n \quad (2.70)$$

in the short time limit, and

$$E_1 = jmE_q \frac{E_D + jE_{\text{ph}}}{E_q + E_D + jaE_{\text{ph}}} \quad (2.71)$$

under stationary conditions. The normalized grating formation time is equal to Re(p), which yields from eqn (2.68)

$$\tau_{\text{nr}} = \frac{E_M}{E_q} \frac{E_q + E_D}{E_M + E_D} \quad (2.72)$$

As an example of a photorefractive material with a large photovoltaic field we shall choose iron-doped LiNbO$_3$. The actual value of E_{ph} may vary considerably from sample to sample but a typical value may be 10 kV cm^{-1}. Taking

$$\varepsilon_s = 29\varepsilon_o, \quad \gamma_R = 5 \times 10^{-14} \text{ m}^3 \text{ s}^{-1}, \quad N_A^- = 4 \times 10^{22} \text{ m}^{-3},$$
$$\mu = 8 \times 10^{-5} \text{ m}^2 \text{ V}^{-1} \text{ s}^{-1} \tag{2.73}$$

for the parameters, and considering a grating spacing of $\Lambda = 3$ μm, we find that E_M and E_q are practically the same, about 120 kV cm^{-1}. The space charge field is then approximately equal to $E_{ph}t_n$ in the short time limit and to E_{ph} under stationary conditions. The normalized rise time is about unity.

2.5 Characteristic lengths and characteristic times

When discussing space charge fields we have so far described them in terms of characteristic electric fields introduced by the Kiev group (see Section 4.3). Alternatively, one might use characteristic lengths and characteristic times as is often done in the literature. In fact, we have already briefly mentioned two characteristic lengths, the diffusion length, eqn (2.52) (well known in semiconductor theory), and the Debye screening length, eqn (2.59) (well known in plasma theory). Two further lengths may be defined by

$$L_E = \mu \tau_e E_o \tag{2.74}$$

and

$$L_{ph} = \mu \tau_e E_{ph} \tag{2.75}$$

where L_E is known as the drift length and L_{ph} as the photovoltaic transport length.

E_D and E_q can be expressed in terms of the characteristic lengths as follows:

$$E_D = \frac{K L_D^2}{\mu \tau_e}, \quad E_q = \frac{1}{\mu \tau_e K} \frac{L_D^2}{L_{Deb}^2} \tag{2.76}$$

We could now rewrite all our previous expressions with the aid of characteristic lengths but they would look complicated and not particularly instructive. It is much preferable to consider here only the simple cases when one of the effects dominates, i.e. we shall look at the photovoltaic, diffusion, and drift dominated regimes separately.

We are in the photovoltaic regime when $E_o = 0$ (short-circuited case) and $E_{ph} \gg E_D$. In terms of characteristic lengths the condition is

$$L_{ph} \gg K L_D^2 \tag{2.77}$$

If we take a crystal that does not display the photovoltaic effect but we apply a field, then the condition to be in the drift regime is $E_o \gg E_D$ (and vice versa for the diffusion dominated regime). The new criterion is

$$L_E \gg K L_D^2 \tag{2.78}$$

Short time limit. In the photovoltaic region

$$E_1 = -m \frac{L_{ph}}{\mu \tau_e} \tau_n \tag{2.79}$$

In the diffusion-dominated region

$$E_1 = j \frac{m}{\mu \tau_e} \frac{KL_D^2}{1 + (KL_D)^2} \tau_n \tag{2.80}$$

from which it may be immediately seen that there is an optimum value of K equal to $1/L_D$.

In the drift-dominated region

$$E_1 = j \frac{m}{\mu \tau_e} \frac{L_E}{1 + jKL_E} \tau_n \tag{2.81}$$

which shows that the maximum is reached for long drift lengths.

Stationary solution. In the photovoltaic region

$$E_1 = -m \frac{L_{ph}}{\mu \tau_e} \tag{2.82}$$

In the diffusion-dominated region

$$E_1 = j \frac{m}{\mu \tau_e} \frac{KL_D^2}{1 + (KL_{Deb})^2} \tag{2.83}$$

It shows immediately that the optimum value of K is equal to L_{Deb}^{-1}.

In the drift-dominated region

$$E_1 = -\frac{m}{\mu \tau_e} \frac{L_E}{1 + j(L_{Deb}/L_D)^2 KL_E} \tag{2.84}$$

We can see that the maximum is reached for large drift length and that the magnitude depends on the ratio of the diffusion length to the Debye length.

Normalized rise time. In the photovoltaic region

$$\tau_{nr} = 1 \tag{2.85}$$

In the diffusion-dominated region

$$\tau_{nr} = \frac{1 + K^2 L_D^2}{1 + K^2 L_{Deb}^2} \tag{2.86}$$

It may be seen that $\tau_{nr} > 1$ if $L_{Deb} < L_D$, which is the usual case. It may also be seen that in the limit of large K

$$\tau_{nr} = \left(\frac{L_D}{L_{Deb}}\right)^2 \tag{2.87}$$

In the drift-dominated region

$$\tau_{\mathrm{nr}} = \frac{1 + (KL_{\mathrm{E}})^2}{1 + (L_{\mathrm{Deb}}/L_{\mathrm{D}})^2 (KL_{\mathrm{E}})} \tag{2.88}$$

which shows again that the normalized rise time is larger than unity whenever $L_{\mathrm{Deb}} < L_{\mathrm{D}}$. It may also be seen that $\tau_{\mathrm{nr}} = 1$ in the limit of small K.

Characteristic times. We have already used, quite extensively, two characteristic times: the rise time and the dielectric relaxation time. Two further characteristic times have been introduced by Valley (1983b) which are used mostly in time-varying problems. They are the diffusion time

$$\tau_{\mathrm{D}} = \frac{1}{DK^2} = \frac{\tau_{\mathrm{e}}}{(L_{\mathrm{D}}K)^2} \tag{2.89}$$

and the drift time

$$\tau_{\mathrm{E}} = \frac{1}{\mu E_{\mathrm{o}} K} \tag{2.90}$$

2.6 Figures of merit

We have introduced a number of new concepts and have shown the variation of the space charge field under various conditions. In general, however, the space charge field itself is not a useful measure for an application. We are interested in the achievable change in the index of refraction and for that the electro-optic coefficient in the particular geometrical configuration must be taken into account.

The only quantity that we have calculated so far which can be taken directly as a figure of merit is the rise time. We did indeed compare BSO and $BaTiO_3$ on that basis in Section 2.3. We shall consider a few further figures of merit, but before we do so we will need to quote two simple equations which will appear only later in Section 3.1. The first one gives the change in the index of refraction in the form

$$\Delta n = \tfrac{1}{2} n_{\mathrm{r}}^3 r_{\mathrm{eff}} E_1 \tag{2.91}$$

where n_{r} is the unperturbed index of refraction and r_{eff} is the effective electro-optic coefficient.

The second equation we need is more controversial. For some reason, presumably because of its clear conclusions and simple formulae, the analysis by Kogelnik (1969) has gained widespread acceptance, even in cases when it is not valid. In most papers concerned with diffraction efficiency in photorefractive materials, the formula quoted is the sine square variation given by eqn (1.20) and plotted in Fig. 1.2. As will be shown in Section 3.7, the sine square expression has a rather limited range of validity. Unfortunately, the correct formula is not suitable for incorporation in simple definitions of figures of merit. Hence, rather reluctantly, we shall give here the expression widely used:

$$\eta_D = \exp(-\alpha d/\cos\theta)\sin^2\left(\frac{\pi\Delta n}{\lambda}\frac{d}{\cos\theta}\right) \quad (2.92)$$

which includes the effect of losses. The argument of the sine square function is the same as in eqn (1.19) but expressed this time with the change in the index of refraction[4]. Armed with eqns (2.91) and (2.92) we can now proceed with our definitions.

Photorefractive sensitivity. It is defined as the change in the index of refraction per absorbed energy per unit volume. It is a measure of how well the material uses a given amount of optical energy in the initial stages of grating development. It may be written as

$$S = \frac{|\Delta n|}{\alpha I_0 t} \quad (2.93)$$

Considering eqn (2.91) for Δn and substituting for E_1 from the short time limit (eqn 2.46), we obtain

$$S = \frac{m}{2}n_r^3\frac{r_{\text{eff}}}{\varepsilon_s}e\mu\tau_e\frac{\eta_q}{\hbar\omega}|q| \quad (2.94)$$

The quantum efficiency η_q and the photon energy $\hbar\omega$, have come from eqn (2.2), and q was given in eqn (2.34). The value of n_r is roughly the same for all photorefractive crystals; $r_{\text{eff}}/\varepsilon_s$ has received some attention in the past as the polarization-optic coefficient (see e.g. Wemple et al. 1968). Its maximum value varies only moderately from crystal to crystal. The three quantities together may constitute a figure of merit and may be defined as

$$Q = n_r^3\frac{r_{\text{eff}}}{\varepsilon_s} \quad (2.95)$$

Its value for a number of photorefractive materials together with some other parameters is given in Table 2.1.

The quantity in eqn (2.94) that can vary widely within orders of magnitude is $|q|$, as witnessed by Figs 2.1(a) and 2.2(a) (note that $|q| = (1/m)\text{d}|E_1|/\text{d}t_n$) plotted for BSO and BaTiO$_3$, respectively. For other materials we can of course always use eqn (2.94) in conjunction with eqn (2.46) provided all the parameters are known.

---

[4]The coupling coefficient, κ, is expressed in eqn (1.16) with the aid of $\varepsilon_{r1}/\varepsilon_{ro}$. Since the relative permittivity is equal to the square of the index of refraction we can define Δn by the relation $n_r + \Delta n = \sqrt{\varepsilon_{ro} + \varepsilon_{r1}} = \sqrt{\varepsilon_{ro}}\sqrt{1 + \varepsilon_{r1}/\varepsilon_{ro}} \approx \sqrt{\varepsilon_{ro}}(1 + \varepsilon_{r1}/2\varepsilon_{ro})$, whence $\Delta n/n = \varepsilon_{r1}/2\varepsilon_{ro}$.

2.6 Figures of merit

Table 2.1 Parameters of some photorefractive materials (after Yeh 1987a)

Materials	λ (μm)	r (pm V^{-1})	n	ϵ/ϵ_o	Q^a (pm/Vϵ_o)	Q (MKS)
BaTiO$_3$	0.5	$r_{42} = 1640$	$n_e = 2.4$	$\epsilon_1 = 3600$	6.3	0.71
SBN	0.5	$r_{33} = 1340$	$n_e = 2.3$	$\epsilon_3 = 3400$	4.8	0.54
GaAs	1.1	$r_{12} = 1.43$	$n_e = 3.4$	$\epsilon_3 = 12.3$	4.7	0.53
BSO	0.6	$r_{41} = 5$	$n = 2.54$	$\epsilon = 56$	1.5	0.17
LiNbO$_3$	0.6	$r_{33} = 31$	$n_e = 2.2$	$\epsilon_3 = 32$	10.3	1.16
LiTaO$_3$	0.6	$r_{33} = 31$	$n_e = 2.2$	$\epsilon_3 = 45$	7.3	0.83
KNbO$_3$	0.6	$r_{42} = 380$	$n = 2.3$	$\epsilon_3 = 240$	19.3	2.2
GaP	0.56	$r_{41} = 1.07$	$n = 3.45$	$\epsilon = 12$	3.7	0.41

As a special case we shall now work out the sensitivity of BSO under the usually valid condition $E_o \gg E_M \gg E_D$. In that case (see eqn 2.46)

$$q = E_M = \frac{1}{\mu \tau_e K} \tag{2.96}$$

and eqn (2.94) takes the form

$$S = \frac{1}{4\pi} m Q \frac{e}{\hbar \omega} \eta_q \Lambda \tag{2.97}$$

In order to have the greatest sensitivity we shall choose unity beam ratio ($m = 1$) and unity quantum efficiency ($\eta_q = 1$). We are left then with the material parameters, the optical wavelength and the grating spacing (interestingly, μ and τ_e no longer appear in the equation). Taking

$$\lambda = 0.5\,\mu\text{m}, \ \Lambda = 10\,\mu\text{m}, \ n_r = 2.54 \text{ and } r_{\text{eff}} = 3.4 \times 10^{-12}\,\text{m V}^{-1} \tag{2.98}$$

we obtain

$$S = 35.6\,\text{cm}^3\,\text{kJ}^{-1} \tag{2.99}$$

which is close to Glass's (1978) estimate (although for a different grating spacing) of 100 cm^3 kJ^{-1} for the ultimate sensitivity.

We have to realize however that for any given crystal Q may very well depend on the geometrical configuration since both the electro-optic coefficients and the dielectric constants are tensor quantities. This is particularly true for BaTiO$_3$ which has three different elements in the electro-optic tensor, and the dielectric constant has transverse and longitudinal values. The usual arrangement for recording a grating in BaTiO$_3$ is shown in Fig. 2.7. The beams are incident with an interbeam angle of 2θ, but the bisector is now tilted at an angle β. The corresponding equations for r_{eff} and ε are given in Appendices A and B. Our intention here is to show the variation of Q and r_{eff} for one particular grating spacing ($\theta = 30°$) and concentrate on the variation with β, the angle subtended by

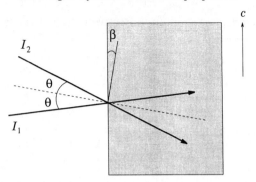

Fig. 2.7 Usual experimental arrangement for recording holograms in BaTiO$_3$.

the grating vector with the polar axis. In may be seen in Fig. 2.8 that there is an optimum value of β around 20°. Changing q does not affect this curve to any great extent although the maximum does become broader for smaller θ. The dashed line in Fig 2.8 represents the concurrent variation in r_{eff}. Recalling that the change in the refractive index (eqn 2.91) and hence the diffraction efficiency is proportional to r_{eff}, we note that the angle β which gives optimum diffraction is not necessarily the same as that which leads to greatest energy transfer in photorefractive two-wave mixing.

Another, related, measure is the amount of *incident energy per unit area* needed to record a grating of 1% diffraction efficiency in a 1 mm thick crystal. Using eqn

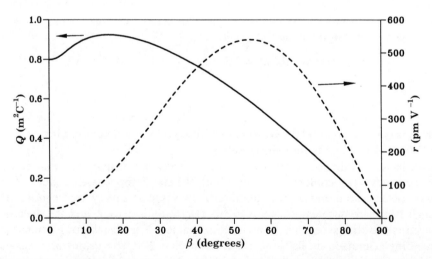

Fig. 2.8 The variation of the photorefractive figure of merit, Q, with angle of incidence, β, for the crystal of BaTiO$_3$ depicted in Fig. 2.7. The incident waves are of extraordinary polarization. The dashed curve (right-hand scale) represents the concurrent variation in effective electro-optic coefficient, r_{eff}.

(2.92) for diffraction efficiency (we can use the argument of the sine function) and eqn (2.91) for Δn we obtain

$$I_o t = \frac{2}{\pi} \frac{\sqrt{\eta_D}}{\eta_q} \frac{1}{Q} \frac{hc}{e} \frac{1}{\mu \tau_e \alpha dm |q|} \quad (2.100)$$

where c is the velocity of light, $\eta_D = 0.01$, and $d = 10^{-3}$ m. It may be seen that nearly the same parameters come in as in our previous measure of photorefractive sensitivity. Assuming again $q = E_M$ for BSO and taking the values given in eqn (2.98) and the further parameter values $\alpha = 1.5\,\text{cm}^{-1}$, $\varepsilon_s = 56\varepsilon_0$, we obtain $I_o t = 0.76\,\text{J m}^{-2}$.

A third measure is the *total available change in the index of refraction*, Δn_{tot} which we get from eqn (2.91) by taking the saturation field for E_1 from eqn (2.58). That will enable us to determine the value of Δn_{tot} for any material under any geometrical configuration. In what follows we shall work out its value under the rather optimistic assumption that we are able to reach a saturation space charge field equal to E_q. We obtain then

$$\Delta n_{\text{tot}} = \frac{1}{2} n_i^3 r_{\text{eff}} E_q = \frac{Q}{4\pi} e N_A^- (1-a) \Lambda \quad (2.101)$$

The new feature appearing is the acceptor density, N_A^-. In principle one should be able to influence it by adding the right impurities at the crystal growing stage. Doping LiNbO$_3$ has indeed been a success story but for most crystals this important parameter is not under control yet. It is also possible to interfere with the properties of the crystal after crystal growth as shown for BaTiO$_3$ by Ducharme and Feinberg (1986). They treated the crystal at 650 °C in oxygen at different partial pressures. Both the density of traps and the sign of the photocarrier (from electrons to holes and vice versa) could be changed.

2.7 The very short time limit

In the previous sections we have presented the development of the space charge field for the short time limit. We have shown the rise time to be proportional to the dielectric relaxation time. Hence we could conclude that by increasing the input intensity we could decrease the rise time at the same rate. Is that true? It is not true. We assumed the electrons to have been excited *before* the grating formation starts. In other words we assumed that the lifetime of the electron is much less than the dielectric relaxation time. If we increase the intensity, a stage will be reached when the dielectric relaxation time becomes *shorter* than the lifetime. Then the physical separation of the phenomena into a stage of excitation and a stage of space charge development is no longer possible. The linear relationship between rise time and dielectric relaxation time is bound to break down.

Mathematically, the low intensity assumption appeared in Section 2.1 where we assumed that the spatially periodic electron density is small relative to the ionized donor density, and failed to include it in eqn (2.9). If we do include the electron density in Poisson's equation then the derivation is much more complicated;

instead of a first order temporal differential equation with constant coefficients for the space charge field (eqn 2.33) we obtain a second-order differential equation with time-varying coefficients. We shall return to these problems in Sections 5.5 and 5.2.10. Our intention in the present section is to look at the simple diffusion-limited case (no photovoltaic field, no applied electric field) and show the variation of rise time and steady-state space charge field as a function of input intensity (see Fig. 2.9). For parameters we shall take those of BSO under the obviously imperfect assumption that the same physical model is applicable independently of the input intensity. This exercise will, however, give an indication of the orders of magnitude involved.

It may be seen in Fig. 2.9 that the slope of the decrease in rise time with increasing intensity has a break-point at around 10^6 W m^{-2}. Looking at the actual figures it may be deduced that the variation changes at that point from an I_0^{-1} to an $I_0^{-1/2}$ relationship. The saturation space charge field remains constant for a remarkably wide range and then starts to decline.

We have of course left many questions unanswered, e.g. could we create the carriers by a short pulse and then wait for the grating to appear in the dark, a question that will be considered in Section 5.5. We have however shown that the photorefractive effect is not inherently slow, that at high enough intensities the nanosecond and even the picosecond ranges are within reach.

2.8 Enhancement mechanisms

Is there a way to obtain space charge electric fields in excess of those calculated in Section 2.3? The answer is yes. Enhancement is possible whenever the coefficient p

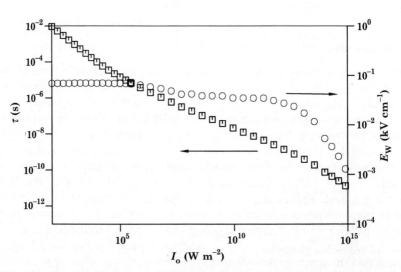

Fig. 2.9 The intensity dependence of rise time (squares, left-hand axis) and imaginary part of the steady-state space charge field (circles, right-hand axis) for a crystal of BSO.

2.8 Enhancement mechanisms

has an imaginary part. There are two ways to do it: either detune one of the beams (Huignard and Marrakchi 1981a; Refregier et al. 1985) or apply an AC voltage instead of a DC voltage (Stepanov and Petrov 1985; K. Walsh et al. 1990).

Detuning. Two coherent optical beams incident at an angle will produce a standing interference pattern as we have seen in eqn (1.37). What happens when we detune one of the beams? Incorporating the phase angles in the complex amplitude we may write the total optical electric field due to both beams:

$$\mathcal{E} = \tfrac{1}{2}[\mathcal{E}_1 \exp j(\omega_1 t - \mathbf{k}_1 \cdot \mathbf{r}) + \mathcal{E}_2 \exp j(\omega_2 t - \mathbf{k}_2 \cdot \mathbf{r}) + \text{c.c.}] \qquad (2.102)$$

where c.c. stands for complex conjugate.

The intensity (i.e. the interference pattern) varies then as[5]:

$$I = I_o\left[1 + \frac{m}{2}\exp j[(\omega_1 - \omega_2) - (\mathbf{k}_1 - \mathbf{k}_2)\cdot \mathbf{r}] + \text{c.c.}\right] \qquad (2.103)$$

where

$$I_o = |\mathcal{E}_1|^2 + |\mathcal{E}_2|^2 \quad \text{and} \quad m = \frac{2\mathcal{E}_1 \mathcal{E}_2^*}{I_o} \qquad (2.104)$$

The difference of the wave vectors has been defined as the grating vector $\mathbf{K}$ (see eqn 1.3). The new relationship is the difference of the frequencies, which is usually referred to as the detuning frequency and is denoted by

$$\Omega = \omega_1 - \omega_2 \qquad (2.105)$$

---

[5] A little reflection may be in order now. We introduced optical electrical fields with eqn (1.6) where R and S were complex amplitudes. This technique is widely used both in circuit theory (where the quantities are supposed to vary periodically in time) and in field theory where the variation is both in time and space and the differential equations are linear. It is a technique so often used that it is easy to forget a very important tacit assumption. When we write an electric field in the complex form we always mean the real part. And when we finish our calculations we have to take the real part of the final answer. The beauty of this technique is not only that it significantly reduces mathematical labour but it also provides a new kind of physical intuition when we imagine (or actually plot) the phase and amplitude in the complex plane. The intensity is then simply given by the modulus square of the electric field which is actually equivalent (apart from a factor 2 which plays no role) with taking the time average. This choice also means that the impedance of the medium may be taken as a constant for all photorefractive phemonena.

We also introduced the excitation in complex form in eqn (2.28) and gave there a warning that it is the real part that should be understood. The differential equations were by that time linear so everything was all right. In the present section however we have changed tack. By introducing the complex conjugate the optical electric field in eqn (2.102) is real. We still use the complex amplitudes but we are now no longer restricted to linear differential equations – and that is a good thing because the differential equations for the optical field coming in the next chapter will be non-linear.

Can we take now intensity as being proportional to the time average of the square of the electric field? Not quite. We can take an average of the fast-varying phenomena (which vary at the optical frequency) but not of the slowly varying phenomena (which vary at the difference of the optical frequencies) because the photorefractive material is capable of responding to the latter one. This is why the intensity is taken in the form of eqn (2.103).

Do we need to take into account the complex conjugate in eqn (2.103) when considering the excitation? No, in the presence of detuning we may still rely on the linearized differential equations and the complex conjugate may be disregarded.

Writing now the interference pattern in the form $\exp[j(\Omega t - \mathbf{K} \cdot \mathbf{r})]$ it is clear that we have an interference pattern that moves with the velocity, $v = \Omega/K$. What is the difference between the previous and the present formulation? In the excitation term we have to replace $\exp(-jKz)$ (remember that for symmetric beam incidence $\mathbf{K}$ has only a z component) by $\exp[j(\Omega t - Kz)]$. Hence the new term is $\exp(j\Omega t)$, which means that on the right-hand side of eqn (2.33) we have to replace mq by $mq \exp(j\Omega t)$. From the point of view of the differential equation this signifies a harmonic excitation at a frequency Ω. But, the differential equation itself has oscillatory eigensolutions when $\text{Im}(p) \neq 0$. *Resonance* will occur when the eigenfrequency agrees with the impressed frequency,

$$\Omega = -\text{Im}(p) \tag{2.106}$$

From eqn (2.34) this happens when

$$\Omega = \frac{1}{\tau_d} \text{Im}\left[\frac{1 + jE_o/E_q}{1 + jE_o/E_M}\right] \tag{2.107}$$

where diffusion has been neglected. Under the assumption

$$E_M \ll E_o \ll E_q \tag{2.108}$$

valid under usual conditions for BSO, eqn (2.107) reduces to

$$\Omega = \frac{1}{\tau_d} \frac{E_M}{E_o} \tag{2.109}$$

The corresponding maximum value of the imaginary part of the space charge field will come to (for the calculation see Section 5.4.2)

$$\text{Im}[E_1] = m \frac{E_o^2 E_q}{E_o^2 + E_M E_q} \tag{2.110}$$

For $E_o = 10 \text{ kV cm}^{-1}$ and $\Lambda = 20 \text{ μm}$, using our usual parameters for BSO, the corresponding value of $\text{Im}(E_1)$ comes to 37 kV cm^{-1} in contrast to about 1 kV cm^{-1} which can be read from Fig. 2.3(b), an increase by a factor of 37. There is in fact a range of grating spacings for which significant enhancement is possible. This is shown in Figs 2.10(a) and (b) where the modulus and the imaginary part of the space charge field are plotted against detuning, again for BSO. The applied electric field is taken as 7 kV cm^{-1}. The solid line shows the maximum space charge field possible at each detuning frequency. The maximum space charge field as a function of grating spacing may be easily obtained from eqn (2.110) by differentiation with respect to Λ. The maximum occurs at a grating spacing of (Refregier *et al.* 1985)

$$\Lambda_{\text{opt}} = \frac{2\pi E_o}{N_A^-} \left(\frac{\varepsilon_s \mu}{e \gamma_R}\right)^{1/2} \tag{2.111}$$

2.8 Enhancement mechanisms

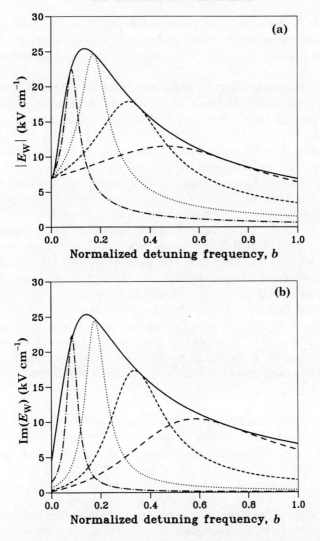

Fig. 2.10 Variation of (a) $|E_w|$ and (b) $\text{Im}(E_w)$ with normalized detuning frequency ($b = \Omega \tau_d$) for BSO for $E_o = 7\,\text{kV cm}^{-1}$ and $\Lambda = 7\,\mu\text{m}$ (-·-), $15\,\mu\text{m}$ (····), $30\,\mu\text{m}$ (- - - -), and $60\,\mu\text{m}$ (– – – –). The solid line shows the maximum space charge field possible at each detuning.

and the corresponding space charge field is

$$\left|\frac{E_{s1}}{m}\right| \simeq \text{Im}\left(\frac{E_{s1}}{m}\right) = \frac{E_{qo}}{2}$$
$$= \frac{E_o}{2}\left(\frac{e\mu}{\varepsilon_s \gamma_R}\right)^{\frac{1}{2}} \quad (2.112)$$

It is interesting to note that this maximum depends on the ratio $\mu/\gamma_R \varepsilon_s$. It may also be seen from Figs 2.10(a) and (b) that at optimum detuning, the imaginary part and the modulus of the space charge field are approximately equal.

We shall go into a lot more detail in Section 5.4 about the merits and demerits of detuning, and we shall in Section 5.4 say more about the topic of space charge waves, which we shall briefly introduce here on the basis of physical considerations.

Waves in physical configurations involving charge carriers drifting with a certain average velocity have been studied in many branches of physics and engineering, particularly in the discipline of plasmas and in the branch of engineering concerned with the design of microwave oscillators and amplifiers. It is not unreasonable to assume that the configuration of electrons, ionized donors, and ionized acceptors will also permit the appearance of waves which we shall call space charge waves. If they are waves they should have dispersion characteristics, that is, their wavenumbers and frequencies will be related to each other. Our moving interference pattern on the other hand could also be regarded as a wave, with Ω and K representing its frequency and wavenumber respectively. Obviously, the moving interference pattern is an impressed wave where we choose Ω by the amount of detuning, and K by the inter-beam angle of the optical beams. The space charge wave is an eigenwave of the system. When the impressed wave and the eigenwave have the same frequency and wave number, then resonance occurs and the resulting space charge field is very much increased.

Applied AC field. We have discussed the application of a DC voltage leading to an applied electric field in the interior of the crystal. Can we do better with a time-varying periodic field? Yes, we can, particularly when the time variation takes the form of a square wave, in which the direction of the electric field reverses every half period. Why does it enhance the space charge field? The basic principle can essentially be attributed to an augmentation of the diffusion process. When electrons are optically excited from the valence band into the conduction band, they will experience a force from the applied electric field which will lead to a preferential drift of electrons in the direction opposite to the field. If the field is varied with a repetition frequency that is much less than the inverse of the electron recombination time (i.e. $\Omega_r \ll 1/\tau_e$), the electrons will experience a constant force during their lifetimes. This force will alternate periodically in direction. If, in addition, the grating formation time takes many periods of field oscillations (i.e. $\Omega_r \gg 1/\tau_r$), the net result will be that there is no preferential direction given to charge redistribution. The grating formed will consequently resemble that formed by diffusion only, and will be shifted 90 degrees with respect to the interference pattern.

Let us look again at eqn (2.33) which describes the temporal variation of the space charge field in the presence of an applied electric field. If the applied electric field is time-varying, then the coefficients of eqn (2.33) will also become time-

2.8 Enhancement mechanisms

varying. It is, however, still a first-order differential equation which has an exact solution. Thus the whole time variation can be obtained from a closed expression, as shown by Grunnet-Jepsen *et al.* (1994a). We shall further explore the AC transients in Section 5.4.6 but just to give an idea of how the space charge field varies, we shall plot in Fig. 2.11 the time variation of $\mathrm{Im}(E_1)$ for square wave amplitude of $E_{oo} = 7\,\mathrm{kV\,cm^{-1}}$, intensity of $10\,\mathrm{mW\,cm^{-2}}$, grating spacing of $15\,\mu\mathrm{m}$ and for a frequency of $10\,\mathrm{Hz}$. The rest of the parameters are those of BSO. It may be clearly seen that an oscillatory steady state is reached in which the basic period is one half of the applied frequency.

Can we find the average value of the space charge field in the steady state? Yes, there is a simple mathematical technique due to Stepanov and Petrov (1985), who made the first experiments as well. The argument is that since the period T of the applied field is small relative to the grating formation time, it is sufficient to consider the *average* effect of the applied field. Hence we may attempt the solution by integrating eqn (2.33) term by term for one period from nT to $(n+1)T$ where n is large enough so that steady state has been reached. Let us do the integration term by term. The first one will yield

$$\int_{nT}^{(n+1)T} \frac{\mathrm{d}E_1}{\mathrm{d}t}\,\mathrm{d}t = E_1[(n+1)T] - E_1(nT) = 0 \tag{2.113}$$

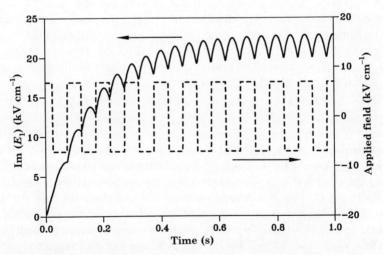

Fig. 2.11 The temporal evolution of $\mathrm{Im}(E_w)$ (solid line, left-hand axis) for an applied AC square wave electric field of frequency $f = 10\,\mathrm{Hz}$ and amplitude $E_o = 7\,\mathrm{kV\,cm^{-1}}$ (dashed line, right-hand axis). $I_o = 10\,\mathrm{mW\,cm^{-2}}$, $\Lambda = 15\,\mu\mathrm{m}$, and the parameters for BSO have been used.

The integral is obviously zero because E_1 is a periodic function. Integrating the second term on the left-hand side we obtain

$$\int_{nT}^{(n+1)T} pE_1 dt \approx E_1 \int_{nT}^{(n+1)T} p\, dt \qquad (2.114)$$

We can take the electric field out of the integral sign if we can assume that it is very nearly constant during the period. Hence the space charge field is given by the expression

$$E_1 = m \frac{\int_{nT}^{(n+1)T} q\, dt}{\int_{nT}^{(n+1)T} p\, dt} \qquad (2.115)$$

If we perform the operations for a square wave we get exactly the same formula as for enhancement by detuning, namely eqn (2.110).

The impression given in this section is that it is easy to obtain large enhancement of the space charge electric field by either of the two enhancement methods. In reality, it is not that easy. The detuning method suffers from the disadvantage that the optimum detuning is intensity-dependent (see the dependence on τ_d in eqn 2.36); hence in the presence of attenuation the optimum condition will be satisfied only at a single position inside the crystal, and consequently the total improvement is much less than the ideal one (see Section 6.8 for more details). The main difficulty with the AC method is that it is very sensitive to the waveform of the applied field. If it differs from the ideal square shape then the enhancement will be considerably reduced (Walsh et al. 1990). For an ideal square wave the enhancement factor is equal to P where

$$P = \frac{KL_E}{1 + L_D^2 K^2} \qquad (2.116)$$

If the waveform is of a trapezoidal instead of a square shape (Fig. 2.12) then a reduction occurs even for modest values of R which is a measure of deviation from the square shape. The new enhancement factor, η_e, will then depend on P as shown in Fig. 2.13. For $R = 0$, obviously, $\eta_e = P$, but for a value of $P = 10$ and $R = 0.1$ the enhancement factor reduces to 40% of its original value. This is rather important in practice because perfect square waves can never be realized. For higher frequencies (say above 500 Hz) the slew rate can no longer be ignored. If we use a sinusodial applied field instead of a square waveform the reduction is even more significant; η_e is slightly below 1 for $P = 10$.

The condition for enhancement by an AC applied field, as mentioned before, is that $\tau_g^{-1} \ll \Omega \ll \tau_e^{-1}$. In fact, this is not the only frequency region in which

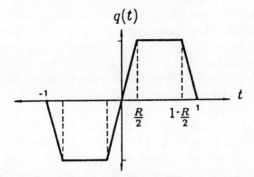

Fig. 2.12 Slew-rate-limited square waveform, normalized to the amplitude of the applied field.

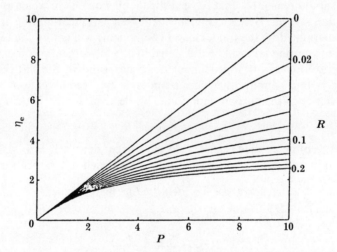

Fig. 2.13 Enhancement factor, η_e, versus the P parameter for varied values of the slew-rate parameter R (after Walsh *et al.* 1990).

enhancement is possible. Under certain conditions there is also a high frequency resonance, to which we shall return in Section 5.4.6.

2.9 Large modulation effects

One of the crucial approximations in Section 2.2 that enabled us to obtain analytical solutions was that m, the modulation of the interference pattern, is small. In practice this condition is usually (albeit not always) satisfied when the crystal is used as an amplifier (i.e. weak signal and strong pump). When the aim is to have a high diffraction efficiency then the best choice is undoubtedly to have incident beams of equal intensity, i.e. a modulation of unity. Is our theory still valid?

What is still true is that the space charge field must be a periodic function of space, and that the fundamental period is given by Λ as worked out from the inter-beam angle. If the space charge field is periodic it can be expanded into a Fourier series. The fundamental component will vary as $\cos Kz$ (in general with some phase angle relative to the interference pattern) and that is exactly what we have assumed so far about the spatially varying space charge field. Hence the questions we should ask are: (i) will the fundamental component remain unchanged when we take into account non-linearity, and (ii) how large will be the harmonic content? The answer to the first question is very important. If we obtain only one third of the expected space charge field, then so much smaller is the useful refractive index modulation. The answer to the second question is less important because the different Fourier components do not interfere with each other. We should remember that we investigate phenomena in which Bragg selectivity is always present. Hence a beam incident on a particular grating at the Bragg angle (as determined by the fundamental component) will ignore the higher components of the grating and will be diffracted by the fundamental component only.

Let us first look at an approximate treatment valid in the diffusion regime under steady-state conditions which assumes linear recombination but retains the non-linearity in the current equation. The electron density is then given by

$$n(z) = sI_o(N_D - N_A^-)\tau_e(1 + m\cos Kz) \tag{2.117}$$

In the absence of an applied electric field the current density is zero and therefore the conduction and diffusion currents must cancel each other, yielding

$$e(1 + m\cos Kz)E(z) = k_B TK \sin Kz \tag{2.118}$$

whence the space charge field can be obtained as

$$E(z) = \frac{mE_D \sin Kz}{1 + m\cos Kz} \tag{2.119}$$

The dependence on m is clearly non-linear but it still reduces to our previous result when m is small (in the present trigonometric formulation the 90 degree phase shift is implied by the sinusoidal variation). The profile of the space charge field for one period, calculated from eqn (2.119), is shown in Fig. 2.14 with m as a parameter. It may be clearly seen that for larger values of m the profile is far from sinusoidal. The magnitude of the fundamental component may be obtained from a Fourier expansion (the singularity at $m = 1$, $Kz = \pi$ occurs because we neglected thermal effects but it does not affect the expansion) (Hall et al. 1985):

$$f_1(m) = \frac{2}{m}[1 - \sqrt{1 - m^2}] \tag{2.120}$$

It is plotted in Fig. 2.15. In the limit of small m we find that $f(m)$ reduces to m and it is equal to 2 at $m = 1$. The emerging picture is that large modulation effects are favourable, we can obtain a higher space charge field. This is definitely true in the

2.9 Large modulation effects

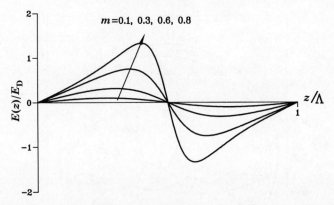

Fig. 2.14 The spatial profile of the space charge field for diffusion recording. The space charge field becomes increasingly non-sinusoidal as the intensity modulation, m, approaches 1.

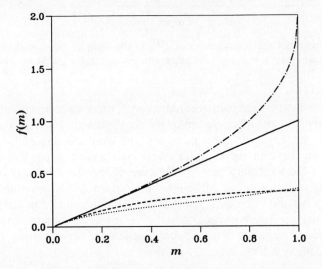

Fig. 2.15 Phenomenological functions to account for large modulation effects: $f_o = m$ (——), f_1 (-··-), f_2 for $a = 2.8$ (- - - -), and f_3 for $a = 7.6$ (·····).

diffusion dominated region as the experimental results indicate (Hall *et al.* 1985). However, when the space charge field is increased using the enhancement methods mentioned in Section 2.8 then large modulation effects can be shown to be harmful (Au and Solymar 1988a, 1990; Brost 1992). To show what happens in the detuning case as the modulation increases, we plot in Fig. 2.16 the normalized space charge field, $w_1/m = E_w/E_q$ in the complex plane as a function of normalized detuning ($y = \Omega\tau_e$) for a grating spacing of $\Lambda = 20\,\mu\text{m}$ and for the usual BSO parameters. For low modulation it is a circle (see Solymar and Ringhofer 1988).

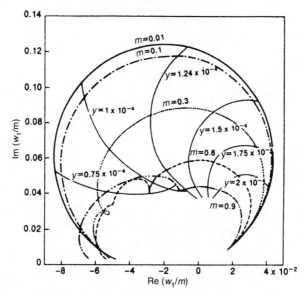

Fig. 2.16 The normalized space charge field E_w/E_q in the complex plane with m and the normalized frequency detuning, $y = \Omega\tau_e$, as parameters (after Au and Solymar 1988a).

The curve for low modulation is in accordance with what we have said about the effect of detuning. At optimum detuning the maximum of $\text{Im}(w_1/m)$ is about equal to w_1/m. As the modulation increases, the available space charge field decreases and the curve in the complex plane changes its shape. For $m = 0.9$ the maximum of the imaginary part of the space charge field reduces to about 30% of its low modulation value. Although the proof for this large reduction was provided only in Au and Solymar (1988a), it was actually guessed in Refregier *et al.* (1985) by setting up a phenomenological model. It was claimed there that large modulation effects at optimum detuning can be accounted for by simply replacing the modulation, m, by a function

$$f_2(m) = \frac{1}{a}[1 - e^{-am}] \tag{2.121}$$

where the value of the parameter a was chosen at $a = 2.8$. Note that for small enough modulation, $f_2(m) = m$. As m increases, the variation is now sublinear as shown in Fig. 2.15.

For AC excitation the reduction in gain due to large modulation effects was first shown by Stepanov and Sochava (1987). The theory both for square wave and sinusoidal applied fields was worked out by Brost (1992), who introduced a phenomenological function of the form

$$f_3(m) = \frac{1}{a}[1 - e^{-am}]e^m \tag{2.122}$$

2.9 Large modulation effects

which is also plotted in Fig. 2.15 for $a = 7.6$. Brost also concluded that under large modulation conditions the space charge field cannot significantly exceed the applied field. For a square wave temporal variation of $10\,\text{kV}\,\text{cm}^{-1}$ amplitude the spatial variation of the space charge field for one grating period for $m = 1$ turns out to be very close to a square waveform as shown in Fig. 2.17 for $a = 8.7$. The amplitude of the fundamental space charge field is then about 30% higher.

We have so far talked about the fundamental component only. Will there be higher harmonics? Obviously, if the profile of the space charge field is non-sinusoidal then higher harmonics are bound to appear. In fact, for the moving grating case, the higher harmonics have a very significant effect on the fundamental. Au and Solymar (1988a) had to take into account up to 30 harmonics in order to find a converging solution for the fundamental. Can we find these higher harmonics experimentally? The experiment is quite easy if we consider that the nth harmonic means a grating vector of $n\mathbf{K}$ magnitude. If that grating is present it will diffract a probe beam incident at the Bragg angle at a different wavelength. The usual situation is that the fundamental grating is written in the green and the probe beam is at a longer wavelength in the red. A schematic diagram of the experimental set-up of Vachss and Hesselink (1988a) may be seen in Fig. 2.18. They detuned one of the beams by a piezoelectric mirror and found the detuning for the second and third harmonic which gave maximum diffraction of the red beam. The diffraction of the red beam by the second harmonic grating (grating vector $2\mathbf{K}$) is shown schematically in Fig. 2.19.

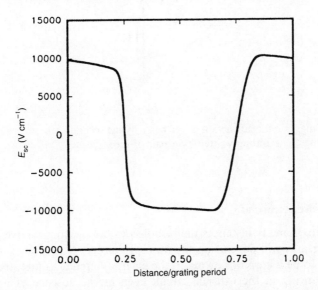

Fig. 2.17 The spatial profile of the space charge field in BTO for an applied square wave AC-field of amplitude $10\,\text{kV}\,\text{cm}^{-1}$, modulation $m = 1$, and $a = 8.7$ (after Brost 1992).

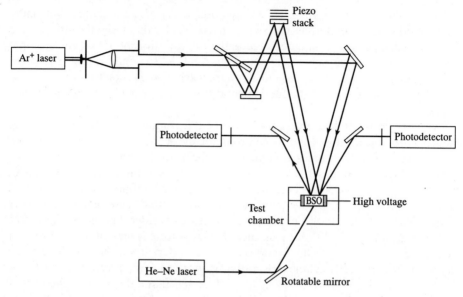

Fig. 2.18 Experimental set-up for probing the higher harmonic gratings (after Vachss and Hesselink 1988a).

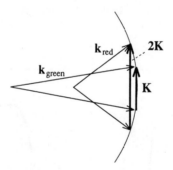

Fig. 2.19 Ewald diagram for diffraction of red beam by the second harmonic component of a grating written by a pair of green beams.

2.10 External circuit current

As we have seen, in many instances the response of the photorefractive crystal to incident light beams will depend on the applied voltage. If a voltage is applied, there is bound to be a current flowing in the external circuit. In fact, there might be a current flowing in the external circuit even in the absence of an applied voltage, provided of course that there is a conducting path between points A and B in Fig. 1.7, the simplest case being a short-circuit. A current will certainly flow when the crystal exhibits the photovoltaic effect, but currents can flow in the

2.10 External circuit current

external circuit even without the photovoltaic effect. A moving interference pattern for example would give rise to a current. We can imagine it as the electrons being given a gentle push in one direction by the moving interference pattern. If the interference pattern were to move in one direction for a while and then reverse its motion, then one could expect an AC current to flow in the external circuit.

The problem was investigated in quite some detail by Stepanov and Petrov and co-workers in the Ioffe Institute. The following analysis is based on a paper by Petrov et al. (1990). The starting point is the equation for the total current which we introduced with eqn (2.6) and reproduce below:

$$J_{\text{total}} = J + \varepsilon \frac{\partial E}{\partial t} \quad (2.123)$$

This is the current that must be the same at every cross-section[6]. It is independent of space but it can depend on time.

How is the current in the external circuit related to the current in the material? Wherever current flows, it will contribute to current in the external circuit. The total contribution may be obtained by taking the average of the current at each cross-section. Hence

$$J_{\text{external}} = \frac{1}{d} \int_0^d J_{\text{total}} dx = \frac{1}{d} \int_0^d \left[e\mu n(x,t) E(x,t) + k_B T \mu \frac{\partial n(x,t)}{\partial x} + \varepsilon \frac{\partial E}{\partial t} \right] dx \quad (2.124)$$

Integration of the second term,

$$\int_0^d \frac{\partial n(x,t)}{\partial x} dx = n(d,t) - n(o,t) = 0 \quad (2.125)$$

yields zero since the charge density varies periodically and is assumed to be the same at the beginning and at the end of the crystal. Integration of the third term gives

$$\varepsilon \frac{\partial}{\partial t} \int_0^d E dx = \varepsilon \frac{\partial V_0}{\partial t} \quad (2.126)$$

which is the capacitive current. We can ignore it because we shall not be concerned here with time-varying applied voltages. Hence the current in the external circuit will be determined solely by the average of the conduction current in the material.

---

[6]In our equations it is always the current density that appears and not the current. This is because we tacitly assumed that the cross-section of the crystal is constant. Thus we talk about current when, strictly speaking, we should talk about current density. We hope that it does not cause any confusion.

The time-varying interference pattern was achieved by Petrov et al. by sinusoidal phase modulation of one of the input beams. This means that m, the modulation of the interference pattern, takes now the form

$$m(t) = m\exp(j\Delta \cos \Omega t) \tag{2.127}$$

which is reduced to

$$m(t) = m(1 + j\Delta \cos \Omega t) \tag{2.128}$$

for small values of Δ. Thus the first term is a constant (leading to a static grating) and the second term corresponds to two travelling waves moving in opposite directions. Consequently, the space charge electric field can be determined from eqn (2.33) where the right-hand side contains now the three driving functions at frequencies 0, $+\Omega$ and $-\Omega$. Having determined the electric field we can obtain the corresponding electron charge densities from eqn (2.31) in a similar form, and finally we can substitute both the space charge electric field and the electron density in eqn (2.124) for the external current. The component of the current that varies with Ω is obtained by Petrov et al. for $E_o = 0$ in the simple form

$$J_{\text{external}} = \frac{m^2 \Delta}{2} \frac{e\mu n_o E_D}{1 + (E_D/E_M)^2} \frac{-j\Omega/\Omega_o}{1 + j\Omega/\Omega_o} \tag{2.129}$$

where

$$\Omega_o = 1/[\tau_d(1 + (E_D/E_M)^2)] \tag{2.130}$$

It may be seen that the current saturates for sufficiently high frequency. The current measured by them for three different intensities is plotted in Fig. 2.20. It shows the expected behaviour with frequency.

In the presence of an applied voltage Petrov et al. show that the external current is highest when the resonance condition given by eqn (2.109) is satisfied.

2.11 Electrons and holes

Up to now we have made the assumption that the photo-excited carriers were of one particular kind, namely electrons. There is no reason of course that in some of the materials the dominant carriers should not be holes. Some of the parameters would then be different like mobility and lifetime, and the change in the polarity of the charge is bound to make some other signs reverse (e.g. the direction of amplification would change), but the basic phenomena would remain unaltered. However, more drastic changes may be expected when both types of carriers are present. One would expect electrons and holes to work against each other because, under the effect of diffusion, both types move in the same direction along the concentration gradient. This is the same kind of thing that happens in the Hall effect when electrons and holes are simultaneously present. The Hall voltage is then reduced owing to the fact that both types of carriers are deflected in the *same* direction by the applied magnetic field.

2.11 Electrons and holes

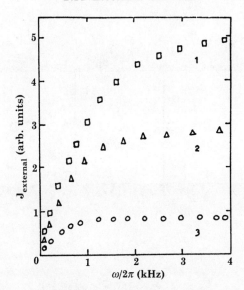

Fig. 2.20 Amplitude of non-steady-state photo-e.m.f. signal as a function of modulation frequency for (1) $I_o = 16\,\text{W}\,\text{cm}^{-2}$, (2) $I_o = 8\,\text{W}\,\text{cm}^{-2}$, (3) $I_o = 3.2\,\text{W}\,\text{cm}^{-2}$. BSO, $\lambda = 633\,\text{nm}$, $\Lambda = 20\,\mu\text{m}$, $m = 0.9$ (after Petrov et al. 1990).

The first report on the deleterious effect of having simultaneously both electrons and holes present is due to Orlowski and Krätzig (1978), who measured the diffraction efficiency for LiNbO$_3$ crystals containing different amounts of Fe$_2^+$/Fe$_3^+$ which determines the ratio of electrons and holes. The diffraction efficiency was high when the mobile carriers were either holes or electrons but declined when both carriers were present. Similar experiments were conducted by Ducharme and Feinberg (1986), who, as mentioned in Section 2.6, changed the polarity of the dominant carriers in BaTiO$_3$ by a heat treatment.

How could we incorporate holes into our models? The first, rather concise, analysis by Stepanov (1982) has been consistently ignored by Western literature. The most popular models are those of Valley (1986) and Strohkendl et al. (1986), shown in Fig. 2.21(a) (model 1) and (b) (model 2).

In model 1 we have a single recombination centre which we may still regard as our donor atom. In the absence of thermal generation the rate equation is given by eqn (2.3) when only electrons are involved. The new feature in model 1 is that ionization and deionization may also occur via holes. What we need to know is that a hole falls spontaneously upwards from the valence band into a neutral atom and makes it positive. Thus recombination of holes will create a positive charge. Alternatively, a hole is excited downwards out of an ionized atom and makes it neutral. Hence the rate of change of the positively ionized donors is now

$$\frac{\partial N_D^+}{\partial t} = s_e I(N_D - N_D^+) - \gamma_e n_e N_D^+ - s_h I N_D^+ + \gamma_h n_h (N_D - N_D^+) \qquad (2.131)$$

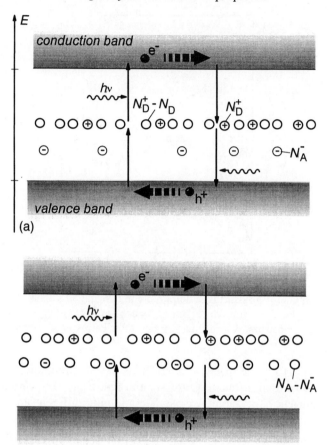

Fig. 2.21 (a) Model 1 and (b) model 2 of photorefractivity incorporating bipolar photoconductivity.

where we introduce the subscripts h and e for holes and electrons respectively. Note that we are still assuming here the presence of ionized acceptor atoms which compensate for the ionized donors in the dark. In model 1 all acceptor atoms are ionized all the time, as in the simple model.

In model 2 the acceptors of density N_A also participate in the process of creating charge carriers. The main feature of model 2, as may be seen in Fig. 2.21(b), is a division of responsibilities: donors cater for electrons, acceptors cater for holes. The rate equation for ionized donors is then still given by eqn (2.3) but we have a separate rate equation for ionized acceptors which takes the form

$$\frac{\partial N_A^-}{\partial t} = s_h I (N_A - N_A^-) - \gamma_h n_h N_A^- \qquad (2.132)$$

2.11 Electrons and holes

The density of negatively charged ionized acceptors *increases* when a hole is excited, i.e. a positive charge leaves the neutral atom. Recombination means that a hole moves upwards, making the negatively charged acceptor atoms neutral, i.e. reducing the number of negatively ionized acceptors[7].

The rest of the equations for both models and some more detailed conclusions will be presented in Section 5.2. One interesting conclusion is that model 1 leads to one exponential decay time whereas model 2 will yield two distinct exponential decay times. Another interesting relationship may be deduced for the steady state space charge field from model 1 in the form

$$E_{s1} = jm\xi(K) \frac{E_D E_q}{E_D + E_q} \quad (2.133)$$

where

$$\xi(K) = \frac{1-C}{1+C} \quad (2.134)$$

and

$$C = \frac{\tau_{de}}{\tau_{dh}} \frac{1 + E_D/E_{Me}}{1 + E_D/E_{Mh}} \quad (2.135)$$

$$E_{Mh} = \frac{1}{\mu_h \tau_h K}, \quad E_{Me} = \frac{1}{\mu_e \tau_e K}, \quad \tau_h = \frac{1}{\gamma_h(N_D - N_A)}, \quad \tau_e = \frac{1}{\gamma_e N_A} \quad (2.136)$$

$$\tau_{dh} = \frac{\varepsilon}{e\mu_h n_{ho}}, \quad \tau_{de} = \frac{\varepsilon}{e\mu_e n_{eo}}, \quad n_{ho} = s_h I_o N_A \tau_h, \quad n_{eo} = s_e I_o (N_D - N_A) \tau_e$$

$$(2.137)$$

where τ_{dh} is now the dielectric relaxation time of holes and E_{Me} and E_{Mh} are the expressions for E_M (eqn 2.37) for electrons and holes respectively.

The function $\xi(K)$ was introduced by Strohkendl *et al.* (1986,1987,1989) as an electron–hole competition factor. It may be seen from eqn (2.134) that for electrons only (when the hole density is zero and the hole dielectric relaxation time tends to infinity) $\xi(K) = 1$ and for holes only, by similar argument, $\xi(K) = -1$, i.e. the space charge field changes sign. It is also easy to see that when the holes and electrons have exactly the same densities, and all their parameters are identical, then $\xi(K) = 0$, there is no space charge field at all. In the general case there is a reduction in the strength of the space charge field which turns out to be grating spacing dependent.

The desirability of avoiding having *both* types of mobile carriers present simultaneously can now be clearly appreciated.

---

[7]In the present case it may be easier to imagine that as a response to the incident optical field an electron is excited upwards from the valence band, making the neutral atom negative and creating a hole in the valence band. Recombination is then the process in which the electron falls back into the valence band and the acceptor atom becomes neutral again. However, if one wishes to avoid confusion it is probably best to stick to the physical picture we used for model 1 in which only holes are considered when the valence band is involved in any carrier movement.

3
HOW MATERIAL PROPERTIES INFLUENCE LIGHT

3.1 The change in the dielectric constant

In Chapter 2 we answered, in quite some detail, the question: how will the incident optical beams affect the material? In the present chapter we shall try to answer the complementary question: how will the changes in the material affect the optical beams?

Let us briefly recapitulate what we have already said in Section 1.3. As a result of the input optical beams there will be a periodic electric field (static or slowly varying) in the photorefractive material. But the material posesses the crucial property, called the linear electro-optic effect, that an electric field E will cause a proportional change in the dielectric constant. It may be written in a simplified form as

$$\Delta \varepsilon = -\varepsilon_{ro}^2 r_{eff} E \tag{3.1}$$

where r_{eff} is the electro-optic coefficient. We shall call ε_{ro} the unperturbed relative dielectric constant. It is the same thing as the background relative dielectric constant of eqn (1.1). It needs to be emphasized here that this is the *optical* dielectric constant related to the index of refraction as $\varepsilon_{ro} = n_r^2$, in contrast to the static dielectric constant appearing in Chapter 2. Their values can be drastically different. For BSO the static value is 56 and the value at the optical frequency is 6.45. We may then write the relative dielectric constant as

$$\varepsilon_r = \varepsilon_{ro} + \Delta\varepsilon \tag{3.2}$$

In general, r is a tensor quantity. The tensorial relationships are discussed in detail in Appendix B and they will be properly taken into account in Chapter 6 where the full field equations will be presented. A particular component of the r tensor can be positive or negative depending on the crystal orientation. For the purpose of this section we shall take r as a scalar, denote it by r_{eff} and take it positive.

The electric field in eqn (3.1) is the actual electric field that is the sum of the applied electric field and of the space charge field. The change in the dielectric constant due directly to the applied electric field will matter under certain circumstances and it will be discussed in Chapter 6. For our present purpose it is sufficient to take into account the space charge field. Instead of writing it in the exponential form and meaning the real part (see footnote on p. 49) we shall take it here as a real quantity:

$$E = \tfrac{1}{2}[E_1 \exp j(\Omega t - \mathbf{K}\cdot\mathbf{r}) + \text{c.c.}] \tag{3.3}$$

remembering that

$$E_1 = mE_w \tag{3.4}$$

and that the modulation is defined as (see eqn 2.104)

$$m = \frac{2\mathcal{E}_1 \mathcal{E}_2^*}{I_o} \tag{3.5}$$

the change in dielectric constant may be obtained by substituting eqns (3.3) to (3.5) into eqn (3-1) to yield

$$\Delta\varepsilon = F \exp j(\Omega t - \mathbf{K}\cdot\mathbf{r}) + \text{c.c.} \tag{3.6}$$

where

$$F = -\varepsilon_{ro}^2 \frac{\mathcal{E}_1 \mathcal{E}_2^*}{I_o} r_{\text{eff}} E_w \tag{3.7}$$

We are going to investigate now what effect the above change in dielectric constant will have upon the fields.

3.2 The wave equation

The mathematical treatment of this interaction goes back to the first few decades of the century, to the times when X-ray diffraction by crystals and the diffraction of light by acoustic waves was described. The method initiated by Raman and Nath (1935) is to treat the problem as that of coupling between waves and derive the corresponding coupled differential equations. We need to start with the wave equation which we have already written in Section 1.2. We shall now derive it from Maxwell's equations

$$\nabla \times H = \varepsilon_o \varepsilon_r \frac{\partial \boldsymbol{\mathcal{E}}}{\partial t} \tag{3.8}$$

and

$$\nabla \times \boldsymbol{\mathcal{E}} = -\mu_o \frac{\partial H}{\partial t} \tag{3.9}$$

where H is the magnetic field at the optical frequencies.

We have of course used Maxwell's equations in the previous chapter as well when discussing static or very low frequency field problems. Equation (3.8) is essentially the same as eqn (2.4), only the notations are different to emphasize that we are now concerned with optical quantities. Further differences are that there is no current varying at the optical frequencies and that the dielectric constant is the one at optical frequencies and not the static one (notice the absence of the subscript s).

We shall now follow the time-honoured method of taking the curl of eqn (3.9) to obtain

$$\nabla \times (\nabla \times \mathcal{E}) = \mu_0 \frac{\partial}{\partial t}(\nabla \times H) = -\mu_0 \varepsilon_0 \varepsilon_r \frac{\partial^2 \mathcal{E}}{\partial t^2} \tag{3.10}$$

use further the vectorial relationship

$$\nabla \times (\nabla \times \mathcal{E}) = \nabla(\nabla \cdot \mathcal{E}) - \nabla^2 \mathcal{E} \tag{3.11}$$

and take[1]

$$\nabla \cdot \mathcal{E} = 0 \tag{3.12}$$

to arrive at the wave equation

$$\nabla^2 \mathcal{E} - \mu_0 \varepsilon_0 \varepsilon_r \frac{\partial^2 \mathcal{E}}{\partial t^2} = 0 \tag{3.13}$$

The electrical field is of course a vectorial quantity, and in many configurations its vectorial character is of great significance. Nevertheless, when our aim is to present the simplest description of the interaction between the two optical waves it will suffice to consider an electric field that is always in the direction perpendicular to the grating vector. Following our previous notations the electric field will be assumed to be along the x direction.

3.3 Beam coupling: derivation of the coupled wave equations

We shall now analyse from the point of view of the field quantities the physical configuration shown in Fig. 1.7 with which we started our treatment of photorefractive interactions: there are two waves incident upon the crystal. The analysis will be similar to that of Section 1.1, looking for a solution in which the amplitudes of both beams vary as a function of the x coordinate only. There is though one major difference: the two beams are assumed to have slightly different frequencies. Thus the solution of the wave equation will be assumed in the form (the notations are a little different from those in Section 1.1, we prefer to use here vector quantities in the exponents)

$$\mathcal{E} = \tfrac{1}{2}[\mathcal{E}_1(x) \exp j(\omega_1 t - \mathbf{k}_1 \cdot \mathbf{r}) + \mathcal{E}_2(x) \exp j(\omega_2 t - \mathbf{k}_2 \cdot \mathbf{r}) + \text{c.c.}] \tag{3.14}$$

---

[1] For further explanation of this assumption see Section 6.2.

3.3 Beam coupling: derivation of the coupled wave equations

where **r** is a radius vector and $\mathcal{E}$ is a scalar quantity. When we substitute the above equation into the wave equation we shall again have a coupling between the waves arising from the term $\varepsilon\Delta\varepsilon$:

$$\begin{aligned}
\Delta\varepsilon\mathcal{E} &= \tfrac{1}{2}([F\exp j(\Omega t - \mathbf{K}\cdot\mathbf{r}) + F^*\exp -j(\Omega t - \mathbf{K}\cdot\mathbf{r})] \\
&\quad \times [\mathcal{E}_1(x)\exp j(\omega_1 t - \mathbf{k}_1\cdot\mathbf{r}) + \mathcal{E}_2(x)\exp j(\omega_2 t - \mathbf{k}_2\cdot\mathbf{r}) + \text{c.c.}] \\
&= \tfrac{1}{2}F\mathcal{E}_1\exp j[(\omega_1 + \Omega)t - (\mathbf{k}_1 + \mathbf{K})\cdot\mathbf{r}] \\
&\quad + \tfrac{1}{2}F^*\mathcal{E}_1\exp j[(\omega_1 - \Omega)t - (\mathbf{k}_1 - \mathbf{K})\cdot\mathbf{r}] \\
&\quad + \tfrac{1}{2}F\mathcal{E}_2\exp j[(\omega_2 + \Omega)t - (\mathbf{k}_2 + \mathbf{K})\cdot\mathbf{r}] \\
&\quad + \tfrac{1}{2}F^*\mathcal{E}_2\exp j[(\omega_2 - \Omega)t - (\mathbf{k}_2 - \mathbf{K})\cdot\mathbf{r}] + \text{c.c.}
\end{aligned} \quad (3.15)$$

But, it needs to be remembered that

$$\Omega = \omega_1 - \omega_2 \quad \text{and} \quad \mathbf{K} = \mathbf{k}_1 - \mathbf{k}_2 \qquad (3.16)$$

hence the second term in eqn (3.15) reduces to

$$\tfrac{1}{2}F^*\mathcal{E}_1(x)\exp j(\omega_2 t - \mathbf{k}_2\cdot\mathbf{r}) \qquad (3.17)$$

and the third term reduces to

$$\tfrac{1}{2}F\mathcal{E}_2(x)\exp j(\omega_1 t - \mathbf{k}_1\cdot\mathbf{r}) \qquad (3.18)$$

The first and fourth terms represent higher-order diffracted beams with wave vectors $\mathbf{k}_1 + \mathbf{K}$ and $\mathbf{k}_2 - \mathbf{K}$. They will be disregarded here but we shall briefly comment on them in Section 3.10.

Having established that the waves may swap their exponential arguments, we can now get down to performing the remaining mathematical operations. Continuing to use operators we determine

$$\nabla[\mathcal{E}_1\exp j(\omega_1 t - \mathbf{k}_1\cdot\mathbf{r})] = \exp j(\omega_1 t - \mathbf{k}_1\cdot\mathbf{r})(\nabla\mathcal{E}_1 - j\mathbf{k}_1\mathcal{E}_1) \qquad (3.19)$$

and

$$\begin{aligned}
\nabla^2[\mathcal{E}_1\exp j(\omega_1 t - \mathbf{k}_1\cdot\mathbf{r})] &= \nabla\cdot\nabla[\mathcal{E}_1\exp j(\omega_1 t - \mathbf{k}_1\cdot\mathbf{r})] \\
&= \nabla(\exp j(\omega_1 t - \mathbf{k}_1\cdot\mathbf{r}))(\nabla\mathcal{E}_1 - j\mathbf{k}_1\mathcal{E}_1) \\
&= \exp j(\omega_1 t - \mathbf{k}_1\cdot\mathbf{r})(\nabla^2\mathcal{E}_1 - 2j\mathbf{k}_1\cdot\nabla\mathcal{E}_1 - k_1^2\mathcal{E}_1)
\end{aligned} \qquad (3.20)$$

and of course we have the same expressions with subscript 2.

The next step is the assumption of slow variation, mentioned in Section 1.2 as the slowly varying envelope approximation, which leads to the neglect of the second spatial derivatives. It may then be seen that substituting eqn (3.14) into eqn (3.13) will yield the equation

$$\exp j(\omega_1 t - \mathbf{k}_1 \cdot \mathbf{r})(-2j\mathbf{k}_1 \cdot \nabla \mathcal{E}_1 - k_1^2 \mathcal{E}_1)$$
$$+ \exp j(\omega_2 t - \mathbf{k}_2 \cdot \mathbf{r})(-2j\mathbf{k}_2 \cdot \nabla \mathcal{E}_2 - k_2^2 \mathcal{E}_2)$$
$$+ k_o^2 [F^* \mathcal{E}_1 \exp j(\omega_2 t - \mathbf{k}_2 \cdot \mathbf{r}) + F \mathcal{E}_2 \exp j(\omega_1 t - \mathbf{k}_1 \cdot \mathbf{r})]$$
$$+ k_1^2 \mathcal{E}_1 \exp j(\omega_2 t - \mathbf{k}_2 \cdot \mathbf{r}) + k_1^2 \mathcal{E}_2 \exp j(\omega_1 t - \mathbf{k}_1 \cdot \mathbf{r}) + \text{c.c.} = 0 \qquad (3.21)$$

where

$$k_o = 2\pi/\lambda \qquad (3.22)$$

We may now recognize that eqn (3.21) can be satisfied if the coefficients of $\exp j(\omega_1 t - \mathbf{k}_1 \cdot \mathbf{r})$ and those of $\exp j(\omega_2 t - \mathbf{k}_2 \cdot \mathbf{r})$ vanish separately. Hence we obtain the coupled differential equations

$$-2j\mathbf{k}_1 \cdot \nabla \mathcal{E}_1 + k_o^2 F \mathcal{E}_2 = 0 \qquad (3.23)$$

and

$$-2j\mathbf{k}_2 \cdot \nabla \mathcal{E}_2 + k_o^2 F^* \mathcal{E}_1 = 0 \qquad (3.24)$$

The wave vectors $\mathbf{k}_1$ and $\mathbf{k}_2$ can again be expressed in coordinates as in eqn (1.2):

$$\mathbf{k}_1 = k(\mathbf{i}_x \cos\theta + \mathbf{i}_z \sin\theta), \quad \mathbf{k}_2 = k(\mathbf{i}_x \cos\theta - \mathbf{i}_z \sin\theta) \qquad (3.25)$$

where we have further assumed that

$$\mathbf{k}_1^2 = \omega_1^2 \mu_o \varepsilon_o \varepsilon_{ro} \approx \omega_2^2 \mu_o \varepsilon_o \varepsilon_{ro} = \mathbf{k}_2^2 = k^2 \qquad (3.26)$$

Equations (3.23) and (3.24) may be rewritten in the form

$$\frac{\partial \mathcal{E}_1}{\partial x} + j \frac{F k_o}{2\varepsilon_{ro}^{1/2} \cos\theta} \mathcal{E}_2 = 0 \qquad (3.27)$$

and

$$\frac{\partial \mathcal{E}_2}{\partial x} + j \frac{F^* k_o}{2\varepsilon_{ro}^{1/2} \cos\theta} \mathcal{E}_1 = 0 \qquad (3.28)$$

which bear strong resemblance to the differential equations (1.17) derived for static gratings. It may be clearly seen that the field $\mathcal{E}_1$ is coupled to the field $\mathcal{E}_2$ and vice versa. The resemblance though is less startling if we look at the symbol F as given by eqn (3.7). It depends on the field amplitudes. Thus in contrast to the coupled wave equations for static gratings the present equations are *non-linear*.

Equations (3.27) and (3.28) may be written in a number of forms. Remembering the phase angle Φ of the space charge field, we may write

$$E_w = |E_w| \exp j\Phi \qquad (3.29)$$

3.3 Beam coupling: derivation of the coupled wave equations

Introducing further the constant

$$\Gamma = \frac{\varepsilon_{ro}^{3/2} k_o r |E_w|}{\cos \theta} = \frac{2\pi}{\lambda} \frac{n_r^3 r |E_w|}{\cos \theta} \quad (3.30)$$

the differential equations take the form

$$\frac{\partial \mathcal{E}_1}{\partial x} - j \frac{\Gamma}{2} \exp(j\Phi) \frac{|\mathcal{E}_2|^2}{I_o} \mathcal{E}_1 = 0 \quad (3.31)$$

and

$$\frac{\partial \mathcal{E}_2}{\partial x} - j \frac{\Gamma}{2} \exp(-j\Phi) \frac{|\mathcal{E}_1|^2}{I_o} \mathcal{E}_2 = 0 \quad (3.32)$$

We are now in a position to prove one of the important effects occurring in photorefractive materials, namely that one of the beams may be amplified at the expense of the other one. Let us assume that beam 1 (pump beam) is much stronger than beam 2 (signal beam). Assuming further an undepleted pump beam, eqn (3.31) is no longer needed. If $\mathcal{E}_2$ is small the equation just confirms that $\mathcal{E}_1$ is constant. Denoting the intensity of beam 1 by I_1 and its value at the input by I_{10} we may then claim that

$$|\mathcal{E}_1|^2 = I_1 \approx I_{10} \approx I_o \quad (3.33)$$

and we can integrate eqn (3.32) to give

$$\mathcal{E}_2 = \mathcal{E}_{20} \exp\left[\frac{\Gamma}{2}(\sin \Phi + j \cos \Phi)x\right] \quad (3.34)$$

where $\mathcal{E}_{20}$ is the value of the electric field associated with beam 2 at $x = 0$, the input of the crystal. It may now be clearly seen that amplification depends on the phase angle Φ.[2] When the space charge electric field is in phase with the interference pattern, Φ is equal to zero and only the phase of $\mathcal{E}_2$ is affected. But as soon as there is an out-of-phase component, i.e. $\Phi \neq 0$, the magnitude of $\mathcal{E}_{20}$ is also affected. It is either amplified or attenuated while crossing the crystal, depending on the sign of Φ. In our present configuration it has to be positive in order to lead to amplification for beam 2. The maximum amplification occurs when $\Phi = 90°$. Since it is the intensity and not the electric field that is measured in an experiment, it is desirable to rewrite eqn (3.34) in terms of intensities. For $\Phi = \pi/2$ we have then the simple expression

$$I_2 = I_{20} \exp \Gamma x \quad (3.35)$$

From eqn (3.35) the coefficient Γ appears as the intensity gain factor per unit length under optimum coupling conditions.

---

[2] Remember that the $\sin \Phi$ relationship was also predicted in Section 1.1 as the condition of power transfer for static gratings.

Let us take a concrete example. For BSO we quoted in Section 2.8 a figure of $37\,\text{kV}\,\text{cm}^{-1}$ for E_w under optimum detuning. For the usual paraxial case $\cos\theta = 1$ and for the other parameters we take $\lambda = 514\,\text{nm}$, $n_r = 2.54$, and $r_{\text{eff}} = 3.4\,\text{pV}\,\text{m}^{-1}$. We obtain $\Gamma = 25.1\,\text{cm}^{-1}$. In practice, no one has ever achieved that kind of gain factor. The maximum reported for BSO (Refregier et al. 1985) was $\Gamma = 12\,\text{cm}^{-1}$ for a 1.27 mm thick crystal. The effective gain factor reported in the same paper for a 1 cm thick crystal was $7.3\,\text{cm}^{-1}$, leading to an amplification of 1480.

3.4 Solution of the coupled wave equations

We shall now find a solution of the coupled wave differential equations (3.31) and (3.32). We are interested both in the modulus and in the phase of the optical electric field. Hence we shall introduce the new variables A and Ψ by the relationship

$$\mathcal{E}_{1,2} = A_{1,2}\exp j\Psi_{1,2} \tag{3.36}$$

where both A and Ψ are real. Substituting eqn (3.36) into eqns (3.31) and (3.32), and noting that $A_{1,2}^2 = I_{1,2}$, we obtain after a number of straightforward mathematical operations the following set of differential equations:

$$\frac{dI_1}{dx} + \Gamma\frac{I_1 I_2}{I_1 + I_2}\sin\Phi = 0 \tag{3.37}$$

$$\frac{dI_2}{dx} - \Gamma\frac{I_1 I_2}{I_1 + I_2}\sin\Phi = 0 \tag{3.38}$$

$$\frac{d\Psi}{dx} + \frac{\Gamma}{2}\frac{I_1 - I_2}{I_1 + I_2}\cos\Phi = 0 \tag{3.39}$$

where

$$\Psi = \Psi_1 - \Psi_2 \tag{3.40}$$

By adding eqns (3.37) and (3.38) it may be seen that

$$\frac{d}{dx}(I_1 + I_2) = 0 \tag{3.41}$$

i.e. the sum of I_1 and I_2 is independent of the spatial variable. This is of course obvious: the total power in the two beams must be constant for physical reasons. However, when one uses so many approximations it is nice to know that at the end we have not violated a fundamental law. (It should be borne in mind that once we make the further approximation of the undepleted pump beam we do abandon power conservation.)

What other conclusions can we draw? As may be expected, it is still true that maximum amplification for beam 2 is obtained when $\Phi = 90°$. What will happen for $\Phi = 0$? As may be seen, the intensities will become uncoupled. There is no

3.4 Solution of the coupled wave equations

power transfer between the two beams[3]. There is however phase transfer (also referred to as phase coupling), maximum at $\Phi = 0$ as may be seen from eqn (3.39).

In principle, this effect provides a mechanism for controlling the phase of one beam with another. However, despite there being many applications for phase modulators, there is not much interest in using the photorefractive effect for this purpose, mainly because the photorefractive response time is much too slow for such applications.

Equations (3.37) and (3.38) are non-linear differential equations offering no guarantee for an analytical solution. It turns out however that they can be integrated out by routine methods to give

$$I_1 = I_{10} \frac{1+\beta}{\beta + \exp(\Gamma_B x)} = \frac{\beta I_o}{\beta + \exp(\Gamma_B x)} \qquad (3.42)$$

and

$$I_2 = I_{20} \frac{1+\beta}{1 + \beta \exp(-\Gamma_B x)} = \frac{I_o}{1 + \beta \exp(-\Gamma_B x)} \qquad (3.43)$$

where β is the input beam ratio, defined as

$$\beta = \frac{I_{10}}{I_{20}}, \quad \text{and} \quad \Gamma_B = \Gamma \sin \Phi \qquad (3.44)$$

When the input beam ratio is sufficiently high it may be easily seen from eqn (3.43) that we obtain the simple exponential growth for the signal beam that we have already derived in eqn (3.35). For the general case we plot the intensities of the beams in Fig. 3.1 against $\Gamma_B x$, normalized length, for a range of beam ratios. With our choice of the direction of power transfer from beam 1 to beam 2 the cases of interest are those when β is equal to or larger than unity. For the largest value, $\beta = 100$, the pump beam may be seen to remain constant at low values of $\Gamma_B x$ and the signal beam increases exponentially. The exponential increase clearly has to stop when the beam intensities become comparable. At the end, i.e. for sufficiently large normalized length, all the power is transferred from the pump beam to the signal beam whatever the value of β.

The phase difference between the optical waves, Ψ, can also be obtained in closed form by substituting eqns (3.42) and (3.43) into eqn (3.39) and integrating:

$$\Psi = \Psi_o + \frac{1}{2} \cot \Phi \ln \left(\frac{(1+\beta)^2 \exp(-\Gamma_B x)}{[\beta + \exp(-\Gamma_B x)]^2} \right) \qquad (3.45)$$

The variation of Ψ as a function of normalized distance may be seen in Fig. 3.2(a) and (b) for $\beta = 100$ and 0.01 respectively. Most of them are approximately linear

---

[3] Note that this implies that two-wave mixing gain cannot be obtained in *steady state* using non-linear mechanisms which exhibit a local response, such as interaction via the third order nonlinear susceptibility, $\chi^{(3)}$.

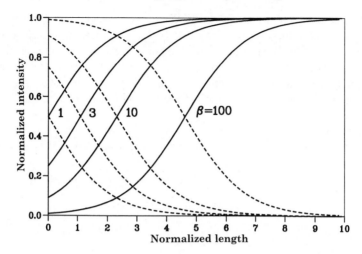

Fig. 3.1 Two-wave mixing energy transfer showing the normalized intensities of the signal (solid lines) and pump (dashed lines) beams versus the normalized interaction length, $\Gamma_B x$, for beam ratios $\beta = 1, 3, 10$, and 100.

curves. For the $\Phi = 0$ case the curve is exactly linear, which could actually be obtained from eqn (3.39) in the form

$$\Psi = \frac{\Gamma x}{2} \frac{1-\beta}{1+\beta} \tag{3.46}$$

(The same expression can also be obtained from eqn (3.45) but one needs to work out the limit as $\Phi \to 0$.)

The position of the maxima (or for that matter the minima) of the interference pattern will of course depend on Ψ (see eqn 1.37) as $\cos(Kz - \Psi)$. Hence the fringes will no longer be straight lines at half the inter-beam angle. They might of course be straight lines at some other angle, but in general their shape, as follows from eqn (3.45), would be quite a complicated function. Choosing our previously taken values of $\beta = 0.01$ and $\Phi = 45°$ (when there is considerable curvature of the fringes), the fringe pattern is shown in Fig. 3.2(c).

3.5 Direction of power transfer

We have just described a process in which one beam can gain power at the expense of another. It is appropriate then to ask what determines the direction of power transfer. Could such power transfer occur in a perfectly isotropic material? Obviously not. The beams would not 'know' which way to transfer power. The sign of the carrier, whether an electron or a hole, does affect the sign of the space charge field and through that the value of the phase angle, Φ. When Φ changes sign, the direction of power transfer will also change, as follows from eqns (3.42)

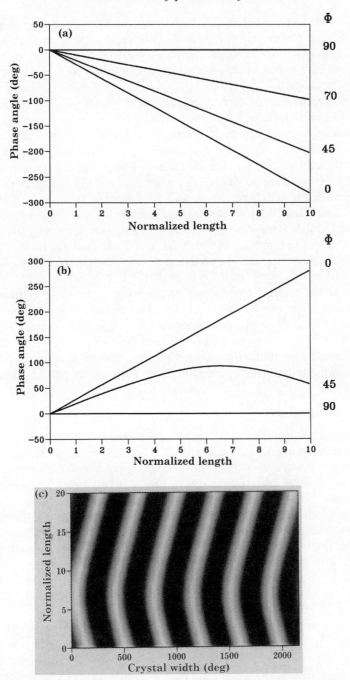

Fig. 3.2 Variation of phase difference, ψ, between the optical waves as a function of normalized distance for beam ratios (a) $\beta = 100$ and (b) $\beta = 0.01$. Curves are plotted for various values of the phase shift, ϕ, between the interference pattern and the space charge field. (c) When $\phi \neq 90°$ there will be an exchange of phase which will result in a bending of the fringe pattern as shown here for $\beta = 0.01$ and $\Phi = 45°$.

and (3.43). However, the mere fact that carriers of different signs exist would not, in itself, make power transfer possible. We need something that is direction dependent, in other words we need an anisotropic material. The relevant anisotropy in our case appears in the sign of the electro-optic coefficient (see Appendix B for further details).

Will the direction of the applied field affect the transfer of power? One might be forgiven to think that here is a prima-facie case for direction dependence. It turns out however that changing the direction of the applied field will change only the sign of the real part of the space charge electric field. The imaginary part remains unchanged (see eqn 2.58) and the direction of power transfer remains unchanged as well. By the same argument an AC applied field will not change the direction of power transfer every half period. Again, what matters is the sign of the electro-optic coefficient and the sign of the charge carrier.

When a DC field is used in conjunction with frequency detuning, the situation is more complex. The direction in which the interference pattern moves depends on the sign of the detuning, and the direction in which the electrons move will depend on the sign of the applied electric field. Since the enhancement mechanism depends on the interaction between the moving interference pattern and the space charge wave (which travel in the same direction as the electrons) it is obvious that both the sign of detuning and the sign of the applied electric field must be considered. If we change one of them the direction of power transfer will change. If we change simultaneously both the direction of the applied field and the direction of detuning then the direction of power transfer remains unaffected.

3.6 Reflection gratings

Up to now all our illustrations have shown the beams incident from the same side of the crystal, known as the transmission configuration. It is the one used most often but not always. The light beams could just as well be incident from the opposite sides of the crystal as shown in Fig. 3.3. This looks very similar to Fig. 1.5 of Section 1.2, the difference being that in the present case the incident angles are θ and $\pi - \theta$. In Section 1.2 we called it a reflection grating. In the present section we shall call this the reflection configuration. Will the physics change in any way in this new configuration? Well, the incident light beams will still create

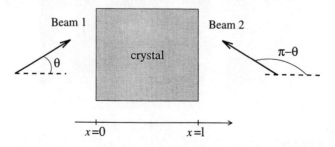

Fig. 3.3 Experimental arrangement for two-wave mixing in a 'reflection' configuration.

3.6 Reflection gratings

an interference pattern, electrons will still be generated by the optical intensity, and they will still diffuse, drift (if an electric field is applied along the crystal), and set up a space charge electric field. Will the space charge field cause a change in the dielectric constant? Yes, if we have the right electro-optic coefficients. For some crystal orientation a dielectric grating appears, and there is no grating for some others, as will be discussed in Appendix B. We are obviously interested in the case when there is a grating. The main difference is then that the grating we obtain is a reflection grating which will reflect the two beams into each other. The physics is otherwise the same.

Is the mathematics the same? In the field equations we assumed that two waves are incident symmetrically at angles $+\theta$ and $-\theta$. That is no longer true. The beams are now incident at angles θ and $\pi - \theta$. What else is different? In the transmission configuration the space charge electric field varied periodically in the z direction and the optical electric fields varied in the x direction. This is no longer so. For the reflection configuration both types of electric fields vary in the x direction but, as it happens, our equations are not affected by the direction in which the space charge field has a periodic variation. In fact, when deriving the coupled wave differential equations in Section 3.3 we did not even assume any specific incident angles. We did the derivation in terms of wave vectors $\mathbf{k}_1$ and $\mathbf{k}_2$. The transmission configuration came in specifically only with eqn (3.25) where we changed to a coordinate representation. With our new incident angles of θ and $\pi - \theta$ the wave vector $\mathbf{k}_1$ is still the same but $\mathbf{k}_2$ is of the form

$$\mathbf{k}_2 = k(-\mathbf{i}_x \cos\theta + \mathbf{i}_z \sin\theta) \tag{3.47}$$

resulting in the new coupled wave equations

$$\frac{\partial \mathcal{E}_1}{\partial x} + j \frac{F k_o}{2\varepsilon_{ro}^{1/2} \cos\theta} \mathcal{E}_2 = 0 \tag{3.48}$$

and

$$\frac{\partial \mathcal{E}_2}{\partial x} - j \frac{F^* k_o}{2\varepsilon_{ro}^{1/2} \cos\theta} \mathcal{E}_1 = 0 \tag{3.49}$$

Hence the only change is in the sign of the second term of our second differential equation.

We may now proceed and introduce the modulus and the phase of the optical electric field as in eqn (3.36), leading to differential equations in terms of intensity and phase angle. We obtain

$$\frac{\partial I_1}{\partial x} + \Gamma \frac{I_1 I_2}{I_1 + I_2} \sin\Phi = 0 \tag{3.50}$$

$$\frac{\partial I_2}{\partial x} + \Gamma \frac{I_1 I_2}{I_1 + I_2} \sin\Phi = 0 \tag{3.51}$$

$$\frac{\partial \Psi}{\partial x} + \frac{\Gamma}{2} \frac{-I_1 - I_2}{I_1 + I_2} \cos \Phi = 0 \qquad (3.52)$$

We should notice now the differences. The sign of the second term in eqn (3.51) differs from that in eqn (3.38). An obvious consequence is that

$$\frac{d}{dx}(I_1 - I_2) = 0 \qquad (3.53)$$

i.e. instead of $I_1 + I_2$ the quantity conserved is now $I_1 - I_2$. It may also be seen that the sign of I_1 is different in eqn (3.52) from that in eqn (3.39). But, of course, $(-I_1 - I_2)/(I_1 + I_2) = -1$ and eqn (3.52) can be integrated to give

$$\Psi = \frac{\Gamma}{2} x \cos \Phi \qquad (3.54)$$

i.e. for the reflection case the difference in the phase angles always varies linearly.
The solution for the intensities is

$$\frac{I_1(x)}{I_{10}} = \frac{1}{2\beta}\{\beta - 1 + |[(\beta - 1)^2 + 4\beta \exp(-\Gamma_B x)]^{1/2}|\} \qquad (3.55)$$

$$\frac{I_2(x)}{I_{10}} = \frac{1}{2\beta}\{1 - \beta + |[(\beta - 1)^2 + 4\beta \exp(-\Gamma_B x)]^{1/2}|\} \qquad (3.56)$$

Note that β still denotes the intensity ratio of beam 1 to beam 2 at $x = 0$. In the present case, however, when beam 1 is incident from the left and beam 2 is incident from the right, β is a rather artificial quantity. It makes much better sense to use the intensity ratio of the incident beams, that is

$$M = \frac{I_1(0)}{I_2(l)} \qquad (3.57)$$

with which β can be expressed as

$$\beta = \frac{M}{1+M}[1 + M \exp(-\Gamma_B l)] \qquad (3.58)$$

Other quantities of interest are the output to input intensity ratios of the beams, i.e.

$$u_1 = \frac{I_1(l)}{I_1(0)} \quad \text{and} \quad u_2 = \frac{I_2(0)}{I_2(l)} \qquad (3.59)$$

From eqns (3.57) and (3.58) we find their values to be

$$u_1 = \frac{1+M}{M + \exp(\Gamma_B l)} \qquad (3.60)$$

$$u_2 = \frac{1+M}{1 + M\exp(-\Gamma_B l)} \qquad (3.61)$$

3.6 Reflection gratings

One more thing we need to check is power conservation. It is no longer true, as in the transmission configuration, that the total beam power is conserved as the two beams move along the x axis. However, it must still be true that the total input power must be equal to the total output power, i.e.

$$I_1(0) + I_2(l) = I_1(l) + I_2(0) \tag{3.62}$$

With our previously introduced quantities eqn (3.62) reduces to the equation

$$M(1 - u_1) = u_2 - 1 \tag{3.63}$$

which can be easily proved to be true.

Let us see now how the beam intensities vary as a function of distance. For Γ_B positive we find that beam 1 declines in power and beam 2 gains power. Taking $\Gamma_B l = 3$, $I_1(0) = 1$, and $I_2(l) = 0.1$ (i.e. $M = 10$), we can see from Fig. 3.4 (solid lines) that I_1 declines from 1 to 0.37, and I_2 increases from 0.1 to 0.73. The difference between the two intensities is always 0.27. Clearly, we have a non-reciprocal window. One may 'look' through it in one direction but not in the other direction. The amount of 'isolation' increases for larger $\Gamma_B l$ and M.

An interesting special case arises when

$$I_1(x) = I_2(x) \tag{3.64}$$

for all values of x. It may be clearly seen from eqns (3.55) and (3.56) that this is satisfied when $\beta = 1$. Equation (3.58) yields then

$$M = \exp(\Gamma_B l/2) \tag{3.65}$$

For the above chosen value of $\Gamma_B l = 3$ the corresponding value of M is 4.48. The two coinciding intensity curves are also shown in Fig. 3.4 as a dashed line.

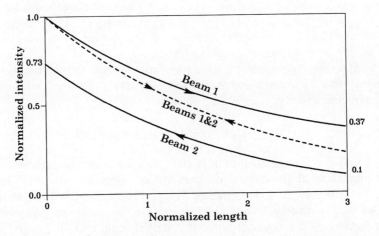

Fig. 3.4 Two-wave mixing energy transfer for the reflection configuration. The normalized intensities of the signal (beam 2) and pump (beam 1) versus normalized interaction length are plotted for $\Gamma_B l = 3$ and $M = 10$ (solid line) and $M = 4.48$ (dashed line).

Non-reciprocal phase shift. We have seen that both for transmission and reflection gratings, phase transfer may take place, which means that the optical path may be different for the transmitted and for the diffracted rays. When these two rays propagate in different directions, as for transmission gratings, such behaviour does not appear to be particularly striking (after all, this is what happens in an anisotropic material). In the reflection case however the two rays may travel in exactly opposite directions. Phase transfer now means that the phase difference of a wave propagating from A to B is not the same as that propagating from B to A, and that is certainly not an everyday occurrence. A photorefractive reflection grating can be non-reciprocal concerning both its intensity and its phase variation. There is no device at the moment that utilizes these properties, but the potential is there for producing an isolator or using these properties in some other non-reciprocal device like a fibre gyroscope.

3.7 Diffraction efficiency

In many of the applications of photorefractive materials the aim is to write a dielectric grating. We may have for example an object wave containing some useful information, say a digital data page in which the information is encoded with black spots for 0's and white spots for 1's. This object wave may then be interfered in the photorefractive material with a plane reference wave to write a hologram. When the hologram is illuminated by the reference wave the object beam is reconstructed. The question is: what is the intensity of the object beam for a given reference beam intensity, or in other words what is the diffraction efficiency? In this case we need to know the diffraction efficiency for illumination by the *same* reference beam. In some other case we may need to know the diffraction efficiency of the recorded grating for a range of input waves. A simple example is a device in which a reflection grating is recorded by two counterpropagating waves, and used as a frequency filter. The filter is designed to reject all frequencies within a narrow band and transmit other frequencies. The application may be as a laser cavity mirror or as a channel dropping filter.

Let us first do the calculation for the transmission case. Two beams incident at the angles θ and $-\theta$ interact and reach the steady state. The amplitude and phase of the refractive index grating is then equal to

$$\varepsilon_{r1}(x) = -\varepsilon_{ro}^2 \frac{r_{\text{eff}}|E_w|}{I_o} (I_1 I_2)^{1/2} \exp j(\Psi + \Phi) \qquad (3.66)$$

We shall now suddenly interrupt one of the beams and ask the question: what will be the amplitude of the diffracted beam at that particular moment? This is a problem of ordinary, static holography. The grating is given, we have an incident wave, we need to determine the amplitude of the diffracted wave. It is true that in the derivation of the coupled wave differential equations in Section 1.2 we assumed that the grating has constant amplitude whereas in the present case both the amplitude and the phase of the grating are dependent on the x coordinate. But, if we look carefully at the derivation we find that at no place did we use

3.7 Diffraction efficiency

that assumption. Hence the differential equations (1.17) are still valid for our present purpose. All we need to do is to replace ε_{r1} by the expression in eqn (3.66). The variation of ε_{r1} can be easily obtained since I_1, I_2, and Ψ have already been given as functions of x by eqns (3.42), (3.43), and (3.45). When one takes the square roots and the exponents and multiplies all of them together they look terribly complicated. To solve them analytically is a genuine *tour de force*, and that's what Kukhtarev *et al.* (1979b) accomplished. They provided an analytical expression although they forgot to mention how they did it. One can get some idea of how to do it from a similar calculation by Vahey (1975) but it is still a hard slog, a very hard slog, in fact. The final result is

$$\eta = \frac{2\beta}{1+\beta} \frac{\exp(-\Gamma_B l/2)[\cosh(\Gamma_B l/2) - \cos(\Gamma_A l/2)]}{1+\beta\exp(-\Gamma_B l)} \quad (3.67)$$

(where $\Gamma_A = \Gamma\cos\Phi$), valid only for the case when the replay beam is the same as one of the recording beams (i.e. same wavelength, same angle of incidence). It needs to be emphasized however that this does not mean that the Bragg conditions are satisfied. If $\Phi = 0$ the fringes are bent and it is impossible to satisfy the Bragg conditions at every point in the grating. The situation is actually even worse. There is no reason why the diffraction efficiency should be highest for reconstruction by the same beam. As will be shown later, reconstruction by a beam incident at a slightly different angle would give higher efficiency. For $\Phi = \pi/2$ it is true of course that reconstruction by the original recording beam gives maximum efficiency. For this particular case, $\Gamma_A = 0$ and $\Gamma_B = \Gamma$. The diffraction efficiency as a function of Γl with the beam ratio as a parameter is shown in Fig. 3.5(a). It may be seen that for the high beam ratio of 100 a 100% diffraction efficiency is achievable. For the lower beam ratios the achievable diffraction efficiency is less and it cannot be increased by making the normalized length larger. The explanation follows from Fig. 3.1. For lower beam ratios there is only a narrow region of space in which the modulation (proportional to the product of the amplitudes) is high. The condition of total power transfer is soon reached and after that the modulation is of course zero. Thus making the crystal longer cannot possibly help.

For $\Phi = 0$ we have $\Gamma_B = 0$, which may be substituted into eqn (3.67) to yield

$$\eta_D = \frac{4\beta}{(1+\beta)^2}\sin^2(\Gamma x) \quad (3.68)$$

There is now no beam coupling, therefore the initial beam ratio will remain constant. Remembering that the relationship between beam ratio and modulation may be written as

$$m = \frac{2\sqrt{I_{10}/I_{20}}}{1+I_{10}/I_{20}} = \frac{2\sqrt{\beta}}{1+\beta} \quad (3.69)$$

eqn (3.68) reduces to

$$\eta_D = m^2 \sin^2(\Gamma x) \quad (3.70)$$

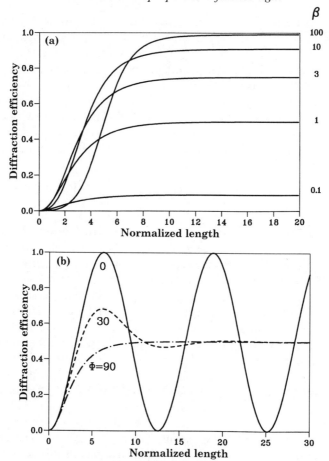

Fig. 3.5 Diffraction efficiency versus normalized length for (a) $\Phi = 90°$ with $\beta = 0.1, 1, 3, 10, 100$, and for (b) $\beta = 1$ with $\Phi = 0°, 30°, 90°$.

a simple expression indeed. For unity modulation we have exactly Kogelnik's formula. The maximum diffraction efficiency is 100% and it varies periodically with distance. What happens when the incident beams are of unequal intensity? Is the similarity to Kogelnik's formula a pure coincidence? Yes, it is a coincidence. The physical situation is quite different.

In the present case ($\Phi = 0$), Ψ varies linearly, which makes the grating slanted. The incident beam is still at the same angle, hence the Bragg conditions are, strictly speaking, not satisfied. However, there is still in some sense a cumulative interaction which makes the diffraction efficiency vary periodically with distance, but this cumulative interaction is not perfect, hence 100% efficiency is not possible.

When Φ is neither 0 nor $\pi/2$ then we have both power transfer and phase transfer. The diffraction efficiency in general varies in a quite complicated

3.7 Diffraction efficiency

manner. We shall show here the results in Fig. 3.5(b) for $\beta = 1$ only. For zero angle the solution is periodic but as soon as Φ differs from zero a periodic solution no longer exists. The diffraction efficiency declines in an oscillatory manner but the behaviour is still different from that in a static grating. The diffraction efficiency does not tend to zero; it tends to 0.5.

Let us imagine now that the grating is somehow fixed (see Chapter 10 on how to do it in practice) and we measure the diffraction efficiency for a range of incident angles. In that case the coupled wave equations must be set up in the manner outlined in Section 1.2 and solved numerically as was done by Heaton et al. (1984). For illustration we shall show here some of their curves. For $\Gamma_B l = 3$, $\beta = 1, 3,$ and 5, a recording angle of $10°$, and phase angles $\Phi = 0$ and $\Phi = \pi/2$, they are shown in Figs. 3.6(a) and (b). It may be clearly seen from Fig. 3.6(a) that

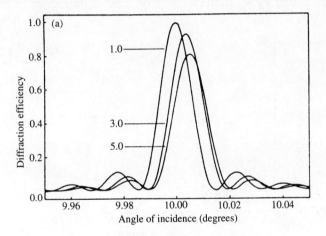

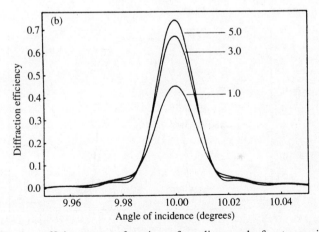

Fig. 3.6 Diffraction efficiency as a function of reading angle for transmission grating recorded with an inter-beam angle of $10°$ and with phase angles of (a) $\Phi = 0°$ and (b) $\Phi = 90°$. $\Gamma_B l = 3$, $\beta = 1, 3,$ and 5 (after Heaton et al. 1984).

for $\beta = 1$ the incident angle maximizing the diffraction efficiency is different from 10°. It makes also good sense that the maximum diffraction efficiency is higher when $\Phi = 0$. Although the fringes are bent, there is no beam coupling, which makes the grating amplitude independent of distance.

We also show here the theoretical results for a reflection grating using our previous set of parameters, in Fig. 3.7(a) and (b). Again, for $\Phi = 0$ the original recording angle gives less than maximum diffraction efficiency and, again, the curves for $\Phi = 0$ give higher efficiency than those for $\Phi = \pi/2$.

3.8 Higher diffraction orders

In our derivation of the basic equations of grating diffraction in Section 1.2 we allowed two waves only, an incident wave and a diffracted wave, excluding

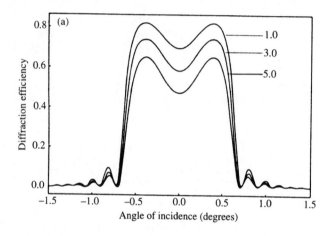

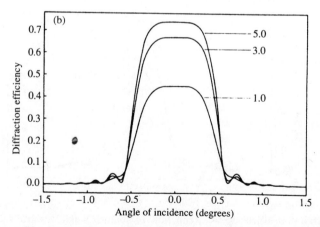

Fig. 3.7 Same as Fig. 3.6 but for a reflection grating recorded with counterpropagating beams (after Heaton et al. 1984).

3.8 Higher diffraction orders

thereby higher diffraction orders. The condition for the absence of higher diffraction orders (the distinction between optically thick and optically thin gratings) is given by Solymar and Cooke (1981) as

$$Q = 2\pi \frac{\lambda L}{\Lambda^2} \gg 1 \tag{3.71}$$

and

$$\Omega = \left(\frac{\lambda}{\Lambda}\right)^2 \frac{n_r}{\Delta n} \gg 1 \tag{3.72}$$

where L is the length of the interaction region. For gratings in photorefractive materials both conditions are usually satisfied but not always. Let us take a concrete example: $L = 5$ mm, $\Lambda = 10\,\mu$m, $\lambda = 514$ nm, $n_r/\Delta n = 5 \times 10^4$, which yields $Q = 40$ and $\Omega = 33$. Neither value seems to be particularly large. Had we taken a larger grating spacing, say 50 μm, then neither condition would have been satisfied. So, using considerations valid for static gratings we must conclude that higher diffraction orders cannot be simply disregarded. Is the situation in any way different for dynamic than for static gratings? As a matter of fact, it is. For static gratings a higher diffraction order is always off-Bragg. For a dynamic grating it is off-Bragg to start with but as soon as the diffracted beam appears, a new grating vector is generated for which the Bragg conditions are satisfied.

The process (not to scale, the inter-beam angle is much exaggerated) is shown with the aid of an Ewald diagram in Fig. 3.8. Two beams incident with wave vectors $\mathbf{k}_o$ and $\mathbf{k}_{-1}$ (notations are changed just for this section) generate a grating vector $\mathbf{K}$ in the photorefractive material. If the inter-beam angle is small enough then higher order beams appear by the construction shown in Fig. 3.8(a). However, as soon as a beam with (e.g.) wave vector $\mathbf{k}_1$ appears, a grating vector $\mathbf{K}_{o1} = \mathbf{k}_o - \mathbf{k}_1$ (see Fig. 3.8b) will also be generated. So in any further interaction beam 1 (with wave vector $\mathbf{k}_1$) is coupled to beam 0 (with wave vector $\mathbf{k}_o$) both off-Bragg and on-Bragg, and of course the same applies to the other beams.

The situation is even more complicated if we consider the large modulation effects discussed in Section 2.9. In that case, owing to non-linearities, grating vectors, $2\mathbf{K}$, $3\mathbf{K}$, etc. may also appear. Hence the beam with wave vector $\mathbf{k}_1$ may also be coupled to beam $\mathbf{k}_{-1}$ via the $2\mathbf{K}$ grating vector as shown in Fig. 3.8(c). The mathematics would indeed be very complicated if one wanted to take into account all the various coupling possibilities.

The appearance of higher diffraction orders may easily be shown if one happens to possess a crystal in which the opposing faces are not parallel but subtend a small angle α. Then a single incident beam will give rise, by double reflection, to another beam, the angle between the two beams being 2α. Experiments with such a crystal were reported by Erbschloe et al. (1989). The interesting feature of such an experiment is that the higher diffracted orders appear after some delay, reach a maximum, and then decline as shown in Fig. 3.9. The transient effects are obviously quite strongly present.

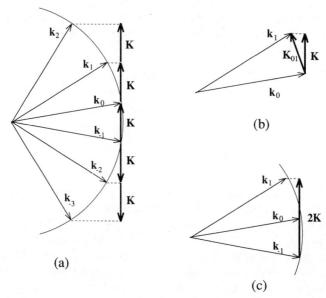

Fig. 3.8 (a) Ewald diagram illustrating the appearance of higher diffraction orders. (b) These higher-order diffracted beams will write new gratings. For example, $\mathbf{k}_1$ and $\mathbf{k}_o$ will record a grating of wave vector $\mathbf{K}_{o1}$. (c) The non-linearity of the material will give rise to grating vectors of higher orders (e.g. $2\mathbf{K}$) which may diffract beams in new directions.

Since a dynamic grating may transfer power from one beam to another beam, it follows that under favourable conditions it is possible to obtain quite a substantial amount of power in a higher diffraction order at the expense of the incident beams (Au and Solymar 1988b).

A further interesting effect that is worth mentioning is the ability of a higher diffraction order to influence the input beams. Let us remember Section 1.3.2 where we definitely stated that two-wave gain is only possible when the phase angle between the interference pattern and the refractive index distribution is different from zero, the optimum occurring for a phase shift of 90°. Well, if we take higher orders into account this is no longer true, as shown by Au and Solymar (1988c). Amplification happens to be possible for zero phase angle as well. The analysis was motivated by the experimental results of Sanchez *et al.* (1988) in liquid crystals in which the non-linearity is a purely local effect, and consequently, the phase angle must be zero (see Section 7.3 for further details).

3.9 Transients for optical beams

Up to now we have only calculated the stationary values of the optical fields. Could we work out how they vary as a function of time? To answer this question let us rewrite eqns (3.30) and (3.31) in a slightly different form:

3.9 Transients for optical beams

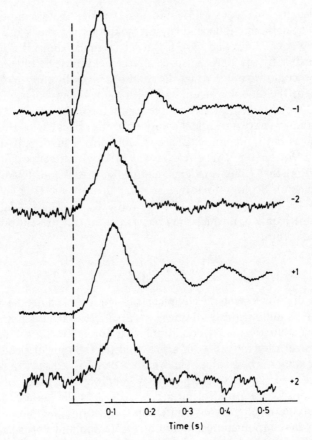

Fig. 3.9 Experimental intensity of the −2, −1, +1, +2 diffraction order beams as a function of time for a crystal with non-parallel opposing faces. The dashed line represents the instant the voltage ($V_o = 4\,\text{kV}$) is applied to the crystal of BSO (after Erbschloe *et al.* 1989).

$$\frac{\partial \mathcal{E}_1}{\partial x} - j\kappa E_w \frac{|\mathcal{E}_2|^2}{I_o} \mathcal{E}_1 = 0 \tag{3.73}$$

$$\frac{\partial \mathcal{E}_2}{\partial x} - j\kappa E_w^* \frac{|\mathcal{E}_1|^2}{I_o} \mathcal{E}_2 = 0 \tag{3.74}$$

where

$$\kappa = \frac{\pi}{\lambda} \frac{n_r^3 r}{\cos \theta} \tag{3.75}$$

and (just as a reminder) E_w is the space charge field divided by the modulation, m. It is a complex quantity.

We managed to solve eqns (3.31) and (3.32) under the assumption that the space charge field is time-independent, the solutions for the transmission case being given by eqns (3.42) and (3.43). But let us think about it a little further. Did we ever explicitly say that the space charge field must be time-independent? We did not. So could we not say that the solutions given by eqns (3.42) and (3.43) are still valid for the time-dependent case with Γ_B being now dependent on time? In Section 2.3 we obtained a fairly simple solution for the variation of E_1/m, so can we claim that we have a transient solution for the optical fields and intensities? This argument is not quite correct. It represents only a first approximation (see Erbschloe and Wilson 1989). The reason why it is not correct is that eqns (3.42) and (3.43) assume that E_1/m is independent of the spatial coordinate, and that is not true in general. So how to solve the general problem? It is very simple in principle. Besides eqns (3.73) and (3.74) we must also have a partial differential equation in time (eqn 2.33), which may be written for our present purpose in the more convenient form

$$\frac{\partial E_1}{\partial t_n} + pE_1 = \frac{q}{I_o} 2\mathcal{E}_1 \mathcal{E}_2^* \tag{3.76}$$

It is clear now that for solving the complete transient problem for the optical fields we need to solve simultaneously all three partial differential equations, eqns (3.73) to (3.76), subject to the initial conditions that for $t = 0$, (i) the space charge electric field is zero for all values of space, and (ii) the optical electric fields are equal to their unperturbed values (that is their values without coupling) everywhere in space, and subject to the boundary conditions that the optical fields of both incident beams are given for all times at $x = 0$ at the input of the crystal.

Analytical solutions of the above differential equations were given by Solymar and Heaton (1984) and numerical solutions by Heaton and Solymar (1985). The results for the build-up of the grating showed quantitative differences but qualitatively the same kind of curves were obtained. At erasure, however, there are considerable differences between the transient solution given by eqn (2.43) (a simple exponential decay) and that relying on the complete differential equations. If we interrupt one of the recording beams, then strictly speaking the right-hand side of eqn (3.76) is zero and therefore, one may argue, eqn (2.43) must be valid. However, when we think of the complete problem it is obvious that the beam which is still incident will be diffracted by the grating and a new grating will be formed. The total effect is that the grating strength might initially increase before it declines. A good example is given by Au and Solymar (1989), who calculated the temporal variation of diffraction efficiency under the conditions when one of the recording beams is blocked. Using the simple solution (material equations only) the diffraction efficiency declines as shown in Fig. 3.10(b). However, when the complete set of differential equations is solved, then the diffraction efficiency may initially increase for large enough coupling constant (Fig. 3.10a). For a small coupling constant $\Gamma_B l = 0.5$ the two solutions may be seen to be identical.

Figure 3.10 shows only the diffraction efficiency, which in a sense integrates the effect of the whole grating. One should really be a little more curious and ask the

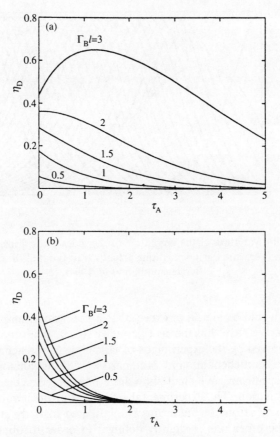

Fig. 3.10 Temporal variation of diffraction efficiency during erasure as calculated using (a) the exact model and (b) the approximate model, for various values of beam coupling, $\Gamma_B l$ (after Au and Solymar 1988c).

question: how does the index grating vary as a function of time and space? A very interesting result due to Jeganathan *et al.* (1994, 1995) is that for a sufficiently long crystal the index grating travels along the crystal as a soliton. A numerical solution for a particular set of parameters is shown in Fig. 3.11.

Our conclusion is that we must be on our guard. Very often the simple solutions of the material equations give the wrong answer both quantitatively and qualitatively. For further discussion see Section 6.11.

3.10 Temporal modulation

We have discussed at length how the beams interact. In the next two sections we shall look at two simple cases when we do something to the beams. In this section we ask the question of what happens when the beam to be amplified is amplitude-

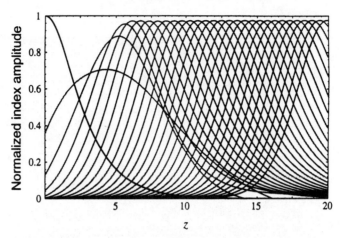

Fig. 3.11 Temporal variation of the amplitude of the index grating inside a long crystal, showing that during erasure the index grating actually travels out of the crystal as a soliton (after Jeganathan et al. 1994).

modulated[4]. It would be logical to expect that if the beam is amplified, so will be the modulation. We have thus means to amplify signals in a very wide band. The principle was proved by the experiments of Hamel de Montchenault et al. (1987) and Hamel de Montchenault and Huignard (1988) on analogue and digital modulation, respectively. We shall show here one of their experimental results for digital modulation. The experimental set-up is that for two-wave amplification. The intensity of one of the beams is modulated digitally at a frequency of 10 kHz, and the other one is slightly detuned in order to obtain higher gain. Oscilloscope traces of the modulated input and output beams may be seen in Fig. 3.12.

The effect of modulation can be understood from an argument well known in telecommunications: amplitude modulation leads to the appearance of sidebands, i.e. the modulation of a carrier frequency ω by a signal at a frequency of $\Delta\omega$ will lead to frequencies of $\omega \pm \Delta\omega$. For small values of ω this is actually the same thing as detuning. Assuming that the optimum detuning in the absence of modulation is Ω_{opt}, we would come to the conclusion that peaks in the gain curve (meaning the gain of the modulated signal) would occur at values of detuning Ω_{opt} and $\Omega_{opt} \pm \Delta\omega$. For a square wave modulation of 20 Hz the experimentally obtained gain against detuning (Webb and Solymar 1990b) and the theoretical results are shown in Fig. 3.13. Optimum detuning in the absence of modulation was 3 Hz. In the presence of amplitude modulation there are gain peaks at 17 Hz and also at 23 Hz.

---

[4]Up to now 'modulation' has always meant the periodic variation of the imposed interference pattern. In this section we had to use the same term to mean the changing intensity of one of the input beams. We hope it will not lead to any confusion.

3.10 Temporal modulation

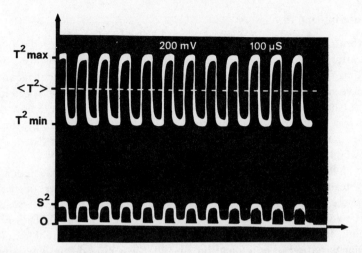

Fig. 3.12 Experimental demonstration of two-wave mixing by modulating the intensity of one of the beams with a square wave of frequency $f = 10\,\text{kHz}$. Lower trace = incident signal intensity, upper trace = transmitted signal intensity. BSO, $E_o = 6\,\text{kV}\,\text{cm}^{-1}$, $\Lambda = 20\,\mu\text{m}$, $I_o = 10\,\text{mW}\,\text{cm}^{-2}$ (after Hamel de Montchenault et al. 1987).

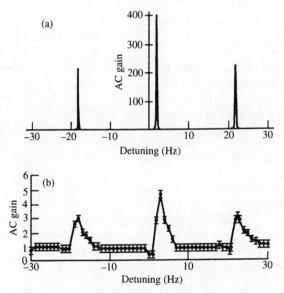

Fig. 3.13 (a) Theoretically and (b) experimentally obtained AC gain in BSO as a function of detuning frequency for a signal beam amplitude modulated by a square wave at 20 Hz (after Webb and Solymar 1990b).

3.11 Scattering and resonators

No photorefractive crystal is perfect. If it is not perfect it will scatter light. Some of the scattering centres will scatter light in some preferred direction, but let us assume that they scatter isotropically in all directions. Would that mean that we can observe scattering in all possible directions? If the crystal is not too imperfect this cannot happen. The scattering in most directions would be below detection level. So, in which direction can we expect to see scattered light? In the direction in which it is amplified. In other words there will be observable scatter in the directions in which two-wave gain is high.

As we have seen in Chapter 2, the imaginary part of the space charge wave, which is responsible for amplification, depends on a number of parameters including the inter-beam angle. We may then claim that a beam scattered in that particular direction will be observable provided the amplification is sufficiently high. In other words, the angular structure of scattered light should reflect the angular variation of amplification, and that was indeed found experimentally (Voronov et al. 1980). Will the scattered beam have the same frequency as the input beam? If the amplification is only due to the diffusion mechanism, then the frequency of the scattered beam will be the same as that of the input beam. However, when an electric field is applied, and the gain is higher for a slight detuning, then the scattered beam will be detuned by the same amount. Maximum scattering will be in the direction of the highest possible two-wave gain. Where would that additional frequency come from? From a space charge wave of the appropriate frequency, that is present in the crystal in the form of noise. We shall show examples of this type of scattering in Section 8.2.3 where we shall discuss other types of scattering too.

Next, we shall discuss a particular consequence of scattering. If the wave vectors of the incident and scattered beams are $\mathbf{k}$ and $\mathbf{k}_s$ as shown in Fig.3.14(a) and we set up a ring resonator so that the propagation direction of one of the beams coincides with the scattering direction, then we shall find a beam going round in the resonator as shown in Fig.3.14(b). We can find observable light intensity in the resonator even in the case when the scattering itself is not observable, because the resonator provides a feedback path. The most effective form of the resonator is

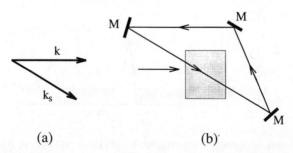

Fig. 3.14 (a) Light of wave vector $\mathbf{k}$ will scatter in directions of high gain, forming $\mathbf{k}_s$. (b) The ring resonator works by providing a feedback path for the amplified scattered beam.

the ring cavity (see the experiments of Rajbenbach and Huignard (1985) and Rajbenbach *et al.* (1989b) in BSO and in GaAs respectively). However, if scattering is large, other types of resonators are also possible. In fact, the first observation of a resonating beam (Feinberg and Hellwarth 1980) was obtained with a single mirror, which we shall discuss in the next section concerning phase conjugators.

3.12 Phase conjugation

Phase conjugation is one of the wonders of the world. It sounds so unlikely when one first comes across it, and sounds so obvious when one has got used to it. The device that does the phase conjugation is known as a phase conjugate mirror. For the difference between an ordinary mirror and a phase conjugate mirror we shall first give a simple example following AuYeung and Yariv (1979). In both cases a plane wave is incident and part of the wavefront (large in comparison with the wavelength) passes through a piece of glass placed in the middle of the wavefront (Fig. 3.15a). As is well known, the velocity of propagation is lower in the glass

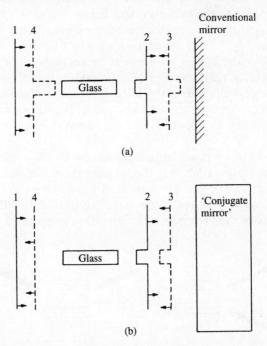

Fig. 3.15 (a) A plane wave (1) incident on a distorting element (a glass cylinder) emerges with a 'bulge' (2). When this wave is reflected (3) by an ordinary mirror and traverses the cylinder in reverse, the result is a doubling of the bulge. (b) A phase conjugate mirror reflects light so that its wavefront is identical to that of the incident wave, i.e. $2 = 3$. When the reflected wave traverses back through the cylinder, all the phase distortions are undone (after AuYeung and Yariv 1979).

than in free space, therefore the wave through the glass will be retarded. This is shown by the wavefront denoted by 2. The conventional mirror will reflect the wavefront as it comes in: first come, first served. After reflection, the middle of the wavefront is still retarded as shown by wavefront 3. After passing through the glass once more, wavefront 4 will show twice the retardation.

Let us now consider the phase conjugate mirror of Fig. 3.15(b). It 'conjugates' or in other words it reverses the phase: last in, first out. The middle of wavefront 3 now shows phase advance instead of retardation. Passage through the piece of glass once more gives again a nice, undistorted wavefront. Thus the role of phase conjugate mirrors is to undo phase distortions.

How can such a phase conjugate mirror be realized? The basic idea is provided by holography. If two waves, a reference wave and an object wave, record a hologram and the developed hologram is illuminated by the reference beam, then according to the rules of holography, the object beam will dutifully spring into existence. If the same hologram is illuminated by the conjugate reference beam (a beam which propagates in the opposite direction, e.g. if the reference beam is a slightly divergent beam, its conjugate is a slightly convergent beam) and it is incident from the opposite direction upon the hologram, then the output beam will be opposite to the original object beam.

The relationships may be appreciated perhaps even better if we consider a volume hologram. Fig. 3.16(a) shows the fringes in a volume hologram recorded by waves 1 and 4. According to our previous terminology, wave 1 is the reference wave and wave 4 is the object wave. If A_2, the conjugate of the reference wave, is incident from the opposite side of the volume hologram, then A_3, a wave propagating in a direction opposite to our object wave, is generated as shown in Fig. 3.16(b). Propagating in an opposite direction means that it is a phase conjugate wave.

To clarify the process further, let us imagine now that the object wave contains some pictorial information. We may then expand the object wave into its Fourier components, i.e. instead of one strong incident wave we may think of a very large number of weak waves incident, all propagating in slightly different directions. We may then invoke the principle of superposition (mentioned in Section 1.2 as

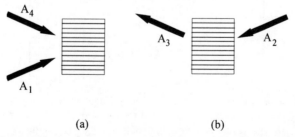

(a)   (b)

Fig. 3.16 Holographic method for producing phase conjugate beams. If a hologram is recorded by beams 4 and 1, then the phase conjugate of beam 4 (or 1) may be generated by illuminating the hologram with the phase conjugate of beam 1 (or 4).

3.12 Phase conjugation

applying to volume holograms), leading to the superposition of a large number of weak holograms. Each of these holograms is then reconstructed by the conjugate reference beam, generating waves which propagate in the direction opposite to all the weak input beams. Hence, however complicated the object beam might be, one can produce the phase conjugate wave of each one of them.

So far we have been talking about volume holograms which need to be developed. If the material is self-developing, i.e. the gratings appear while the radiation is on, then all the above-mentioned phenomena may occur simultaneously, and we can talk about real-time holography. Photorefractives may, of course, be described as real-time holographic materials.

The first demonstration of phase conjugation in a photorefractive material was, to our knowledge, done by Huignard *et al.* (1979). The experimental set-up is shown in Fig. 3.17. The input signal beam, A_4, is represented by an illuminated object slide from which the radiation propagates through some phase-disturbing medium to the BSO crystal to which a DC voltage is applied. The conjugate reference beam, A_2, is supplied simply by reflecting the plane wave input reference beam. Since the conjugate reference beam generates the phase conjugate of the input object beam, it will retrace its path through the phase-distorting medium and the undistorted image will appear behind the beam splitter as indicated in Fig. 3.17. A photograph of the image viewed through the aberrating glass may be seen in Fig. 3.18(a). It is highly distorted. Fig. 3.18(b) shows the image quality achieved by this phase conjugation technique. Finally, about tolerances. As mentioned before, reversal of the wavefront can be achieved if the hologram is illuminated by its conjugate reference beam, but how accurately should that conjugate reference beam be aligned? A misalignment of the mirror by 1.5 mrad results in the image shown in Fig. 3.18(c).

Phase conjugation in photorefractive materials has become quite a big subject. We shall discuss it in somewhat more detail in Section 7.6. Here we wish to mention only two further developments, *self-phase conjugation* and the *double phase conjugate mirror*.

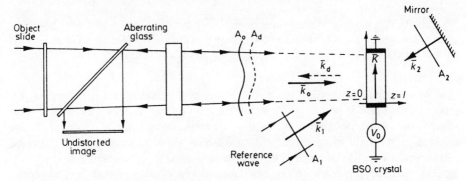

Fig. 3.17 Experimental setup to demonstrate real-time phase conjugation and aberration correction by four-wave mixing in a crystal of BSO (after Huignard *et al.* 1979).

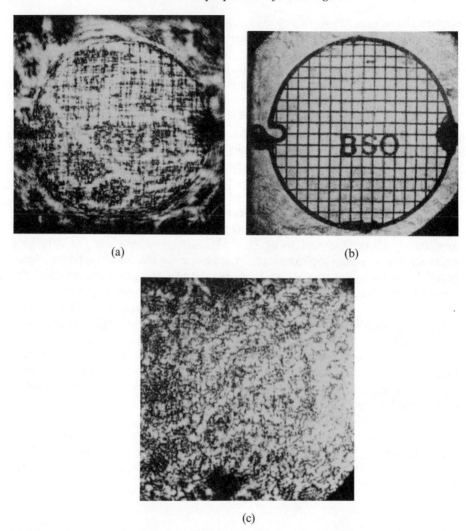

Fig. 3.18 Demonstration of aberration correction: (a) direct imaging through aberrating glass; (b) imaging via real-time phase conjugation (i.e. after beamsplitter); (c) imaging with a vertical misalignment of the mirror by 1.5 mrad (after Huignard et al. 1979).

One of the most interesting advances is self-phase conjugation discovered by Feinberg (1982b). It is easy to describe the experiment. A single beam carrying some pictorial information is incident at an angle upon a $BaTiO_3$ crystal. If the angle is in the right range, then miraculously a phase conjugate beam appears, retracing the rays of the original beam and producing a phase conjugate image. A simple explanation given by K. R. MacDonald and Feinberg (1983) will be reproduced here. Since scattering is known to be strong in $BaTiO_3$, there can be waves propagating in a number of different directions. If these beams can establish some

3.12 Phase conjugation

mutual interactions, then they can reinforce each other and eventually a steady state will be reached. According to MacDonald and Feinberg (1983) the steady-state wave distribution may be described by Fig. 3.19. The incident beam 4 partly continues as 4′ and partly gives rise to 1, which after suffering two edge reflections turns into 2′. The wave 4′ gives rise to 1′, which after the same edge reflections turns into 2. The two pump waves are now 1′ and 2′. The grating is produced by 4′ and 1′ and is read by 2′, producing the phase conjugate 3′. Similarly, 1 and 4 generate the grating which is read by 2 and produces 3. Under the right conditions 3 and 3′ reinforce each other and we obtain the observed phase conjugate. Is this a correct description of what is happening? It is difficult to tell. It certainly has the merit of simplicity. A numerical model by Zozulya et al. (1994) is probably closer to the truth but does not easily lend itself to physical explanation.

The double phase conjugate mirror, invented by Sternklar et al. (1986), may be understood by reference to scattering. The physical arrangement is shown in Fig. 3.20(a). Two beams are incident upon a $BaTiO_3$ crystal from opposite sides with wave vectors k_2 and k_4. Both input beams scatter in all possible directions. They generate gratings and erase gratings. In the long run all gratings will be erased, with the exception of the one that is generated by k_2 and its scattered wave k_3, and by k_4 and its scattered wave k_1, as shown in Fig. 3.20(b). In other words, only the grating which is mutual or *shared* will remain. We have double phase conjugation in the sense that k_3 is the phase conjugate of k_4, and k_1 is the phase conjugate of k_2, i.e. this particular physical configuration produces the phase conjugate of *both* input beams.

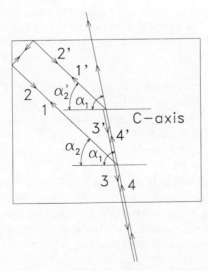

Fig. 3.19 Self-phase conjugation via the 'cat' geometry showing how fanning and total internal reflection work to set up two separate four-wave mixing interaction regions for generation of a phase conjugate beam.

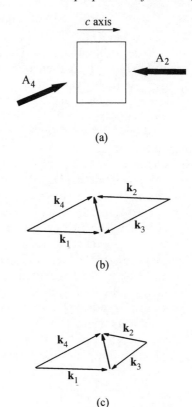

Fig. 3.20 (a) Double phase conjugate mirror. (b) The two incident waves interact via a shared noise grating. (c) The interaction may occur even when the incident beams are of different colour, e.g. k_2 = red and k_4 = blue.

Interestingly, the interaction does not need coherent waves. In fact, it is easier to obtain the interaction between the input beams if they are *not* coherent. The wavelength difference between the input beams may be quite substantial. One might be blue and the other one red as shown by Sternklar and Fischer (1987). Let us assume that red light is incident from the right, and blue light from the left, i.e. $k_4 > k_2$. Then, naturally, $k_4 = k_1$ and $k_3 = k_2$. The geometrical construction to give the same grating vector is shown in Fig. 3.20(c). Note that k_1 and k_2, and k_4 and k_3, are no longer parallel with each other. Consequently, the reflected scatter beam is no longer the exact phase conjugate of the input beam.

It is also possible to have interactions between three colours (blue, green, and red) as shown by Kaczmarek *et al.* (1994).

PART II
Discussion of the physics

GENERAL INTRODUCTION

The main aim in this Part is to provide a comprehensive coverage of the literature, describe the various models, offer experimental evidence, and try to describe the great variety of phenomena which belong to the photorefractive family.

We start in Chapter 4 with a history section. We believe that the subject is mature enough to have a history and that the views about the way the subject developed are not too divergent. We believe there is a rough consensus as to what were the more important advances though, naturally, every history written is bound to be subjective, and we do not think that ours is an exception. We round off the history section with some statistics which show the development of the subject in figures.

Chapter 5 covers the response of the material to the incident light, which represents an extremely wide range of phenomena. Not surprisingly, Chapter 5 is the longest. We have as many as 16 models whose properties we duly describe. We write a lot about our favourite subject, the enhancement of the space charge field, and we treat, in quite some detail, responses to pulsed excitation.

Chapter 6 is on the field equations for two-wave mixing, which extend the calculations of Chapter 3 by including the vectorial character of the optical field and some further properties (besides the electro-optic effect) of anisotropic materials like optical activity and piezoelectricity. We also include there experimental verification of the theoretical models.

Chapter 7 is on multi-wave mixing, in particular on forward three-wave and on forward four-wave mixing. This is where we should have included the vast range of effects related to reverse phase conjugation but it would have made the book too unwieldy. So we decided on no more than three brief sections.

Chapter 8 is devoted to the problem when some radiation appears out of the blue, when the output beams show no strict relation to the input beams. We include under this heading both scattering and instabilities (giving perhaps too much space to another favourite subject of ours, subharmonics) and refer to both of them as causing the appearance of spurious beams.

The last chapter of Part II contains topics which somehow could not find their way into any of the previous chapters. One of these topics is the photovoltaic effect, which has been around since the very beginning of photorefractive studies. The other five have come to prominence in fairly recent times. They are: photorefractive polymers, band-edge photorefractivity, multiple quantum well structures, stratified holographic elements, and solitons.

4
A HISTORY OF THE PHOTOREFRACTIVE EFFECT

4.1 Introduction

A book written on a mature subject in the applied physical sciences should, by custom, have a historical review that should be objective, describe briefly and concisely all the major advances, name the people who made the advances, and, possibly, evaluate the state of the art.

What is a mature subject? A general consensus on the definition is unlikely to exist but most people would agree that the following considerations play a role: (i) age; (ii) was it difficult to solve the arising problems?; (iii) have most of the major problems been solved?; (iv) have potential applications been demonstrated?; (v) how many major laboratories all over the world have been involved?; (vi) what is the number of publications?

Firstly, age. A subject that has been around for a couple of years can hardly be called a mature subject. The unit of time to be used is obviously not the year but the decade. A subject that can boast of a history amounting to one decade starts to become respectable. With two decades behind a subject, maturity can no longer be doubted. With three decades – and photorefractivity has nearly got that far – maturity is assured.

Secondly, the difficulty of the problems. It's no use for a subject to have three decades behind it if all the problems were foreseen and all the solutions were foreseen as well. Then the three decades merely illustrate sluggish progress owing to lukewarm interest. Hard-won advances, on the other hand, show that great efforts were invested because the importance or the interest of the subject demanded it.

In the third place, maturity requires the solution of most major problems. If the advances are won with great effort but understanding is still lacking, then it is quite likely that the subject will die in its adolescence and will never reach maturity. Even the most enlightened and most benevolent financial directors will be forced at a certain stage to draw the purse-strings.

In the fourth place, potential applications. There should obviously be at least one major application, but preferably many. If the subject is, or becomes, void of applications it cannot survive for long in today's commercial atmosphere.

In the fifth place, the number of laboratories involved. If the subject is of any importance then several laboratories should clamber upon the bandwagon. If all the research is done in one particular laboratory then it is more likely to be one man's hobby-horse.

In the sixth place, publications. In a subject in which military interests play only a minor role, advances will always be linked to publications. It is possible to imagine that a corporation will allow no publications on a sensitive subject for a limited time, but not for long. No corporation will ever be able to keep a good applied scientist if publication is forbidden. Glory is more important than money. What is the number of publications that characterize a mature subject? A few hundred does not sound enough. A thousand will perhaps be a sufficient proof of maturity.

It would be possible to give a few more criteria, but six are sufficient for our pupose, namely to prove that the physics and applications of photorefractive materials is a subject that has reached maturity. It has been around for 29 years, it has been a great one to be involved with if one wanted to solve difficult problems, it has reached the stage at which all effects have been qualitatively and most effects have been quantitatively understood, it can boast of a score of potential applications, it has been pursued by a good many laboratories around the world, and the number of publications is well over one thousand.

There is just one blot on this beautiful landscape: materials. The properties of materials grown in different batches and even of those grown in different laboratories have been roughly the same, but only roughly. With the possible exception of SBN (strontium barium niobate), a major candidate for applications as a memory element, there has been very little painstaking work on growing the crystals and particularly in attempting to tailor the properties of the crystals. In the majority of the cases the identity of traps capturing photocarriers has still not been unambiguously determined. The actual values of the mobilities can only be very roughly predicted and what determines the mobilities is nearly entirely unknown. The hope is however strong that whoever writes the next historical review will have a lot to say about materials.

Before making the actual attempt at this review it seems desirable to ask the question: 'can one write an *objective* historical review?' and give the answer that 'the chances are vanishingly small'. Nobody who has ever participated in an event (whether small or great, social or political) has ever been able to give an objective description of it. A participant will always give a coloured history. The contribution of the group from which the historical assessment comes is bound to be overestimated. What will be the likely reaction of readers who themselves appear in that history? They will find that their contribution has been drastically underestimated. They will be of the opinion that some of their papers which first expounded a theory or which clearly laid the foundations of such-and-such applications had been unreasonably and inexplicably undervalued or simply omitted from the list. That is a reaction absolutely inevitable. Those who participated in bringing the subject to maturity are bound to view the advances in a certain physical and mathematical framework of their own. In that framework some advances will seem trivial, other advances may seem immensely significant. So whether an advance is regarded as great or not will depend on the reference frame of the contributor. Since advances made by oneself are best understood, the complaint might be entirely legitimate that the authors of a review did not sufficiently understand the advances made by other researchers. We cannot hope that any contributor to the subject will find his or her own work sufficiently represented. We do hope however that most of the contributors will be happy with the general picture emerging of many laboratories and many individuals making a number of significant contributions. There were indeed a number of significant contributions but there were no giant leaps forward. Having offered up our apologies in advance for what we have suggested must be a job poorly done, we shall commence our review.

4.2 Early history

It is worth noting that the story of photorefractivity has a clear kinship with some stories that have been influential in forming our Civilization. The subject of photorefractivity has an unusual birth, like Moses or Jesus, although for photorefractivity, the story is better documented: the first paper was published in *Applied Physics Letters* in 1966. The seven authors, Ashkin, Boyd, Dziedzic, Smith, Ballman, Levinstein, and Nassau, worked for Bell Telephone Laboratories. The title was 'Optically induced refractive index inhomogeneities in $LiNbO_3$ and $LiTaO_3$'. They wrote in the first paragraph: 'The effect, although interesting in its own right, is highly detrimental to the optics of nonlinear devices based on these crystals'. For the purpose they had in mind, frequency doubling, the effect was clearly harmful they called it therefore 'optical damage'. A new material, KTN (potassium tantalate niobate), with similar behaviour was reported by F. S. Chen (1967) soon afterwards. The main difference was that the damage occurred only in the presence of an applied electric field. A qualitative description of the origin of the effect was given as follows: 'The process of the space charge build-up can be explained by postulating that electrons are photoexcited from traps by the laser beam into the conduction band and drift toward the positive electrode leaving behind positive stationary space charges. These electrons may be retrapped and reexcited out of the traps as they eventually drift out of the laser spot and are retrapped there'. This is as good a qualitative description as any.

Applications of the effect were quickly conceived. In the scheme of F. S. Chen *et al.* (1968a) the damaged portion of the material ($LiNbO_3$) could be used as a memory element. A little later came the realization by F. S. Chen *et al.* (1968b) that the 'damage' reproduces the original intensity variation in the form of a varying dielectric constant, hence it is suitable for holographic recording. The first experiments yielded resolution in excess of 1600 lines mm^{-1}. The authors also showed that the effect was enhanced when the number of available deep traps was increased. Three new materials came soon: Lin (1968) reported the effect in $Bi_4Ti_3O_{12}$, Thaxter (1969) in SBN and Townsend and LaMacchia (1970) in $BaTiO_3$. Further experimental and thoretical results were reported by F. S. Chen (1969) on $LiNbO_3$ and $LiTaO_3$. The theory contained rate equations both for electrons and traps[1], and it was suggested that the linear electro-optic effect was responsible for the change in the index of refraction.

In the early 1970s action was transferred to the RCA Laboratories in Princeton. Amodei (1971) showed that diffusion can be responsible for the observed effects in $LiNbO_3$. The model used relied on the instantaneous generation of the electrons in response to the imposed interference pattern. It was a non-linear model in the sense that the resulting space charge electric field was not a pure sinusoidal one but allowed the existence of higher harmonics. The author also realized that the maximum achievable space charge field could not be described by his model because trap density was not included. The relative significance of the applied field and the diffusion field and the pertaining phase angles were determined by Amodei and Staebler (1972). Beam coupling effects were observed at about the same time (Staebler and Amodei 1972a). Their model, based on coupled wave differential equations, did indeed account for transfer of power from one beam to the other beam. However, they obtained an incorrect oscillatory solution (energy going from one beam to the other one and back again if the interaction length is long enough) since they disregarded the dependence of the coupling constant on the optical fields. Transients were worked out by Alphonse et al. (1975). On the application side, Staebler et al. (1975) recorded and fixed simultaneously as many as 500 holograms in $LiNbO_3$:Fe. For a more complete review of the early work at RCA, including storage applications, see Staebler (1977).

Early work on applications was also carried out at the Thomson-CSF Laboratories in Orsay on fast random-access holographic memories in $LiNbO_3$ (d'Auria et al. 1973). A particularly interesting result of theirs was to be able to erase a single bit from a data sheet when a large number of data sheets were recorded (Huignard et al. 1976a). They also had remarkable results in real-time (Huignard and Herriau 1977) and in time-average (Huignard et al. 1977) holographic interferometry and in demonstrating edge enhancement (Huignard and Herriau 1978).

Progress was of course made at a number of other laboratories as well in the early and middle 1970s. Short-time transients were derived by Young et al. (1974), who showed that the phase angle depended on time. Conical scattering in $LiNbO_3$ was observed by Magnusson and Gaylord (1974). The same authors (1976) derived a time-varying theory of beam coupling based on the formulation of Ninomiya (1973). A non-linear theory, in which both the electron density and the space charge field were expanded into a Fourier series, was devised by D. M. Kim et al. (1976). The problem of recording gratings taking into account the gaussian envelope of the beams was treated by Moharam and Young (1976).

4.3 The emergence of a new set of equations, and some related thoughts

Where do the Middle Ages end and where does the Modern Age start? When drawing lines between various periods in history it is customary to refer to some

---

[1] Rate equations for traps and charge carriers were already known from photoconductivity studies (see e.g. Rittner 1956).

4.3 The emergence of a new set of equations, and some related thoughts

cataclysmic event which heralds the arrival of a new age. The history of photorefractivity cannot quite claim a Columbus who discovers the New World, but there is a fairly sharp dividing line between using 'early' theories and the full-fledged apparatus developed by the group working in Kiev at the Institute of Physics of the Ukrainian Academy of Sciences. Their first paper published in the West (Kukhtarev *et al.* 1977) was very general. It was the biggest single step forward since the discovery of the photorefractive effect. In order to arrive at their final results they started off with the full set of (what we call now) the material equations and the field equations, and they used the two crucial approximations: small modulation of the interference pattern in solving the materials equations and the slowly varying envelope approximation for solving the field equations. The paper had however one major disadvantage. It was entirely incomprehensible. The authors failed to mention where they started from and what were the approximations they used. It is true that for details they gave a reference to a paper by Vinetskii *et al.* (1977) which was published in the *Bulletin of the Soviet Academy of Sciences of the USSR, Physical Series.* Unfortunately that was not a journal readily accessible in the West. Those few who might have taken the trouble to find the paper would have found there another innocent sentence saying that the authors omitted the lengthy calculations, which however could be found in Preprint No. 14, Institute of Physics, Ukrainian Academy of Sciences. We do not know whether Preprint No. 14 was ever consulted by Western scientists. What is known however is that the Kukhtarev *et al.* (1977) paper went unnoticed. Two years later came two further papers by Kukhtarev *et al.* (1979a,b). They gave the starting points of their physical model, a definite advance on the 1977 paper, but there was very little given about the mechanics of the solution, and particularly which equations were valid under what approximations (antiquated units, a factor of 4π in Poisson's equation did not help either). Although many of the formulae were derived under the approximation of small modulation, that approximation was not mentioned at all.

The 1977 paper was still entirely ignored but the two 1979 papers slowly entered Western literature. Huignard *et al.* (1980b) made use of some of their expressions, although the parameter E_q was misquoted owing to the factor of 4π mentioned above. The journey across the Atlantic took longer. Feinberg *et al.* (1980) were clearly unaware of those new comprehensive theories. Had they known about them, they would have probably not bothered with setting up their alternative 'hopping model'. However, the 'Kiev' theories were adopted as soon as the work on phase conjugation in photorefractive materials gathered momentum. Fischer *et al.* (1981) for example used the 'Kiev' expression for the relationship between the perturbed index of refraction and the space charge field.

There is a consensus of opinion among workers in the field of photorefractivity that the papers by the Kiev group provide the best foundation for any theoretical work. There is however no consensus about the thorny problem of what to call these theories. Should we call them by the neutral name of band transport model?

After all, very few theories are called by the name of their originators. Should we give priority to Kukhtarev on account of his 1976 paper or should we reward him because his name is first among the five authors using the Roman alphabet. It is of course different if we rely on the Cyrillic alphabet. The paper already mentioned, Vinetskii *et al.* (1977), was first published in Russian and Vinetskii (presumably) is the first because V is close to the beginning of the Cyrillic alphabet. Or as a compromise solution should we call them the equations of Vinetskii and Kukhtarev, considering that the papers by Vinetskii and Kukhtarev (1975,1976) did already display all the basic elements of the theory? But then it turns out that there is actually an earlier paper from the Kiev group which bears the name of neither Kukhtarev nor Vinetskii. It was written by Deigen, Odoulov, Soskin and Shanina (1975) and gives the rate equations, etc. So should they be called the equations of Deigen *et al.*? We believe the historically correct description would be to call them the 'Kiev' equations. The alternative is to refer to the paper with which consensus came, and refer to the equations as those of Kukhtarev *et al.* (1979a).

Before going on to the contribution of other groups it seems worthwhile to give a little thought to the *mésentente* East–West. Why did it take a good 6 or 7 years before the theories of the Kiev group were quoted, digested and further developed in the West? The first Western paper which whole-heartedly adopted the approach of the Kiev group and made the small modulation assumption explicit for everyone to understand was written by Valley and Klein (1983). Scientific news did not always travel so slowly. Oersted's paper on the effect of electric current on a magnetic pole, written in August 1820 in Copenhagen, was fully evaluated and understood by Ampère a month later in Paris. We wish to argue that the fundamental reason was in the nature of the Soviet system. It was a system that greatly encouraged the development of science. People in the Party apparatus may have risen farther but they could also have spectacular falls. Scientists represented the only stable elite, and the ultimate rewards were high. A scientist's salary was more than doubled when elected to the Academy of the Soviet Union and there were many privileges going with it. (This is in contrast to what happens in the UK where someone elected to the Royal Society pays a yearly subscription.) A corollary of high esteem is strong competition. Hence the ideal solution is to publish, thus ensuring priority, but not to give enough information for a competitor to be able to use the results. The aim was *not* to keep out Western scientists. The aim, whether explicitly formulated or not, was to keep out everybody. If one asks scientists from the former Soviet Union why they published so many obscure articles they usually answer that printing paper was scarce and they had to be very concise. That was probably a contributory reason but could not have possibly accounted for the consistent omission of starting points and details of derivations and for not mentioning relevant approximations. Should we conclude that we in the West are perfect? Well, we also have our vanities, jealousies, and ambitions, but we are in the field for the fun of it. There is no money – not even much prestige – for those engaged in the pursuit of science in the West.

4.4 Modern history

We have already mentioned Bell Laboratories where the first discoveries were made, and RCA which took over the lead. Gradually, a number of new laboratories joined the race. The major centres in the United States were Caltech, Hughes Research Laboratories, the University of Southern California, the University of California in San Diego, and the National Bureau of Standards in Colorado. Apart from the Institute of Physics in Kiev the prime place in Europe should be given to the Thomson-CSF Laboratories in Orsay which, in addition to their early studies on storage and interferometry, have produced the most consistent and most versatile output for well over 20 years. Other well known laboratories in the game have been the Ioffe Institute in Leningrad (resurrected as St Petersburg), the Eidgenössische Technische Hochschule in Zürich, the Université de Paris, the Universities of Osnabrück and Darmstadt in Germany, the Universidad Autónoma in Madrid, King's College, London and the University of Oxford in the UK, to name a few. What were the main advances made in these laboratories? We shall take now a few topics and follow their development.

On the materials side the most significant advance was probably the explanation of the photovoltaic effect (called photogalvanic in the Russian-language literature) by Glass *et al.* (1974) who realized that the local asymmetry of the $Nb-Fe^{2+}$ distances in the $+c$ direction are responsible for the appearance of an asymmetric current. An extension to space-oscillating currents was made by Belinicher (1978) and a thorough study of the fundamental theoretical concepts was published later by Belinicher and Sturman (1980). New materials also appeared. Huignard and Micheron (1976) found two new high sensitivity photorefractive crystals, $Bi_{12}SiO_{20}$ and $Bi_{12}GeO_{20}$ (a different chemical formula from the bismuth germanate reported by Lin 1968). Potassium niobate was first introduced by the group in Zürich (Günter *et al.* 1976). A major step forward was the discovery of the photorefractive effect in semi-insulating GaAs and InP by Glass *et al.* (1984), and in undoped GaAs by M. B. Klein (1984) (the traps being provided by the EL2 defect). More recently, Larkin *et al.* (1993) succeeded in growing pure BSO by hydrothermal growth, which has opened the way towards controlled doping.

Phase conjugation by degenerate four-wave mixing was first proposed by Hellwarth (1977); the equivalence of four-wave mixing and real time holography was shown by Yariv (1978). The first phase conjugation in photorefractive materials was demonstrated more or less simultaneously by Huignard *et al.* (1979) in BSO, by Kukhtarev and Odoulov (1979a,b) in $LiNbO_3$, and by Khizhnyak *et al.* (1979) in $LiTaO_3$. Phase conjugation in $BaTiO_3$ was reported by Feinberg *et al.* (1980). A little later Feinberg and Hellwarth (1980) obtained c.w. phase conjugation with higher than 1 reflectivity in $BaTiO_3$, which in Western literature has been regarded as a first. In fact, Khizhnyak *et al.* (1979) reported a reflectivity of 5 a year before, which should not have been ignored, considering that it was published in the SPIE Proceedings. The spatial frequency dependence of phase conjugation was demonstrated by Huignard et al. (1980b). The condition for self-oscillation in the phase conjugate geometry was found by Fischer et al. (1981). A particularly interesting self-phase conjugation effect was discovered by Feinberg (1982b) in $BaTiO_3$. A very comprehensive paper on the theory and on a large number of phase conjugate configurations was published by Cronin-Golomb et al. (1984). Phase conjugation in the infrared was reported by Cronin-Golomb and Lau (1985), and coherent coupling of diode lasers by Cronin-Golomb et al. (1986b). The double phase conjugate mirror yielding phase conjugation in both directions was conceived by Weiss et al. (1987).

Coming now back to the simplest effect demonstrated, two-wave mixing, the main advances were made by devising enhancement methods and refining our understanding of the phenomena. Gain enhancement by detuning was first reported by Huignard and Marrakchi (1981a). An early explanation was given by Stepanov et al. (1982) in terms of space charge waves (although they did not call them by that name). This explanation, unfortunately, shared the fate of other papers from the former Soviet Union. It was disregarded in the West, where a later paper by Refregier et al. (1985) has usually been quoted as the fundamental one on the consequences of detuning. A later explanation in terms of space charge waves was given by Furman (1987). A new intensity resonance in InP:Fe was reported by Picoli et al. (1989a). An alternative method of enhancing gain by the application of a time-varying voltage was proposed by Stepanov and Petrov (1985). A new resonance at high repetition frequency for the AC voltage and at high detuning was found by Pauliat et al. (1990) and Grunnet-Jepsen et al. (1995). Significant results for the temporal variation of the space charge field were obtained by Valley (1983a,b) for short pulse conditions and for two photo-active species respectively. An analytical solution for the combined materials and field equations as a function of time was found by Solymar and Heaton (1984). The amplification of modulated signals was investigated by Hamel de Montchenault et al. (1987) for digital and by Hamel de Montchenault and Huignard (1888) for analogue modulation. The effect of induced birefringence and optical activity was investigated by Stace et al. (1989) for AC applied fields (who found an approximate analytical solution) and by Webb and Solymar (1991a) for detuning. The latter authors made a detailed comparison between theory and experiments showing the crucial effect of absorption.

High modulation effects were already treated in the papers of the RCA group at the beginning of the 1970s in a fairly simple manner. The problem however became much more complex when enhancement methods were used for maximizing the fundamental component of the space charge field. The solution was given by Au and Solymar (1988a) in the presence of detuning and by Brost (1992) for AC modulation of the applied field.

The deleterious effect of holes was first shown by Orlowski and Krätzig (1978). The first theoretical account was again a brief paper in the Russian literature, Stepanov (1982), which was ignored in the West where references are usually made to the practically simultaneously published papers of Strohkendl et al. (1986) and Valley (1986).

4.4 Modern history

The anisotropy of some of the photorefractive materials, particularly LiNbO$_3$ and BaTiO$_3$, makes possible some additional diffraction effects, as first realized by Stepanov et al. (1977). One of the many consequences is a fine structure in the Bragg selectivity of the grating, as shown by Pencheva et al. (1982) in the absence and by Vachss and Hesselink (1987a) in the presence of an electric field. Anisotropic scattering, that is, scattering due to the anisotropic properties of the material, also became a big subject. A particularly interesting type, conical anisotropic scattering (quite different from the scattering cone found by Magnusson and Gaylord 1974), was shown and explained by Odoulov et al. (1985). The understanding of scattering through this and many other papers led to schemes for the reduction of scatter noise. A successful solution was found by Rajbenbach et al. (1989a).

Multi-wave forward mixing was initiated by Khizhnyak et al. (1984) in the form of forward phase conjugation. Three-wave mixing was studied theoretically by Ringhofer and Solymar (1988a), showing that optimum gain occurs when the interference pattern and the refractive index distribution are in phase. High gain by forward three-wave mixing was measured by D. C. Jones and Solymar (1989). Two-wave gain for a local grating (phase angle zero) was shown to be possible by Au and Solymar (1988c) provided a higher diffracted wave is present.

A diagnostic method for measuring photorefractive parameters based on two-wave interaction was initiated by Mullen and Hellwarth (1985). The phase modulation of fringes and subsequent measurement of the external current for diagnostic purposes was introduced by Petrov et al. (1990).

Subharmonic instabilities were discovered by Mallick et al. (1988). The models accounting for the threshold conditions were developed by Au et al. (1990) and Sturman et al. (1992a). Comparisons between a simplified theory and experimental results were made by Kwak et al. (1992). An interesting domain structure (shown on the front cover) which bears some similarity to domains in ferromagnetic materials was found by Grunnet-Jepsen et al. (1993).

We have already mentioned some early applications of the photorefractive effect, storage being the first to be suggested. More recently, the recording of many holograms with the same diffraction efficiency has been carried out using an incremental recording technique (Anderson and Lininger 1987). In 1990, Brady et al. suggested another approach to indefinite storage involving refreshing the stored image. In 1993, Tao et al. reported the storage of 750 holograms using combined spatial and angular multiplexing. Mok improved the total number of holograms that he could store from 500 in 1991 (Mok et al. 1991) to 5000 in 1993 (Mok 1993). Heanue et al. (1994) have demonstrated a storage system with a readout rate of 6.3×10^6 pixels s^{-1}.

Another obvious application of photorefractive materials that suggests itself is the amplification of an image bearing beam. In 1979, Kukhtarev et al. (1979b) demonstrated amplification by a factor of 10 in LiNbO$_3$; in 1983 a gain of 4000 was obtained in BaTiO$_3$ (Tschudi et al. 1986); amplification of signal beams in the picowatt range was reported in 1991 by Rajbenbach et al. and a gain of 11 000 was obtained in $BaTiO_3$ in 1991 using a pulsed read-out scheme to overcome problems due to beam fanning (Joseph et al. 1991a).

The use of photorefractives for the correlation and convolution of optical images was first demonstrated by White and Yariv (1980). This is another area that has witnessed steady development over the years and a prototype engineered system was demonstrated by Rajbenbach et al. (1992b). Closely related to correlators are associative memories based on photorefractive crystals, which are able to recover a complete stored image from a corrupted or partial input image. Again there has been considerable effort devoted to this area since the first demonstrations of all optical associative memories around 1986 (Soffer et al. 1986; Yariv et al. 1986).

There have been a number of other interesting developments in the general area of image processing using photorefractives. In 1981, Levenson et al. demonstrated lensless imaging for projection lithography using a photorefractive phase conjugate mirror. Photorefractive crystals may be used in a variety of ways as spatial light modulators, possibly the earliest approach being suggested by Kamshilin and Petrov (1980). A photorefractive novelty filter (which only transmits the parts of a scene which are changing) was described by Anderson et al. (1987).

Other interesting areas for applications that have been suggested include: interferometry using phase conjugate mirrors (Feinberg 1983); dynamic holographic interconnections (Wilde et al. 1987); optical neural networks (Psaltis et al. 1988) and self-organizing circuits (Saffman et al. 1991). Finally, the first paper describing a commercial device incorporating a photorefractive crystal, an extremely narrow band optical filter, was published in 1993 (Rakuljic and Leyva 1993).

4.5 A review of the reviews

In a subject as appealing as photorefractivity, one would expect to see many reviews written in the course of years, and that is indeed the case. Moving chronologically, the first review paper is probably that of Staebler (1977) on the photorefractive applications of ferroelectric crystals, containing 74 references. The review by Glass (1978) is also written from the materials point of view and has 60 references. A broader review containing about 200 references was published by Günter (1982) but the theories mentioned are mostly those of the pre-Kiev-group type. A review that does give a good summary of the Kiev group theory is that of Hall et al. (1985) but it is much more selective, giving only 82 references. The expansion of the field then became so fast that the reviews following no longer aimed at full coverage, they were specialist reviews, like Huignard et al. (1985) on wave mixing in BSO, Fainman and Lee (1988) on signal processing, Feinberg (1988) on self-phase conjugators, Valley et al. (1988) on materials, Fischer et al. (1989) on photorefractive oscillators, Owechko (1989) on associative memories, Anderson and Feinberg (1989) on novelty filters, Yeh et al. (1989) on optical computing, Pepper et al. (1990) for the layman in Scientific American, Rajbenbach et al. (1990) on signal processing, Cheng et al. (1991) on image processing with semiconductors, Yu and Yin (1991) on signal processing,

Pauliat and Roosen (1991) on holographic memories, Hong (1993) on neural network applications, Hesselink and Bashaw (1993) on optical memories, Moerner and Silence (1994) on photopolymers, and Solymar *et al.* (1994) on forward wave interactions.

Coming now to the books: the first one was written in Russian by Petrov *et al.* (1983), followed by another book in English, Petrov *et al.* (1991). In the latter book they covered a wide field. There is a particularly good chapter on anisotropic diffraction. An introduction to the subject is available from Yeh (1993) but its coverage is not broad and there are very few references in it. There are also two volumes available with Günter and Huignard (1988,1989) as editors, and one collection of papers (Yeh and Gu 1993) on applications. The last three are anthologies which have the advantage that each subject is treated by an expert in the field but, in our opinion anyway, this is offset by the disadvantage of not having a clear structure and by the lack of cross-references between the chapters.

4.6 Some statistics

For entertainment purposes, we present here the results of a short statistical survey of publications covering the photorefractive effect. The survey was carried out using the *Science Citation Index* running from the beginning of 1981 until the end of 1994. The survey should not be considered in any way exhaustive, as many journals from the former Soviet Union are not included, but it does illustrate the development of the field.

To begin with, Fig. 4.1 shows the total number of publications dealing with the photorefractive effect for each calendar year[2]. The rapid growth in number of publications in the field is quite marked: from 13 in 1981 to 342 in 1994. Of course, this rapid growth has to be seen in the light of the increasing total number of publications in all fields worldwide. For example the number of papers in the *Journal of the Optical Society of America* has risen from roughly 250 in 1981 to about 650 in 1994. This trend is repeated to a greater or lesser extent in other journals. In Fig. 4.2 we attempt to take this factor into account by choosing the three journals[3] publishing the most number of papers on the photorefractive effect – *Optics Letters, Journal of the Optical Society of America* (parts A and B), and *Optics Communications* – and calculating the percentage of papers each year dealing with the subject. We can see an increase from 0.3% to 6.5% over the time period of this study, indicative we would claim of the increasing importance of the field.

Finally, we have looked at the different journals in which people choose to publish this work. The three journals listed above account for almost 40% of the

---

[2] Care should be taken in a search of this kind as there are a great number of papers in other fields dealing with photorefactoriness and photorefractive keratectomy. Our search was done using photorefracti* combined using logical AND with NOT(keractect*), where * is the wild card character.
[3] These are the journals each containing over 10% of the total number of photorefractive publications, though of course we are cheating a little in counting the *Journal of the Optical Society of America* parts A and B as one journal.

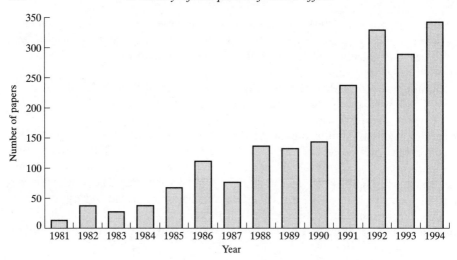

Fig. 4.1 Number of publications for each calendar year, 1981–1994.

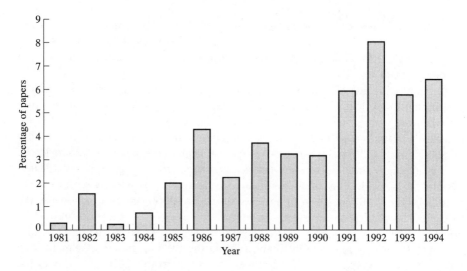

Fig. 4.2 Percentage of publications dealing with the photorefractive effect in the journals *Optics Letters*, *JOSA* and *Optics Communications*.

photorefractive literature over the period of this survey. However, the remaining 60% of papers are spread thinly over many journals; including the next three most popular journals – *Applied Optics, Applied Physics Letters*, and *Journal of Applied Physics* – still leaves 44% of the papers unaccounted for (see Fig. 4.3). In fact when the thirteen most popular journals are included (see Fig. 4.4), there are still about 20% of the papers lying elsewhere.

4.6 Some statistics

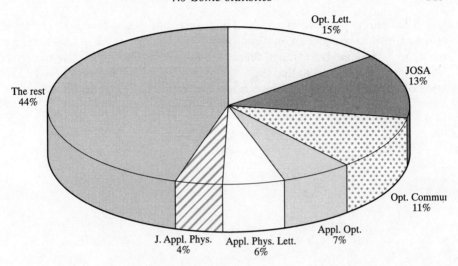

Fig. 4.3 Percentage of photorefractive papers by journal, 1981–1994.

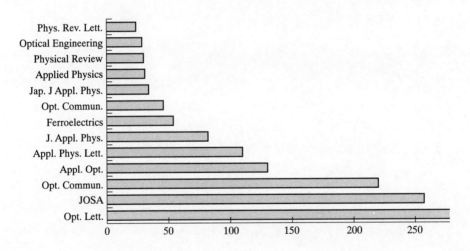

Fig. 4.4 Number of photorefractive papers by journal title, 1981–1994.

What conclusions can be drawn from these figures? Well, it is clear that measured by the sheer volume of work, the importance of the photorefractive effect as a research field has increased dramatically since 1980 (for example, in 1994, papers dealing with the photorefractive effect accounted for over 9% of the number of publications in the JOSA journals). We believe this growth is due to a combination of two factors: firstly, the complex nature of the physics underlying the photorefractive effect, and secondly, the large number of applications suggested for photorefractive materials. The former has meant that a great deal of effort has had to be spent on (and continues to be applied to) the understanding of the many interesting behavioural features exhibited by photorefractive materials. The latter factor has meant that there has been interest from industry based on the hope of commercial exploitation: a hope that is just beginning to be realized.

5
THE MATERIALS EQUATIONS

5.1 Introduction

We have already discussed a simple model of charge transport in Section 1.3 and two somewhat more complicated models in Section 2.11 where we considered both electrons and holes. Our aim is now to present further models which were used at some time or other in the literature. The complications that come are quantitative rather than qualitative. One may have more species of impurities (or traps as they are often called) and more transitions between them, but each process can still be understood by a modest amount of solid state theory.

Unfortunately, there are no standard notations, no standard diagrams, and no attempts at a neat classification. We shall try to introduce here the various models slowly, using a common notation. Donors and acceptors are well known from semiconductor theory. A neutral donor atom will turn into an ionized donor by donating an electron. The electron becomes then available for conduction in the conduction band. A neutral acceptor atom will turn into an ionized acceptor by accepting an electron. The electron leaves a hole behind in the valence band which becomes available for conduction. The trouble with this terminology of 'donating' and 'accepting' is that it is 'electron-centric', meaning by this that an electron is donated or accepted and holes have no equivalent status. This lack of symmetry causes difficulties when an atom may communicate with both the conduction and the valence band.

We think it would be easier to relate the various models to each other if we use consistently a slightly different terminology. We shall distinguish neutral donors and ionized donors, and neutral acceptors and ionized acceptors. An ionized donor has a positive charge and an ionized acceptor has a negative charge. So far there is nothing new. However, instead of talking of donating and accepting we shall talk about excitation. An electron can be excited out of a neutral donor, and a hole can be excited out of a neutral acceptor. Similarly, and that is important for symmetry, a hole can be excited out of an ionized donor (turning it into a neutral donor) and an electron can be excited out of an ionized acceptor (turning it into a neutral acceptor). Excitation may of course mean two things: either photoexcitation (which is of most interest in the physics of photorefractives) or thermal excitation.

Recombination is the inverse process. An electron and an ionized donor may recombine to produce a neutral donor. A hole and a neutral donor may recombine to produce an ionized donor. An electron and a neutral acceptor may recombine to produce an ionized acceptor. A hole and an ionized acceptor may

118   *The materials equations*

recombine to produce a neutral acceptor. In Fig. 5.1 we summarize how these processes may lead to a redistribution of charge among (a) donors, and (b) acceptors.

Starting with these building blocks let us build up a few models.

In Fig. 5.2 we depict a number of different band transport models which are frequently used in explaining the photorefractive effect. Consider first model I. It

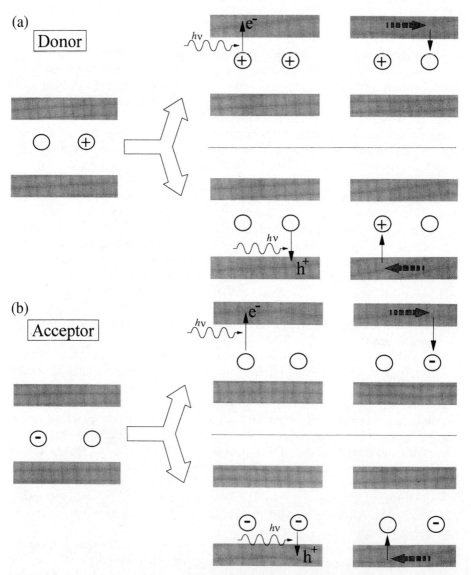

Fig. 5.1 Redistribution of charge via photoexcitation of electrons or holes from (a) donor atoms and (b) acceptor atoms.

5.1 Introduction

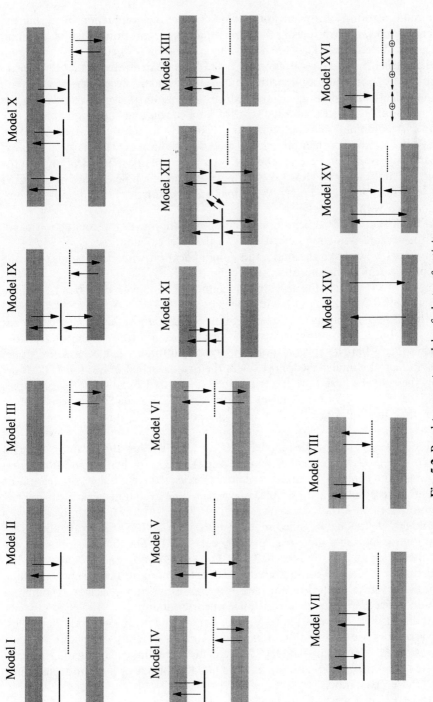

Fig. 5.2 Band transport models of photorefractivity.

is a good introduction to our notations. We have the conduction band and the valence band, and two lines in between, which represent donors (solid line) and acceptors (dotted line). They are not in communication with either the conduction or the valence band. The condition for overall charge neutrality is that the density of ionized donors must agree with the density of ionized acceptors.

In model II the donors are connected with arrows to the conduction band. The arrow upwards denotes the excitation of an electron and the downward arrow denotes recombination. The acceptors do not communicate with either band. They are 'inactive' but their presence ensures that we have overall charge neutrality.

Model III is the converse of model II. Acceptors are in communication with the valence band (the arrow down shows the excitation of a hole, the arrow up shows the recombination of a hole with an ionized acceptor); donors are not in communication with either band.

In model IV we have the donors in communication with the conduction band, and the acceptors in communication with the valence band.

In model V we have the donors in communication with both bands, and the acceptors not in communication with either band.

In model VI we have the acceptors in communication with both bands, and the donors not in communication with either band.

Model I served only as an introduction to our notations. Such a case can occur in a solid but it is of no interest to us since it does not lead to photorefraction. Models II to VI are however all good models of photorefractive behaviour. Model II is, in fact, the model used most often, the one discussed in Section 1.3. Model III is the converse. We have not mentioned it in Part I because all the principles could be explained with the aid of varying donors. For $BaTiO_3$ however, more often than not holes are the dominant charge carrier, so this is the model that applies.

Model IV is what we called model 2 in Section 2.11. Now the acceptors are also of two varieties, they are either ionized or not. We need then an additional rate equation. The density of ionized acceptors increases when a hole is excited out of a neutral acceptor, and it decreases when a hole recombines into an ionized acceptor (see eqn 2.132).

Model V is the one we called model 1 in Section 2.11. The corresponding rate equation for ionized donors was given by eqn (2.131). The ionized donor density increases when an electron is excited out of a neutral donor, and it decreases when a hole is excited out of an ionized donor. Furthermore, it increases when a hole recombines into a neutral donor, and it decreases when an electron recombines with an ionized donor and turns it into a neutral donor.

Model VI is the same as model V, only donors and acceptors swap their roles. The acceptors are now in communication with both bands. With this we have exhausted the 'elementary' models. More complicated models will either be constructed by putting together more of these building blocks or by introducing some new kind of excitation.

Model VII is a generalization of model II to two different species of donors. Model VIII is a generalization of model II to the case when the acceptor level is in communication with the conduction band. Model IX is the slightly more complicated case where the donors are in communication with both the conduction and valence bands. We could of course present further models by adding any pleasing number of donors or acceptors. We shall desist from doing so but we shall exceptionally add one more since it has been discussed in the literature. Model X contains three species of donors in communication with the conduction band, and one species of acceptors in communication with the valence band.

Next, we shall permit communication between the impurity levels. Model XI is a generalization of model II for the case when an electron can be excited from a lower donor level into a higher donor level and, of course, the electron can drop back and recombine. A somewhat more complicated arrangement is shown in model XII. It adds two features to model V. Instead of one species of donors there are now two species, and the two can communicate with each other.

At the beginning of our classification we mentioned that excitation can take two forms: photoexcitation and thermal excitation. When talking of photoexcitation we tacitly assumed that one photon is incident giving rise (with a certain probability that defines the quantum efficiency) to an electron or a hole. It is customary to use a different notation for the case when two photons are needed to accomplish the excitation, called, quite logically, a two-photon process. This is shown in model XIII, which is a generalization of model II. When exciting an electron into the conduction band we now permit excitation by two photons, shown duly by two subsequent arrows pointing upwards.

We have so far relied on various kinds of impurities to lead to a space charge field. That is the way photorefraction is usually realized, but that is not the only way. After all, electrons and holes are charged particles, so they can be responsible for setting up a space charge field without the assistance of impurities. This possibility is shown in model XIV for a single photon process. This is then extended in model XV to include the recombination to donor levels, but without any photoexcitation from them.

Finally, we shall show a model taking into account a new piece of physics. There are two kinds of mobile carriers: one is positive and the other one is negative but the positive carriers are not holes. They are positive ions. This model XVI differs from model II by adding the positive ions. The horizontal arrows mean that they are mobile. This can happen at high temperature. The model is useful for discussing fixing (see Chapter 10 for details).

Having introduced a fair number of models we shall now turn to the literature and try to summarize how these models can improve our understanding of what goes on in photorefractive materials.

5.2 Models and equations

The first question is inevitably: why are there so many models? When devising a model one always attempts to find the simplest model which can reproduce the experimentally observed effects. As we have seen, the photorefractive effect relies on the redistribution of charges in impurities[1], and as such the choice of model essentially boils down to a question of what kind and how many impurities participate in the photorefractive effect. In fact, electrons can be 'trapped' without the presence of impurities by intrinsic defects, as for example in the so-called EL2 defect in 'pure' GaAs (see e.g. Valley *et al.* 1989) or the silicon vacancy complex in BSO (Hou *et al.* 1973). Hence from now on we shall adopt the more general terminology of referring to any defect level in the forbidden band gap as a trap.

Obviously, it is highly desirable to control the type of traps and their concentrations, and that can be more easily done by doping. Thus, LiNbO$_3$ is often doped with Fe, GaAs with Cr, InP with Fe, CdTe with V, and BaTiO$_3$ with Fe and Co, to name just a few of the many combinations which have been tried in the past. It is here important to note that when a material is doped with acceptors or donors these may in principle reside anywhere in the forbidden energy band. Thus, even though conventional wisdom has acceptors placed near the valence band and donors near the conduction band, that is only a reflection on the position of donors and acceptors in most applications of semiconductor technology (Conwell 1958; Sze and Irvin 1968).

So far, we have noted that the photorefractive properties of a material are likely to depend very much on the concentration of donor and acceptor traps. Another important factor is the energy or position of these trap levels within the forbidden band. This is what determines the spectral dependence of the photorefractive effect. Obviously the photon energy has to be larger than the energy difference between the trap level and the relevant band in order for the photon to be absorbed. More specifically, if we consider the traps as hydrogenic centres from which electrons or holes may be excited into parabolic conduction or valence bands, then the electron and hole photoexcitation cross sections, s_e and s_h respectively, are given by (Bashaw *et al.* 1994; Jaros 1982):

$$s_e = \frac{8}{\lambda_{e0}^3} s_{e0} \lambda^{3/2} (\lambda_{e0} - \lambda)^{3/2} \tag{5.1}$$

and

$$s_h = \frac{3.86}{\lambda_{h0}^3} s_{h0} \lambda^{5/2} (\lambda_{h0} - \lambda)^{1/2} \tag{5.2}$$

where λ_{e0} and λ_{h0} are the cutoff wavelengths corresponding to the energy depths of the centres and s_{e0} and s_{h0} are the maximum excitation cross sections which occur at 1/2 and 5/6 of the cutoff wavelength for electron and hole excitation, respectively. To give an example, in Fig. 5.3 we plot the spectral dependence of the absorption cross sections for BSO and BGO (bismuth germanium oxide) as measured by Baquedano *et al.* (1989). In short, we see that even when several trap levels are present in the material, for any given wavelength only a few of them will be actively contributing to the photorefractive effect.

---

[1] With the exception of model XIV.

5.2 Models and equations

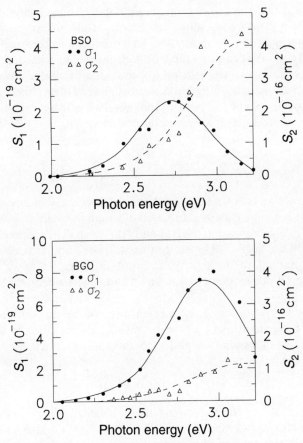

Fig. 5.3 Spectral dependence of photoexcitation cross sections, $s_1(\lambda)$ (circles, left-hand scale) and $s_2(\lambda)$ (triangles, right-hand scale), for a crystal of (a) BSO and (b) BGO (after Baquedano *et al.* 1989).

Another consideration is whether the traps are 'deep-level' traps situated near midgap or whether they are 'shallow traps' located close to the conduction or valence bands. In the latter case, thermal excitation (and optical intensity, as we shall see later) can play a major role in determining whether a particular level is active. The thermal emission, β, of carriers from a trap level which is E_t away from a band follows the form (see e.g. Cheng and Partovi 1986):

$$\beta \propto T^{3/2} \exp\left(-\frac{E_t}{k_B T}\right) \qquad (5.3)$$

where k_B is the Boltzmann constant, and T is temperature. Note that any trap which is very close to a band will be completely ionized by thermal excitations.

Having argued that the choice of material, doping, optical wavelength, optical intensity, and temperature are all crucial factors which determine which photorefractive model is most appropriate, we shall now turn to the main topic of this section, namely the traits and features of each model. In what follows, we shall attempt to proceed in a logical fashion to show what prompted various researchers to propose new models. In the process we shall introduce effects which cannot be explained by the standard model, and we shall attempt to provide a clear picture of the physical processes based on new models.

5.2.1 Standard model

We commence with a quick resumé of the most elementary model, the 'standard band-transport model' which was discussed at length in Chapter 2. In our present terminology this encompasses models II and III. We shall here develop this model in a building-block fashion. Hopefully this will not only make it more readily digestible, but also allow for the easy expansion of the model as we proceed later to include such effects as multiple trap levels and simultaneous electron and hole conduction.

We shall start by considering the concentration of electrons in the conduction band, n_e. The net generation rate of electrons, G_e, will be given by the difference between the rates of generation and recombination of electrons to the donor level:

$$G_e = (\beta_e + s_e I)(N_D - N_D^+) - \gamma_e n_e N_D^+ \tag{5.4}$$

where s_e, γ_e, and β_e are the photoionization coefficient, the recombination constant, and the thermal excitation rate for electrons. N_D and N_D^+ are the total concentration of donors and the concentration of ionized donors, respectively. Having accounted for how the electrons are generated, we now need to consider the movement, or current, of electrons in the conduction band. The electrons will be driven by three forces: (1) they will drift due to an applied electric field; (2) they will be preferentially ejected in certain crystal directions as a result of the photovoltaic effect; and (3) they will diffuse from regions of high electron concentration to regions of low concentration. Restricting the variation to one dimension (here z), the electron current may consequently be written as

$$J_e = e n_e \mu_e E + \mu_e k_B T \frac{\partial n_e}{\partial z} + p_n (N_D - N_D^+) I \tag{5.5}$$

where μ_e is the electron mobility, p_n is the photovoltaic constant, I is optical intensity, and E is the total electric field. Equations (5.4) and (5.5) can subsequently be combined in a continuity equation to describe the variation of the electron concentration with time:

$$\frac{\partial n_e}{\partial t} = G_e + \frac{1}{e} \frac{\partial J_e}{\partial z} \tag{5.6}$$

5.2 Models and equations

Having described the behaviour of electrons, we turn now to the donors. The continuity equation for the density of ionized donors is given by

$$\frac{\partial N_D^+}{\partial t} = G_e \tag{5.7}$$

Note that in this 'standard model', traps are not allowed to move. Finally, the electric field in the crystal is related to the net charge density by Poisson's equation:

$$\frac{\partial E}{\partial z} = \frac{e}{\varepsilon_s}(N_D^+ - N_A^- - n_e) \tag{5.8}$$

where ε_s is the static dielectric permittivity, which is assumed to be independent of position. It is important to note that the acceptor density, N_A, is completely ionized in this model, i.e. $N_A = N_A^-$, and hence does not play any role in charge redistribution. Nevertheless, it has to be included in eqn (5.8) in order to account for the fact that the photorefractive material is not electrically charged.

These coupled non-linear materials equations describe the response of a photorefractive material to light. We have already discussed in detail in Chapter 2 the various aspects and predictions of this standard model, and shall therefore not elaborate much on that here. We shall, however, expand on one aspect, namely the steady-state solution for the general case of an applied DC electric field. In Chapter 2 we presented mainly simplified solutions to special cases. To remedy that, we shall here include simultaneously the photovoltaic effect and thermal excitation, while making no simplifying assumptions about the electron or ionized donor densities.

Taking a sinusoidal variation in light intensity:

$$I = I_o(1 + m\cos Kz) \quad \text{where} \quad m = I_1/I_o \tag{5.9}$$

we shall assume that the intensity modulation represents only a small perturbation (i.e. $m \ll 1$). This simplification allows us to assume that the variables, J_e, n_e, E, and N_D^+, will respond in a similar fashion, taking on a sinusoidal variation of the form $X(z,t) = X_o(t) + 1/2[X_1(t)\exp(jKz) + \text{c.c.}]$. Thus, by linearizing the materials equations in this manner, the solution for the fundamental Fourier component of the space charge field can be found to be (Solymar et al. 1984)

$$E_1 = -\frac{m}{D}[E_{ph}(E_e[1-a\nu] + E_q) + \nu E_q(E_o - jE_D)] \tag{5.10}$$

$$D = E_D + E_e + E_q + j(E_o + aE_{ph}) \tag{5.11}$$

where

$$E_q = \frac{eN_{\text{eff}}}{\varepsilon_s K}, \quad E_{\text{ph}} = \frac{p_n I_o(N_D - N_{Do}^+)}{e\mu_e n_{eo}}, \quad E_e = \frac{en_{eo}}{\varepsilon_s K} = \frac{\tau_e}{\tau_d} E_M, \quad E_D = \frac{k_B T}{e} K,$$

$$E_M = \frac{1}{\mu_e \tau_e K},$$

$$E_o = \frac{J_{eo}}{e\mu_e n_{eo}} - E_{\text{ph}}, \quad N_{\text{eff}} = \frac{N_{Do}^+}{N_D}(N_D - N_{Do}^+), \quad a = \frac{N_{Do}^+}{N_D}, \quad \nu = \frac{s_e I_o}{\beta_e + s_e I_o}$$

$$\tau_e = \frac{1}{\gamma_e N_{Do}^+}, \quad \tau_d = \frac{\varepsilon_s}{e\mu_c n_{co}}, \quad N_{Do}^+ = n_{eo} + N_A^-$$

(5.12)

and the background electron density, n_{eo}, may be obtained as the positive root of the quadratic equation

$$n_{eo}^2 + n_{eo}\left[N_A^- + \frac{s_e I_o + \beta_e}{\gamma_e}\right] + \frac{s_e I_o + \beta_e}{\gamma_e}(N_A^- - N_D) = 0 \quad (5.13)$$

In the limit of $n_{eo} \ll N_{\text{eff}}$ and no photovoltaic effect, eqn (5.10) reduces to

$$E_1 = -\nu m_1 \frac{E_q(E_o - jE_D)}{E_D + E_q + jE_o} \quad (5.14)$$

Note that the reduction factor, ν, did not appear in the treatment of Part I, where it was assumed that thermal excitation was negligible. Thus, in the previous chapter the space charge field was seen to be completely independent of intensity. This conclusion now has to be revised slightly. Equation (5.14) reveals a very important feature of the standard model, namely that two-wave mixing gain is independent of intensity, provided the optical excitation dominates over thermal excitation (i.e. $\nu = 1$).

We have so far concentrated on the steady state. Before proceeding to more complex models, it is important to emphasize two key features concerning the transient behaviour in the standard model. Firstly, in Chapter 2 it was shown that for relatively low intensities the response rate of the photorefractive effect scales linearly with the average optical intensity. Thus, even though the magnitude of the photorefractive effect is independent of the illumination, the intensity does matter where the 'speed' of the photorefractive effect is concerned. Secondly, the photorefractive gratings were predicted to build up and decay exponentially with a *single* time constant.

This standard model, which was extensively developed by Kukhtarev et al. in the late 1970s, is by far the most commonly used model. It has been surprisingly successful at describing the photorefractive effect in many of the original mainstream photorefractive materials such as BSO, $LiNbO_3$, and in many instances $BaTiO_3$. However, by the mid-1980s experimental results were amassing which could not be explained by this simple model.

5.2.2 Bipolar transport

One of the primary deficiencies of the standard model is that it does not account for bipolar transport, i.e. the simultaneous involvement of electrons and holes in the charge transport process. By the mid-1980s evidence was mounting which implicated electron–hole competition as the principal culprit responsible for reducing the strength of the photorefractive effect. In 1978 Orlowski and Krätzig demonstrated experimentally that the relative contributions of electron and hole transport in $LiNbO_3$:Fe depended on the concentration of oxidized Fe impurities, as shown in Fig. 5.4. Similar results were obtained by Ducharme and Feinberg (1986), who showed that the electron and hole conduction in $BaTiO_3$ could be altered by heating the crystal in oxygen at different partial pressures. Beam coupling experiments were also turning up peculiar results. M. B. Klein and Valley (1985) found that in one particular 'anomalous' crystal of $BaTiO_3$ the signal gain could be seen to change sign with grating period (i.e. inter-beam angle), as shown in Fig. 5.5. This is definitely not possible with the standard model. In short, there was clearly a need to extend the standard model to include both electron and hole conduction.

When extending the standard model to include both electrons and holes, there are basically two ways to proceed. The first is to assume that holes and electrons can be excited to and from a single impurity band, e.g. donors. This involves retaining the assumption that only one trap level is active while the other level, in this case the acceptors, is present only to ensure overall charge neutrality. This approach was adopted independently by Valley (1986) and Strohkendl et al. (1986) (for earlier work see Stepanov 1982). In Fig. 5.2 we have termed this model V (or model VI when the active level is an acceptor). The other method is to have electrons in communication with the donor (or acceptor) level while

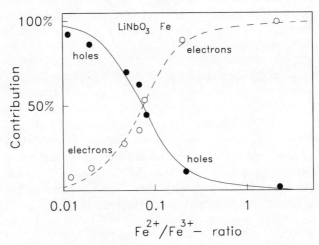

Fig. 5.4 Electron (○) and hole (●) contributions to photoconductivity versus Fe^{2+}/Fe^{3+} ratio in $LiNbO_3$:Fe crystals (after Orlowski and Krätzig 1978).

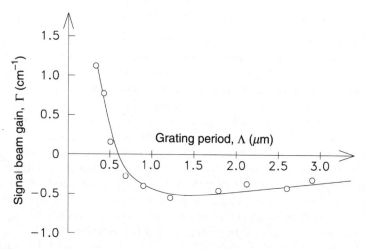

Fig. 5.5 Two-wave mixing gain as a function of grating period for an 'anomalous' crystal of BaTiO$_3$ (after M. B. Klein and Valley 1985).

holes are in communication with the acceptor (or donor) level. This model, here termed model IV, was developed by Valley (1986). In the present section we shall concentrate on model V, deferring model IV to the next section which deals with the general implications of multiple active trap levels.

Returning to the 'one-species electron–hole competition model', in this model we still only have excitation and recombination to and from the one donor level, but we have now included communication with both the conduction and valence bands. First we consider the concentration of electrons, n_e, and holes, n_h. Clearly the net generation rate of electrons, G_e, and holes, G_h, is given by the difference between the generation and recombination rates to the trap level. Thus, for this model they will be given by

$$G_e = (\beta_e + s_e I)(N_D - N_D^+) - \gamma_e n_e N_D^+ \tag{5.15}$$

$$G_h = (\beta_h + s_h I)N_D^+ - \gamma_h n_h (N_D - N_D^+) \tag{5.16}$$

where we have introduced the notation that the subscripts e and h refer to electron and hole processes, respectively. We note that electrons are excited from filled traps while holes are excited from empty traps. The movement of the electrons in the conduction band and the holes in the valence band are governed by the continuity equations:

$$\frac{\partial n_e}{\partial t} = G_e + \frac{1}{e}\frac{\partial j_e}{\partial z} \tag{5.17}$$

5.2 Models and equations

$$\frac{\partial n_h}{\partial t} = G_h - \frac{1}{e}\frac{\partial j_h}{\partial z} \tag{5.18}$$

where the electron and hole current densities are defined as:

$$J_e = en_e\mu_e E + \mu_e k_B T \frac{\partial n_e}{\partial z} \tag{5.19}$$

$$J_h = en_h\mu_h E - \mu_h k_B T \frac{\partial n_h}{\partial z} \tag{5.20}$$

The time variation of the concentration of ionized donors, which are immobile in this model, can be expressed as

$$\frac{\partial N_D^+}{\partial t} = G_e - G_h \tag{5.21}$$

Before we proceed to the solutions, it is worthwhile to develop a physical picture for how the inclusion of holes might influence the build-up of space charge. During diffusion both carrier types will move in the same direction. However, since they carry an opposite charge they will effectively be working against one another. This is referred to as electron–hole competition.

We shall make the usual assumptions in deriving a solution for the space charge field. From now on we restrict ourselves to low illumination intensities and small intensity modulation, unless stated otherwise. Searching for the steady-state solution for the case of no applied field, no thermal excitation, and no photovoltaic effect, we may obtain the results of eqns (2.133). This formulation of the result by Strohkendl *et al.* (1986) neatly separates the space charge field into a product of the result expected from the standard model and a new term, $\xi(K)$, which describes the electron–hole competition. We find that the electron–hole competition has the following effects: (1) it reduces the space charge field (and hence beam coupling); (2) it leads to a variation of the relative contribution of electrons and holes with grating spacing; and (3) the sign of the space charge field depends on differences in photoconductivity for large grating spacings and absorption coefficients for small grating spacings, i.e.

$$\xi(K \to 0) = \frac{\sigma_e - \sigma_h}{\sigma_e + \sigma_h}, \quad \xi(K \to \infty) = \frac{\alpha_e - \alpha_h}{\alpha_e + \alpha_h} \tag{5.22}$$

where the electron and hole photoconductivity and absorption are respectively defined as $\sigma_e = e\mu_e n_{eo}$ and $\sigma_h = e\mu_h n_{ho}$, and $\alpha_e = s_e(N_D - N_{Do}^+)\hbar\omega$ and $\alpha_h = s_h N_{Do}^+ \hbar\omega$ (Strohkendl *et al.* 1989).

This one-species electron–hole competition model has been extremely successful in explaining the photorefractive effect in many materials. Strohkendl *et al.* (1986), were able to explain the 'anomalous' beam coupling results of Fig. 5.5. Medrano *et al.* (1988), and Rytz *et al.* (1990), had similar success with describing the photorefractive effect in various samples of KNbO$_3$ and BaTiO$_3$:Co, respectively, with Medrano *et al.* confirming as well that the grating spacing at which the cross-over in coupling direction occurs will change with wavelength. However, at the same time there were several indications that this was not the 'do-all' model of photorefractivity. The foremost concern was that many experiments were showing photorefractive gratings which would decay with two or more time constants instead of the single time constant predicted by this model (see e.g. Strohkendl and Hellwarth 1987; Strohkendl et al. 1989). It was clear that more active species, or trap levels, were called for.

5.2.3 Two or more impurity species

In this section we shall take a look at how including more than one trap level will affect the photorefractive behaviour. By first and foremost emphasizing those effects which are new and unique to multiple level systems, we hope to demonstrate why these slightly more complex models often need to be invoked to fully explain the photorefractive effect.

We shall begin by considering a system of two photo-active levels, each of them in communication with the conduction band. This 'monopolar two-species model' is depicted in Fig. 5.2 as models VII and VIII, depending on whether the photo-active levels are two donor levels or an acceptor and a donor level, respectively. Both models actually behave in exactly the same way. The only difference seems to be that model VII requires an inactive level in order to ensure overall charge neutrality.

So how do we expect this monopolar two-species model to behave? It is actually very easy to develop an intuitive feel for what will happen. We already know that the photorefractive effect relies on the photo-assisted redistribution of charge to set up a space charge field, which in turn alters the refractive index. With only one active species, the charges will only be redistributed within one trap level. If we now increase the number of trap levels to two, any light incident on the material will redistribute the charges in both levels, in effect creating two in-phase superimposed gratings. Moreover, since the trap densities, the photoionization rates, and the recombination times are likely to be different for each level, as we have remarked earlier, we would expect each grating to build up and decay with different time constants. This is the type of behaviour which was observed for example by Strohkendl and Hellwarth (1987), who were able to fit the decay rate of a grating in BSO to a simple double-exponential function for a range of grating spacings, as shown in Fig. 5.6. However, two-trap-level systems should not be expected always to behave in such a simple way. In general the individual gratings cannot be considered to be uncoupled, because when charges in the separate levels form and evolve, they do so simultaneously. The upshot is of course that the gratings will influence each other, and the grating erasure (or writing) will no longer be just a sum of two exponential decays, but will take on a decidedly non-exponential behaviour.

The mathematical framework for handling the monopolar two-species model was laid by Valley (1983b). Later extensions to the model were made by Carrascosa and Agulló-López (1988), who included the photovoltaic effect, Jariego and Agulló-López (1991), who introduced more than two species, and

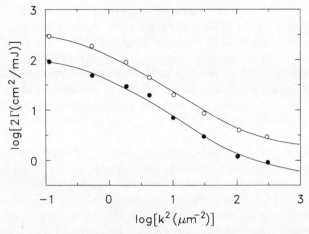

Fig. 5.6 Erasure data for a crystal of BSO showing the fast (○) and the slow (●) exponent of the normalized decay rates as a function of spatial frequency. The curves were fitted (solid lines) by the sum of two exponential decays (after Strohkendl and Hellwarth 1987).

Rustamov (1993), who studied diffraction efficiency by including the coupling to the optical field equations. The materials equations for this model (model VIII), as derived by Valley, are very similar to the set of equations describing the standard model. The first difference is that since electrons now communicate with two trap levels, we have that the overall net electron generation rate, $G_e = G_{eA} + G_{eD}$, where the net generation rates from the donor, G_{eD}, and acceptor levels, G_{eA}, are

$$G_{eD} = s_{eD}I(N_D - N_D^+) - \gamma_{eD}n_e N_D^+ \tag{5.23}$$

$$G_{eA} = s_{eA}IN_A^- - \gamma_{eA}n_e(N_A - N_A^-) \tag{5.24}$$

Here the additional subscripts A and D refer to the processes related to acceptors and donors, respectively. The second difference is that we now have to account for the variation of ionized traps in both levels:

$$\frac{\partial N_D^+}{\partial t} = G_{eD}, \quad \frac{\partial N_A^-}{\partial t} = -G_{eA} \tag{5.25}$$

and with that the model is established.

It is important to note that in this monopolar multiple-species model the gratings which are set up in the different trap levels are all in-phase, contributing constructively to set up a refractive index grating. This is true irrespective of whether the active traps are all donors or whether they are donors and acceptors. This is perhaps somewhat surprising; after all, the ionized donors and acceptors do carry an opposite sign. It is nevertheless true as may be realized by considering the photo-induced charge redistribution once again. Roughly speaking, when a photoconductive material is illuminated, the electrons which are trapped in (negatively charged) ionized acceptors or (neutral) un-ionized donors will be ejected into the darker regions of the material. In both cases a more positive charge will be left behind. The opposite would be true if one grating was set up by the redistribution of electrons, and the other grating by the redistribution of holes. In this case, the two gratings will be 180 degrees out of phase with each other and are consequently commonly referred to as *complementary gratings*. The ability to set up complementary gratings is what forms the backbone of most concepts of 'fixing' of holograms for long term storage, as we shall see in later sections.

Complementary gratings yield some very interesting effects. As the total space charge field in the crystal will be the sum of the fields created by the spatial distribution of carriers in each trap level, for complementary gratings the net amplitude of the space charge field will be smaller than the separate contributions. Moreover, the net space charge field will be reduced compared with the case of only one type of carrier. This is how electron–hole competition manifests itself in this model. We also find that if the initial amplitude of the 'slower' grating is smaller than that of the fast-decaying grating, then net space charge field will change sign during erasure. This type of behaviour was for example observed by Attard and Brown (1986) in their study of trapping levels in BSO. The results of Fig 5.7, which were obtained by Miteva and Nikolova (1986) for the recording and erasure of holograms in a BTO (bismuth titanium oxide) crystal, can quite easily be explained in the context of this complementary grating model. During writing of the hologram, the fast trap level builds up first, giving the initial quick rise in the diffraction efficiency. At this stage only one grating is present in the crystal. If the hologram is erased at this time (i.e. B in Fig. 5.7) then the diffraction efficiency will follow a single exponential decay in accordance with the standard

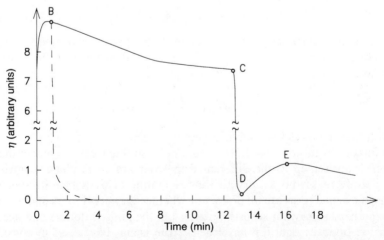

Fig. 5.7 Dependence of diffraction efficiency on time for a crystal of BTO (after Miteva and Nikolova 1988).

model, as seen by the dashed line in Fig. 5.7. However, if the hologram is written for a long time the slower complementary grating will have time to build up, in the process reducing the net space charge field and hence the diffraction efficiency. Erasing the hologram at this stage (C in Fig. 5.7) will lead to a characteristic behaviour whereby the diffraction efficiency falls to zero and then increases again before eventually decaying completely.

The bipolar two-species model, or model IV in Fig 5.2, was originally proposed by Kukhtarev *et al.* (1984a) and followed by a more detailed treatment by Valley (1986). Subsequent work by Zhivkova and Miteva (1990), and by Bashaw *et al.* (1990a,b, 1992) provided the framework for the interpretation of the recording, the revelation, and the time evolution of complementary gratings. Zhivkova (1992) compared the steady-state space charge field for this model for the cases of an applied DC, an AC sine wave, and an AC square wave electric field.

The basic materials equations for the bipolar two-species model are only slightly different from those of the bipolar one-species model of Section 5.2.2. The only difference is that responsibilities have now been divided so that one trap level caters only for electrons while the other caters only for holes. Thus, the net generation rates for electrons and holes are now

$$G_{eD} = s_{eD} I (N_D - N_D^+) - \gamma_{eD} n_e N_D^+ \tag{5.26}$$

$$G_{hA} = s_{hA} I (N_A - N_A^-) - \gamma_{hA} n_h N_A^- \tag{5.27}$$

and the continuity equations for the traps are

$$\frac{\partial N_D^+}{\partial t} = G_{eD}, \quad \frac{\partial N_A^-}{\partial t} = G_{hA} \tag{5.28}$$

Having introduced the model, we shall without further ado jump directly to the steady-state solution for the fundamental component of the space charge field (Zhivkova and Miteva 1990):

$$E_1 = jm \frac{E_{qD} - E_{qA}}{1 + j\dfrac{E_{qD}}{E_o + jE_D} - j\dfrac{E_{qA}}{E_o - jE_D}} \tag{5.29}$$

where

$$E_{qD} = \frac{e}{\varepsilon_s K} \frac{(N_D - N_{Do}^+) N_{Do}^+}{N_D}, \quad E_{qA} = \frac{e}{\varepsilon_s K} \frac{(N_A - N_{Ao}^-) N_{Ao}^-}{N_A}, \quad N_{Ao}^- = N_{Do}^+ \tag{5.30}$$

E_{qD} and E_{qA} are the saturation fields for the space charge in the donor and acceptor levels, respectively. Note that if the acceptor traps are completely ionized, i.e. $N_A = N_{Ao}^-$, then $E_{qA} = 0$ and this model simplifies to the standard model, as expected.

There are several important remarks concerning this model which should be made at this stage. First, eqn (5.29) reveals that the sign of the space charge field (and hence energy transfer) depends on the relative concentration of donor and acceptor traps, rather than relative response rates as was the case for the single-species electron–hole competition model. In other words, this model cannot lead to a change in direction of the steady-state energy transfer with grating spacing. On the other hand, the direction of energy transfer can change with time, as complementary gratings are written or revealed. For example, referring back to Fig. 5.7 we note that as we go from point C to E the space charge field actually changes sign as the diffraction efficiency (which is proportional to the square of the space charge field) recovers from the dip, revealing the slower complementary grating.

The existence of complementary gratings is perhaps the most characteristic feature of this model. Taking for example eqn (5.29), the space charge field is in fact a superposition of the gratings set up in each of the two trap levels. The steady-state amplitudes of these gratings are

$$N_{D1}^+ = m \frac{\varepsilon_s K}{e} \frac{E_{qD}(E_D + 2E_{qA})}{E_D + E_{qD} + E_{qA}}, \quad N_{A1}^- = m \frac{\varepsilon_s K}{e} \frac{E_{qA}(E_D + 2E_{qD})}{E_D + E_{qD} + E_{qA}} \quad (5.31)$$

where for simplicity we have set $E_o = 0$. Consider now the consequences of having a crystal with equal trap densities, i.e. $N_A = N_D$. In this case $E_{qD} = E_{qA}$ and, as may be seen from eqn (5.29) the space charge field vanishes. However, the same is not true of the individual trap densities. The steady-state space charge field resulting from each trap level actually reaches its maximum possible value of E_{qD} because there is no electric field which hampers or halts their growth. Will we ever be able to 'see' these gratings? Yes: if the response rates are different for each grating, then we may 'reveal' one of the gratings by erasing the other[2]. Unfortunately, Nature has conspired to limit the space charge field that may be revealed. Zhivkova and Miteva (1990) showed that in the extreme case where, for instance, the donor grating is fixed while the acceptor grating is erased by thermal excitation (e.g. if $\beta_{eD} \ll \beta_{hA}$), the acceptor grating will not be completely erased, but will decay to a steady state of

$$N_{A1}^- = \frac{N_{D1}^+}{1 + E_D/E_{qA}} \quad (5.32)$$

revealing a net space charge field of

$$E_1 = \frac{jeN_{D1}^+}{\varepsilon_s K} \frac{E_D}{E_D + E_{qA}} \quad (5.33)$$

For $E_{qD} = E_{qA}$, this is equivalent to the space charge field which is written in a material with only one active donor species, as described by the standard model.

---

[2] The trap gratings can in fact be seen by non-electro-optic means which rely on the fact that the optical absorption is proportional to $N_D - N_D^+$ and $N_A - N_A^-$. Knyaz'kov and Lobanov (1985), Pierce et al. (1990), and Cudney et al. (1991) have described and measured contributions to two-wave mixing gain from absorptive coupling which acts to reduce or increase the absorption experienced by both beams.

5.2 Models and equations

Before concluding this section, we shall return briefly to the response rates. We have already emphasized that having two active trap levels leads in general to a non-exponential decay of the space charge field. However, for high applied fields or small grating spacings the electron and hole processes decouple and the two gratings build up and decay separately. Moreover, it is interesting to note that the relative magnitudes of the response rates can change with grating spacing. Figure 5.8 illustrates an example by Bashaw *et al.* (1992), showing how the response rates might be expected to behave in a sillénite-like crystal. The response rates are seen to cross over at a particular spatial frequency. This means that on one side of the cross-over the diffraction efficiency will decay with the characteristic 'dip' depicted in Fig. 5.7, while on the other side it will rise initially before it eventually decays to zero. In other words, the temporal position of the dip, or mimimum, in diffraction will change with grating spacing.

5.2.4 Shallow trap models and the influence of temperature

We shall now investigate various models which include 'shallow' traps, meaning traps from which electrons or holes may be excited thermally. The term 'shallow' is used rather loosely, because at high temperature the trap levels need not be particularly close to the valence or the conduction band for thermal excitation to play a significant role. We have already investigated the effect of thermal excitation on the standard model, having shown that the uniform thermal excitation serves to reduce the effective modulation in much the same way as uniform illumination would. The effect of thermal excitation is here quite simple, and consequently easy to understand. Much more complex and peculiar behaviour

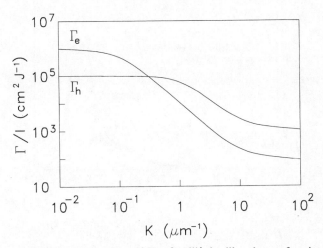

Fig. 5.8 Uncoupled response rates Γ_e and Γ_h of a sillénite-like photorefractive medium as a function of spatial frequency (after Bashaw *et al.* 1992).

may be observed if thermal excitation is included in models with more active trap levels or with simultaneous electron and hole conduction. In the following we shall for example show how shallow traps can lead to screening, photoinduced absorption, sublinear photoconductivity, intensity-dependent gain, fixing, and intensity–temperature resonance-like phenomena.

We shall start by looking at the 'multiple defect screening model', represented by model X in Fig. 5.2, which is a variation on the complementary grating model of the previous section. This model was introduced by Nolte *et al.* (1989) to explain phenomenologically the temperature dependence of the photorefractive effect in InP:Fe. In this model, there is only one photoactive trap level from which holes may be excited. Nolte *et al.* argued that if additional defects are introduced from which electrons may be excited purely thermally, then the electrons will redistribute in these shallow traps so as to cancel the fields built up by the photoactive trap level. Using this model, Nolte *et al.* were able to quite successfully explain their diffraction efficiency measurements of Fig. 5.9. Note that for very high temperatures, all trap levels are completely emptied, including the photoactive trap level, and there is no resulting space charge field. On lowering the temperature, the photorefractive effect builds up as photoionization starts to dominate over thermal ionization for the deep-level photoactive trap. This behaviour is explained by the standard model, eqn (5.14). On further lowering of temperature, the additional shallow traps are no longer completely emptied and charge will consequently redistribute in these levels in response to the space charge field and will eventually completely screen the space charge field as the reservoir of charge in the additional shallow traps becomes large enough. This screening

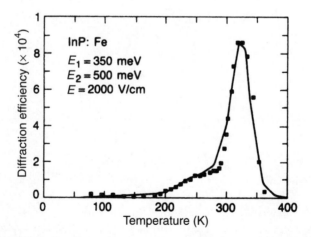

Fig. 5.9 Diffraction efficiency as a function of temperature for a crystal of InP:Fe for $E_o = 2\,\text{kV}\,\text{cm}^{-1}$ and $I_o = 40\,\text{mW}\,\text{cm}^{-2}$. The solid lines are fits assuming the presence of two shallow trap levels with concentrations 5 times that of the Fe concentration, and with binding energies of $E_1 = 350$ meV and $E_2 = 500$ meV (after Nolte *et al.* 1989).

model is, however, not the only model which can explain the peak in gain versus temperature of Fig. 5.9, as we shall see later.

We shall now turn to perhaps the most thoroughly developed and frequently used, shallow trap model to date, the 'monopolar shallow- and deep-trap model' due largely to Brost et al. (1988), Holtmann (1989), Mahgerefteh and Feinberg (1990), Jariego and Agulló-López (1991), Tayebati (1991), Tayebati and Mahgerefteh (1991), and Garrett et al. (1992). This model is an extension of the monopolar two-species model (models VII and VIII) of the previous section, to describe the case where one of the trap levels can be excited thermally as well as optically (i.e. shallow traps), while the other can only be excited optically (i.e. deep traps). Letting N_A^- represent filled shallow traps, the materials equations (5.23) to (5.25) still apply, with the change that electrons may also be excited thermally from these levels. Poisson's equation now takes the form

$$\frac{\partial E}{\partial z} = \frac{e}{\varepsilon_s} (N_D^+ - N_A^- - n_e - N_C^-) \tag{5.34}$$

where N_C^- is an additional inactive acceptor level which we introduced as a slight modification of model VIII. In the low-modulation and low-intensity approximation, the mean density of filled shallow traps at steady state may be obtained by combining eqns (5.23) to (5.25) and (5.34). In the special limit of low density of filled shallow traps, i.e. $N_{Ao}^- \ll N_C^-, N_A$, Tayebati and Mahgerefteh (1991) obtained the simple analytical solution

$$N_{Ao}^- = \frac{1}{1 + \beta_{eA}/s_{eA}I_o} \frac{s_{eD}\gamma_{eA}N_A(N_D - N_C^-)}{s_{eA}\gamma_{eD}N_C^-} \tag{5.35}$$

and $N_{Do}^+ = N_{Ao}^- + N_C^-$. Thus, the uniform components of the densities of ionized donors and acceptors change with illumination. This is one of the key features of this model. Brost et al. (1988) used this behaviour to explain photoinduced absorption in BaTiO$_3$ by noting that the total optical absorption depends on the densities of filled traps:

$$\begin{aligned}\alpha_{\text{traps}} &= [s_{eD}(N_D - N_{Do}^+) + s_{eA}N_{Ao}^-]\hbar\omega \\ &= [s_{eD}(N_D - N_C^-) + (s_{eA} - s_{eD})N_{Ao}^-]\hbar\omega\end{aligned} \tag{5.36}$$

It follows that if N_{Ao}^- is intensity-dependent, so is the absorption. Specifically, eqn (5.36) predicts photoinduced absorption if $s_{eD} > s_{eA}$, and photoinduced transparency if $s_{eD} < s_{eA}$. We shall return to this in more detail in Section 5.5, which is concerned with various effects of high optical intensity.

The intensity dependence of the uniform trap densities has even wider consequences. For instance, the electron recombination times, which are given by

$$\tau_{eD} = \frac{1}{\gamma_{eD}N_{Do}^+}, \quad \tau_{eA} = \frac{1}{\gamma_{eA}N_{Ao}^-} \tag{5.37}$$

will now depend on intensity. Thus, if the light intensity increases, the uniform trap densities increase, leading to a decrease of the electron recombination times. Furthermore, the electron concentration in the conduction band, which is given by

$$n_{eo} = s_{eD} I_o (N_D - N_{Do}^+) \tau_{eD} \qquad (5.38)$$

will also depend on intensity. Holtmann (1989) and Mahgerefteh and Feinberg (1990) showed that the electron density and hence the photoconductivity ($\sigma_e = e\mu_e n_{eo}$) will depend sublinearly on the intensity, taking the approximate functional form I_o^x where $0.5 < x < 1$. Finally, the grating formation time will also be directly affected by the intensity variation of the uniform trap densities. Since the grating formation rate is proportional to the inverse of the dielectric relaxation time ($1/\tau_{de} = \sigma_e/\varepsilon_s$), the 'speed' of the photorefractive effect is expected to follow the same sublinear dependence on intensity as the photoconductivity.

Proceeding to the space charge gratings set up in each trap level, Tayebati and Magherefteh (1991) found that in the absence of an externally applied electric field the amplitude of the steady-state grating is

$$N_{D1}^+ = m \frac{\varepsilon_s K}{e} \left[\frac{E_{qD} E_D}{E_D + E_{qA} + E_{qD}} + S \right] \qquad (5.39)$$

in the deep traps, and

$$N_{A1}^- = m \frac{\varepsilon_s K}{e} \left[\frac{1}{1 + \beta_{eA}/s_{eA} I_o} \frac{E_{qA} E_D}{E_D + E_{qA} + E_{qD}} - S \right] \qquad (5.40)$$

for the shallow traps. E_{qA} and E_{qD} have been defined in eqn (5.30). The gratings have here been decomposed into an in-phase grating and a screening grating, S, given by:

$$S = \frac{1}{1 + s_{eA} I_o/\beta_{eA}} \frac{E_{qA} E_{qD}}{E_D + E_{qA} + E_{qD}} \qquad (5.41)$$

Note that although the screening grating may cause the shallow trap grating to change size and sign with intensity or time, Brost et al. (1988) came to the conclusion from numerical studies that the net space charge field will never actually switch sign. The steady-state space charge field may be found by combining the above trap gratings in Poisson's equation, yielding

$$E_1 = jm\chi(I_o) \frac{E_D E_q}{E_D + E_q} \qquad (5.42)$$

with the intensity-dependent factor

$$\chi(I_o) = \frac{1}{E_q} \left(E_{qD} + \frac{E_{qA}}{1 + \beta_{eA}/s_{eA} I_o} \right) \qquad (5.43)$$

where the saturation space charge field has been defined as $E_q = E_{qA} + E_{qD}$. By expressing the solution in this manner, Tayebati and Mahgerefteh (1991) have introduced an intensity-dependent factor, $\chi(I_o)$, which lies between 0 and 1. Note that in this model none of the aspects which depend on intensity also depend on grating spacing. This includes the intensity-dependent reduction factor. Thus, $\chi(I_o)$ should be contrasted with the reduction factor, $\xi(K)$, of the single species electron–hole competition model (models V and VI), where the reduction factor was shown to depend on the grating spacing and to lie between -1 and $+1$.

Before we conclude our discussion of this model, it is important to make a few additional observations concerning grating decay. In this shallow trap model, it is logical that for thermal erasure in the dark, the shallow trap grating will be erased quite quickly. However, what is equally important is that the deep trap grating will also be partially or completely erased, because the charge which is released in the depopulation of shallow traps can lead to the erasure of the grating in the deep traps. This quick erasure of net space charge field on the time scale of the thermal erasure of the shallow traps is referred to as *coasting* (Mahgerefteh and Feinberg 1988). To illustrate the effects of coasting, we plot in Fig. 5.10 normalized diffraction efficiency decay curves for BSO, as obtained by Tayebati (1991). The gratings have here been written at different intensities, after which their decay in the dark has been monitored. The shallow trap model readily explains the observed behaviour in the context of coasting. For low intensities, shallow traps play no role and the grating decays slowly. For recording at higher intensities, the shallow traps become more and more populated, and as a consequence grating erasure by coasting plays an increasing role.

Finally, it is now necessary to revise slightly a previous statement. In Section

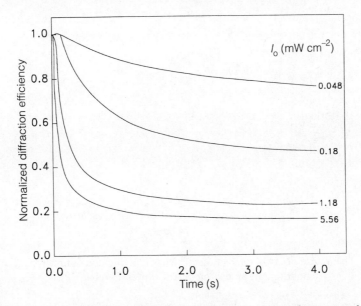

Fig. 5.10 Normalized diffraction efficiency as a function of time for a crystal of BSO, showing the role of coasting in grating erasure. The gratings were recorded with a grating spacing of $\Lambda = 1.9\,\mu\text{m}$ and at different intensities I_o (after Tayebati, 1991).

5.2.3 we suggested that in order to explain the grating decay shown in Fig. 5.7, a bipolar two-species or 'complementary grating' model was needed. As hinted then there is another explanation: Tayebati (1991) and Jariego and Agulló-López (1991) have shown that this monopolar shallow trap model can under certain circumstances lead to similar behaviour, as may be seen in Fig. 5.11, which shows a theoretical prediction of diffraction efficiency (Tayebati 1991). However, two differences should be emphasized. First, the direction of energy transfer changes sign during grating erasure for the complementary grating model, but remains unchanged in the shallow trap model. Second, in the complementary grating model the diffraction efficiency should dip to zero[3], whereas this is not necessary for the shallow trap model.

In the shallow trap model discussed so far, we have seen how temperature and intensity can have a profound effect on the population of the trap levels, resulting in such interesting effects as intensity-dependent gain and photoinduced absorption. We shall now turn to another model for which the role played by temperature is also far from trivial.

In what follows, we return to the one-species electron–hole competition model (introduced in Section 5.2.2 as models V and VI) with the intention of examining how the photorefractive properties of this model are altered by the inclusion of thermal emissions of both electrons and holes. Starting with the simplest case of diffusion recording (i.e. no applied field), it turns out that a reduction factor, ν, must be multiplied to the space charge field in eqn (2.124). This reduction factor

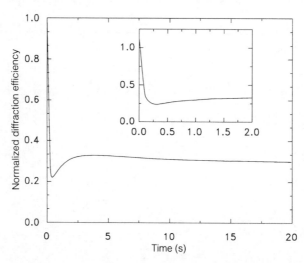

Fig. 5.11 Theoretical decay of diffraction efficiency as a function of time for the monopolar shallow trap model, model VII (after Tayebati 1991).

---

[3]In experiments, however, vibrations during hologram recording can lead to a slight phase-shift in the complementary gratings so that they do not cancel at the dip (Bashaw et al. 1990b; Zhivkova and Miteva 1991). Similarly, the simultaneous appearance of photochromic or absorption gratings can explain a non-zero dip in diffraction efficiency (Bernado et al. 1990).

reflects the comparative influence of photoconductivity and dark conductivity, as was the case for the standard model in eqn (5.14). So far, there is seemingly nothing untoward about the effects of temperature. However, to describe how the space charge field varies with temperature we need to consider the temperature dependence of all the terms. Thus, the thermal emission varies exponentially with temperature, as seen from eqn (5.3), and the diffusion field, E_D, varies linearly with temperature. Moreover, the electron–hole competition factor, $\xi(K)$, is now also dependent on temperature through diffusion. In Fig. 5.12 we illustrate the results of Földvári et al. (1993a), who used this model to explain the temperature dependence of the diffraction efficiency in BGO. Note that the diffraction efficiencies have been normalized to their room temperature values. The apparent decrease of diffraction efficiency with intensity is merely an illusion, the true (unnormalized) diffraction actually increases with intensity as expected. Nevertheless, there is clearly an optimum temperature, in this case near 120 K, which yields a maximum in diffraction.

The situation becomes even more complex once an external field is applied to the crystal, and some surprising intensity–temperature resonance effects spring into action. To place this development in its correct historical perspective, it is interesting to note that Gravey et al. (1989a,b) observed some unusual characteristics when measuring gain in InP:Fe. They were surprised to find that (1) there was such a high gain for applied DC fields, and (2) there was a very distinct maximum gain as a function of intensity (some indication of such optimum

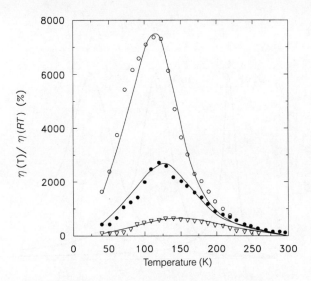

Fig. 5.12 Normalized diffraction efficiency, $\eta(T)/\eta(T = 300\,\text{K})$, as a function of temperature for a crystal of BGO. $\lambda = 488$ nm, $2\theta = 18°$, and the write intensities are $I_o = 25\,\text{mW cm}^{-2}$ (○), $80\,\text{mW cm}^{-2}$ (●), and $800\,\text{mW cm}^{-2}$ (▽). The solid lines are fits using the one-species electron–hole competition model, model V (after Földvári et al. 1993a).

intensity was shown somewhat earlier by Mainguet (1988). Picoli et al. (1989a,b) promptly proffered an explanation based on the one-species electron–hole competition model incorporating thermal excitations. The authors showed that if E_{Me}, $E_{Mh} \ll E_D$, E_q, then the space charge field simplifies to

$$E_1 = -jmI_o \frac{s_h N_{Do}^+ - s_e(N_D - N_{Do}^+)}{(N_D - N_{Do}^+)(\beta_e + s_e I_o)\left(\frac{1}{E_q} + \frac{1}{E_D + jE_o}\right) + N_{Do}^+(\beta_h + s_h I_o)\left(\frac{1}{E_q} + \frac{1}{E_D - jE_o}\right)} \tag{5.44}$$

This equation contains all the relevant physics. Figure 5.13 shows the theoretical two-wave mixing gain in InP:Fe as a function of pump intensity for temperature $T = 290$, 300, and 305 K. The resonant character of the gain is clearly evident. To understand how the gain is affected by temperature and intensity it is necessary to look at the imaginary part of the space charge field:

$$\mathrm{Im}(E_1) = m\xi_R \frac{A}{A^2 + D^2 B^2} \tag{5.45}$$

where

$$A = \frac{1}{E_q} + \frac{E_D}{E_o^2 + E_D^2}, \quad B = \frac{E_o}{E_o^2 + E_D^2} \tag{5.46}$$

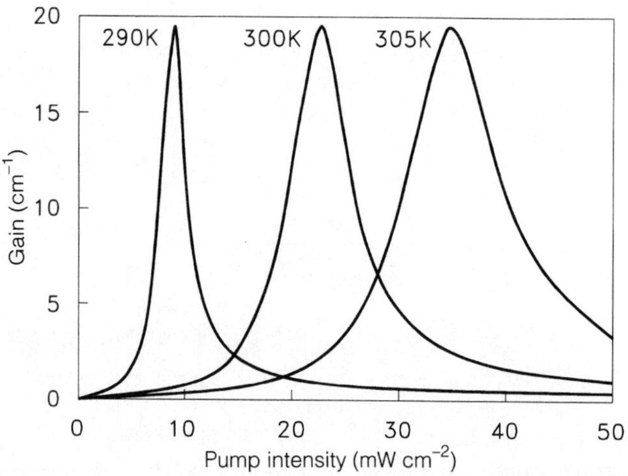

Fig. 5.13 Theoretical predictions of the intensity–temperature resonance model (model V) showing two-wave mixing gain in InP:Fe versus pump intensity for $T = 290$, 300, and 305 K. $\Lambda = 10\,\mu\mathrm{m}$, $E_o = 10\,\mathrm{kV\,cm^{-1}}$, and $\lambda = 1.06\,\mu\mathrm{m}$ (after Picoli et al. 1989a).

$$\xi_R = \frac{[s_h N_{Do}^+ - s_e(N_D - N_{Do}^+)]I_o}{(\beta_e + s_e I_o)(N_D - N_{Do}^+) + (\beta_h + s_h I_o)N_{Do}^+},$$

$$D = \frac{(\beta_e + s_e I_o)(N_D - N_{Do}^+) - (\beta_h + s_h I_o)N_{Do}^+}{(\beta_e + s_e I_o)(N_D - N_{Do}^+) + (\beta_h + s_h I_o)N_{Do}^+}$$

(5.47)

The origin of the intensity–temperature resonance now becomes clear. From eqns (5.45) to (5.47) we note that if $E_o \neq 0$ then $B \neq 0$, in which case the space charge field is influenced by the term $D^2 B^2$. If the weak dependence of ξ_R on I_o is neglected, then the imaginary part of the space charge field is maximized when

$$(\beta_e + s_e I_o)(N_D - N_{Do}^+) = (\beta_h + s_h I_o)N_{Do}^+ \qquad (5.48)$$

In other words, when $E_o \neq 0$ the space charge field can be optimized and made purely imaginary by controlling I_o! Inspection of relation (5.48) reveals that when thermal emission dominates for electrons (i.e. $s_e I_o \approx 0$) and photoionization dominates for holes (i.e. $\beta_h \approx 0$), or vice versa, then there is a pronounced optimum intensity which increases with temperature, as seen in Fig. 5.13. Moreover, the optimum intensity is independent of both grating spacing and the applied field.

Before concluding the discussion of this model, a few words should be devoted to the implementation of this method for improving two-wave mixing gain. By its very nature, this new resonance method is extremely sensitive to intensity and temperature variations throughout the crystal. Unfortunately, the incident light will inevitably decay exponentially into the crystal due to absorption. Picoli *et al.* (1989b) suggested that this could be remedied by having an incoherent optical beam incident from the opposite side of the crystal in order to make the total intensity more uniform. Alternatively, the gain could be increased by creating a temperature gradient to ensure that the temperature–intensity resonance condition is met everywhere in the crystal. Both techniques have been successfully demonstrated for InP:Fe (Picoli *et al.* 1989b; Ozkül *et al.* 1990, 1991; Mainguet 1991).

The intensity-resonance model has been developed further by Mainguet *et al.* (1990) to include the effect of detuning one of the beams by a frequency Ω. The authors showed that the resonance condition in eqn (5.48) is now modified slightly to read

$$(\beta_e + s_e I_o)(N_D - N_{Do}^+) - (\beta_h s_h I_o)N_{Do}^+ = \Omega E_o \frac{\varepsilon_s K}{e} \qquad (5.49)$$

Both the shallow trap model and the single-species bipolar models have been quite successful in explaining various effects of temperature and intensity which could not be accounted for by the standard model. Recently, these two models have been taken one step further by uniting them in a so-called 'shallow-trap electron–hole competition model', or model IX in Fig. 5.2. Tayebati (1992), deriving analytical solutions in low intensity approximation, showed that the zero cross-over of the space charge field becomes dependent on intensity. Barry *et al.* (1994) extended the analysis to higher intensities to explain the intensity dependence of the beam coupling and response time in $BaTiO_3$. Moisan *et al.* (1994) found that the model could reproduce some of the experimentally observed traits of beam coupling in photorefractive CdTe:V under applied alternating electric field, in particular, the optimum in gain for low AC frequencies.

5.2.5 Electron–ion 'fixing' model

In the models discussed so far, the holograms which are stored in the photorefractive crystal are erased if the readout laser beam is of the same wavelength as the one used for recording. While this may be an advantage for real time holography, other applications, such as holographic memories, would benefit greatly from long term storage capabilities and the possibility for non-destructive readout of information. Section 12.3 summarizes the field of 'fixing', whereby holograms can be made semi-permanent by means of thermal, electrical, or two-photon techniques. In this section we concentrate on understanding the physics of thermal fixing. To summarize briefly, this is the process whereby thermally activated ions are present inside the crystal. For low temperatures, these ions are immobile or fixed. However, at sufficiently high temperature these secondary charge carriers become mobile and will subsequently (1) drift under the effect of the space charge field set up by the primary charge carriers so as to neutralize the field, and (2) diffuse in their own concentration gradient. The idea is then that when the crystal is cooled back down to room temperature, the ions remain fixed in their new positions. Erasing the primary grating will subsequently reveal the ionic grating, in effect leaving an imprint of the primary grating.

Much work has been devoted to understanding the process of thermal fixing (Kulikov and Stepanov 1979; Meyer et al. 1979; Hertel et al. 1987; Carrascosa and Agulló-López, 1990; Montemezzani et al. 1993a). The model which best describes thermal fixing is depicted in Fig. 5.2 as model XVI. This is essentially the standard model to which mobile ions have been added. Note that these ions are optically inactive. The materials equations are the same as for the standard model with the exception that there are now mobile ions (here with positive charge) of concentration, N_I, which are free to move, generating a current, J_I:

$$\frac{\partial N_I}{\partial t} = -\frac{1}{e}\frac{\partial J_I}{\partial z} \tag{5.50}$$

$$J_I = e\mu_I N_I E - \mu_I k_B T \frac{\partial N_I}{\partial z} \tag{5.51}$$

In addition, Poisson's equation has to be modified to include the ions:

$$\frac{\partial E}{\partial z} = \frac{e}{\varepsilon_s}[(N_D^+ - n_e - N_A^-) + (N_I - N_{Io})] \tag{5.52}$$

where N_{Io} is the average background concentration of ionic charges.

The steady-state solution may be obtained in the usual small modulation c.w. illumination approximation. Neglecting the photovoltaic effect and setting the applied field to zero, the net space charge field simplifies to

$$E_1 = j\nu m \frac{E_D E_{qD}}{E_D + E_{qD} + E_{qI}} \tag{5.53}$$

which is made up of a contribution from the redistribution of charge in the deep traps, E_{e1}, and from the ionic space charge field, E_{I1}:

$$E_{e1} = j\nu m \frac{E_{qD}(E_D + E_{qD})}{E_D + E_{qD} + E_{qI}}, \quad E_{I1} = -j\nu m \frac{E_{qD}E_{qI}}{E_D + E_{qD} + E_{qI}} \quad (5.54)$$

The limiting field for the ions is $E_{qI} = eN_{Io}/\varepsilon_s K$. Note that all the space charge fields are purely imaginary, but that E_{I1} and E_{e1} are in antiphase. The very fact that the ions merely respond to the space charge field set up by the deep traps means that E_{I1} cannot exceed E_{e1}. However, if the effective density of ions is sufficiently large (i.e. $E_{qI} \gg E_{qD}$) then the ionic charge will nearly completely compensate the field in the deep traps, yielding a negligible net space charge field.

We shall not here go into more detail, but will return to this model in Section 12.3 to discuss the properties and dynamics of hologram fixing. We shall however make one last comment. Most of the features of this 'mobile ion' model are mirrored by the complementary grating model (model IV) for which one of the trap level is only excited thermally (e.g. $s_h I_o \ll \beta_h$). Both models give identical steady-state solutions for the space charge field, but the dynamics are slightly different (Montemezzani et al. 1993a). Furthermore, formally the fixing in the mobile ion model is based on a thermally activated mobility of the Arrhenius type:

$$\mu_I(T) = \mu_{Io} \exp\left(-\frac{\Delta E}{k_B T}\right) \quad (5.55)$$

where ΔE is the activation energy. For the complementary grating model the same temperature dependence is obtained via the thermal excitation, β, as given by eqn (5.3)

5.2.6 Two-photon writing model

Until this stage, our survey of photorefractive models has been restricted to models based on photoexcitation of electrons and holes from traps. In 1974 von der Linde et al. suggested and performed experiments to demonstrate that it is also possible to write holograms by two-photon absorption. This was seen as a significant development toward the goal of high-volume optical storage in photorefractive materials. The attraction is that hologram writing and erasure can only occur if two beams of different wavelengths are present. In the case of $LiNbO_3$ and $LiTaO_3$, von der Linde et al. (1974, 1975, 1976), and Vormann and Krätzig (1984) recorded gratings by illuminating the crystals simultaneously with beams of wavelengths 1.06 μm and 0.53 μm. The gratings could then be read repeatedly with only one wavelength present, with negligible erasure of the hologram. In other words, a two-photon photorefractive effect offers the possibility of non-destructive readout of information, and therefore eliminates the need for fixing and maintains the versatility of optical erasure.

The model which probably best describes this type of behaviour is model XIII in Fig. 5.2. This 'two-photon absorption model' describes a system whereby electrons are either generated directly by two-photon absorption or by two consecutive single-photon absorptions via an excited state. The standard model therefore applies, with the change that generation of electrons becomes proportional to the product of the intensities of the writing beams of one colour and the 'sensitizing' beam of a different colour. One result of the two-photon process is that the refractive index now changes quadratically with incident intensity, in contrast to the linear dependence of the conventional one-photon process. This behaviour is illustrated in Fig. 5.14 for $LiNbO_3$:Cu. Note that quite high energies are needed, as would be expected from a multiphoton process.

5.2.7 Quantum well models

In all the models discussed so far, the charges in the deep traps are spatially redistributed by optically exciting electrons (and/or holes) out of traps, so that they may diffuse and drift in the conduction (or valence) band before they recombine with ionized traps elsewhere. Photorefractive multiple quantum wells (MQW), however, attest to the fact that photoexcitation from traps is not a

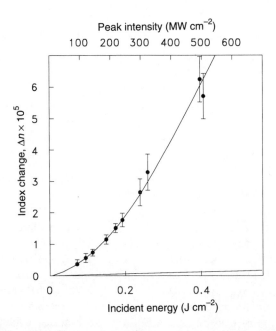

Fig. 5.14 Birefringence change $\triangle n$ for $LiNbO_3$:Cu^{2+} as a function of energy (lower scale) and pulse intensity (upper scale). The curve drawn through the experimental points was calculated assuming two-photon absorption. The straight line at the bottom is $\triangle n$ obtained with a c.w. argon ion laser, i.e. one-photon absorption (after von der Linde et al. 1974).

prerequisite for charge redistribution. In the MQW model, or model XV in Fig. 5.2, there is no photoexcitation of charges out of traps; instead the necessary free electrons and holes are generated by direct band-to-band absorption. The charges in the traps are subsequently redistributed due to the repeated recombination of both electrons and holes with the traps.

Theoretical models to account for the steady-state photorefractive response of MQWs have been developed by Q. N. Wang et al. (1992, 1994a,b), Carrascosa et al. (1992), and Magana et al. (1994). MQWs rely for their operation on wavelengths tuned to the sharp absorption features of quantum confined excitons. This means that photogeneration of electron–hole pairs occurs predominantly by interband transitions for which the absorption is usually several orders of magnitude larger than for trap photoionization. The net generation rates of electrons and holes are consequently given by

$$G_e = g_{dir}I - \gamma_{dir}n_e n_h - \gamma_e n_e N_D^+, \quad G_h = g_{dir}I - \gamma_{dir}n_e n_h - \gamma_h n_h(N_D - N_D^+) \quad (5.56)$$

where γ_{dir} is the recombination coefficient for direct recombination across the band gap, and $g_{dir} = \alpha_{dir}/\hbar\omega$, where α_{dir} is the absorption coefficient for direct band-gap excitation. Otherwise the model is exactly the same as the single-species bipolar model detailed in Section 5.2.2.

Q. N. Wang et al. (1992) showed that for negligible direct recombination, weak illumination, and well below the trap limit (i.e. $E_1 \ll E_q$), the steady-state space charge field simplifies to

$$E_1 = -m \frac{E_{Mh}(E_o + jE_D) + E_{Me}(E_o - jE_D)}{2E_D + E_{Me} + E_{Mh}} \quad (5.57)$$

where $E_{Me} = 1/\mu_e\tau_e K$ and $E_{Mh} = 1/\mu_h\tau_h K$. The feature to note is that the diffusion field in the denominator serves to limit the space charge field.

This model has been quite successful at describing the photorefractive effect in GaAs/AlGaAs multiple quantum wells for low values of applied field. However, it is at a loss when it comes to explaining the surprising experimental observation that the phase shift of the space charge field approaches $\pi/2$ for fields above $4\,\text{kV}\,\text{cm}^{-1}$. Q. N. Wang et al. (1994a,b), recently suggested that this peculiarity could be attributed to hot-electron transport. This relates to the fact that under high electric fields, electrons in GaAs are promoted to an auxiliary conduction band valley with a smaller mobility, meaning that the effective electron mobility becomes a function of applied field[4]. Furthermore, the electron temperature also becomes a function of electric field, and will in fact exceed the lattice temperature by several hundred degrees for electric fields of $10\,\text{kV}\,\text{cm}^{-1}$. By modifying the electron current to read

$$J_e = e\mu_e(E)n_e E + k_B \frac{\partial}{\partial z}[n_e T_e(E)\mu_e(E)] \quad (5.58)$$

---

[4]This is the same effect which, for high electron densities, leads to non-linear differential resistance and the Gunn effect (Ridley and Watkins 1961; Gunn 1963).

Wang et al.(1994b) obtained the curves in Fig. 5.15, indicating that hot-electron transport is the probable origin of the phase shift. It should be noted that eqn (5.58) is not restricted to multiple quantum wells, but should also be applied to the photorefractive effect in bulk GaAs.

5.2.8 Trapless photorefractive model

In all the previous models we have concentrated on the space charge field which is set up by the ionized traps. In actuality, free charges in the conduction and valence bands also contribute to the net space charge field, but under normal conditions of low illumination the concentration of electrons and holes is negligible compared with the concentration of ionized traps. Nevertheless, the fact remains that the photorefractive effect can occur in materials without the participation of photoactive traps, provided the density of free carriers can be made sufficiently large. One obvious way to achieve this is to use large optical intensities. A potentially more attractive alternative is to use photon energies larger than the energy gap so as to benefit from the tremendous increase in absorption. Montemezzani et al. (1993b, 1994) recently demonstrated interband photorefractive effects in $KNbO_3$ by illuminating the crystal with ultraviolet light. In this way they were able to write gratings with a diffraction efficiency of 60%, and to obtain two-beam coupling gains of 10 cm^{-1} for optical intensities of a few $mW\,cm^{-2}$.

The 'trapless model' of photorefractivity, model XIV, is described by the same equations as the MQW model of the previous section, with the change that terms containing traps are omitted. Montemezzani et al. (1994) showed that for no applied field the steady-state space charge field simplifies to

$$E_1 = -jm \frac{E_D(E_{Rh} - E_{Re})E_{qf}}{(E_D + 2E_{qf})(E_D + E_{Rh} + E_{Re})} \qquad (5.59)$$

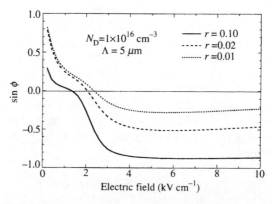

Fig. 5.15 Phase shift of the space charge field versus the applied electric field, obtained by taking into account hot electron transport. The trap density of each of the three curves corresponds to the ratio $r = N_D/N_A$ (after Wang et al. 1994b).

where

$$E_{qf} = \frac{en_{eo}}{\varepsilon_s K}, \quad E_{Re} = \frac{\gamma_{dir} n_{eo}}{\mu_e K}, \quad E_{Rh} = \frac{\gamma_{dir} n_{ho}}{\mu_h K}, \quad n_{eo} = n_{ho} = \sqrt{\frac{g_{dir} I_o}{\gamma_{dir}}} \quad (5.60)$$

Here E_{qf} is the limiting space charge field that can be set up by the free charge carriers. Furthermore, this model has a response speed which is proportional to the conductivity, meaning that the grating erasure time constant should vary as $I_o^{-1/2}$. This intensity dependence was confirmed experimentally by Montemezzani et al., as may be seen in Fig. 5.16.

This trapless photorefractive effect is both very new and very interesting. With such prospective benefits as high speed and sensitivity, it is definitely an effect which deserves further investigation.

5.2.9 Charge exchange between species (trap intercommunication)

In our survey so far, we have progressed in a step by step fashion from the simple standard model to increasingly more complex models so that the significant features of each new addition may be better understood. Naturally, we can also envision more general models which can exhibit a combination of the effects we have discussed. For instance, Rana et al. (1992) suggested that a system of two impurity levels in communication with both the conduction and valence bands was best suited for understanding the photorefractive effect in InP:Fe in order to take into account the presence of the excited state Fe^{2+*}. Moreover, the authors proposed that traps could be in direct communication with each other. The merits of this model, which we refer to here as model XII, have been investigated by Ozkül et al. (1994) and Bashaw et al. (1994). Jermann and Otten (1993) extended

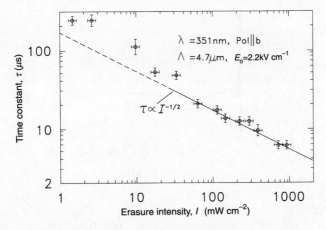

Fig. 5.16 Intensity dependence of the exponential erasure time measured in a thin sample of $KNbO_3$ (after Montemezzani et al. 1993b).

the equations to include the photovoltaic effect. Essentially, model XII opens up the possibility for combining in one model such traits as electron–hole competition, multiple decay rates, and complementary gratings. The role of inter-trap communication is more complicated and we shall not here attempt to discuss it in the context of this complex model. Instead we turn to very recent work on certain aspects of the photorefractive effect in polymers.

Silence et al. (1994a) found that grating erasure in their new photorefractive polymers was not behaving as predicted by any of the existing models: while a grating would be erased if read with high intensity light, for a low intensity reading beam the grating would not experience any appreciable decay. This behaviour may be seen from Fig. 5.17(a) where the lower curve is the decay for a reading beam of intensity 0.1 W cm^{-2}, and the upper curve is for a readout intensity of 20 mW cm^{-2}. Figure 5.17(b) shows that in the latter case the grating is still present after 24 h. Silence et al. proposed that this surprising, yet welcome, ability for quasi-non-destructive readout could be explained using a 'series model' for which

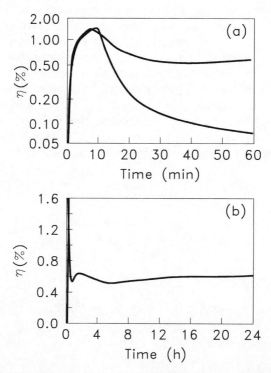

Fig. 5.17 (a) Diffraction efficiency, η, as a function of time for a photorefractive polymer (PMMA:DTNBI:C$_{60}$). Writing beams are turned on at $t = 0$ and blocked at $t = 10$ min. The gratings are subsequently read by a beam of intensity $I_o = 0.1$ W cm^{-2} (lower curve) and $I_o = 2 \times 10^{-5}$ W cm^{-2} (upper curve). The long-term readout of the grating read at $I_o = 2 \times 10^{-5}$ W cm^{-2} is shown in (b) (after Silence et al. 1994a).

two active trap levels are connected sequentially, as shown in model XVI in Fig. 5.2. The exact features of this model still remain to be investigated.

5.2.10 Valley's second-order differential equation

In this section we return briefly to the standard model to pave the way for the understanding of the photorefractive effect under short optical pulses, high intensity, and high frequencies (AC field or detuning frequencies). Valley (1983a) solved the materials equations in the so-called quasi-c.w. illumination approximation, where it is assumed that the average electron density reaches its steady-state value before significant drift or diffusion occurs. In this case the space charge field is governed by the following second-order differential equation (for $m \ll 1$):[5]

$$\frac{\partial^2 E_1}{\partial t^2} + \frac{\partial E_1}{\partial t}\left[\frac{1}{\tau_d} + \frac{j}{\tau_E} + \frac{1}{\tau_D} + \frac{1}{\tau_+} - \frac{j}{E_D + jE_o}\frac{dE_o}{dt}\right]$$
$$+ E_1\left[\frac{1}{\tau_I}\left(\frac{1}{\tau_D} + \frac{j}{\tau_E}\right) + \frac{1}{\tau_d\tau_+} - \frac{j}{\tau_d}\frac{1}{E_D + jE_o}\frac{dE_o}{dt}\right] \qquad (5.61)$$
$$= -\left[\frac{1}{\tau_E} - \frac{j}{\tau_D}\right]\frac{seI_1}{\varepsilon_s K}(N_D - N_A - n_o)$$

where

$$\tau_d = \frac{\varepsilon_s}{e\mu n_o}, \quad \tau_E = \frac{1}{\mu K E_o}, \quad \tau_D = \frac{e}{k_b T \mu K^2},$$
$$\tau_+ = \frac{1}{\gamma_e N_A + 2\gamma_e n_o + sI_o}, \quad \tau_I = \frac{1}{sI_o + \gamma_e n_o} \qquad (5.62)$$

We shall make use of this equation in Section 5.5, concerned with high powers and short optical pulses, and in Sections 5.4.5 and 5.4.9 which deal with high frequency resonances.

5.3 Space charge waves

5.3.1 Introduction

We have already touched upon this subject in Section 2.8. We said that drifting electrons can carry such waves and we also said that space charge waves played an important role in the design of microwave tubes. In those devices the electrons drift in vacuum and the space charge waves carried by them are either modulated by the electromagnetic field of a cavity (as in a klystron) or are interacting continuously with electromagnetic waves (as in a travelling wave tube). Their operation is made possible (for a detailed treatment see Beck 1958) by the transfer of

---

[5]This equation differs from that of Valley only in that we allow for the temporal variation of the applied field, E_o. As a consequence, terms containing dE_o/dt have been included.

power from the space charge waves to the electromagnetic waves. The source of power is the power supply which initially accelerates the electrons.

It makes good sense that electrons drifting in vacuum possess a lot of kinetic energy and that energy could somehow be tapped if the designer is clever enough. Is it possible to use electrons drifting in a solid as sources of power? The common-sense answer is no. The electrons suffer collisions so frequently that they can't possibly have enough energy to transfer it to some other kind of wave capable of propagation in the solid. This common-sense view turned out to be incorrect. In the so-called acoustic wave amplifier (invented by Hutson *et al.* 1961) an acoustic wave could be amplified by interaction with electrons drifting in the solid, piezo-electricity providing the coupling mechanism. Somewhat later this phenomenon was reinterpreted by Bløtekjaer and Quate (1964) as interaction between space charge waves and acoustic waves. Thus a clear proof was provided that electrons which suffer collisions at a high rate are still capable of transferring their energy to another wave, namely to an acoustic wave.

If transfer of power to an acoustic wave in a solid is possible, then surely it is also possible to transfer power to an electromagnetic wave. A solid state travelling wave amplifier based on this principle was proposed by Solymar and Ash (1966). Unfortunately, neither the acoustic wave amplifier nor the solid state travelling wave amplifier reached the glory of commercial exploitation. The subject was slowly forgotten in the West but, fortunately, kept alive in the Russian literature with occasional articles appearing. It was shown by Kazarinov *et al.* (1972), Suris and Fuks (1975) and Fuchs *et al.* (1977) that in a material with traps some new kind of weakly damped waves may appear. (These waves were referred to as 'trap exchange waves' but they were only representations of space charge waves under somewhat more general conditions.) Experiments by Zhdanova *et al.* (1978) showed that in the impedance of a thin slice of compensated semiconductor (gold-doped germanium) resonances occurred whenever the thickness of the sample was equal to an integral number of wavelengths of the space charge wave excited.

The enhancement of the gain found by Huignard and Marrakchi (1981a) in photorefractive materials was reinterpreted by Stepanov *et al.* (1982) as a resonance phenomenon when the imposed travelling interference pattern interacts with the eigenwave of the system. These eigenwaves were space charge waves although the authors did not use that terminology.

Further extensions to the optical range came with two theoretical papers by Alimpiev and Gural'nik (1984, 1986). In the first paper they discussed the excitation of space charge waves by inhomogeneous optical illumination in a compensated semiconductor sample. In the second paper they analysed the conditions for the appearance of instabilities. Space charge waves in photorefractive materials were considered by Furman (1987, 1988a,b) in a series of theoretical papers. He showed that space charge wave instabilities may exist due to the photovoltaic effect (Furman 1988a) and reinterpreted the experimental results of Lemeshko and Obukhovskii (1985) on scattering as due to stimulated space charge wave scattering (in analogy with stimulated Brillouin scattering). The relationship

5.3 Space charge waves

between space charge waves and the photovoltaic effect has been further investigated by Sturman et al. (1995b). They showed that weakly attenuating space charge waves may occur both for small and large values of the mobility lifetime product. Some other related effects were discussed by Solymar and Shamonin (1992) (power conservation between optical and space charge waves) and by Sturman et al. (1993c) (excitation of subharmonics). Space charge waves also play a role in inducing an AC current in a crystal as shown by Sokolov and Stepanov (1993). For some more general comments about space charge waves see also the review article by Solymar et al. (1994).

5.3.2 Derivation of the dispersion equation

Any set of linear differential equations which contain derivatives both in time and space and have constant coefficients will lead to wave solutions. All one needs to do is to assume a solution in the form of $\exp j(\Omega t - Kx)$, substitute it into the differential equations, and obtain a set of simultaneous linear equations in the variables. The dispersion relationship (between frequency Ω and wavenumber K) is then obtained by the condition that the determinant vanishes.

Our aim is to find the space charge–wave dispersion relationship. The question arises as to which model to choose. The obvious answer is to use first the simplest one possible, i.e. the standard model. The relevant equations are (5.4) to (5.8). Should we use any approximations? Should we neglect for example the electron density in Poisson's equation as we did in Part I? The answer is no. If we used that approximation, that would exclude what we are going to call the high frequency branch. Can we then proceed to the wave solution? Not yet, because the differential equations are not linear. We can however linearize them as explained in Section 2.2 and use then the procedure outlined above to find the dispersion equation. In fact, there is an even shorter way to the dispersion equation. We have already done the linearization etc. when we derived the second-order temporal differential equation (5.61) for the fundamental component of the electric field. Taking the right-hand side equal to zero and substituting $d/dt = j\Omega$ we obtain

$$-\Omega^2 + j\Omega \left[\frac{1}{\tau_d} + \frac{j}{\tau_E} + \frac{1}{\tau_D} + \frac{1}{\tau_+}\right] + \left[\frac{1}{\tau_I}\left(\frac{j}{\tau_E} + \frac{1}{\tau_D}\right) + \frac{1}{\tau_+\tau_d}\right] = 0 \qquad (5.63)$$

If desired, the above equation could also be rewritten as a polynomial in K in which the coefficients are dependent on Ω. The usual practice is to assume either a complex K and real Ω or a complex Ω and a real K, depending on the physical situation.

Let us first look at erasure. If at zero time we switch off one of the beams, then the excitation in the form of an interference pattern (whether moving or not) suddenly stops and the grating decays. In that case the wavenumber should be regarded as real and the imaginary part of the frequency is responsible for the decay. However, there should still be a space charge wave moving with a velocity

$v_s = \Omega/K$ where K is the wavenumber imposed previously and Ω is the frequency of the space charge wave calculable from the dispersion equation.

The relevant experiment was done by Kandidova *et al.* (1988) in iron-doped lithium niobate using beams with extraordinary polarization. They monitored the decay of the grating in the dark by illuminating it by two equal-frequency very low intensity beams (at the same angles as the original recording beams). The diffracted beam was Doppler-shifted whereas the transmitted beam did not change its frequency. Hence, by measuring the intensity in either of the output beams they could observe beating at the frequency Ω. They found the dispersion relationship for a range of inter-beam angles from about 2 to 12 degrees. Similar experiments during erasure were conducted by Hamel de Montchenault *et al.* (1986) in a BSO crystal. They showed that the diffracted beam was frequency-shifted and the amount of frequency shift was equal to that required for optimum beam coupling (i.e. the one given by the dispersion equation).

Next, we shall assume that Ω is real and K is complex, and look at the properties of the space charge waves. The decay of the wave is then given by the imaginary part of K. Solving eqn (5.63) numerically, we do actually obtain four roots in Re(K) which, in principle, correspond to four different waves. We believe that three of them, having very high attenuation, are of no physical significance. We shall assume that only the low attenuation wave is of any practical interest.

Just to give an idea how the real part of K varies with frequency we plot it in Fig. 5.18 for BSO for $I_o = 10\,\text{mW cm}^{-2}$ and for $E_o = 4$ (solid line), 7 (dotted line), and 10 kV cm^{-1} (dashed line) using the parameters given in eqns (2.54). The figure itself shows the low frequency behaviour on the scale of a few hertz whereas the high frequency behaviour on a megahertz scale is shown in the inset. Obviously, the dispersion curve depends on the applied electric field but its general shape

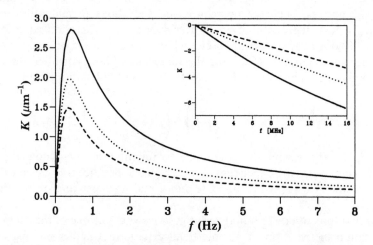

Fig. 5.18 Dispersion curves for space charge waves in photorefractive BSO, showing Re(K) as a function of frequency for $I_o = 10\,\text{mW cm}^{-2}$ and $E_o = 4\,\text{kV cm}^{-1}$ (———), 7 kV cm^{-1} (····), 10 kV cm^{-1} (- - - -). The inset shows the corresponding curves for high frequencies.

5.3 Space charge waves

remains unchanged: it is positive for low frequencies, it has a range of frequencies for which it represents a backward wave (i.e. the phase and group velocities are in opposite directions), it then crosses the axis becoming again a forward wave, and for high enough frequencies (as shown in the inset) the variation is linear. Before discussing further the significance of the dispersion curve, let us look at both the real and imaginary parts of K. It turns out that there are only two regions in which this wave (we called it earlier the only wave with low attenuation) has genuinely low attenuation. These regions are shown in Fig. 5.19(a) and (b).

It may be easily shown from eqn (5.63) that $\text{Re}(K)$ and $\text{Im}(K)$ are odd and even functions of Ω respectively. This is the reason why one of our plots is for negative frequency (it is actually a convenient representation, it might mean a wave propagating in the opposite direction or, simply, negative detuning). In any case, Fig. 5.19(a) and (b) show a high frequency and a low frequency region with low attenuation. What is the significance of these low attenuation regions? In order to answer this question let us return to wave vector diagrams introduced in Figs 1.1 and 1.4.

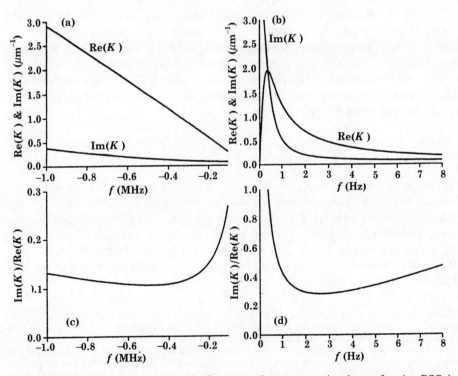

Fig. 5.19 The dispersion relationship for space charge waves in photorefractive BSO in the (a) high frequency and (b) low frequency regions; (c) and (d) are the corresponding curves for the loss per wavelength of the space charge waves.

For two incident waves of wave vectors $\mathbf{k}_1$ and $\mathbf{k}_2$ and for a grating vector $\mathbf{K}$ the construction is shown in Fig. 1.1(b). Let us now reinterpret this diagram as showing parametric interaction between three waves. $\mathbf{K}$ is now not the grating vector but the wave vector of the space charge wave. Then, according to the rules of parametric wave interactions (see e.g. Yariv 1991), the effect is maximum when the two conditions

$$\mathbf{k}_1 - \mathbf{k}_2 = \mathbf{K} \quad \text{and} \quad \omega_1 - \omega_2 = \Omega \tag{5.64}$$

are satisfied[6]. The space charge wave appears now as a wave in its own right. Can it carry its own power? It can, because it has at its disposal a power source in the form of the power supply which imposes the applied electric field. That leaves open the intriguing possibility that power could be transferred from the space charge wave to one of the optical waves.

We obtain the low frequency dispersion curve by neglecting first Ω^2. If we further assume diffusion to be negligible and take $\tau_I = \infty$ and $\tau_+ = \tau_e$ the following relationship may be derived for the complex K:

$$K = \frac{1}{\tau_e \tau_d v_o \Omega} + j \frac{1}{v_o} \left(\frac{1}{\tau_e} + \frac{1}{\tau_d} \right) \tag{5.65}$$

where $v_o = \mu_e E_o$ is the drift velocity of the electrons. The region of validity of this equation starts a little beyond the maximum of the $\text{Re}(K)$ curve, as may be seen in Fig. 5.19(b).

For high frequencies it is sufficient to retain in eqn (5.63) the quadratic and linear terms in Ω and disregard the constant term. Using the same approximations as above, we obtain the relationship

$$K = -\frac{\Omega}{v_o} + j \frac{1}{v_o} \left(\frac{1}{\tau_e} + \frac{1}{\tau_d} \right) \tag{5.66}$$

The slope of the dispersion curve is $-1/v_o$, which we have already shown in the inset of Fig. 5.18. The wave may be seen to travel with the drift velocity of the electrons. Its attenuation depends on the lifetime of the electron and on the dielectric relaxation time. This is nearly the same expression as that of Bløtekjaer and Quate (1964) in the absence of diffusion. The difference is that they did not need to take recombination into account because their electron lifetime was infinitely long.

---

[6]Parametric interactions between waves can also be explained in terms of particles. In an optical parametric amplifier for example we might say that two photons are annihilated and a third photon at the sum frequency is created. We could use similar terms here. There has been no incentive to define a particle corresponding to a space charge wave, because that is not a useful concept in the microwave region. In a recent review article however (Solymar et al. 1994) the plunge was taken and the particle was called a *spastron*. We may then look at the three-wave interaction as collision between three particles, two photons and a spastron, in which both momentum and energy are conserved. If, for example, there is a power transfer from beam 1 to beam 2 then we can claim that a photon at ω_1 and a spastron at Ω have been annihilated and a photon at ω_2 has been created.

5.3.3 Effect of holes

So far we have concentrated on the standard model when electrons, excited from donor atoms, are the mobile carriers. We shall now briefly comment on what happens to the space charge waves when we allow the presence of holes according to models IV and V (Aubrecht *et al.* 1995a). We shall not give here the modified dispersion equations, as they are far too complicated. It turns out that nothing interesting happens on the basis of model V. However, model IV (what we called model 1 in Section 2.10) predicts a new low attenuation region at certain hole densities. Again, a low attenuation space charge wave leads to resonance with the moving interference pattern. The corresponding variation of the space charge electric field will be discussed in Section 5.4.4.

The effect of holes on the dispersion curves in the presence of a photovoltaic field was investigated by Sturman *et al.* (1995). They concluded that under certain conditions the presence of holes has only a minor influence on space charge wave loss.

5.4 Time-varying interference patterns and applied voltages

5.4.1 Introduction

The main aim of this section is to discuss the various mechanisms which serve to enhance the space charge electric field, but that is not the only aim. One might wish to apply a periodic voltage in order to study the currents flowing in the external circuit or, for example, to measure the mobility.

We have already discussed enhancement mechanisms in Section 2.8 consisting of the application of (i) a DC voltage accompanied by detuning of one of the beams, and (ii) a time-varying voltage. We shall discuss them here in a little more detail with particular reference to high frequency resonance. In addition, we shall look at a few other phenomena like the 'intensity resonance' and the effect of holes, and we shall also look at a combination of AC fields and detuning. Finally, we shall discuss phase modulation of the interference pattern and its effect on circuit currents.

5.4.2 DC applied field plus detuning: low frequency resonance

We shall give here some further information about the derivation of the optimum detuning and the corresponding maximum value of $\text{Im}(E_1)$, and we shall also endeavour to interpret the result in terms of space charge waves.

In the presence of detuning, the excitation term of eqn (2.28) modifies to

$$I_\text{p} = I_1 \exp j(\Omega t - Kz) \tag{5.67}$$

which follows also from eqn (2.104). The solution is sought in the form of

$$X_\text{p} = X_1(t) \exp j(\Omega t - Kz) \tag{5.68}$$

The linearized differential equations may then be solved by the simple technique outlined in Section 2.8 to yield (Refregier et al. 1984, 1985)

$$\text{Im}(E_1) = m \frac{E_M^2}{E_o} \frac{A-b}{(b-C)^2 + (B-bF)^2} \tag{5.69}$$

where

$$A = \frac{E_o}{E_q} + \frac{E_D}{E_o}\left(1 + \frac{E_D}{E_q}\right), \quad B = \frac{E_M}{E_q}$$
$$C = \frac{E_M}{E_o}\left(1 + \frac{E_D}{E_q}\right), \quad F = \frac{E_D}{E_o} + \frac{E_M}{E_o}, \quad b = \Omega\tau_d \tag{5.70}$$

The value of b that maximizes the imaginary part of the space charge field may then be calculated by differentiating eqn (5.69) with respect to b, a very simple operation, leading though to a rather complicated expression. However, if we neglect diffusion and assume that

$$E_M \ll E_o \ll E_q \tag{5.71}$$

we end up with the expression for the optimum detuning already given in eqn (2.109):

$$\Omega = \frac{1}{\tau_d} \frac{E_M}{E_o} \tag{5.72}$$

Substituting the value of E_M from eqn (2.37), the above equation reduces to

$$\Omega = \frac{1}{\tau_d} \frac{1}{\tau_e \mu_e E_o K} \tag{5.73}$$

which we may recognize as the dispersion equation (eqn 5.65) in our chosen low frequency, low attenuation region. The conclusion is that for a given inter-beam angle the optimum detuning frequency is the one which may be obtained from the dispersion equation by substituting into it the right value of the grating spacing. We can talk about resonance because the eigenwave and the impressed wave have identical frequencies and wavenumbers. Alternatively, we might say that resonance occurs when the impressed wave velocity agrees with the phase velocity of the space charge wave with the same K vector.

It may be noticed that the optimum detuning frequency depends on the dielectric relaxation time, which depends on the average electron density, which depends on the intensity. If the total intensity is the same at every point in the crystal then this dependence leads to no disadvantage. However, in any practical case the input optical beams suffer a certain amount of attenuation. If the beams are attenuated, then their intensity varies along the crystal. If the intensity varies, it can have its optimum value only at one particular point in the crystal. But that means that the velocity of the interference pattern is non-optimum at all other points. Hence the optimum two-wave gain as worked out from the lossless case is an overestimate. The available gain will be considerably less. We shall return to

5.4 Time-varying interference patterns and applied voltages

this problem in Section 6.8 where we shall show some relevant experimental results and compare them with the theoretical predictions.

5.4.3 DC field: variation of space charge field with intensity

We discussed in Section 5.2.4, among the shallow trap models, how the space charge field may be optimized by a certain intensity. We adopted the terminology of Picoli et al. (1989a), calling the phenomenon an 'intensity resonance'. Is then 'intensity resonance' a special feature of the shallow trap model under conditions given in Section 5.2.4? The answer is no. A similar dependence of the space charge field on intensity can also be obtained from the standard model when one of the beams is detuned. Let us recall eqn (5.72) giving the expression for the frequency of optimum detuning. It depends on intensity through the value of the dielectric relaxation time, τ_d. Hence for a given detuning a certain intensity will maximize the space charge field. We do not think this should be called an 'intensity resonance'. We should rather say that the condition of resonance (when the moving grating velocity agrees with the velocity of the space charge wave) happens to be satisfied for that particular value of the intensity.

It needs to be admitted however that no such intensity maximum follows from the standard model for zero detuning, whereas the shallow trap model predicts an optimum intensity even in the absence of detuning. Further argument in favour of 'intensity resonance' is that the space charge field is purely imaginary when it occurs. Could we possibly generalize our concept of resonance so as to include the shallow trap model at zero detuning? One could argue that when the parameters do not permit the existence of a propagating space charge wave, then the stationary interference pattern must be the optimum. However, different values of the parameters may represent different amount of 'mismatch'. The optimum intensity might be the one which causes the least mismatch. This is an interesting question on which, we hope, further work will be done.

5.4.4 DC field plus detuning: bipolar low frequency resonance

As mentioned in Section 5.3, it was found theoretically by Aubrecht et al. (1995a) that there is a new region in which low attenuation space charge waves exist when the electron density is accompanied by a certain hole density. The space charge electric field for that case can be found in closed form. Taking our usual BSO parameters for electrons and making the assumption that the hole parameters (mobility, lifetime, photoionization coefficient) are equal to the electron parameters, the imaginary part of the space charge field is plotted in Fig. 5.20 for n_h/n_e (hole density/electron density) equal to 0, 0.02, 0.05, 0.1, 0.2, 0.4, and 0.7. It may be seen that the curves have maxima for finite values of n_h/n_e, and the maximum is reached around the value of 0.1.

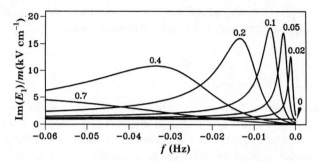

Fig. 5.20 Imaginary part of the space charge field as a function of frequency for $n_h/n_e = 0, 0.02, 0.05, 0.1, 0.2, 0.4$, and 0.7, as obtained using photorefractive model IV.

5.4.5 DC field plus detuning: high frequency resonance

As mentioned in Section 5.3, there is also a low attenuation, high frequency branch in the dispersion curve which one may expect to be suitable for our particular kind of three-wave interaction, namely the interaction between the two optical waves and the space charge wave.

The steady state space charge field may be derived from eqn (5.61) by assuming that one of the beams is detuned. The complete expression is given by Grunnet-Jepsen *et al.* (1995). Under resonance conditions, i.e. when Re(K) and Ω satisfy the dispersion relationship given by eqn (5.66), its maximum value may be obtained in the form

$$E_1 = -\mathrm{j}m \frac{en_{eo}}{\varepsilon_s K} \tag{5.74}$$

The space charge field against detuning for this high frequency resonance is shown in Fig. 5.21 using again BSO parameters (Grunnet-Jepsen *et al.* 1995) and an intensity of $I_o = 10\,\mathrm{mW\,cm^{-2}}$ at a grating spacing of 29 μm. It needs to be noted that for the high frequency resonance the space charge field is mainly set up by electrons, which explains why the magnitude of the space charge field and detuning frequency have opposite signs to that of the low frequency resonance, where the space charge field is due to positively charged ionized traps.

It may further be seen from eqn (5.74) that the maximum space charge field increases with the density of electrons and hence with illumination. However, under the normally used low illumination levels of a few $\mathrm{mW\,cm^{-2}}$, the space charge field at high frequency resonance in BSO is a few orders of magnitude smaller than that attainable at the low frequency resonance. It is also worth noting that while the frequency of the low frequency resonance is intensity-dependent, that of the high frequency resonance depends only on the grating spacing and on the electron velocity.

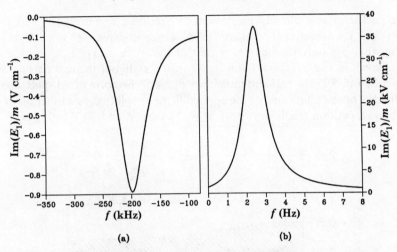

Fig. 5.21 Im(E_w) versus detuning frequency, showing the (a) high and (b) low frequency resonances for BSO with $E_o = 10\,\text{kV cm}^{-1}$, $\Lambda = 29\,\mu\text{m}$, and $I_o = 10\,\text{mW cm}^{-2}$ (after Grunnet-Jepsen *et al.* 1995).

5.4.6 AC applied field

We have already discussed the effect of AC applied fields upon the space charge field in Section 2.8 and presented the integrating method of Stepanov and Petrov (1985). It turns out that the maximum achievable space charge field obtainable by applying an AC field is equal to that which may be obtained by detuning one of the beams and applying a DC field of the same amplitude. The main difference between the two methods is that the detuning method is based on a resonance and therefore depends strongly on the actual value of the detuning frequency (see the resonance curves of Fig. 2.10) whereas the AC method is *not* resonant. There is a wide range of frequencies for which the AC method gives the same high value of the space charge field. The condition to be satisfied is that the frequency of the AC field must be large relative to the inverse rise time of the grating and small relative to the inverse lifetime of the electrons. Which method is better? The disadvantage of the 'DC plus detuning' method is that the optimum grating velocity depends on intensity; hence in an absorptive crystal there is only one point at which the velocity is optimum, as mentioned before. The disadvantage of the AC method, as pointed out by Stepanov and Petrov (1985) and Walsh *et al.* (1990), is that enhancement of the space charge field is large only for a perfect square wave. Any deviation from that perfect shape results in the decline of the available space charge field, as was already shown in Fig. 2.13. For a detailed experimental comparison of the two methods see Sochava *et al.* (1993b). Some results are shown in Section 6.7.

The temporal variation of the space charge field in this low frequency region may be determined from the first order differential equation (2.33) with time-varying coefficients as already mentioned in Section 2.8. For the closed form solution and its subsequent evaluation see Grunnet-Jepsen et al. (1994a). It turns out that not only the imaginary part of the space charge field (which was already shown in Fig. 2.11) oscillates in the steady state but the real part as well. In Figs 5.22 and 5.23 the real and imaginary parts of the normalized space charge field (divided by the amplitude of the applied field) are plotted as a function of the normalized repetition frequency for $E_{oo} = 1\,\text{kV}\,\text{cm}^{-1}$ and $7\,\text{kV}\,\text{cm}^{-1}$, both for

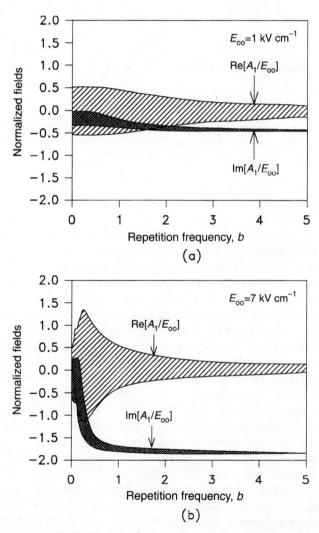

Fig. 5.22 Envelope of oscillation amplitudes of the steady-state space charge field for an applied AC square wave electric field as a function of normalized frequency, for an applied field amplitude of (a) $E_{oo} = 1\,\text{kV}\,\text{cm}^{-1}$ and (b) $E_{oo} = 7\,\text{kV}\,\text{cm}^{-1}$ for a BSO crystal (after Grunnet-Jepsen et al. 1994a).

5.4 Time-varying interference patterns and applied voltages

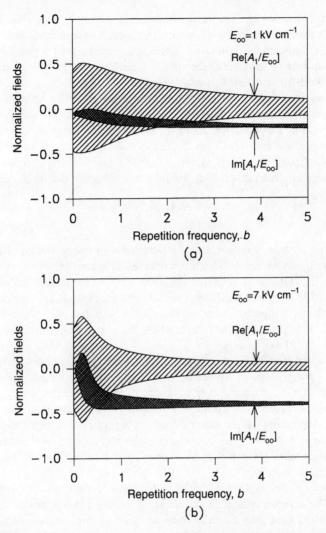

Fig. 5.23 Same as Fig. 5.22 but for an applied AC sine wave electric field.

square wave and sinusoidal applied fields. Firstly, it may be seen that the envelope of the oscillations becomes narrower as the frequency increases, as already indicated in Section 2.8. The second conclusion one may draw is that the real part, that one hardly ever mentions, has in fact larger amplitude than the imaginary part. Thirdly, one may see the confirmation that the sinusoidal applied field leads to much smaller space charge fields than the square wave.

We should note here that there is a solution of the differential equation (2.33) for a sinusoidal applied field, as shown by Kumar et al. (1987a). They assumed the solution for the space charge field as a Fourier series and obtained the first few coefficients by equating the respective harmonics. Unfortunately, the method is valid only for the case when E_M/E_o is small (in which case the AC field technique can give little enhancement on its own, although enhancement is possible when an additional detuning is applied, as discussed in Section 5.4.7), otherwise too many terms are needed for the series to converge.

For experimental results on the temporal variation of the gain in two-wave mixing (which is of course related to the space charge field) see Wolffer and Gravey (1991) and Grunnet-Jepsen et al. (1994c). The dependence of the two-wave mixing gain on the repetition frequency was investigated by Besson et al. (1989) and Vachss (1994). They showed that the gain declines both at low and high frequencies for the reason that the inequality, mentioned in Section 2.8,

$$1/\tau_r \ll \Omega \ll 1/\tau_e \tag{5.75}$$

is not satisfied.

Before going further, we have to stop here for a moment and discuss in more detail what we mean by low and high frequencies. Our terminology is based on the inequality (5.75). Therefore the frequencies in the low megahertz region used by Kumar et al. (1987a,b) in experiments with GaAs:Cr crystals are regarded as low because the electron lifetime in those crystals is about 1 ns. On the other hand a frequency of 10 kHz is definitely high when the experiments are done in BGO (Besson et al. 1989).

We need to mention here the experiments of Belaud et al. (1994), who showed that for high enough applied square wave field the two-wave mixing gain declines whenever the applied square wave field or the repetition frequency exceeds a certain threshold value. The frequency threshold varied from a few to a few tens of hertz, depending on the electric field (at higher electric fields the threshold frequency is lower). We believe the reason is the appearance of the subharmonic instability which takes away power from two-wave amplification. We shall return to this topic in Section 8.3.

Let us now consider high frequencies in the sense defined above. According to Vachss (1994) the space charge field declines. However, this is not necessarily so. The space charge field may actually increase towards higher frequencies, as was demonstrated by Pauliat et al. (1990). In order to make it easy to exceed in frequency the electron lifetime, Pauliat et al. had a BGO crystal grown with a long lifetime, a little over 1 ms. This meant that frequencies in the kilohertz region which were experimentally available could be employed. They claimed that for square waves maximum interaction occurs when the distance an electron drifts in one half period of the applied voltage is equal to one half of the spatial period of the grating, i.e.

$$\mu E_o T/2 = \Lambda/2 \tag{5.76}$$

where $T = 2\pi/\Omega$. This is equivalent to the condition

$$\Omega = K v_o \tag{5.77}$$

5.4 Time-varying interference patterns and applied voltages

which we have seen before as the resonance condition when a DC voltage is applied and one of the beams is detuned by Ω. For a sinusoidal voltage the drift length was calculated by integrating the velocity for half a temporal period, leading to the resonance condition

$$\Omega = (2/\pi)Kv_o \tag{5.78}$$

Thus from a knowledge of the inter-beam angle and from the measurement of the resonance frequency (with either square wave or sinusoidal voltage), the drift velocity and from that the mobility could be calculated. Hence one of the applications of this high frequency resonance is the measurement of the drift mobility.

It has been recently shown (Grunnet-Jepsen et al. 1995) that eqns (5.77) and (5.78) are in error and the correct condition for the square wave resonance is

$$\Omega = Kv_o/2 \tag{5.79}$$

meaning that the drift length in one half period of the applied voltage must be equal to one grating spacing. It was also shown that the resonance frequency for the sinusoidal case cannot be calculated by the integration outlined above.

Pauliat et al. (1990) found experimentally that the two-wave mixing gain against frequency curve had some other maxima besides the main maximum. Their experimental curve for an applied AC sine wave electric field for $\Lambda = 30\,\mu\text{m}$ and the theoretical match, calculated numerically by Grunnet-Jepsen et al. (1995), taking the parameters of the BGO as quoted by Pauliat et al. ($E_o = 3.1\,\text{kV}\,\text{cm}^{-1}$, $\varepsilon_s = 46\varepsilon_o$, $\mu = 2 \times 10^{-6}\,\text{m}^2\,\text{V}^{-1}\,\text{s}^{-1}$, $N_D = 10^{25}\,\text{m}^{-3}$, $\tau_e = 0.525\,\text{ms}$) are shown in Fig. 5.24(a) and (b). The agreement is remarkably good. Fig. 5.24(b) shows in addition the corresponding curve for a square wave applied field, which also has multiple peaks. It is interesting to note that the sinusoidal field is much more efficient at high frequencies than at low frequencies.

Let us first consider the position of the main peak for the sine wave in Fig. 5.24(b). We made it agree with the experimental result by choosing the mobility. The agreement between theory and experiment for the other peaks shows the success of the theory. Turning now to the square wave, we may see that the frequency of the main peak, obtained from the heuristic argument leading to eqn (5.79), agrees with that given by the numerically calculated curve of Fig. 5.24(b). With our choice of mobility it should occur at a frequency of 9 kHz, and that is indeed where the main peak lies (see also eqn C.13 in Appendix C). The positions of the additional peaks follow a simple rule. The second peak is at $\Omega_o/2$, the third peak is at $\Omega_o/3$, the fourth peak is at $\Omega_o/4$, etc., where $\Omega_o = Kv_o/2$. Where does this relationship come from? It can be derived by a modification of the qualitative argument used above, leading to eqn (5.76). The nth resonance occurs when the electrons move a distance of $n\Lambda$ in half a period of the applied field whence the resonant frequency, Ω_n, may be obtained from

$$\Omega_n = 2\pi/T = Kv_o/2n = \Omega_o/n \tag{5.80}$$

Is the space charge wave concept still useful when discussing AC resonances? We hope that subsequent research will prove it. Meanwhile we offer a simple

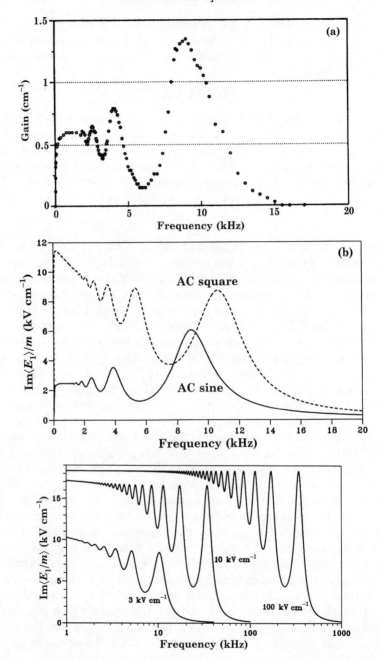

Fig. 5.24 (a) Experimental two-wave mixing gain as a function of frequency of an applied AC sine wave electric field (after Pauliat *et al.* 1990). (b) The average theoretical value of the imaginary part of the space charge field for $I_o = 10\,\text{mW cm}^{-2}$, $\Lambda = 29\,\mu\text{m}$ (after Grunnet-Jepsen *et al.* 1995). (c) The average theoretical value of the imaginary part of the space charge field for BGO for an applied square wave field, $E_{oo} = 3\,\text{kV cm}^{-1}$, $10\,\text{kV cm}^{-1}$, $100\,\text{kV cm}^{-1}$ and $I_o = 10\,\text{mW cm}^{-2}$, $\Lambda = 29\,\mu\text{m}$.

5.4 Time-varying interference patterns and applied voltages

derivation of eqn (5.80) from space charge wave arguments. We shall start by mentioning some numerical calculations of ours which concluded that at the nth peak the strongest temporal component of the space charge field (when expanded in a Fourier series) is the $2n$th component. Consequently the space charge field may be written in the form $\exp j(2n\Omega t - Kx)$, leading to a velocity $v = -2n\Omega/K$. For a given K the space charge wave dispersion relation gives $\Omega_s = -Kv_0$. Resonance occurs when the two waves move with the same velocity, i.e. $v = v_0$, leading to

$$2n\Omega/K = v = v_0 = \Omega_s/K \tag{5.81}$$

and from there

$$\Omega = \Omega_s/2n. \tag{5.82}$$

The lowest resonance occurs at $\Omega = \Omega_s/2$ and we can find the nth resonance by dividing further by n, in agreement with our numerical calculations.

What happens as we increase E_{oo}? We may expect the space charge field to increase and the number of resonances also to increase. This is indeed so, as may be seen in Fig. 5.24(c) where the imaginary part of the space charge field is plotted against repetition frequency for E_{oo} = 3, 10, and 100 kV cm^{-1} for BGO parameters. As E_{oo} increases, the space charge field gets nearer and nearer to the limiting value of E_q. For E_{oo} = 100 kV cm^{-1} (not a practical figure of course) all the peaks reach E_q. How many resonances are there? With a little effort one can count about 30 resonances. How many should be expected to be there? A simple argument would suggest that the resonances vanish when the frequency lifetime product is equal to 1. Remembering that in plotting Fig. 5.24(c) we took $\tau_e = 0.525$ ms, we would expect the lowest resonance to be at $1/\tau_e = 1.9$ kHz, which corresponds to about the 180th resonance. This cannot of course be seen in Fig. 5.24(c) by the naked eye but a peak could still be identified there by computer analysis.

For a square wave the nth resonant frequency is an exact subharmonic of the main resonant frequency. Is there a similar simple relationship for sinusoidal fields? An empirical relationship was indeed derived by Grunnet-Jepsen et al. (1995) but not from first principles.

We shall finish this fascinating section with the question of whether trap density limitations can be overcome, i.e. is it possible for the space charge field to exceed E_q? But if not trap density, what would then determine the maximum achievable gain? Obviously, the mobile carriers. In fact, in the acoustic wave amplifier, as mentioned in Section 5.3, the energy that was transferred to the acoustic waves came from the space charge field of the mobile electrons without any help from traps. The same message is offered by eqn (5.74). The space charge field is determined by the electron density. However, if we return to practical realities under normal c.w. operation, then it is obvious that the electron densities will always be much less than the trap densities, so that is not the way in which a space charge field in excess of E_q can be obtained. Could particular temporal variations of the applied field be responsible for exceeding E_q? Mathey et al. (1991) came to the conclusion that a particular waveform consisting of periodic pulses may lead to such large space charge fields, but this was refuted by Aubrecht et al. (1995b).

5.4.7 Detuning plus AC applied field

What happens when the AC applied field is asymmetric? A particular type of asymmetry was investigated by Doogin and Zel'dovich (1993). They assumed the time-varying field in the form shown in Fig. 5.25 when the time average over a period is still equal to zero. They found that under these conditions the enhancement of the space charge field is smaller than under symmetric fields. However, if one of the beams is detuned by the appropriate amount (which turns out to be always smaller than that for the DC field case), they found that the same enhancement can be obtained as either by the DC plus detuning method or by the pure AC symmetric field

We may obtain major enhancement of the space charge field when the inequality (2.108) (repeated in this section as eqn 5.71) is valid, which is equivalent to the requirement of having a high $\mu\tau$ product, which is available in sillénites and semiconductors. However, when the $\mu\tau$ product is not large, as in ferroelectric materials, these mechanisms are of little use. To overcome this problem Zel'dovich and co-workers proposed a solution in which a symmetric AC voltage is applied and one of the beams is detuned by the same frequency as that of the AC voltage. Under these conditions a strong static grating can be recorded (Ilinykh et al. 1989). The method is known as phase-locked detection.

The mathematical solution for the stationary space charge field may be obtained from the differential equation (5.61). We may neglect the second-order term since the frequencies employed are relatively small, but we need to take into account the time derivative of the applied voltage in the coefficient of the zero-

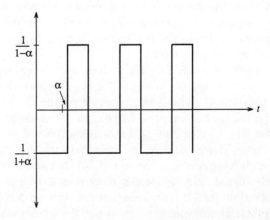

Fig. 5.25 Time-varying applied field assumed by Doogin and Zel'dovich (1993).

order term. In addition the modulation of the interference pattern m must be replaced by $m \exp j\Omega t$ owing to detuning. It may be shown (Ilinykh et al. 1991a) that when Ω is much larger than the grating formation time, then the rapidly oscillating part of the space charge field is negligible and the stationary solution can be obtained in a quite simple analytical form. For materials with small $\mu\tau$ the available space charge field may be shown to be equal to the applied field for all the available grating spacings. For a discussion of the various special cases see Zel'dovich and Nestiorkin (1991).

Analyses are also available for the case when the detuning frequency is n times as large (n being an integer) as the repetition frequency (Doogin et al. 1992). For generalization to phase modulation see Doogin et al. (1993). For experimental results see Ilinykh et al. (1991b,c) and for a theoretical paper taking into account the interaction between the beams see Nestiorkin and Zel'dovich (1992). An extension to four-wave mixing is given by Ilinykh et al. (1991d).

5.4.8 External circuit current due to phase modulation

A vibrating interference pattern induces a current in the external circuit of a photorefractive material, as was already mentioned in Section 2.10, where the basic principles have been explained. The topic was investigated in detail in a series of papers by Petrov et al. (1986), Trofimov and Stepanov (1986, 1988), Trofimov et al. (1987), Sokolov et al. (1989), Stepanov et al. (1990), Petrov et al. (1990), Sokolov and Stepanov (1990,1993), Sochava et al. (1993a), and Bryksin et al. (1993).

The treatment based in Section 2.10 on the work of Petrov et al. (1990) was further generalized by Sokolov and Stepanov (1993). The validity of the expression for the external current was extended to permit high modulating frequencies and the experimental investigation using BSO and BTO crystals went up to 50 kHz. They showed that besides the low frequency maximum the current has a high frequency maximum as well when the relationship $\Omega = Kv_0$ is satisfied. In common with the method proposed by Pauliat et al. (1990) this measurement of current can be used, and was used by Sokolov and Stepanov (1993), for determining the drift mobility. An extension to the bipolar case was recently done by Korneev et al. (1995). Measurements were also carried out in GaAs in the presence of a magnetic field in which a transverse Hall current was observed (Sochava and Stepanov 1994).

It is interesting to note that the low frequency and the high frequency resonances discussed earlier are equally important in the investigations of the external current. This is not altogether surprising, because the current in the external circuit is related to the space charge field.

The sign of the external current depends on the sign of the mobile carrier; hence it is a powerful method for determining the dominant charge carrier. Such experiments were done recently by Davidson et al. (1994b) on GaAs:Cr and InP:Fe, showing that, at a wavelength of 1.06 μm, the dominant charge carriers are electrons and holes respectively.

5.5 Optical pulses and high powers

5.5.1 Introduction

One of the main attractions of the photorefractive effect is that it does not necessitate the use of extremely high laser powers in order to exhibit an appreciable non-linear effect. In fact, in its simplest manifestation involving charge redistribution of one type of carriers and a single deep-trap level, the magnitude of the photorefractive effect was in Chapter 2 seen to be entirely independent of intensity once the photoionization exceeded the thermal ionization. However, the 'speed' of the effect (the inverse of the response time) is another matter. In addition to depending on material parameters, the speed was seen to increase linearly with incident optical intensity. For example, for BSO, which is one of the faster or more 'sensitive' photorefractive materials, the steady state is reached after only a few milliseconds for green light of intensity 10 mW cm^{-2}. This video frame rate may be fast enough for many real-time applications, but it does not quite hold a candle to the nearly instantaneous response in $\chi^{(3)}$ materials. In this chapter we shall take a more in-depth look at the speed of the photorefractive effect. We shall investigate what really happens at increasing optical intensities. Are there any limits on the speed? Is the optical energy required to write gratings at high intensity the same as for low intensity? What other effects start to play a role when the intensity is increased?

It is true that to a first approximation the speed of response of a photorefractive material increases linearly with intensity, as can be seen from eqn (2.60) (the dielectric relaxation time being inversely proportional to intensity). However, this relationship must break down when the build-up time approaches the material characteristic times, i.e. the diffusion and recombination times. This much should be clear. The photorefractive effect relies on the setting up of a space charge field in the crystal. For this to occur we must have photoexcitation of electrons (or holes), movement in the conduction (or valence) band by diffusion and drift, and recombination with traps. Clearly the speed of the photorefractive effect should ultimately be limited by the smallest of these time constants.

5.5.2 Predictions of the standard model

In the following we shall commence by looking at how the photorefractive effect performs under short optical pulses. This is not only for its instructive value, but also because experimentally pulsed lasers are needed in order to achieve high intensities. The photorefractive effect during pulsed illumination has been studied in a number of different materials including BSO (Hermann *et al.* 1981; Le Saux and Brun 1987; Partanen *et al.* 1990; Pauliat and Roosen 1990; Jones and Solymar 1991a; Partanen *et al.* 1991; Nouchi *et al.* 1992; Okamato *et al.* 1993), BaTiO$_3$ (L. K. Lam *et al.* 1981; Smirl *et al.* 1989b; Yao *et al.* 1990), LiNbO$_3$ (von der Linde *et al.* 1974; C. T. Chen *et al.* 1980), KNbO$_3$ (Zgonik *et al.* 1990; Reeves *et al.* 1991; Zgonik *et al.* 1991; Biaggio *et al.* 1992; Ewart *et al.* 1994], KTN (von der Linde

1975; H. M. Liu et al. 1991a,b), and GaAs (Smirl et al. 1988; Valley et al. 1989; Tomita and Ishii 1994). The seminal work in understanding short-pulse grating formation in photorefractive materials was done by Valley (1983a), based on the standard model. Valley pointed out that a material's response to a pulse can be grouped into two regimes, depending on the relative length of the optical pulse. If the pulse width, τ_p, is much longer than the recombination time, τ_e, then we are in a quasi-c.w. regime in which the charge redistribution takes place while the pulse is on. This regime has already been discussed qualititatively in Chapter 1. The temporal evolution of the space charge field is governed by a second-order differential equation given in Section 5.2.10.

In the reverse case of $\tau_p \ll \tau_e$ we are in the instantaneous illumination regime. Here, the space charge field builds up in the dark *after* the laser pulse ends. To understand how this happens, we note that the 'instantaneous' photoionization generates two superimposed gratings: a mobile electronic grating and a stationary ionic grating. What happens next depends on the relative sizes of the recombination time and the diffusion time, τ_D. If $\tau_e \gg \tau_D$ the 'faster' diffusion process washes out the carrier distribution before the carriers have time to recombine. This leaves only the ionic grating and consequently a space charge field. Steady state is reached when the diffusion current is balanced by the current due to drift of carriers in the space charge field (Pauliat and Roosen 1990). The speed of this process is governed by the diffusion time constant. If, on the other hand, $\tau_e \ll \tau_D$, then only a very small space charge field will appear. This is because with minimal diffusion, charge carriers recombine in nearly the same place from where they were excited. Steady state is consequently reached in the time it takes carriers to recombine. This type of behaviour may be seen in Fig. 5.26 where the temporal variation of the diffraction efficiency is shown. This is a typical experimental

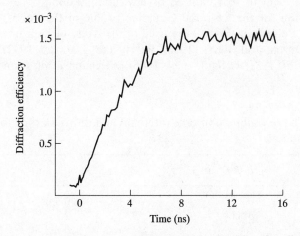

Fig. 5.26. Build-up of a photorefractive grating in reduced $KNbO_3$ after illumination with 70 ps pulses. $\Lambda = 0.8\,\mu m$, the peak intensities of the two writing beams are 8 and 40 MW cm^{-2}. The grating vector is oriented along the c axis (after Biaggio et al. 1992).

result for reduced $KNbO_3$ after illumination with a 70 ps pulse. From this curve a recombination time of $\tau_e = 4\,\text{ns}$ may be deduced.

The build-up of a space charge field in the instantaneous illumination regime is best summarized by the equation derived by Biaggio *et al.* (1990) for the case of low intensities and in the absence of an external applied field:

$$E_1 = \frac{esN_{\text{eff}}\tau_p I_o m}{\varepsilon_s h\nu K} \frac{1}{1+\tau_e/\tau_D}\left[1 - \exp\left(\frac{-t}{\tau}\right)\right], \quad \text{where} \quad \frac{1}{\tau} = \frac{1}{\tau_e} + \frac{1}{\tau_D} \quad (5.83)$$

where $N_{\text{eff}} = (N_D - N_A)N_A/N_D$. We have here for simplicity assumed a rectangular shaped pulse.

Note that in the instantaneous illumination regime photoionization is temporally separated from the processes of charge migration and recombination. This can be a benefit when attempting to characterize a photorefractive material, as we return to in Appendix C.

Which regime we are in depends very much on the pulse length and the photorefractive material. Table 5.1 lists typical values of τ_e and τ_D ($\Lambda = 1\,\mu\text{m}$) for a number of commonly used photorefractive materials. Note that $\tau_D = e/\mu k_B T K^2$ and as such depends on grating spacing. The values in Table 5.1 should be taken only as a guideline. There is a large discrepancy between the values of μ and τ_e reported in the literature, as we discuss in more detail in Appendix C.

So far we have described how pulsed grating formation falls into the two categories of quasi-c.w. and instantaneous illumination. Before continuing, it should be pointed out that it is also possible to simulate c.w. illumination with pulsed light. This may be accomplished by having a train of pulses. Grating build-up may take several pulses, but provided dark decay is negligible between pulses, the steady-state space charge field can for all intents and purposes be considered as being equivalent to that induced by c.w. illumination at the pulsed intensity (with correction for the temporal Gaussian profile of the pulses) (Damzen and Barry 1993). Pulse trains have been used in the study of high intensity effects in $BaTiO_3$ (Damzen and Barry 1993; Barry and Damzen 1992) and in BSO (Hermann *et al.* 1981; le Saux *et al.* 1986). Hermann *et al.* found that with 30

Table 5.1 The recombination time (lifetime) and diffusion time for a few photorefractive materials

	μ ($10^{-5}\,\text{m}^2\,\text{V}^{-1}\,\text{s}^{-1}$)	τ_e	$\tau_D(1\,\mu\text{m})$	
$Bi_{12}SiO_{20}$	0.3	5 μs	0.3 μs	$\tau_e > \tau_D$
GaAs	52 000	32 ns	2 ps	$\tau_e > \tau_D$
$BaTiO_3$	5	1 ns	20 ns	$\tau_e < \tau_D$
Reduced $KNbO_3$	4	4 ns	24 μs	$\tau_e < \tau_D$
$LiNbO_3$	8	1 ns	12 μs	$\tau_e < \tau_D$

5.5 Optical pulses and high powers

ns pulses, less than 10 shots of intensity 50 kW cm^{-2} were required to reach saturation of a diffraction grating in BSO.

As for the energy requirements for writing or erasing gratings, Valley (1983a) concluded that for high irradiances and short pulses the energy needed to write a grating was always greater than for low c.w. illumination. This can essentially be put down to carrier number saturation, which occurs when all the available donors have been ionized.

5.5.3 More complex models

Until now, we have concentrated on understanding how the photorefractive effect would be expected to behave for high intensities for an ideal model of one type of carrier and one deep-trap level. Unfortunately it would be rather naive to think that this is actually what nature presents us with. The application of high intensity to any material is somewhat of a minefield. Almost every conceivable effect rears its head. High irradiance of photorefractive materials is no exception. In the following we will recount briefly some of the effects which have been seen to play important roles in photorefractive materials.

We shall start by looking at the implications of extending the standard models to include the effects of simultaneous electron and hole photoconduction. This model was introduced earlier in this chapter as model IV. Barry and Damzen (1992) showed that with the help of this model it was possible to explain the very strong intensity dependence of two-beam coupling gain which had been observed in BaTiO$_3$. Their experimental results, reproduced in Fig. 5.27, illustrate how gain was seen to decrease with intensity, and that for sufficiently high intensities it was even possible to achieve sign reversal of the gain. Even though carrier saturation effects are expected to play an important role in reducing the gain, the bulk of the

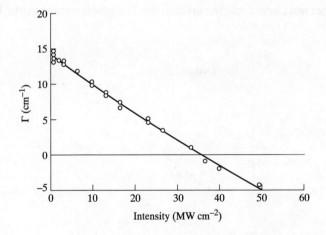

Fig. 5.27 Measured intensity dependence of the two-beam coupling gain coefficient in BaTiO$_3$ (after Damzen and Barry 1993).

effect was attributed to electron–hole competition (Damzen and Barry 1993). This can be understood by noting that a high photoionization rate significantly alters the ratio of the number densities of the photorefractive sites, N_D^+ and $N_D - N_D^+$, from which holes and electrons are excited, respectively. This subsequently leads to a strong intensity dependence in the relative contribution of the electron and hole conductivities. This model neatly explains the sign reversal as being due to a switch from hole-dominated to electron-dominated photoconductivity.

In order to explain the concurrent photoinduced absorption which has been observed in $BaTiO_3$ (Motes and Kim 1987; Ye *et al.* 1991), SBN (Orlov *et al.* 1994), $LiNbO_3$ (C. T. Chen *et al.* 1980; von der Linde *et al.* 1978), and $KNbO_3$ (Reeves *et al.* 1991), the presence of shallow traps had to be invoked. An expression for the photoinduced absorption has already been derived in Section 5.2.4, based on model VIII. We shall give here a simple physical reasoning. At low intensities the shallow traps are empty due to thermal ionization, and the standard model applies. For increasing intensity, however, the concentration of charge carriers in the conduction (or valence) band increases due to the increased photoexcitation from the deep traps. By recalling that recombination is proportional to carrier density, it follows that for sufficiently high intensities a balance will be reached between the rates of recombination to the shallow traps and their thermal re-excitation. The secondary (shallow) traps start to fill. The absorption is affected because we now have a new level which will absorb photons.

Thus, by increasing the intensity we *induce* a change in absorption by making another trap level active. This behaviour can clearly be seen in Fig. 5.28, which depicts a typical experimental result of absorption versus intensity for $BaTiO_3$. It should here be noted that this effect can easily be distinguished from other non-linear absorption effects stemming, for example, from two-photon absorption. Figure 5.29 illustrates an experimental set-up which is often used in this context to study wave interactions for short pulses and high powers. Here the pump and signal pulses intersect inside the crystal, but the much weaker probe beam can be

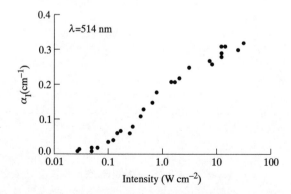

Fig. 5.28 Measured intensity-dependent absorption coefficient as a function of intensity in $BaTiO_3$ (after Brost *et al.* 1988).

5.5 Optical pulses and high powers

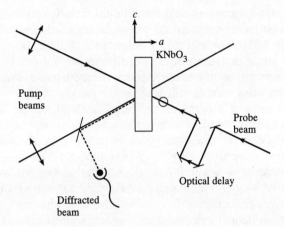

Fig. 5.29 Schematic representation of picosecond pulse-probe experiments (after Reeves et al. 1991).

delayed in time using a delay line (or trombone stage). This method offers the possibility of observing any gratings which are formed during or directly after the pump and signal pulses have crossed. We shall return to this set-up later. For now, it is sufficient to note that the temporal development of photoinduced absorption can be measured by blocking the signal beam and measuring the transmission of the probe in the presence and absence of the pump. Interestingly, Reeves et al. (1991) and Orlov et al. (1994) found that the photoinduced absorption in $KNbO_3$ and SBN, respectively, depended on the polarization of the probe beam but not on the pump beam polarization. The tentative explanation of Orlov et al. was to suggest that if traps are structure defects then their associated wave functions would possess a crystal-imparted symmetry.

There are two consequences of photoinduced absorption which deserve mention. The first is that according to the shallow-trap model of photoinduced absorption it should be possible to 'activate' a crystal for holographic recording at another wavelength. Recall that by populating a trap level which would otherwise be empty, we effectively introduce a new trap level which can be used in holographic recording. Buse et al. (1991) demonstrated that this was in fact possible. They showed that by pre-illuminating $BaTiO_3$ with green light they were able to write photorefractive gratings with infrared pulses.

The second point which must be made about photoinduced absorption is that an interference pattern incident on a crystal will now write an absorption grating in addition to the usual photorefractive grating (Ye et al. 1991; Khromov et al. 1990; Cudney et al. 1991). These gratings are also often referred to as photochromic gratings because the photoinduced absorption can lead to discolouration of the crystal as the absorption characteristics change (Bosomworth and Gerritsen 1968; Wardzynski et al. 1979; Khromov et al. 1990; Brost and Motes 1990; Martin et al. 1991; Fish et al. 1993; Jeffrey et al. 1993). It is well known that absorption gratings will diffract light (recall for instance diffraction from a normal opaque grating mask). The diffraction efficiency, however, cannot exceed about 7.2% (Solymar and Cooke 1981). Furthermore, the nature of the absorption grating dictates that the absorption grating will be in phase with the interference pattern. Such a local response means that an absorption grating set up during two-wave mixing will lead to a transient energy transfer from a strong to a weak beam depending on the beam ratio. This extra contribution to the gain in a photorefractive crystal is independent of crystal orientation and as such can be distinguished from the photorefractive gain. It should be noted, however, that according to K. Walsh et al. (1987) an absorption grating can never produce true gain when offset against the absorption of the crystal. Another feature of absorption gratings is that their formation and decay characteristics are in general quite different from those of photorefractive gratings. Absorption gratings tend to decay much slower. Vainos et al. (1989) demonstrated that this could be used to good effect because it enabled the multiplexing of semi-permanent (absorption grating) and real-time (photorefractive) holograms.

Returning to the shallow-trap level model we find that another consequence of this model is that the photorefractive gain actually increases (see eqn 5.42) with intensity as new traps contribute to the build-up of a space charge field (Tayebati and Mahgerefteh 1991). Figure 5.30 shows the two-beam coupling coefficient (corrected for the absorption) versus grating spacing for different laser intensities, as measured by Brost et al. (1988) for a crystal of $BaTiO_3$. The apparent contradiction with the experimental results of Fig. 5.27 is a testament to the fact that photorefractive crystals can behave very differently depending on the controlled (or uncontrolled) impurity concentrations. In short, there is no knowing for certain which model one should use for any given photorefractive crystal before

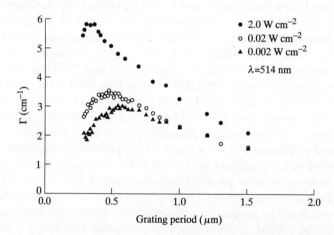

Fig. 5.30 Measured two-beam coupling coefficient, corrected for absorption, as a function of grating spacing for three different intensities (after Brost et al. 1988).

5.5 Optical pulses and high powers

it has been characterized by a series of measurements of gain or diffraction efficiency versus intensity, grating spacing, electric field, etc. To press home this fact, yet another model has been proposed to explain the apparent sublinear photoconductivity of $BaTiO_3$. Measurements by Fridkin and Popov (1977) and Ducharme and Feinberg (1984) had shown that the photoconductivity in $BaTiO_3$ did not scale linearly with intensity, as predicted by the simple model, but instead followed an apparent I^x dependence, where $0.5 < x < 1$. Holtmann (1989), used the aforementioned shallow-trap model to derive a relationship between photoconductivity and light-induced absorption. However, Mahgerefteh and Feinberg (1990) found that they could reproduce the sublinear dependence by using model IV assuming that the acceptor level was close enough to the valence band to allow thermal excitation of holes. Even more complex models with several trapping levels would seem to be needed to correctly explain the behaviour of other materials, in particular at high intensities and short pulses (Cudney et al. 1991; Okamato et al. 1993; Tomita and Ishii 1994). We noted earlier in this chapter that a one-level model leads to a grating which decays as a single exponential, and that the two-level model (e.g. model IV) gives a double exponential decay. Try now to imagine the model that would be needed to explain the measurements of Reeves et al. (1991) shown in Fig. 5.31 for differently doped crystals of $KNbO_3$!

So far we have only mentioned absorption due to photoexcitation from deep traps and shallow traps. These are not the only contributions to absorption. Figure 5.32 illustrates a few of the mechanisms of photon absorption (Valley et al. 1989a,b): (1) band-to-band two-photon absorption (or multiphoton absorption); (2) photoexcitation of electrons from trap levels to conduction band; (3) photoexcitation of electrons from valence band to trap levels (i.e. hole excitation to valence band); (4) free-electron or excited state absorption; and (5) free hole absorption. Note that n-photon absorption is proportional to I^n. These processes will therefore play an increasingly important role at higher intensities (for measurement of two-photon absorption coefficients see e.g. Kim and Hutchinson 1989). For example, the two-photon absorption coefficient of $BaTiO_3$ is 0.1 $cm\,GW^{-1}$ so this process will dominate when the intensity exceeds a few $GW\,cm^{-2}$ (Boggess et al. 1990). Tomita and Ishii (1994) recently examined theoretically the effect of two-photon absorption on the photorefractive effect, and concluded that for large fluences the peak value of the space charge field is enhanced, but the long-lived component is suppressed significantly. The two-photon absorption process can usually be separated from free-carrier and excited state absorption because these tend to have lifetimes longer than the laser pulse duration (Földvári et al. 1993b).

At this stage it is worth returning to the problem of how, with so many different processes, we may isolate the effects that are purely photorefractive in origin. One way of separating the photorefractive effect from the others is to orient the crystal so that no photorefractive gain is possible. Another way is to measure the gain twice, the second time with the crystal rotated by 180 degrees. To see how this works, we note that for small gain or loss the relative change in

178   *The materials equations*

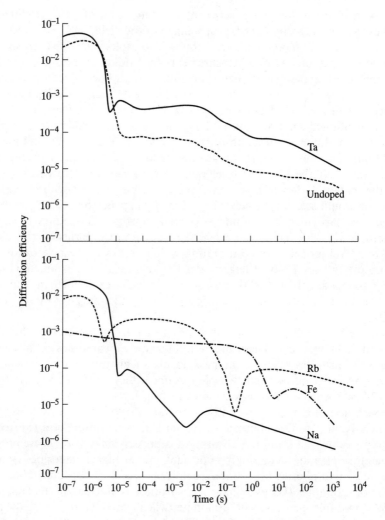

Fig. 5.31 The decay of measured diffraction efficiency as a function of time in doped KNbO$_3$ crystals induced by two single 30 ps writing pulses (after Reeves *et al.* 1991).

transmission of the signal beam in Fig. 5.29 will be (Petrovic *et al.* 1991; Schroeder *et al.* 1991b)

$$\frac{T - T_o}{T_o} = \frac{\exp(-\alpha d - \triangle\alpha d + \Gamma_{pr} d) - \exp(-\alpha d)}{\exp(-\alpha d)} = \exp(-\triangle\alpha d + \Gamma_{pr} d) - 1$$

(5.84)

where T and T_o are the transmissions in the presence and absence of the pump, respectively, and d is the sample length. Here, $\triangle\alpha$ represents the absorption which depends on the pump, i.e. saturable absorption, two-photon absorption, transient

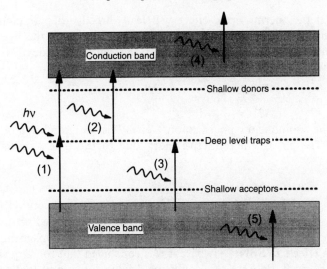

Fig. 5.32 Photon absorption mechanisms.

energy transfer, and absorption gratings. We may therefore write the photorefractive gain coefficient, Γ_{pr}, as

$$\Gamma_{\text{pr}} = \frac{1}{2d} \ln\left(\frac{T^+}{T^-}\right) \qquad (5.85)$$

where the superscript sign refers to the orientation of the crystal. Another way of sorting the photorefractive effect from the other effects is to choose a crystal orientation such that light diffracted from the photorefractive grating will be polarized orthogonal to the incident light. The feasibility of this technique has been demonstrated by Smirl et al. (1987) and Schroeder et al. (1991a) for GaAs.

5.5.4 Non-photorefractive effects

Let us now look briefly at some of the other effects which may play a role at high intensities and short pulses.

Pyroelectric and thermo-optic effects. Exposing a crystal to high intensities must invariably heat up the crystal. In other words, an interference pattern will set up a thermal grating (Eichler et al. 1986). To understand how this can affect optical wave interactions we note that two things will happen: (1) the temperature grating will set up a spatially modulated pyroelectric field which will affect charge redistribution (Vinetskii and Itskovskii 1978; Ducharme

1991), and (2) the thermo-optic effect will lead directly to a refractive index grating which is in phase or 180 degrees out of phase with the light pattern depending on whether heat raises or lowers the refractive index (Nye, 1985). The pyroelectric field is given by

$$E_{\text{pyro}}(t) = -\frac{1}{\varepsilon_s}\frac{\partial P_s}{\partial T}\triangle T(t) \tag{5.86}$$

where P_s is the spontaneous polarization and $\triangle T$ is the amplitude of the thermal grating. The change in refractive index due to the thermo-optic effect is given by

$$\triangle n_{\text{T}} = \frac{\partial n_{\text{a,b,c}}}{\partial T}\triangle T \tag{5.87}$$

Buse (1993) has presented a thorough treatment of the combined action of these two effects in $BaTiO_3$ and $KNbO_3$. This has recently been supplemented with a study of these effects in SBN:Ce, where it was concluded that for very short light pulses pyroelectric fields were the dominant driving force, and that they could significantly enhance the photorefractive effect (Buse et al. 1994). Note, however, that since these effects only occur when there is a thermal grating, they can only be expected to play a role during transient conditions before the heat is redistributed. The effect will be negligible for c.w. experiments.

Piezoelectric and photoelastic effects. What of the influence of the piezoelectric and photoelastic effects on hologram recording by an optical pulse? Do the clamped or unclamped electrooptic coefficients apply? Illuminating a crystal with an interference pattern will excite acoustic oscillations. This may occur due to electrostriction or the piezoelectric effect where the space charge electric field is set up via the photorefractive effect (Deev and Pyatakov 1986; Zelenskaya and Shandarov 1986; Stepanov et al. 1987; Litvinov and Shandarov 1994), or it may stem from thermoelasticity (Nelson and Fayer 1980). Acoustic oscillations will subsequently influence the magnitude of the electro-optic effect, as we discuss in Appendix B. So the answer to the above questions is not trivial. Acoustic oscillations will certainly affect optical wave interactions, but the size of the effect will depend very much on crystal orientation and the magnitude of the piezoelectric and photoelastic coefficients. Litvinov and Shandarov (1994) performed the relevant calculations for BSO and GaAs and concluded that only for GaAs would any substantial effect be seen.

Free-carrier gratings. At high intensities we obtain very high concentrations of electrons and holes. A consequence is that these free carriers can set up a grating directly through the Drude–Lorenz (or band-filling) contribution to the index of refraction (Smirl et al. 1987; Valley et al. 1990). Note that this free-carrier grating does not require diffusion or drift in order to be established. In fact, diffusion is detrimental as it reduces the modulation. Thus, in order to observe transient energy transfer due to free-carrier gratings, the optical pulses have to be shorter than the recombination time and the ambipolar[7] diffusion time. In most semiconductors this means of the order of picoseconds. Valley *et al.* investigated the significance of free-carrier non-linearity on transient energy transfer in GaAs and Si and demonstrated that gains of over 10 could be achieved.

Dember field. The Dember field needs to be taken into consideration for high electron and hole densities. The Dember field is the electric field which arises across the crystal when electrons and holes separate because they have different mobilities (see e.g. Pankove 1971). This field will subsequently affect photorefractive grating formation (Smirl *et al.* 1988; Schroeder *et al.* 1991c; Petrovic *et al.* 1991; Tomita and Ishii 1994).

Higher-order non-linear effects. Finally, we must not forget the higher-order non-linear processes, in particular the effects of the $\chi^{(3)}$ non-linearity. Figure 5.33 illustrates a typical result of four-wave mixing in unreduced $KNbO_3$ for an intensity of about 100 MW cm^{-2} (Zgonik *et al.* 1991). We clearly see a peak in the diffraction efficiency of the probe beam at zero time delay. Similar peaks have been observed in BSO (Ferrier *et al.* 1986; Jonathan *et al.* 1988) and $BaTiO_3$ (Smirl *et al.* 1989b). This peak may either be caused by the above-mentioned

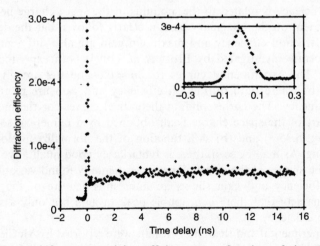

Fig. 5.33 Measured four-wave mixing efficiency as a function of time for unreduced $KNbO_3$ at a grating spacing of 0.8 μm. Apart from the zero time delay peak, the efficiency is approximately constant (after Zgonik *et al.* 1991).

---

[7]The concept of ambipolar diffusion is an approximation whereby the space charge field set up by the separation of the electron and holes is ignored and the electron and hole densities are instead assumed to be governed by the same diffusion equation with an ambipolar diffusion coefficient.

free-carrier grating, or it may be due to a non-dissipative third-order non-linear process stemming from an intraband anharmonic response of bound electrons (Ferrier et al. 1986; Flytzanis 1975). The $\chi^{(3)}$ is also responsible for self-induced non-linear absorption effects. Sylla et al. (1992) recently observed strong non-linear absorption in BSO and BGO with picosecond pulses and intensities of 10^8 W cm^{-2} while demonstrating phase conjugate reflectivities of 2×10^{-3} obtained by degenerate four-wave mixing.

5.6 Other topics

We have discussed a great many problems related to the response of the photorefractive material to light. We put them in neat categories but there are two quite unrelated topics which have not fitted into any of the sections. We shall present them here. One is concerned with the effect of large modulation, discussed already in Section 2.9. We wish to make here a few further comments and add a few more references. The second topic stands entirely on its own. It is for the benefit of those who can easily think in terms of electrical circuits.

5.6.1 Large modulation solutions

It has already been shown in Section 2.9 (see Fig. 2.11) that for large modulation the maximum of the imaginary part of the space charge field does not occur at the same detuning as the maximum of the modulus of the space charge field. Since diffraction efficiency is related to the modulus of the space charge field whereas gain is dependent on the imaginary part, it clearly follows that the detuning for maximum diffraction efficiency and maximum gain are also different. The problem was recently investigated by Brost et al. (1994) both experimentally and theoretically. The experimental curves for $m = 0.6$ and 0.9 are given in Fig. 5.34(a) and (b) where both diffraction efficiency and gain are plotted against detuning frequency. The corresponding theoretical curves for the modulus and imaginary part of the space charge field, obtained by a numerical solution, are plotted in Fig. 5.35(a) and (b) as a function of the normalized velocity of the moving grating. As may be seen, there is remarkably good qualitative agreement concerning the shapes of the curves. The experimentally found variation of both diffraction efficiency and gain shows an interesting feature of the theoretical predictions, namely that there is a double peak in gain but only a single peak in diffraction efficiency.

Further experimental and theoretical results were reported by McClelland et al. (1995) both for gain and diffraction efficiency as a function of detuning. They showed that for the modulus of the space charge field the main peak shifts with higher modulation and an additional peak appears at lower frequencies. Further interesting results were obtained by Serrano et al. (1994a,b) both for the steady state and for transients.

It is worth noting that owing to a strong quadratic electro-optic effect there are, at high modulation, significant second and third harmonic gratings in

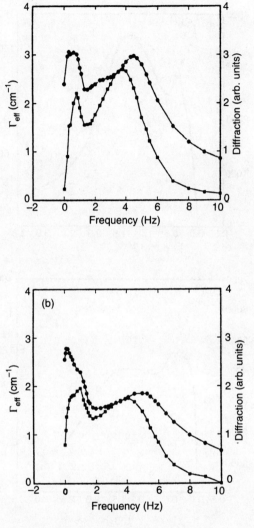

Fig. 5.34 The measured gain (■) and diffraction efficiency (●) as a function of detuning frequency for (a) $m = 0.6$ and (b) $m = 0.9$ (after Brost et al. 1994).

photorefractive quantum wells, as shown by Q. N. Wang et al. (1991) (for a more detailed discussion see Section 9.5).

Finally, we need to mention the analytical solution (the only one available for large modulation) of Swinburne et al. (1989) for low frequency AC enhancement. The curves obtained seem to be in the right range but the derivation is unclear and the introduction of a 'scaling factor' seems entirely unjustified.

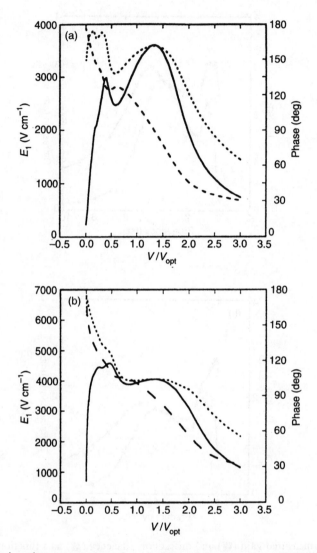

Fig. 5.35 The imaginary component (——), the modulus (- - - -), and the phase (– – –) of the space charge field as a function of normalized grating velocity for (a) $m = 0.6$ and (b) $m = 0.9$ (after Brost et al. 1994).

5.6.2 Equivalent circuits

The space charge field derived from the standard model is given in eqn (2.58). It is a complex quantity: it has a magnitude and a phase. The expression looks like one for impedance in an electrical circuit. The analogy is so close that we may introduce the concept of a 'spatial' circuit or 'equivalent' circuit (Solymar 1987a,b). What is the advantage of doing so? It would enable those familiar with electrical

5.6 Other topics

circuits to make an instant judgement concerning the phase and amplitude of the space charge field, and develop a feel for the variation of the space charge field with K, the spatial frequency.

For the present purpose we shall write eqn (2.58) in the slightly different form

$$-\frac{E_1}{m} = \frac{-jE_q(E_o - jE_D)}{E_o - jE_D - jE_q} \quad (5.88)$$

Recalling eqn (2.37) we may observe that E_D is proportional to K, E_q is inversely proportional to K, and E_o is independent of K. Hence we may introduce an 'inductance', a 'capacitance', and a 'resistance' with the relations

$$L = -\frac{k_B T}{e}, \quad C = \frac{\varepsilon_s}{eN_A^-(1-a)}, \quad R = E_o \quad (5.89)$$

Equation (5.88) may then be rewritten in the form of an 'impedance', denoting the left-hand side simply by Z:

$$Z = \frac{(R + j\omega L)\dfrac{1}{j\omega C}}{R + j\omega L + \dfrac{1}{j\omega C}} \quad (5.90)$$

This may be recognized as two impedances in parallel, as shown in the equivalent circuit of Fig. 5.36. The phase and the magnitude of the impedance correspond to the phase and the magnitude of the space charge field whereas the temporal frequency at which the impedance is driven corresponds to the spatial frequency determined by the inter-beam angle. The analogy is perfect apart from the unfortunate fact that the inductance is negative.

Let us see now some quick conclusions which may be drawn on the strength of the analogy. We shall put side by side the conclusion as reached from the equivalent circuit and the corresponding statement for the space charge field. First, in the absence of the resistance (no applied electric field):

(1) the phase angle is $\pi/2$;
(1') the phase angle is $\pi/2$;

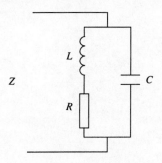

Fig. 5.36 Space charge field 'equivalent circuit'.

(2)   as $\omega \to 0$ the impedance is zero on account of the inductance;
(2') as $K \to 0$ the space charge field is zero on account of the vanishing diffusion field;
(3)   as $\omega \to \infty$ the impedance is zero on account of the capacitance;
(3') as $K \to \infty$ the space charge field is zero on account of $E_q \to 0$.

In the presence of the resistance (electric field applied) the following conclusions may be drawn:

(4)   for low enough resistance the phase angle is $\pi/2$;
(4') for low enough applied field the phase angle is $\pi/2$;
(5)   for high enough resistance the phase angle is $\pi/2$;
(5') for high enough applied field the phase angle is $\pi/2$;
(6)   for high enough resistance the magnitude of the impedance is determined solely by the capacitance;
(6') for high enough applied field the space charge field is determined solely by the value of E_q, i.e. diffusion is irrelevant;
(7)   as $\omega \to 0$ the impedance is equal to the resistance;
(7') as $K \to 0$ the space charge field is equal to the applied field;
(8)   as $\omega \to \infty$ the impedance is zero on account of the capacitance;
(8') as $K \to \infty$ the space charge field is zero on account of $E_q \to 0$.

For an extension of the concept of equivalent circuit to the case of detuning see Solymar and Ringhofer (1988).

6
THE FIELD EQUATIONS FOR TWO-WAVE MIXING

6.1 Introduction

The field equations have already been introduced in Chapter 3 in a simple scalar form. It was hinted there that a more complex treatment is required in order to explain some of the more interesting phenomena observed in experiments on photorefractive materials; that treatment is the subject of this chapter. In addition to looking in more detail at the field equations, we shall also take the opportunity to review the comparisons that have been made between theoretical predictions and experimental results. This will help to differentiate between those situations where simple theory provides a good description and those where more effort is needed if quantitative agreement is required.

As in Chapter 3, we must begin by deriving the wave equation, this time taking into account the fact that the relative permittivity and the electro-optic constant are both tensors. Proper account must also be taken of the vectorial nature of the electric field. From the wave equation we shall obtain the coupled equations relating the complex amplitudes of the electric fields of the two beams interacting within the photorefractive crystal. We shall see that in general these equations are quite complex and we shall look at simplifications that may be made in certain well-defined situations, leading in many cases to analytic solutions.

Most of our treatment will concern beam coupling, though we shall spend some time discussing the diffraction efficiency of gratings recorded in photorefractive media. All electro-optic materials are also piezoelectric and we shall also look briefly at the role the piezoelectric effect can play in photorefractive two-wave mixing. Most of the chapter is concerned with steady-state behaviour but we shall end up by looking at the response of photorefractive materials to input beams which vary with time.

Throughout the chapter we shall restrict ourselves to the case of transmission gratings, where the optical waves may be considered to be propagating in the same general direction through the crystal. Admittedly there are reflection gratings as well, as we have already discussed in Section 3.6. We will briefly return to them in Part III dealing with applications, but since their properties are not drastically different from those of transmission gratings and since they represent a small fraction of the total literature we shall not include them in the present chapter.

6.2 Derivation of the coupled wave equations

The crystal geometry under consideration is shown in Fig. 6.1. The input beams are assumed to be incident on the crystal such that the bisector of the two beams is in the x direction, each beam subtending an angle θ with the bisector. The plane of incidence lies in the x–y plane of the external coordinate system. Note that in general the crystallographic axes may have any orientation with respect to our chosen external coordinate system. The input waves are assumed to be completely polarized but we allow the polarization states to be arbitrary.

We write the total optical field amplitude due to the presence of the two beams, which may be of different frequencies, as

$$\mathcal{E}(\mathbf{r}, t) = \tfrac{1}{2}[\mathcal{E}_1(\mathbf{r}, t)e^{j(\omega_1 t - \mathbf{k}_1 \cdot \mathbf{r})} + \mathcal{E}_2(\mathbf{r}, t)e^{j(\omega_2 t - \mathbf{k}_2 \cdot \mathbf{r})} + \text{c.c.}] \tag{6.1}$$

The resulting intensity pattern within the crystal was given in eqn (2.103). For convenience, we repeat it here:

$$\begin{aligned} I &= I_1 + I_2 + \mathcal{E}_1 \cdot \mathcal{E}_2^* e^{j(\Omega t - \mathbf{K} \cdot \mathbf{r})} + \text{c.c.} \\ &= I_0\left[1 + \frac{m}{2} e^{j(\Omega t - \mathbf{K} \cdot \mathbf{r})} + \text{c.c.}\right] \end{aligned} \tag{6.2}$$

where $\Omega = \omega_1 - \omega_2$ is the angular frequency difference between the two beams (the detuning frequency), $\mathbf{K} = \mathbf{k}_1 - \mathbf{k}_2$ is the grating vector, I_1 and I_2 are the intensities of beams 1 and 2, respectively, and m is the fringe modulation (sometimes called the visibility) in complex notation, defined here as

$$m = \frac{2\mathcal{E}_1 \cdot \mathcal{E}_2^*}{I_0} \tag{6.3}$$

Note that this expression differs slightly from eqn (2.104) in that we have permitted the polarizations of the two beams to be different and therefore must use the scalar (dot) product of their fields. Also we are explicitly assuming that the optical waves are monochromatic. In the vast majority of experimental situations this is an excellent approximation, but some study has been devoted to dynamic holography with broad band light sources (Grousson and Mallick 1980;

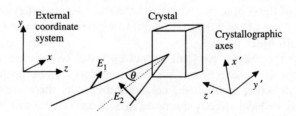

Fig. 6.1 Schematic diagram illustrating the relationship between the external coordinate system, the crystallographic axes, the polarization states of the incident beams, and their directions of propagation.

6.2 Derivation of the coupled wave equations

Rabinovich and Feldman 1991; Rabinovich *et al.* 1991b). Indeed, it has been shown (Rabinovich *et al.* 1991a) that gratings formed with achromatic light can suppress the problem of beam fanning.

We have seen in a previous chapter (Section 3.1) that consideration of the materials equations shows that this intensity pattern will produce an electric field, **E**, within the medium given by eqns (3.3) and (3.4).

$$\begin{aligned}\mathbf{E} &= \mathbf{E}_0 + \tfrac{1}{2}\{\mathbf{E}_1 e^{j(\Omega t - \mathbf{K}\cdot\mathbf{r})} + \text{c.c.}\} \\ &= \mathbf{E}_0 + \tfrac{1}{2}\{\mathbf{E}_w m e^{j(\Omega t - \mathbf{K}\cdot\mathbf{r})} + \text{c.c.}\}\end{aligned} \tag{6.4}$$

where the spatially constant component $\mathbf{E}_0$ is made up of contributions from any externally applied or photovoltaically induced fields. Here we are assuming for the moment that the space charge field amplitude is linear in fringe modulation so that

$$\mathbf{E}_1 = \mathbf{E}_w m \tag{6.5}$$

with $\mathbf{E}_w$, the normalized space charge field, being independent of modulation. This is an assumption that is often only satisfied for small m – see Section 2.9.

We write the relative permittivity of the crystal as

$$\underline{\underline{\varepsilon_r}} = \underline{\underline{\varepsilon_r(0)}} + \underline{\underline{\Delta\varepsilon_r}}(\mathbf{E}) \tag{6.6}$$

where $\underline{\underline{\varepsilon_r(0)}}$ is the relative permittivity in the absence of any electric field and $\underline{\underline{\Delta\varepsilon_r}}$ is the field induced change in relative permittivity. Each of the terms in eqn (6.6) is a tensor of the second rank (indicated by the two underlines), so the optical electric field, $\boldsymbol{\mathcal{E}}$, and its associated displacement field, **D**, are related by

$$\mathbf{D} = \varepsilon_0 \underline{\underline{\varepsilon_r}} \boldsymbol{\mathcal{E}} \tag{6.7}$$

As vectors, $\boldsymbol{\mathcal{E}}$ and **D** may be considered to be tensors of the first rank, though we shall indicate this with bold type rather than by a single underline.

In the following treatment it will sometimes be useful to write products of tensors and vectors using the index notation, in which case eqn (6.7) could be written as

$$\begin{aligned}\mathcal{E}_m &= \varepsilon_0 \sum_{l=1}^{3}(\varepsilon_r)_{ml} D_l \\ &\equiv \varepsilon_0 (\varepsilon_r)_{ml} D_l\end{aligned} \tag{6.8}$$

where the summation is over the three spatial dimensions. In the second line we have dropped the summation symbol by invoking the convention that any indices that are repeated must be summed over (the Einstein summation convention).

Using this convention, in the absence of any field the relative permittivity in general takes the form

$$\varepsilon_r(0)_{il} = \varepsilon'_{il} - j\varepsilon''_{il} + jg_{il} \tag{6.9}$$

where $\underline{\underline{\varepsilon}}'$ is the real part of the relative permittivity tensor, $\underline{\underline{\varepsilon}}''$ is the imaginary part of the relative permittivity tensor, and $\underline{\underline{g}}$ is the optical activity (or gyration) tensor[1]. The first term of eqn (6.9) is responsible for the real part of the refractive index experienced by light propagating within the medium (which of course in general depends on the direction of polarization of that light). The second term describes the absorption of the crystal and $\underline{\underline{\varepsilon}}''$ takes the form

$$\begin{bmatrix} \varepsilon'' & 0 & 0 \\ 0 & \varepsilon'' & 0 \\ 0 & 0 & \varepsilon'' \end{bmatrix} \quad (6.10)$$

where ε'' is related to the intensity absorption constant, α, by

$$\varepsilon'' = \frac{n_r^2 \alpha}{k} = \frac{\alpha}{(\varepsilon')^{1/2} k_o} \quad (6.11)$$

n_r being the refractive index of the crystal.

In the presence of an electric field, the relative permittivity is modified[2] by the term $\underline{\underline{\Delta \varepsilon_r}}(\mathbf{E})$ in eqn (6.6), which is given by

$$(\Delta \varepsilon_r)_{il}(\mathbf{E}) = -\varepsilon'_{im} r_{mqr} E_r \varepsilon'_{ql} \quad (6.12)$$

where $\underline{\underline{r}}$ is the electro-optic tensor. $\underline{\underline{\Delta \varepsilon_r}}(\mathbf{E})$ contains a spatially uniform component due to the applied external field or photovoltaic effect (should they exist) plus a periodic spatial component due to the space charge field.

Proceeding as in Chapter 3, we must solve the wave equation (3.10), which we write using tensor and vector notation as

$$\nabla \wedge (\nabla \wedge \boldsymbol{\mathcal{E}}) - \mu_o \varepsilon_o \frac{\partial^2}{\partial t^2} (\underline{\underline{\varepsilon_r}} \boldsymbol{\mathcal{E}}) = 0 \quad (6.13)$$

In Chapter 3 we were able to simplify things by expanding the first term using the vector identity

$$\nabla \wedge (\nabla \wedge \boldsymbol{\mathcal{E}}) \equiv \nabla(\nabla \cdot \boldsymbol{\mathcal{E}}) - \nabla^2 \boldsymbol{\mathcal{E}} \quad (6.14)$$

and then making the assumption

$$\nabla \cdot \boldsymbol{\mathcal{E}} = 0 \quad (6.15)$$

Now we shall address the question as to just how valid this assumption really is. We can proceed with the investigation by using the constitutive relationship[3]

---

[1] Note that this way of treating the effects of optical activity is not completely general since the components of $\underline{\underline{g}}$ depend on the propagation vector of the optical fields (Yariv and Yeh 1984a). However, the general treatment involves a rank-3 tensor (Landau and Lifshitz 1984), and this added complexity brings no advantages here.
[2] The electro-optic effect is traditionally defined in terms of the change in the impermeability tensor; in Appendix B we show the derivation leading from that definition to eqn (6.12).
[3] Strictly speaking of course, in the spirit of this section the inverse of the relative permittivity in eqn (6.16) should be a second-rank tensor. This would however complicate things unnecessarily and we shall use the scalar notation here for clarity.

6.2 Derivation of the coupled wave equations

$$\varepsilon_r^{-1} \mathbf{D} = \boldsymbol{\mathcal{E}} \tag{6.16}$$

to expand (6.15) to give

$$\nabla(\nabla \cdot \varepsilon_r^{-1} \mathbf{D}) = \nabla(\varepsilon_r^{-1} \nabla \cdot \mathbf{D}) + \nabla(\mathbf{D} \cdot \nabla \varepsilon_r^{-1}) \tag{6.17}$$

We may neglect the first term on the right-hand side of eqn (6.17) since Gauss's theorem tells us that

$$\nabla \cdot \mathbf{D} = 0 \tag{6.18}$$

that is to say, there are no free charges oscillating at optical frequencies. There will be of course a quasi-static space charge within the material which affects static dielectric permittivity; this is included in the materials equations.

The second term on the r.h.s. of eqn (6.17) is more problematic. In the geometry of Fig. 6.1 the grating vector lies in the z direction, so there will be a spatial variation of the dielectric constant in that direction[4]. Furthermore, coupling between the beams will alter the spatial profile of the interference pattern – and therefore the dielectric constant – along the x direction. With planar waves though, the dielectric constant will not vary along the y direction.

Does this information help us? Unfortunately, within the context of this chapter it does not. We are trying to consider waves of arbitrary polarization and so in general we cannot ignore the second term on the r.h.s. of eqn (6.17). We can, however, justify eqn (6.15) as used in Chapter 3, where we were only interested in a scalar treatment. For that equation to be valid we must simply specify that the two beams are linearly polarized in the y direction; then the dot product on the r.h.s. of eqn (6.17) is between $\mathbf{D}$, which we have just specified to be in the y direction, and the gradient of ε_r^{-1}, which as we have just seen must lie in the x–z plane. In this case, both terms on the r.h.s. of eqn (6.17) are equal to zero and assumption (6.15) is valid.

For the purposes of this chapter with arbitrary input polarizations, we shall keep eqn (6.13) as our wave equation and substitute directly into this eqns (6.1) to (6.6), (6.9), and (6.12). We follow the same procedure described in Chapter 3 of gathering together the terms in $\exp(j[\omega_1 t - \mathbf{k}_1 \cdot \mathbf{r}])$ and in $\exp(j[\omega_2 t - \mathbf{k}_2 \cdot \mathbf{r}])$. The results are

$$\nabla(\nabla \cdot \boldsymbol{\mathcal{E}}_1) - \nabla^2 \boldsymbol{\mathcal{E}}_1 - j\mathbf{k}_1 \wedge (\nabla \wedge \boldsymbol{\mathcal{E}}_1) - j\nabla \wedge (\mathbf{k}_1 \wedge \boldsymbol{\mathcal{E}}_1) - \mathbf{k}_1 \wedge (\mathbf{k}_1 \wedge \boldsymbol{\mathcal{E}}_1)$$
$$+ \mu_0 \varepsilon_0 \omega_1^2 \left\{ \underline{\underline{\varepsilon'}} \underline{\underline{r}} \left[\mathbf{E}_w \frac{\boldsymbol{\mathcal{E}}_1 \cdot \boldsymbol{\mathcal{E}}_2^*}{I_0} \underline{\underline{\varepsilon'}} \boldsymbol{\mathcal{E}}_2 + E_0 \underline{\underline{\varepsilon'}} \boldsymbol{\mathcal{E}}_1 \right] - [\underline{\underline{\varepsilon'}} - j\underline{\underline{\varepsilon''}} + \underline{\underline{g}}] \boldsymbol{\mathcal{E}}_1 \right\} = 0 \tag{6.19}$$

---

[4] With certain crystal geometries, it is in fact possible that the space charge field does not give rise to an index grating. These will be geometries for which the appropriate component of the electro-optic tensor is zero. However, since in such cases there can be no coupling between the two beams, these situations are generally uninteresting and we shall ignore them for the time being.

$$\nabla(\nabla \cdot \boldsymbol{\mathcal{E}}_2) - \nabla^2 \boldsymbol{\mathcal{E}}_2 - j\mathbf{k}_2 \wedge (\nabla \wedge \boldsymbol{\mathcal{E}}_2) - j\nabla \wedge (\mathbf{k}_2 \wedge \boldsymbol{\mathcal{E}}_2) - \mathbf{k}_2 \wedge (\mathbf{k}_2 \wedge \boldsymbol{\mathcal{E}}_2)$$
$$+ \mu_0 \varepsilon_0 \omega_2^2 \left\{ \underline{\underline{\varepsilon'r}} \left[\mathbf{E}_w^* \frac{\boldsymbol{\mathcal{E}}_1^* \cdot \boldsymbol{\mathcal{E}}_2}{I_0} \underline{\underline{\varepsilon'}} \boldsymbol{\mathcal{E}}_1 + E_0 \underline{\underline{\varepsilon'}} \boldsymbol{\mathcal{E}}_2 \right] - [\underline{\underline{\varepsilon'}} - j\underline{\underline{\varepsilon''}} + \underline{\underline{g}}] \boldsymbol{\mathcal{E}}_2 \right\} = 0$$
(6.20)

and we can immediately simplify these equations a little by making the slowly varying envelope approximation as in Section 1.2. This says that we can neglect second derivatives of the fields, meaning that the first two terms in each of eqns (6.19) and (6.20) may be dropped, giving

$$-j\mathbf{k}_1 \wedge (\nabla \wedge \boldsymbol{\mathcal{E}}_1) - j\nabla \wedge (\mathbf{k}_1 \wedge \boldsymbol{\mathcal{E}}_1) - \mathbf{k}_1 \wedge (\mathbf{k}_1 \wedge \boldsymbol{\mathcal{E}}_1)$$
$$+ \mu_0 \varepsilon_0 \omega_1^2 \left\{ \underline{\underline{\varepsilon'r}} \left[\mathbf{E}_w \frac{\boldsymbol{\mathcal{E}}_1 \cdot \boldsymbol{\mathcal{E}}_2^*}{I_0} \underline{\underline{\varepsilon'}} \boldsymbol{\mathcal{E}}_2 + E_0 \underline{\underline{\varepsilon'}} \boldsymbol{\mathcal{E}}_1 \right] - [\underline{\underline{\varepsilon'}} - j\underline{\underline{\varepsilon''}} + \underline{\underline{g}}] \boldsymbol{\mathcal{E}}_1 \right\} = 0 \quad (6.21)$$

$$-j\mathbf{k}_2 \wedge (\nabla \wedge \boldsymbol{\mathcal{E}}_2) - j\nabla \wedge (\mathbf{k}_2 \wedge \boldsymbol{\mathcal{E}}_2) - \mathbf{k}_2 \wedge (\mathbf{k}_2 \wedge \boldsymbol{\mathcal{E}}_2)$$
$$+ \mu_0 \varepsilon_0 \omega_2^2 \left\{ \underline{\underline{\varepsilon'r}} \left[\mathbf{E}_w^* \frac{\boldsymbol{\mathcal{E}}_1^* \cdot \boldsymbol{\mathcal{E}}_2}{I_0} \underline{\underline{\varepsilon'}} \boldsymbol{\mathcal{E}}_1 + E_0 \underline{\underline{\varepsilon'}} \boldsymbol{\mathcal{E}}_1 \right] - [\underline{\underline{\varepsilon'}} - j\underline{\underline{\varepsilon''}} + \underline{\underline{g}}] \boldsymbol{\mathcal{E}}_2 \right\} = 0 \quad (6.22)$$

Equations (6.21) and (6.22) describe the evolution of the pump and signal wave in general terms.

6.3 Further simplifications

A great deal of simplification may be obtained by ignoring any longitudinal component of the electric field (Ringhofer et al. 1991), and also by assuming that the two beams subtend small angles to the x axis – the paraxial approximation. This paraxial approximation is satisfied in many situations. For example, when the moving grating technique is used to optimize the space charge field in a crystal of the sillénite family, the greatest enhancement is obtained for inter-beam angles of the order of 1° (Refregier et al. 1985).

Assuming the optical electric field to be perpendicular to the propagation direction allows us to use the following relationship to simplify the third terms in eqns (6.21) and (6.22):

$$\mathbf{k}_i \wedge (\mathbf{k}_i \wedge \boldsymbol{\mathcal{E}}_i) = -k_i^2 \boldsymbol{\mathcal{E}}_i, \quad i = 1, 2 \quad (6.23)$$

The paraxial approximation along with this assumption of a transverse electric field allows the first and second terms in eqns (6.21) and (6.22) to be simplified using

$$\mathbf{k}_i \wedge (\nabla \wedge \boldsymbol{\mathcal{E}}_i) = \nabla \wedge (\mathbf{k}_i \wedge \boldsymbol{\mathcal{E}}_i) = -k_i \cos\theta \frac{d\boldsymbol{\mathcal{E}}_i}{dx}, \quad i = 1, 2 \quad (6.24)$$

Taking these various assumptions into account allows eqns (6.21) and (6.22) to be rewritten as

$$-2j k_1 \cos\theta \frac{d\mathcal{E}_1}{dx} = \mu_0 \varepsilon_0 \omega_1^2 \left\{ \underline{\underline{\varepsilon}}' \underline{\underline{r}} \left[\mathbf{E}_w \frac{\mathcal{E}_1 \cdot \mathcal{E}_2^*}{I_0} \underline{\underline{\varepsilon}}' \mathcal{E}_2 + \mathbf{E}_0 \underline{\underline{\varepsilon}}' \mathcal{E}_1 \right] + [j\underline{\underline{\varepsilon}}'' + j\underline{\underline{g}}] \mathcal{E}_1 \right\} \quad (6.25)$$

$$-2j k_2 \cos\theta \frac{d\mathcal{E}_2}{dx} = \mu_0 \varepsilon_0 \omega_2^2 \left\{ \underline{\underline{\varepsilon}}' \underline{\underline{r}} \left[\mathbf{E}_w^* \frac{\mathcal{E}_1^* \cdot \mathcal{E}_2}{I_0} \underline{\underline{\varepsilon}}' \mathcal{E}_1 + \mathbf{E}_0 \underline{\underline{\varepsilon}}' \mathcal{E}_2 \right] + [j\underline{\underline{\varepsilon}}'' + j\underline{\underline{g}}] \mathcal{E}_2 \right\} \quad (6.26)$$

These equations may be used to describe quite well many of the features observed when two-wave mixing is carried out in a number of the more interesting experimental arrangements.

6.4 Crystal orientation

Equations (6.25) and (6.26) describe the interaction of two waves essentially travelling in the x direction and interfering so as to produce an electric field in the z direction. The next thing to consider is how we should orient the crystallographic axes with respect to our external coordinate system. The tensorial nature of the electro-optic constant is such that a field in one particular direction may or may not produce a change in the refractive index experienced by light of a given polarization state; if there is to be coupling between the two beams it is essential that they are able to experience a modulation in the refractive index produced by the space charge field.

Two geometries are in common usage with these crystals (Marrakchi et al. 1981): the longitudinal geometry (where the field is parallel to the [001] direction) and the transverse geometry (where the field is parallel to the [110] direction). In each case the beams are incident near-normally on the (110) face of the crystal. Appendix B deals with the evaluation of the product $\varepsilon_0 \underline{\underline{\varepsilon}}' \underline{\underline{r}} \mathbf{E} \underline{\underline{\varepsilon}}'$ contained in eqns (6.25) and (6.26) for these two geometries. To summarize: for the longitudinal geometry (eqn B.20), we have

$$\varepsilon_0 \underline{\underline{\varepsilon}}' \underline{\underline{r}} \mathbf{E} \underline{\underline{\varepsilon}}' = \varepsilon_0 \varepsilon'^2 r \begin{bmatrix} E_z & 0 & 0 \\ 0 & -E_z & 0 \\ 0 & 0 & 0 \end{bmatrix} \quad (6.27)$$

while for the transverse geometry (eqn B.21b), the result is

$$\varepsilon_0 \underline{\underline{\varepsilon}}' \underline{\underline{r}} \mathbf{E} \underline{\underline{\varepsilon}}' = \varepsilon_0 \varepsilon'^2 r \begin{bmatrix} 0 & 0 & 0 \\ 0 & 0 & -E_z \\ 0 & -E_z & 0 \end{bmatrix} \quad (6.28)$$

The vast majority of experimental work involving the sillénites and inorganic semiconductors is carried out in these two geometries, though work has been reported where the sides of the crystal are aligned with the crystallographic axes (Yeh 1987b; Cheng and Yeh 1988; T. Y. Chang et al. 1988; Roy and

Singh 1990a,b), where the sides are in the directions [111], [1$\bar{2}$1], and [$\bar{1}$01] (Sugg et al. 1993), and where the crystal is cut as in the transverse geometry, but the angle between the grating and the [110] direction is varied (Strait et al. 1990).

6.5 Jones vector description

The twin assumptions that the optical beams are paraxial and that their electric fields are transverse imply that the x-component of the optical electric field is zero. In that case, the polarizations of the two waves may be described by the two-dimensional Jones vectors (Yariv and Yeh 1984b)

$$\boldsymbol{\mathcal{E}}_1 = \begin{pmatrix} \mathcal{E}_{1y} \\ \mathcal{E}_{1z} \end{pmatrix}; \quad \boldsymbol{\mathcal{E}}_2 = \begin{pmatrix} \mathcal{E}_{2y} \\ \mathcal{E}_{2z} \end{pmatrix} \tag{6.29}$$

In isotropic[5] and cubic materials, the optical activity tensor, $\underline{\underline{g}}$, takes the simple form (Landau and Lifshitz 1984)

$$\underline{\underline{g}} = \gamma \begin{bmatrix} 0 & -k_3 & k_2 \\ k_3 & 0 & -k_1 \\ -k_2 & k_1 & 0 \end{bmatrix} \tag{6.30}$$

where γ is a constant of the material, and k_i are the components of the wave vector of the relevant optical wave. In the paraxial approximation we have $k_2 = k_3 \approx 0$ and $k_1 \approx k$ and so we can write the optical activity tensor as

$$\underline{\underline{g}} = \gamma \begin{bmatrix} 0 & -k \\ k & 0 \end{bmatrix} = \begin{bmatrix} 0 & -g \\ g & 0 \end{bmatrix} \tag{6.31}$$

In the case of the longitudinal geometry, the field-induced change in dielectric constant (6.27) becomes

$$\varepsilon_0 \underline{\underline{\varepsilon}}' \underline{\underline{r}} \mathbf{E} \underline{\underline{\varepsilon}}' = \varepsilon_0 \varepsilon'^2 r \begin{bmatrix} -E_z & 0 \\ 0 & 0 \end{bmatrix} \tag{6.32}$$

while for the transverse geometry eqn (6.28) becomes

$$\varepsilon_0 \underline{\underline{\varepsilon}}' \underline{\underline{r}} \mathbf{E} \underline{\underline{\varepsilon}}' = \varepsilon_0 \varepsilon'^2 r \begin{bmatrix} 0 & -E_z \\ -E_z & 0 \end{bmatrix} \tag{6.33}$$

With the Jones vector notation then, in the longitudinal geometry eqns (6.25) and (6.26) become

$$2j k_1 \cos\theta \frac{d}{dx}\begin{pmatrix} \mathcal{E}_{1y} \\ \mathcal{E}_{1z} \end{pmatrix} = -\mu_0 \varepsilon_0 \omega_1^2 \left\{ \varepsilon'^2 r \left(\begin{bmatrix} -E_w \frac{m}{2} & 0 \\ 0 & 0 \end{bmatrix} \begin{pmatrix} \mathcal{E}_{2y} \\ \mathcal{E}_{2z} \end{pmatrix} + \begin{bmatrix} -E_0 & 0 \\ 0 & 0 \end{bmatrix} \right.\right.$$
$$\left.\left. \begin{pmatrix} \mathcal{E}_{1y} \\ \mathcal{E}_{1z} \end{pmatrix} \right) + j \left(\begin{bmatrix} \varepsilon'' & 0 \\ 0 & \varepsilon'' \end{bmatrix} + \begin{bmatrix} 0 & -g \\ g & 0 \end{bmatrix} \right) \begin{pmatrix} \mathcal{E}_{1y} \\ \mathcal{E}_{1z} \end{pmatrix} \right\} \tag{6.34}$$

---

[5]Optical activity can be present in isotropic materials only if they do not possess a centre of symmetry (Nye 1957a).

6.5 Jones vector description

$$2jk_2 \cos\theta \frac{d}{dx}\begin{pmatrix}\mathcal{E}_{2y}\\\mathcal{E}_{2z}\end{pmatrix} = -\mu_0\varepsilon_0\omega_2^2\left\{\varepsilon'^2 r\left(\begin{bmatrix}-E_w^* \frac{m^*}{2} & 0\\0 & 0\end{bmatrix}\begin{pmatrix}\mathcal{E}_{1y}\\\mathcal{E}_{1z}\end{pmatrix} + \begin{bmatrix}-E_0 & 0\\0 & 0\end{bmatrix}\right.\right.$$
$$\left.\left.\begin{pmatrix}\mathcal{E}_{2y}\\\mathcal{E}_{2z}\end{pmatrix}\right) + j\left(\begin{bmatrix}\varepsilon' & 0\\0 & \varepsilon'\end{bmatrix} + \begin{bmatrix}0 & -g\\g & 0\end{bmatrix}\right)\begin{pmatrix}\mathcal{E}_{2y}\\\mathcal{E}_{2z}\end{pmatrix}\right\}$$
(6.35)

which may be written more compactly as

$$\frac{d}{dx}\begin{pmatrix}\mathcal{E}_{1y}\\\mathcal{E}_{1z}\end{pmatrix} = j\kappa\begin{bmatrix}-E_w \frac{m}{2} & 0\\0 & 0\end{bmatrix}\begin{pmatrix}\mathcal{E}_{2y}\\\mathcal{E}_{2z}\end{pmatrix} + \left\{j\kappa\begin{bmatrix}-E_0 & 0\\0 & 0\end{bmatrix} - \begin{bmatrix}\frac{\alpha}{2} & -\rho\\\rho & \frac{\alpha}{2}\end{bmatrix}\right\}\begin{pmatrix}\mathcal{E}_{1y}\\\mathcal{E}_{1z}\end{pmatrix}$$
(6.36)

$$\frac{d}{dx}\begin{pmatrix}\mathcal{E}_{2y}\\\mathcal{E}_{2z}\end{pmatrix} = j\kappa\begin{bmatrix}-E_w^* \frac{m^*}{2} & 0\\0 & 0\end{bmatrix}\begin{pmatrix}\mathcal{E}_{1y}\\\mathcal{E}_{1z}\end{pmatrix} + \left\{j\kappa\begin{bmatrix}-E_0 & 0\\0 & 0\end{bmatrix} - \begin{bmatrix}\frac{\alpha}{2} & -\rho\\\rho & \frac{\alpha}{2}\end{bmatrix}\right\}\begin{pmatrix}\mathcal{E}_{2y}\\\mathcal{E}_{2z}\end{pmatrix}$$
(6.37)

where κ is the coupling constant introduced in eqn (3.75) and given by

$$\kappa = \frac{\mu_0\varepsilon_0\omega_1^2(\varepsilon')^2 r}{2k_1\cos\theta} \approx \frac{\mu_0\varepsilon_0\omega_2^2(\varepsilon')^2 r}{2k_2\cos\theta} = \frac{k_0(\varepsilon')^{3/2} r}{2\cos\theta} \quad (6.38)$$

ρ is the rotatory power of the crystal given by

$$\rho = \frac{\mu_0\varepsilon_0\omega_1^2 g}{2} \approx \frac{\mu_0\varepsilon_0\omega_2^2 g}{2} \quad (6.39)$$

and α is the intensity absorption constant given in eqn (6.11). Note that if optical activity and absorption may be neglected, and the two interacting waves are polarized in the y direction, then eqns (6.36) and (6.37) reduce to the simple form

$$\frac{d\mathcal{E}_{1y}}{dx} = -j\kappa\left(E_w\frac{I_2}{I_0} + E_0\right)\mathcal{E}_{1y} \quad (6.40)$$

$$\frac{d\mathcal{E}_{2y}}{dx} = -j\kappa\left(E_w^*\frac{I_1}{I_0} + E_0\right)\mathcal{E}_{2y} \quad (6.41)$$

Here, the effect of the external field is merely to introduce a phase shift in both waves proportional to the applied field and to the distance travelled within the crystal. A transformation of the form

$$A_i = \mathcal{E}_{iy}e^{-j\kappa E_0 x}, \quad i = 1, 2 \quad (6.42)$$

reduces eqns (6.40) and (6.41) to

$$\frac{dA_1}{dx} = -j\kappa E_w \frac{I_2}{I_0} A_1 \quad (6.43)$$

$$\frac{dA_2}{dx} = -j\kappa E_w^* \frac{I_1}{I_0} A_2 \tag{6.44}$$

which are of the same form as eqns (3.31) and (3.32) describing two-wave mixing in the scalar approximation when

$$\kappa E_w = -\frac{\Gamma}{2} e^{j\Phi} \tag{6.45}$$

In the absence of any external field, E_0, eqns (6.40) and (6.41) reduce directly to eqns (3.31) and (3.32).

In our discussion of two-wave mixing in Chapter 3 we assumed that the medium was lossless. This is a useful simplification, as the results that we obtained in the absence of absorption still contain the essential physics of the photorefractive process and provide a good qualitative description of beam coupling. For more quantitative calculations though, it is often necessary to account for absorption, which is often quite high, for example $1.1\,\text{cm}^{-1}$ at 514 nm in BSO (Webb and Solymar 1991a)[6]. The presence of such appreciable losses means that the photorefractive gain must be quite high in order simply to overcome losses. In BSO or GaAs, for example, without any field enhancement it is virtually impossible to exceed the losses.

From the viewpoint of someone who is interested in increasing the power of a signal beam, the absorption of the material can be taken into account simply by subtracting the attenuation from the photorefractive gain to obtain the net gain experienced by a signal beam passing through a photorefractive crystal. On the theoretical side, in many cases absorption may be included quite readily. Using eqn (6.45), when absorption is included, eqns (6.43) and (6.44) take the form

$$\frac{dA_1}{dx} = j\frac{\Gamma}{2} e^{j\Phi} \frac{I_2}{I_0} A_1 - \frac{\alpha}{2} A_1 \tag{6.46}$$

$$\frac{dA_2}{dx} = j\frac{\Gamma}{2} e^{-j\Phi} \frac{I_1}{I_0} A_2 - \frac{\alpha}{2} A_2 \tag{6.47}$$

Following the same procedure as in Section 3.4, these may be easily integrated to give

$$I_1(x) = \frac{I_1(0) e^{-(\Gamma_B + \alpha)x}}{1 + I_1(0)\left(\dfrac{\Gamma_B}{I_0(\Gamma_B + \alpha)}\right)(e^{-(\Gamma_B + \alpha)x} - 1)} \tag{6.48}$$

---

[6] Absorption must in fact always be present in a photorefractive material since the absorption of photons is responsible for the generation of the mobile charge carriers that produce the space charge field.

$$I_2(x) = \frac{I_2(0)e^{(\Gamma_B-\alpha)x}}{1 + I_2(0)\left(\frac{\Gamma_B}{I_0(\Gamma_B-\alpha)}\right)(e^{(\Gamma_B-\alpha)x} - 1)} \qquad (6.49)$$

where

$$\Gamma_B = \Gamma \sin \Phi \qquad (6.50)$$

Care must be exercised in deciding when these equations are valid; in particular, they are not valid when the space charge field is enhanced using the moving fringe technique. In that case, the normalized space charge field, E_w, depends on the total intensity at a given location and will therefore vary in the x direction; Γ_B cannot then be treated as a constant for the purposes of integration.

6.6 Degenerate two-wave mixing in the diffusion regime

We shall now see how well the theory presented above is able to explain the results of experiments. We shall choose the simplest theoretical formulation that is able to provide a reasonable agreement with the data, in order to try to convey an understanding of the physical processes involved.

As we have seen in Section 2.3, the materials equations are particularly simple when the interfering beams have the same frequency and there is no externally applied electric field. In this situation, the space charge field is shifted by $\pi/2$ from the optical interference pattern – which is optimum for beam coupling – and the amplitude of the space charge field, obtained from eqn (2.58), is simply

$$E_w = j\frac{E_D E_q}{E_D + E_q} \qquad (6.51)$$

Figure 6.2 shows experimental and theoretical plots of the gain, γ_0 (defined as the output signal beam intensity in the presence of the pump beam divided by the output signal beam intensity when the pump is absent[7]), as a function of the magnitude of the grating K-vector. The experiments were carried out on a 2 mm thick crystal of BSO with light from an Ar^+ laser at 514 nm. The crystal was oriented in the longitudinal geometry with the grating K-vector in the [001] direction and both beams were polarized in the [110] direction and incident on the ($\bar{1}10$) face. For the theoretical curves, eqns (3.42), (3.43), and (6.51) were used along with a single set of material parameters typical for BSO. The electro-optic coefficient was used as the fitting parameter and the value obtained was 2.75 pm V^{-1}, which is somewhat lower than the value determined by other

---

[7]When making measurements of gain in experiments, one often has to guard against (or subtract out) light scattered from the pump beam on to the detector measuring the signal output. Also note that for an ideal lossless material, the definition of gain given above agrees with the simpler one, which states that the signal gain is equal to the output signal power with the pump present divided by the input signal power.

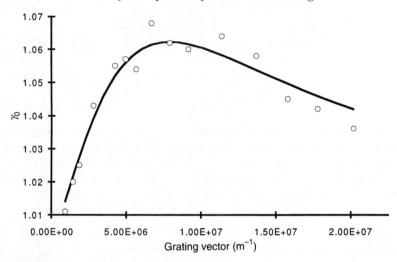

Fig. 6.2 Experimental (circles) and theoretical (solid line) plot of gain as a function of grating vector for a 2 mm thick BSO crystal. The experimental data cover a range of inter-beam angles (marked θ in Fig. 6.1) from 2 to 55° (in air).

means[8]. We shall have more to say on this discrepancy later. Other than that one point, the gain process may be seen to be well described by the theory.

Unfortunately, as is clear from Fig. 6.2, the gain obtained in the diffusion regime with BSO is rather small. Much higher gains may be obtained by using an applied field and either detuning the incident beams (Huignard and Marrakchi 1981a) or modulating the applied field (Stepanov and Petrov 1985) in order to enhance the space charge field. Higher gains are also obtainable using other materials with much higher electro-optic coefficients, such as $Sr_{1-x}Ba_xNb_2O_6$ (SBN), $(K_{1-y}Na_y)_{2A-2}(Sr_{1-x}Ba_x)_{2-A}Nb_2O_6$ (KNSBN) (Ramazza and Zhao 1993), and $BaTiO_3$.

Figure 6.3 shows comparisons between theory and experiment performed by Rak *et al.* (1984) on $BaTiO_3$. Figure 6.3(a) shows the variation of gain with input beam ratio at 514 nm and with a fringe spacing of 1.5 μm. For values of β below about 10^{-4} the gain saturates at nearly 500, corresponding to a value of Γ_B of around 20 cm^{-1}; in this region the undepleted pump approximation becomes valid. Figure 6.3(b) shows the variation in gain with fringe spacing; in both figures the agreement between theory and experiment is reasonable.

Detailed measurements on $BaTiO_3$ and comparison with theory have been carried out by Fainman *et al.* (1986). Figure 6.4(a) shows the geometrical

---

[8] Various values have been reported in the literature: 4.25 ± 0.13 pm V^{-1} (Bayvel *et al.* 1988) at 633 nm, 3.6 ± 0.2 pm V^{-1} (Pellat-Finet 1984), 5.0 ± 0.2 pm V^{-1} (Vachss and Hesselink 1987b) at 514 nm, and 4.1 ± 0.1 pm V^{-1} (Henry *et al.* 1986) at 500 nm. See Appendix C for details of how this parameter may be experimentally determined.

6.6 Degenerate two-wave mixing in the diffusion regime

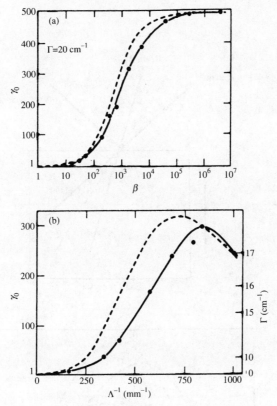

Fig. 6.3 Experimental and theoretical results for two-wave mixing in BaTiO$_3$ (after Rak *et al.* 1984): (a) gain vs. input beam ratio; (b) gain vs. grating spacing.

configuration for their measurements, while Fig. 6.4(b) shows plots of two-wave mixing gain as a function of incident angle, for a range of inter-beam angles. Experimental points are limited to a range of values close to $\beta = 90°$, but here the agreement with theory is good.

With no applied field, the gains obtainable with photorefractive semiconductors tend to be rather small (Fabre *et al.* 1989; Delaye *et al.* 1994). However, in the diffusion regime these materials are interesting because they are able to exhibit pure cross-polarization coupling, where a linearly polarized input beam will be diffracted as the orthogonal polarization state. This arises due to the off-diagonal terms in the expression for the relevant components of the permittivity tensor, eqn (6.28). Pure cross-polarization coupling is not generally possible in naturally birefringent materials, such as LiNbO$_3$ (linear birefringence) or BSO (circular birefringence) (Yeh 1987b; Cheng and Yeh 1988; T. Y. Chang *et al.* 1988; Yeh 1988; Roy and Singh 1990a,b).

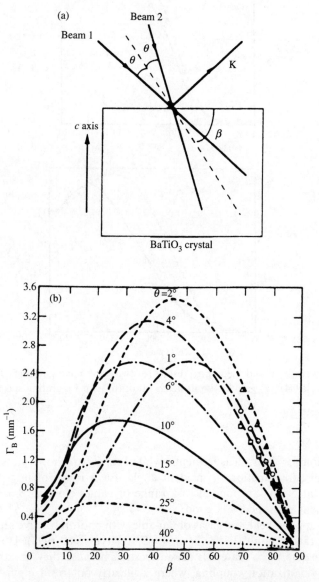

Fig. 6.4 (a) Geometrical configuration for two-wave mixing in BaTiO$_3$. (b) Theoretical gain coefficient (curves) against incident angle, β, for a range of inter-beam angles for BaTiO$_3$. Experimental points are given for inter-beam angles of 1° (○), 2° (△) and 4° (□).

6.7 Space charge field enhancement

As described in Section 2.8 and in Section 5.4, two techniques for the enhancement of the space charge field are in common usage: using an externally applied DC electric field along with moving fringes, and using an applied AC field. Turning our attention first of all to the case of moving fringes: in Section 5.4.2 we derived an expression for the optimum detuning frequency (eqn 5.73), which may be expressed in the following form:

$$\Omega_{opt} = \frac{eI_0 s N_D}{\varepsilon_s K E_0} \qquad (6.52)$$

Clearly, Ω_{opt} ought to be proportional to I_0 and the optimum fringe velocity ($V_{opt} = \Omega_{opt}/K$) ought to be proportional to $1/E_0$. Following on from their earlier experimental work in the field (Refregier et al. 1984), this was experimentally demonstrated by Refregier et al. (1985), their results being reproduced here in Fig. 6.5. They presented further experimental results, showing the variation of optimum gain as a function of grating spacing at different applied fields and with a high beam ratio (10^5). Two crystals were used with lengths 1.27 mm and 10 mm. These data were well matched by theoretical curves using just one set of material parameters (s, N_D, μ, γ_r, and N_A), with the exception that in order to obtain good agreement, a different value for the electro-optic coefficient was required for each crystal: namely 0.95 pm V^{-1} for the short crystal and 0.6 pm V^{-1} for the long. Both of these values are again significantly less than that measured by other means[9].

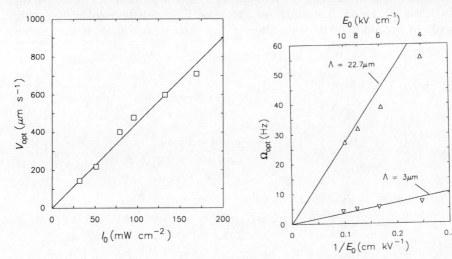

Fig. 6.5 Experimental and theoretical plots of detuning frequency providing optimum gain as a function of I_o and $1/E_o$ (after Refregier et al. 1985).

[9] See Kawata et al. (1992) for recent experimental studies of gain under field enhancement by moving fringes.

It may be remembered that the linearization of the materials equations that was carried out in Section 2.8, in order to obtain the space charge field amplitude, is based on the assumption that the modulation of all the variables is small. A key feature of the later Refregier paper was the realization by the authors that this linearization process was invalid for values of the intensity modulation that are still surprisingly small – particularly when a moving grating is used to enhance the space charge field (this was discussed in Section 2.9). For example, Fig. 6.6. shows their experimental data for the variation of the exponential gain, Γ_B, calculated using eqns. (3.42) and (3.43) as a function of input beam ratio. For a grating spacing of 23 μm (close to optimum), even a beam ratio of 1:1000 was not large enough to obtain the small modulation (i.e. $\beta \to \infty$) value of E_w.

The authors attempted to deal with this problem using a second perturbation approach to the linearization process. The essence of this treatment was to include the second spatial harmonic of all the variables, which were then assumed to have the form

$$X(x,t) = X_0(t) + \tfrac{1}{2}\{X_1(x,t)e^{j(\Omega t - Kx)} + X_2(x,t)e^{2j(\Omega t - Kx)} + \text{c.c.}\} \qquad (6.53)$$

(compare this with eqn 2.12). Having substituted this form into the basic differential equations of Section 2.1, three sets of algebraic equations were obtained by equating terms with coefficients of unity, $\exp(j[\Omega t - Kx])$ and $\exp(2j[\Omega t - Kx])$. From these equations, the amplitude of the second spatial harmonic components could be found in terms of the amplitudes of the fundamental components and this then led to an expression for the second perturbation correction which needs to be added to the amplitude of the space charge field. Unfortunately,

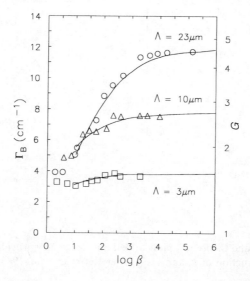

Fig. 6.6 Exponential gain, Γ_B, as a function of input beam ratio (after Refregier *et al.* 1985).

6.7 Space charge field enhancement

the resulting expression was very complicated, the only clear factor emerging being that the correction term was negative and proportional to m^3, implying a sublinear dependence of the space charge field amplitude on the fringe modulation (which is in accordance with the experimental results in Fig. 6.6). Even more unfortunately, the authors estimated that this second perturbation was in any case only valid for $m < 0.03$, and thus invalid for a wide range of beam ratios of interest for many photorefractive applications.

With a third perturbation looking too complex and numerical solution of the material equations appearing to be too time-consuming, the authors were forced to take the phenomenological approach described in Section 2.9 in which the linear dependence of the space charge field amplitude on m was replaced with eqn (2.121). In Fig. 6.6 are also shown theoretical curves which assume this functional form and where the parameter a takes the values 2.8, 1.0, and 0.5 for grating spacings of 23 μm, 10 μm, and 3 μm, respectively.

Space charge field enhancement using detuning plus an applied constant field has also been used to improve the gain obtainable in two-wave mixing with semiconductors. In chromium-doped GaAs at 1.06 μm, without appropriate space charge field enhancement the gain is often insufficient to overcome losses within the material (M. B. Klein 1984; Albanese et al. 1986). A modest increase in photorefractive gain in GaAs has been reported due to the application of an r.f. AC field (Kumar et al 1987a). The use of an applied DC field along with frequency detuning has also been carried out with GaAs (Kumar et al. 1987b) and gains of 6-7 cm^{-1} have been reported (Imbert et al. 1988). Figure 6.7 shows experimental and theoretical plots of Γ_B as a function of fringe velocity and applied voltage across the 5 mm thick crystal at a grating spacing of 18 μm, which provided the largest gains. The theoretical plots were made using eqns (5.69) and (5.70) for the space charge field amplitude. Parameter values typical for GaAs:Cr were used; however, in order to get a reasonable fit to Fig. 6.7(b) it was necessary to assume that the effective electric field within the crystal was of the form

$$E_0 = \frac{V_0}{d_{\text{eff}}} \tag{6.54}$$

where

$$d_{\text{eff}} = CV_0^{1/2} \tag{6.55}$$

and C is a constant. This form was explained as arising due to the formation of a blocking Schottky barrier at the electrode–crystal interface (Sze 1985); indeed it was observed that there was a region of high gain within the crystal whose width varied as $V_0^{1/2}$.

Space charge field enhancement using an applied field and detuned beams has also been applied to GaP at 633 nm (Ma et al. 1991b). Values of Γ_B up to 2.5 cm^{-1} were obtained with an input beam ratio of 1:100, representing an increase of more than eightfold over results reported for a similar crystal with no applied field (Kuroda et al. 1990; Itoh et al. 1990). The data was well described by the theory of Refregier et al. (1985), though the authors had to assume an effective electric

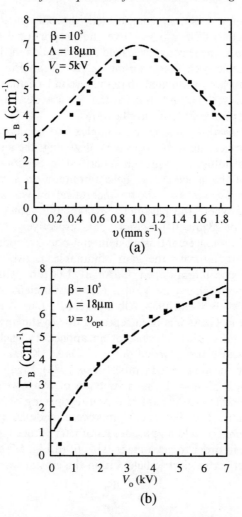

Fig. 6.7 Two-wave mixing gain in GaAs:Cr. Experimental (■) and theoretical (- - -) plots of Γ_B as a function of (a) fringe velocity and (b) applied voltage (after Imbert et al. 1988). $\beta = 10^3$, $\Lambda = 18\,\mu\text{m}$, (a) $V_o = 5\,\text{kV}$, (b) $v = v_{\text{opt}}$.

field within the crystal that was reduced by a factor of 0.42 from the field calculated from the applied voltage and crystal dimensions.

Detuning of the incident beams is not usually carried out in BaTiO$_3$ or SBN. In the optimum geometry, the relevant electro-optic coefficient is so large that high gains may be obtained in the diffusion regime[10]. We have shown above that in this

---

[10]Conversely, detuning has been used to reduce the coupling in SBN in order to prevent pump beam depletion and thereby obtain a large grating amplitude throughout a crystal, which is beneficial for storage applications (Ma et al. 1991a).

6.7 Space charge field enhancement

regime the phase angle Φ may be taken to be $\pi/2$, which is optimum for power transfer. There is evidence though that this is not always strictly true for BaTiO$_3$ (McMichael and Yeh 1987). Figure 6.8 shows a plot of gain as a function of detuning frequency for crystals of SBN and BaTiO$_3$. No field was applied. If the phase angle is $\pi/2$ when no detuning is applied, these curves would all be expected to be symmetric about zero, falling off similarly with positive and negative detunings. Whilst this was the case for the SBN crystal (and three other samples of that material not shown in this figure), there is evidence of asymmetry in the BaTiO$_3$ data, indicating a phase angle slightly different from $\pi/2$.

The use of an AC square wave field to enhance the space charge field and thereby increase the gain was originally proposed by Stepanov and Petrov (1985). Both square wave and sinusoidal waveforms have been used with BSO (Gan et al. 1988) and the use of sinusoidal fields with GaAs has been reported (Kumar et al. 1987b; K. Walsh and Hall 1988). It has been shown that the AC field enhancement technique can lead to a temporal modulation in the intensity, phase, and polarization of the output beams (Pauliat et al. 1989; Stace et al. 1989).

As mentioned in Section 5.4.6, the optimum gain obtainable with enhancement provided by an external AC field or by a DC field plus detuning ought to be the same. Experimental results suggest otherwise: Fig. 6.9 shows the optimum gain as a function of applied field amplitude for the two techniques for a crystal of BSO and two crystals of BTO (Sochava et al. 1993b). It was suggested that the radically different gains obtained with the BTO sample used in Fig. 6.9(c) was related to the short drift length of electrons in that sample.

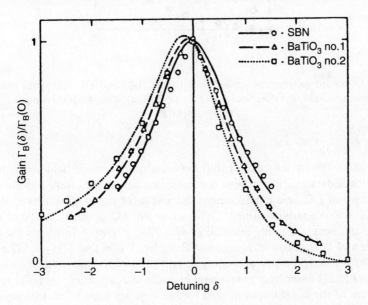

Fig. 6.8 Measurements of two-wave mixing gain as a function of detuning for BaTiO$_3$ and SBN crystals (after McMichael and Yeh 1987).

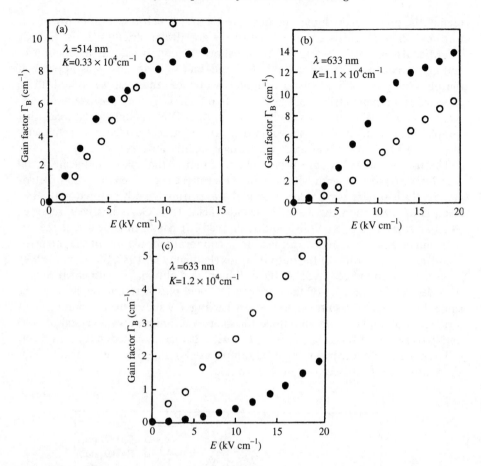

Fig. 6.9 Optimum gain versus applied field for AC (●) and DC (○) space charge field enhancement techniques (after Sochava *et al.* 1993b): (a) BSO; (b) BTO sample 1; (c) BTO sample 2.

Non-linearities in the relationship between space charge field amplitude and fringe modulation, m, for large m (i.e. violation of eqn 6.5) were reported earlier for the case of DC field enhancement. Similar large signal effects have also been observed with vanadium-doped CdTe when an AC applied field was used to enhance the beam coupling (Belaud *et al.* 1994). Figure 6.10 shows the gain as a function of beam ratio at source wavelengths of 1.06 and 1.55 μm. The dashed line is the theoretical curve according to eqn (6.5) whilst the solid curve incorporates eqn (2.121) describing phenomenologically the non-linear material response. The values of the constant, a, needed for the best fits were 2.7 at 1.06 μm and 2.2 at 1.55 μm. Other workers have also studied CdTe under an AC applied field (Ziari *et al.* 1992).

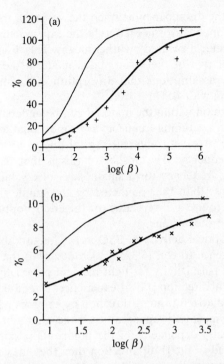

Fig. 6.10 Two-wave mixing in CdTe. Gain as a function of beam ratio at $\lambda =$ (a) 1.06 μm and (b) 1.55 μm. Crosses are experimental points. Upper (thin) curve corresponds to simple theory (space charge field proportional to fringe modulation), lower (thick) curve corresponds to large signal theory (after Belaud et al. 1994). $\Lambda =$ (a) 15 μm, (b) 11 μm, $f =$ (a) 95 Hz, (b) 4 Hz, $E_o = 10\,\text{kV}\,\text{cm}^{-1}$, $I_{\text{inc}} = 3\,\text{mW}\,\text{cm}^{-2}$.

6.8 Limitations of the simple theory

It is clear from the results reported so far that the simple theoretical formulations of the materials equations (presented in Chapter 2) and the field equations (6.43) and (6.44) provide a reasonably good description of the two-wave mixing process. However, they are seriously deficient in a number of areas. We have just seen that the linearization of the materials equations is only valid for quite small values of the modulation when the space charge field is enhanced by the use of an applied electric field and detuning of the beams. This may be acceptable if we are dealing with the amplification of very weak image-bearing beams (Kukhtarev 1979b; Tschudi et al. 1986; Rajbenbach et al. 1991), for example; but there are many applications where a high diffraction efficiency is required and the beam ratio may be close to unity (e.g. holographic interferometry (Huignard and Herriau 1977) or image correlation (White and Yariv 1980)). In these latter situations, an accurate description of the grating amplitude is only possible by resorting to numerical techniques or adopting a phenomenological approach.

Returning now to the discrepancy between the values of the appropriate component of the electro-optic tensor needed to fit the experimental curves of Fig. 6.2 and Fig. 6.6 and the value measured by other means. This discrepancy arises from the limited validity of two of the assumptions made in the derivation of eqns (6.43) and (6.44): the assumptions that absorption and optical activity can be neglected when dealing with BSO.

The effect of absorption within the material can be understood by reference to eqn (6.52); the optimum detuning frequency is proportional to the total intensity, so as the intensity decreases through the crystal due to absorption, so will the optimum detuning frequency. The upshot of this is that it is only possible to optimize the detuning frequency for a small 'slice' of crystal, so in practice the gain will always be less than that predicted by the simple theory. This in part accounts for the need to use a lower value for the electro-optic coefficient for the longer crystal than the shorter one.

The problem with optical activity in BSO in the longitudinal geometry can be understood with reference to eqn (6.32), which describes the space charge field-induced change in the permittivity for light incident paraxially on the ($\bar{1}10$) face. Light polarized in the y direction [110][11] experiences a modulation in the permittivity, whereas light polarized in the z direction experiences no permittivity modulation. It is now immediately apparent that the gain obtained in two-wave mixing must be critically dependent on the polarization state of the beams. If the beams are polarized in the [110] direction then the index modulation experienced by the two beams will be maximized. Conversely, if the two beams are polarized in the direction [001], they experience no index modulation and therefore there can be no beam coupling.

Because of the presence of appreciable optical activity in BSO[12], if the beams are initially polarized in the direction of maximum coupling, after passing through just a few millimetres of the crystal the direction of polarization will have rotated by 90° and there will be no beam coupling. This effect also contributes in part to the smaller electro-optic coefficient needed by Refregier *et al.* (1985) to fit the data from their thicker crystal compared with that for the thinner crystal.

In general it is necessary to use numerical techniques in order to get agreement between experiment and theory with a given sample over a wide range of experimental conditions; however, here we present a simple treatment based on that of Foote and Hall (1986) that allows eqns (6.36) and (6.37) to be used in conjunction with an optically active medium, so long as the gain provided by the crystal is small and the two beams are incident linearly polarized in the same direction. These conditions are often well satisfied when the sillenites are used in the diffusion regime. Figure 6.11(a) shows a schematic diagram of the electric field vectors of the pump and signal beams (E_1 and E_2 respectively) at a given position within

---

[11] In experiments using the longitudinal geometry, this direction is often somewhat loosely referred to as being the vertical direction, since the two beams are usually incident in the horizontal plane of the optical bench. Such a geometry gives rise to a grating vector also in the horizontal plane and hence the [110] direction corresponds to the vertical.

[12] 39° mm^{-1} at 514 nm (Webb and Solymar 1991a).

6.8 Limitations of the simple theory

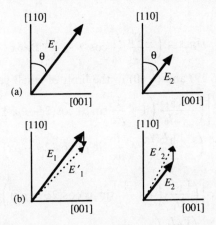

Fig. 6.11 Wave amplitudes (a) before and (b) after traversing a distance δx in a crystal of the sillénite family.

the crystal. Both beams are assumed to be polarized at an angle θ to the [110] direction, where

$$\theta = \phi - \rho x \qquad (6.56)$$

and where ϕ is the polarization angle at which the beams are incident on the crystal.

Figure 6.11(b) shows the effect on the two beams after they have traversed a further small distance δx of the crystal. According to eqns (6.36) and (6.37) the field coupling is determined by the y component (in the [110] direction) of the field, hence the change in the two field amplitudes may be written as

$$\begin{pmatrix}\delta\mathcal{E}_{1y}\\ \delta\mathcal{E}_{1x}\end{pmatrix} = \begin{pmatrix}\left[j\dfrac{\Gamma}{2}e^{j\Phi}\dfrac{I_2}{I_0}-\dfrac{\alpha}{2}\right]\cos\theta \\ -\dfrac{\alpha}{2}\sin\theta\end{pmatrix}|\mathcal{E}_1|\delta x \qquad (6.57)$$

and

$$\begin{pmatrix}\delta\mathcal{E}_{2y}\\ \delta\mathcal{E}_{2x}\end{pmatrix} = \begin{pmatrix}\left[j\dfrac{\Gamma}{2}e^{-j\Phi}\dfrac{I_1}{I_0}-\dfrac{\alpha}{2}\right]\cos\theta \\ -\dfrac{\alpha}{2}\sin\theta\end{pmatrix}|\mathcal{E}_2|\delta x \qquad (6.58)$$

Strictly speaking, coupling of this form will cause the two beams to subtend different angles, θ, with the [110] direction. We can disregard this though if the coupling is small; in that case the change in the magnitude of the two field vectors is given by

$$\delta|\mathcal{E}_1| = \left(\frac{\Gamma_B}{2}\frac{I_2}{I_0}\cos^2\theta - \frac{\alpha}{2}\right)|\mathcal{E}_1|\delta x \qquad (6.59)$$

and

$$\delta|\mathcal{E}_2| = \left(-\frac{\Gamma_B}{2}\frac{I_1}{I_0}\cos^2\theta - \frac{\alpha}{2}\right)|\mathcal{E}_2|\delta x \tag{6.60}$$

Integration of eqns (6.59) and (6.60) in the limit of small gain leads to

$$\begin{aligned}I_1(x) &= I_1(0)\exp\left[\left(-\frac{\Gamma_B I_2}{2I_0}\left(1 + \frac{1}{\rho x}\sin\rho x\cos(2\phi - \rho x)\right) - \alpha\right)x\right] \\ &\approx I_1(0)\left[1 + \left(-\frac{\Gamma_B I_2}{2I_0}\left(1 + \frac{1}{\rho x}\sin\rho x\cos(2\phi - \rho x)\right) - \alpha\right)x\right]\end{aligned} \tag{6.61}$$

$$\begin{aligned}I_2(x) &= I_2(0)\exp\left[\left(\frac{\Gamma_B I_1}{2I_0}\left(1 + \frac{1}{\rho x}\sin\rho x\cos(2\phi - \rho x)\right) - \alpha\right)x\right] \\ &\approx I_2(0)\left[1 + \left(\frac{\Gamma_B I_1}{2I_0}\left(1 + \frac{1}{\rho x}\sin\rho x\cos(2\phi - \rho x)\right) - \alpha\right)x\right]\end{aligned} \tag{6.62}$$

A similar treatment for the transverse geometry leads to the following expressions:

$$\begin{aligned}I_1(x) &= I_1(0)\exp\left[\left(-\frac{\Gamma_B I_2}{2I_0}\left(\frac{2}{\rho x}\sin\rho x\sin(2\phi - \rho x)\right) - \alpha\right)x\right] \\ &\approx I_1(0)\left[1 + \left(-\frac{\Gamma_B I_2}{2I_0}\left(\frac{2}{\rho x}\sin\rho x\sin(2\phi - \rho x)\right) - \alpha\right)x\right]\end{aligned} \tag{6.63}$$

$$\begin{aligned}I_2(x) &= I_2(0)\exp\left[\left(\frac{\Gamma_B I_1}{2I_0}\left(\frac{2}{\rho x}\sin\rho x\sin(2\phi - \rho x)\right) - \alpha\right)x\right] \\ &\approx I_2(0)\left[1 + \left(\frac{\Gamma_B I_1}{2I_0}\left(\frac{2}{\rho x}\sin\rho x\sin(2\phi - \rho x)\right) - \alpha\right)x\right]\end{aligned} \tag{6.64}$$

where for the transverse geometry, ϕ is a measure of the incident polarization angle with respect to the [001] direction.

Figures 6.12 and 6.13 show comparisons between this theory and experimental results obtained with a 2 mm thick crystal of BSO in the diffusion regime (Webb *et al.* 1994). For all the theoretical curves the value 4.5×10^{-12} was used for the electro-optic coefficient. Expanding eqn (6.51) for the normalized space charge field in the diffusion regime gives

$$E_w = -j\frac{E_D E_q}{E_D + E_q} = \frac{\left(\frac{k_B K T}{e}\right)}{1 + \left(\frac{k_B T \varepsilon_s K^2}{e^2 N_A^-}\right)} \tag{6.65}$$

from which it is apparent that the only unknown microscopic parameter is the trap density, N_A^-. This parameter was used to fit the theoretical curve in Fig. 6.12(a) which shows the dependence of gain on the input polarization angle of the beams in the longitudinal geometry. The value returned was $3.5 \times 10^{21}\,\text{m}^{-3}$, which is close to values measured in other crystals. Using the same value, Fig.

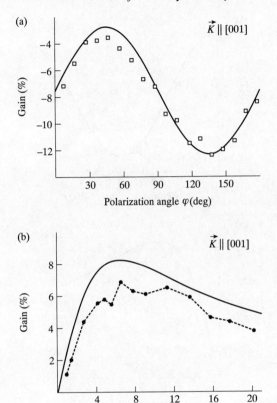

Fig. 6.12 Two-wave mixing in the longitudinal geometry in BSO. (a) Gain as a function of polarization angle. Inter-beam angle = 23°. (b) Gain as a function of grating vector. $\phi = 0°$ (after Webb *et al.* 1994).

6.12(b) shows the gain as a function of grating vector for $\phi = 0$. Figure 6.13 shows similar results for the same crystal in the transverse geometry with the same value for the trap density.

It may be seen that the fit between the sets of results is everywhere quite good with the exception of Fig. 6.12(b), where the experimental results are all consistently lower than the theory. Considering the simple nature of the model (one carrier, one type of donor, small angle approximation) the agreement between theory and experiment is encouraging.

For the class 43m semiconductors (GaAs, InP, CdTe), eqns (6.61) and (6.63) are simplified somewhat, since the lack of optical activity in such materials implies

$$\rho = 0 \tag{6.66}$$

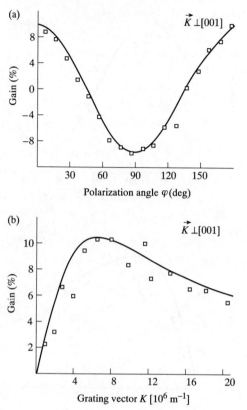

Fig. 6.13 Two-wave mixing in the transverse geometry in BSO. (a), (b) as Fig. 6.12.

and therefore

$$\frac{\sin \rho x}{\rho x} = 1 \tag{6.67}$$

Figure 6.14 shows experimental plots of Γ_B as a function of incident polarization angle for $Cd_{0.9}Mn_{0.1}Te$ in the transverse geometry (Rana et al. 1994) using 799 nm light. Here the angle α_0 is measured relative to the grating direction, and therefore

$$\alpha_0 = \phi - \frac{\pi}{2} \tag{6.68}$$

Maximum gain is predicted and observed at $\alpha_0 = 45°$, when the polarizations are aligned with the crystal eigenaxes. Note the slight asymmetry about the gain of 0.0; this was attributed to the presence of an absorption grating. This asymmetry was also reported in similar work on GaAs (K. Walsh et al. 1987).

When field enhancement is used, numerical methods are required. As an example of what can be done, Fig. 6.15 shows a set of results from a two-wave mixing

6.8 Limitations of the simple theory

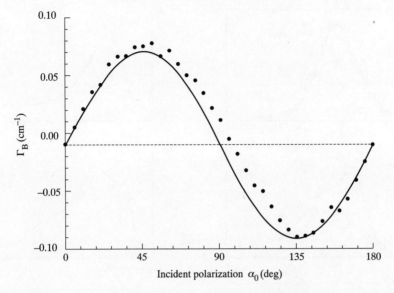

Fig. 6.14 Two-wave mixing in the transverse geometry in CdMnTe. Gain as a function of input polarization angle. Grating spacing = 4 μm, $\lambda = 799$ nm (after Rana *et al.* 1994).

experiment (Webb and Solymar 1991a) with a 1 cm thick crystal of BSO in the longitudinal geometry, alongside theoretical curves obtained by solving eqns (6.36) and (6.37) using the fourth-order Runge–Kutta technique (Press *et al.* 1992). The experimental detuning curves show the variation of gain, G, as a function of detuning frequency at several different wavelengths obtained from an Ar^+ laser. It is clear, particularly at the shorter wavelengths, that the detuning curve contains more structure than the simple resonance peak that would be expected from a variation of the space charge field as indicated in Fig. 2.10.

The explanation of the structure is the following. We have already seen that due to the optical activity of BSO, in this geometry there are positions in the x direction where there is little or no beam coupling (polarization nearly horizontal) and other positions where the coupling is maximized (near vertical polarization states). Furthermore, the presence of absorption means that the space charge field is only optimally enhanced by the frequency detuning at one point along the x axis. Now, when the detuning frequency is changed, this point will move along the x axis, and when the point coincides with a position of maximal coupling, the total gain will be high. Conversely, when the position of optimum detuning coincides with a position of near zero coupling, the total gain is liable to be low. Therefore, as the detuning is varied, we should expect to see a series of maxima and minima in the detuning curve; and this is just what is revealed in the experimental data – particularly for the shorter wavelengths where the absorption is high and so this phenomenon is most pronounced.

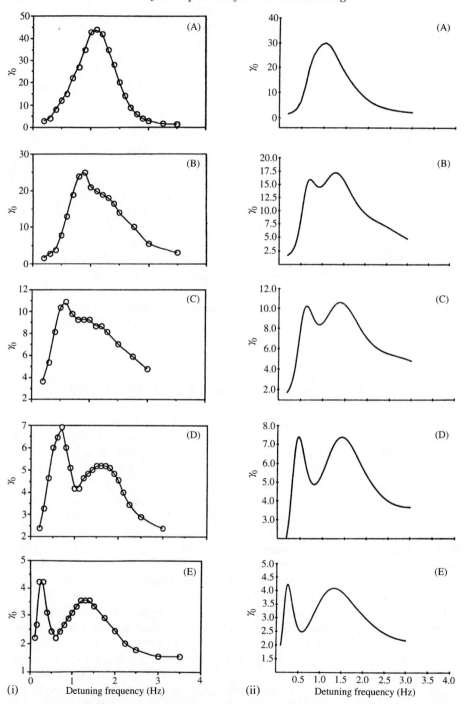

Fig. 6.15 (i) Experimental and (ii) theoretical detuning curves for a 1 cm thick crystal of BSO in the longitudinal geometry. $\lambda =$ (A) 514 nm, (B) 502 nm, (C) 497 nm, (D) 488 nm, (E) 477 nm (after Webb and Solymar 1991a).

6.9 Diffraction efficiency

To obtain the theoretical curves, one set of material parameters was used with values close to those typical for BSO, with the exception that the photoionization constant was varied as a function of wavelength to match the variation in absorption. One discrepancy, though, concerns the electric field. A voltage of 8 kV was applied across the 10 mm wide crystal, which therefore ought to imply a field of 8 $kV\,cm^{-1}$ within the crystal. To obtain the best agreement with the experimental data, it was necessary to use a value of 4.75 $kVcm^{-1}$ for the theoretical plots. It was suggested that this significant difference was possibly due to a potential drop occurring near the crystal electrodes, producing a much smaller field than expected in the body of the crystal. This point is discussed in more detail in Appendix C.

A similar numerical approach has also been taken to the study of gain in BSO when the AC field enhancement technique is used (Krainak and Davidson 1989).

6.9 Diffraction efficiency

So far, we have concerned ourselves with the case of dynamic holography where the optical beams and space charge field affect each other. A simpler situation occurs when we want to investigate the diffraction of incident beams from a fixed grating as was done early in Chapter 3 (see Section 3.7). This is a problem which is of some relevance to the study of photorefractive materials, particularly in applications such as correlation where we are often interested in the diffraction of one beam from a grating written by other beams (White and Yariv 1980). It is also a useful tool for studying the photorefractive response of materials, as it is often possible to probe a grating with a laser whose wavelength has been selected to lie in a region where the photorefractive sensitivity of the material is very low; in this situation the probe beam does not perturb the grating (Pichon and Huignard 1981).

Despite being a simpler problem than that of dynamic holography, diffraction nevertheless has associated with it a considerable volume of literature. This is largely due to the complex behaviour that results when the photorefractive material concerned possesses linear birefringence (either natural or field-induced), circular birefringence (optical activity), or a combination of the two properties.

We begin, however, by looking at diffraction in an isotropic medium. In the simplest case the problem is one central to the field of holography: what happens when we illuminate a uniform, sinusoidal[13] refractive index grating with a plane wave? This problem actually first arose in the field of acousto-optics (Phariseau 1956) but the seminal work is usually acknowledged to be that of Kogelnik (1969), whose treatment deals with transmission and reflection gratings, on- and off-Bragg replay and both index and absorption gratings.

---

[13] As pointed out in Section 2.9, the index grating will not necessarily be sinusoidal – particularly if it is recorded with a high intensity modulation. This can lead to the presence of a number of higher diffraction orders (H.J. Zhang et al. 1993).

The difference between dynamic holography in photorefractive materials and conventional holography is of course that in the former the grating amplitude may be time-dependent. Measuring the diffraction using a read-out beam with a wavelength at which the photorefractive sensitivity (see Section 2.6) is very low allows the grating dynamics to be observed. For example, Fig. 6.16 is taken from a study of the time dependence of grating writing and decay in LiNbO$_3$ by Alphonse et al. (1975)[14]. It illustrates the diffraction efficiency of 633 nm HeNe light Bragg-matched to a grating being recorded by light of wavelength 488 nm at various total incident intensities. As expected, higher incident intensities result in a more rapid rise towards the saturation diffraction efficiency, which was 5.4% in this experiment. A key feature of this work was the investigation of the dependence of both the time evolution of the grating and the maximum grating amplitude on the modulation of the interference fringes (see Section 2.9).

When a stored refractive index grating is read out by light of a wavelength at which the photorefractive sensitivity is appreciable, the situation is more complicated. Typically this situation arises when the read-out wavelength is the same as that used to record the grating in the first place. As early as 1972, Staebler and Amodei (1972a) had realized that the grating recording process itself was complicated by the interference of the undiffracted incident light with light diffracted from the developing grating. Following on from this work, Ninomiya (1973) and Magnusson and Gaylord (1976) investigated the read-out of a stored grating with

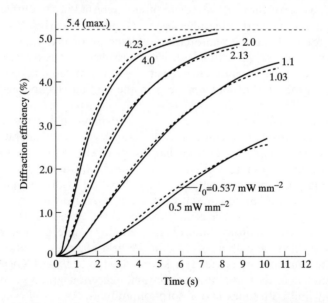

Fig. 6.16 Experimental (——) and theoretical (- - - -) curves showing the time dependence of diffraction efficiency for different intensities incident on LiNbO$_3$ (after Alphonse et al. 1975).

---

[14]Note that LiNbO$_3$ is actually a naturally birefringent material, but with judicious choice of crystal orientation and beam polarizations it is possible to make use of the scalar theory developed for isotropic media.

6.9 Diffraction efficiency

a single beam. In this situation the diffracted beam and incident beam interfere and begin to record their own grating in the medium, as discussed briefly in Section 3.9.

In Magnusson and Gaylord's theory, the complex amplitude of the grating at a given point in the material was allowed to depend on the past history of the incident and diffracted beam amplitudes at that point. Numerical integration of the resulting equations was required, but the theory was able to reproduce a wide range of observed behaviour of the diffracted beams in a number of experimental configurations. Depending in particular on the absorption constant used, the phase angle between the intensity fringe pattern and the resulting grating, and which of the original two writing beams was used to read-out the diffraction grating, the following experimentally observed diffraction efficiency behaviour could be reproduced by this theory: exponential decrease to zero, followed by a slight increase (Staebler and Phillips 1974; Staebler and Amodei 1972b); exponential decay followed by oscillations at low efficiency (Staebler and Phillips 1974); irregular oscillations in diffraction efficiency (Magnusson and Gaylord 1976); initial enhancement of diffraction efficiency (Staebler and Phillips 1974; Gaylord et al. 1973)[15]. The fact that the diffraction efficiency can actually increase during the read-out process (self-enhancement) is of particular interest and has since been studied by others (Kukhtarev 1978; Reinfelde et al. 1984; Shvarts et al. 1987; Au and Solymar 1989; Saxena et al. 1991; Shepelevich and Khramovich 1991).

The theory of Magnusson and Gaylord was also able to describe the angular sensitivity of replay from a stored diffraction grating. For example, Fig. 6.17 shows experimental and calculated plots of diffraction efficiency as a function of incident angle, where the Bragg condition is satisfied at an angle of 5°. At a similar time, Cornish and Young (1975) showed the importance of multiple internal reflections to the diffraction efficiency of holograms stored in $LiNbO_3$. They showed that changes of temperature of around 0.5°C (such as could easily be induced by absorption of the read-out beam) could cause the diffraction efficiency to vary from 19% to 25%, due to the change in the optical length of the crystal and the consequent round-trip phase variations of the multiple reflections. This observation could act as a warning that anti-reflection coatings should be used if repeatable results are required with such an arrangement.

Very high diffraction efficiencies can be obtained from some photorefractive crystals. For example, by using an externally applied electric field to enhance the space charge field amplitude, Herriau et al. (1987) were able to obtain efficiencies of as high as 95% from BGO. As discussed in Sections 3.4 and 3.7, the fringes are bent in the presence of an applied field and the optimum diffraction efficiency is obtained when the angle of the reading beams is slightly different from that of the

---

[15]Note that the most commonly observed behaviour is an approximately exponential decrease in diffraction efficiency as the grating is erased by the read-out process (Staebler and Amodei 1972a; Gaylord et al. 1972; Phillips et al. 1972; Amodei et al. 1972; Ishida et al. 1972; Micheron and Bismuth 1973; Phillips and Staebler 1974; Huignard et al. 1975a; Amodei et al. 1971; Amodei and Staebler 1971; Staebler and Amodei 1972b).

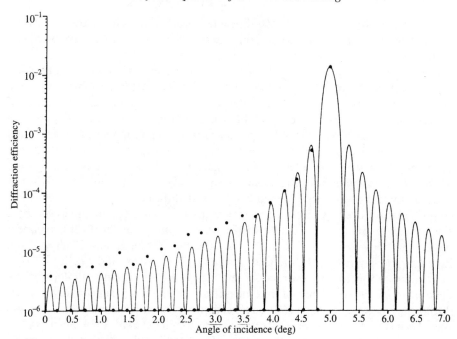

Fig. 6.17 Experimental (●) and theoretical (——) diffraction efficiency as a function of read-out beam angle in 1.66 mm thick LiNbO$_3$ crystal. The Bragg condition is satisfied at 5° (after Magnusson and Gaylord 1976).

writing beams. This effect was investigated in detail by de Vré et al. (1994, 1995). Measurements of the diffraction efficiency as the Bragg condition is increasingly violated have been used to determine the spatial variation of the complex grating amplitude within a crystal of BaTiO$_3$ (Okamura et al. 1993).

As discussed in Section 1.2, for a uniform grating the diffraction efficiency is a sine-squared function of the amplitude of the permittivity variation. For a sufficiently long crystal, when a grating is being recorded and the permittivity amplitude is increasing, it therefore follows that there should be an oscillation of the diffraction efficiency as measured with a separate probe beam. This behaviour has been observed experimentally; Fig. 6.18 shows results obtained with LiNbO$_3$ (Maniloff and Johnson 1993). The decrease in the values of succcesive peaks in the diffraction efficiency was due to the effects of scattered light, which were theoretically modelled (solid line).

Despite the fact that many of the experimental results which have just been referenced were obtained in birefringent materials such as LiNbO$_3$, the diffraction processes occurring are understandable with a simple scalar treatment of the dielectric permittivity. There are however a host of interesting diffraction phenomena that can only be understood when the anisotropic nature of the permittivity is properly taken into account. This field of anisotropic diffraction is a rich

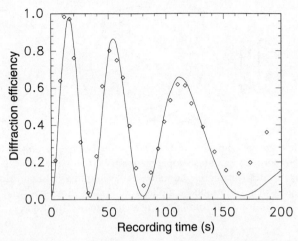

Fig. 6.18 Experimental (◇) and theoretical (——) diffraction efficiency as a function of recording time for an LiNbO$_3$ crystal (after Maniloff and Johnson 1993).

one and there is insufficient space here to do it full justice with an in-depth treatment; instead we shall briefly survey some of the concepts involved and review some of the more important results. For more detail, the reader is referred to Chapter 5 in the book by Petrov *et al.* (1991) and to the review article by Voit (1987).

In general in a birefringent medium, due to the form of the permittivity tensor a particular propagation direction will have two associated polarization eigenstates (as discussed in Appendix A). Light whose polarization matches one of the eigenstates will propagate without a change in the polarization. For example, in a medium like LiNbO$_3$ possessing linear birefringence, the eigenstates will have orthogonal linear polarizations. On the other hand, BSO in the absence of any electrical fields is circularly birefringent (optically active) and the eigenstates of polarization correspond to left and right circular states. The most complicated situation with which we shall deal is when a material is both optically active and linearly birefringent (as is the case with BSO when an electric field is applied); in this case the eigenstates take the form of elliptically polarized states (Nye 1957a).

In such materials, when the Ewald construction is used to illustrate the diffraction process, in general two surfaces (or in two-dimensional representations two arcs) should be drawn, corresponding to the wave vectors of the two eigenstates of polarization (Nye 1957b). Figure 6.19(a) illustrates how the same grating can be involved in Bragg-matched replay involving pairs of beams of the same eigenstates[16]; this has been called intramode coupling (Petrov *et al.* 1991). Note that the

---

[16]We are assuming here that light of both eigenstates of polarization can experience the field-induced grating. Whether or not this is the case actually depends on the form of the dielectric permittitivity tensor and the crystal orientation.

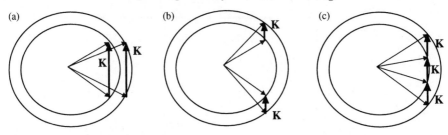

Fig. 6.19 Ewald diagrams showing on-Bragg coupling between beams of light mediated by a single grating recorded in an anisotropic medium. (a) Coupling between beams in similar polarization eigenstates (intramode coupling). (b) Coupling between beams in dissimilar polarization eigenstates (intermode coupling). (c) Interaction involving both inter- and intramode coupling (anisotropic self-diffraction).

two pairs subtend different angles inside the anisotropic medium, though they would subtend the same angle outside in vacuum. In either of these interactions, the transmitted and diffracted light would have the same polarization states.

However, this diffraction process is not the only one that can occur. Instead of coupling between light beams of the same polarization state, it is possible for coupling to occur between beams of light with different eigenstates of polarization; this has been called intermode coupling. This is shown in Fig. 6.19(b). The first extensive studies of intermode coupling were carried out by workers in the former Soviet Union (Stepanov *et al.* 1977; Petrov *et al.* 1979a,b, 1981; Pencheva *et al.* 1981).

Finally, for one particular grating spacing a situation involving both inter- and intramode coupling is possible; see Fig. 6.19(c). Here, typically the innermost pair of beams (which have the same polarization state) would be incident on the crystal and would write the grating. All four beams would be seen to emerge from the crystal, the outermost pair possessing the other polarization eigenstate. This process has been called anisotropic self-diffraction and has been observed in several naturally birefringent crystals: e.g. $LiNbO_3$ (Kukhtarev and Odulov 1980), $BaTiO_3$ (Kukhtarev *et al.* 1984b), $KNbO_3$ (Voit and Günter 1987). High-order anisotropic diffraction and anisotropic self-diffraction have also been reported (Temple and Ward 1986, 1988), where the fundamental grating vector is an integer sub-multiple of the vector needed to span the two shells in the Ewald diagram.

We shall now briefly review anisotropic diffraction in three situations of increasing complexity: in naturally linearly birefringent crystals, in isotropic optically active crystals, and in linearly birefringent, optically active crystals.

Crystals which are naturally linearly birefringent tend to be characterized by a relatively large separation between the two shells illustrated in Fig. 6.19, due to the large difference in the principal refractive indices, 0.1 being a typical figure (Landolt–Börnstein 1979; Baumert *et al.* 1985). As a result of this the Bragg angles for the inter- and intramode processes usually differ considerably. For example, Fig. 6.20 shows the variation in the angle of incidence required for

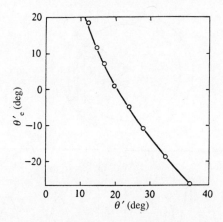

Fig. 6.20 Angle of incidence (in air) needed for on-Bragg replay by intermode diffraction as a function of angle of incidence used to record the grating in $KNbO_3$. Angles measured from normal to surface. Grating is parallel to surface; c axis is normal to incident plane; $\lambda = 442$ nm (after Petrov et al. 1983).

Bragg-matched intermode diffraction as a function of the angle at which the grating was recorded in a crystal of $LiNbO_3$ (Petrov et al. 1983). This property has been utilized to realize a system able to deflect a HeNe read-out beam over a range of 5.7° by changing the wavelength of the pair of Ar^+ beams that were used to write the grating in a crystal of $KNbO_3$ (Voit et al. 1986).

In cubic optically active crystals, such as BSO, in the absence of any applied field, the difference in the refractive indices of the circularly polarized eigenstates is somewhat smaller than was the case for crystals with natural linear birefringence, 0.0001 being typical. The shells in Fig. 6.19 in this case are very close to each other, and beams generated by normal and anisotropic diffraction are not usually well separated angularly. As an example, Fig. 6.21 illustrates the diffraction of linearly polarized light from a grating recorded in BSO (Pencheva et al. 1982). The Ewald diagram of the various interactions is shown in Fig. 6.21(a), and a plot of the diffraction efficiency as a function of deviation from the angle used to record the grating is shown in Fig. 6.21(b), for the crystal in the longitudinal geometry. The explanation for the structure is as follows.

The linearly polarized incident wave may be decomposed into two circularly polarized states of equal amplitude and opposite handedness. When the wave is incident at the same angle as one of the waves used to record the grating (zero deviation), waves 2 and 3 in Fig. 6.21(a), corresponding to the two circular states, are replayed on-Bragg to generate waves 2′ and 3′, which are also circular states of opposite handedness, these states combine to give a linear output state – the central peak in Fig. 6.21(b). Remember that although waves 2 and 3 have different propagation directions inside the crystal, they have the same incident angle from outside. Similarly, waves 2′ and 3′ leave the crystal travelling in the same direction.

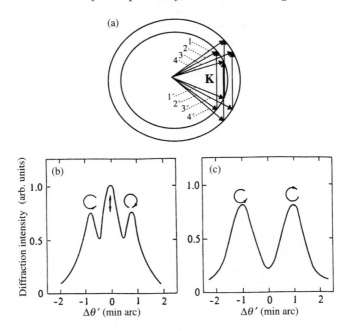

Fig. 6.21 Anisotropic diffraction in BSO (after Pencheva *et al.* 1982): (a) Ewald diagram; (b) **K** ∥ [001]; (c) **K** ∥ [110].

Altering the incident angle slightly in one direction destroys the Bragg-matched condition corresponding to waves 2 and 3 in Fig. 6.21(a). However, the interaction corresponding to wave 1 is soon reached. Here, only one of the circular polarization eigenstates is Bragg-matched (wave 1) and anisotropic diffraction results in an output state of the opposite handedness (wave 1′). Similarly, altering the incident angle in the other direction leads to the interaction corresponding to wave 4, which generates wave 4′, which is of the opposite handedness to wave 1′.

In addition to the anisotropy of the refractive index of the crystal, it is important to remember that the form of the electro-optic tensor and the crystal geometry are also important. Just because two beams are Bragg-matched to a grating does not in itself imply that there will be coupling between them and there is no reason why all of the diffraction processes shown in Fig. 6.21(a) should be equally efficient (Miridonov *et al.* 1978; Petrov *et al.* 1979c, 1981). For example, when BSO is in the transverse geometry, it turns out that the intramode coupling is non-existent (Pencheva *et al.* 1982); in this situation, the equivalent plot to Fig. 6.21(b) does not contain a central linearly polarized peak, but only the two circularly polarized side peaks (see Fig. 6.21c).

When an electric field is applied to the sillénites the situation becomes more complicated, as the changes in the permittivity tensor due to the grating may be comparable with those arising from both the material's natural optical activity and the external field-induced birefringence (Brignon and Wagner 1993; Goff

6.9 Diffraction efficiency

1995). The eigenstates of polarization are in this case elliptical states with orthogonal major axes of opposite handedness (Nye 1957a). Furthermore, diffraction of an incident beam may occur at angles which violate the Bragg condition, and in this case in general the relevant coupled wave equations must be solved numerically (Marrakchi *et al.* 1986a). When the coupling may be considered to be weak – which essentially means that the diffracted beam is much weaker than the transmitted beam – an analytic approach is possible (Vachss and Hesselink 1987a; Pauliat and Roosen 1987). By way of example, Fig. 6.22 shows comparisons between this simple theory and experiments carried out in a crystal of BSO

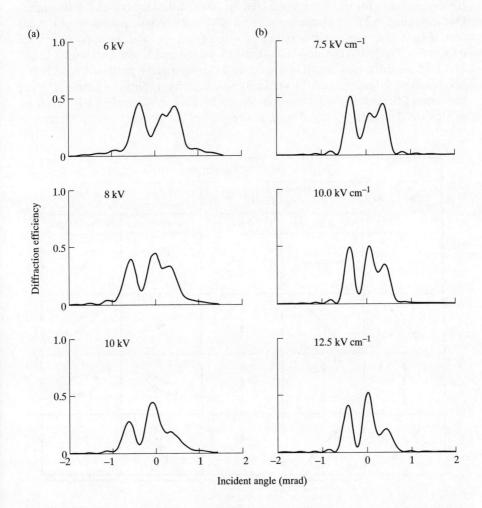

Fig. 6.22 Comparison of (a) experimental and (b) theoretical curves for diffraction efficiency as a function of incident angle with different applied fields in the longitudinal geometry in BSO. Theoretical curves correspond to an analytic expression incorporating a weak coupling approximation (after Vachss and Hesselink 1987a).

in the longitudinal geometry at several values of applied electric field (no detuning was used). It is clear that the theory matches quite closely the experimental results. What is particularly striking about these results is the large amount of structure observable in the figures – this structure may cause experimental difficulties in applications where it is desirable to maximize the diffraction efficiency.

When Bragg-matched diffraction is considered, the theoretical analysis becomes simpler. Figure 6.23 shows solutions to the coupled wave equations describing diffraction in the longitudinal geometry for a 5 mm thick BSO crystal (Marrakchi et al. 1986b). For different input polarization states, the figure shows the output states for the transmitted (readout) and diffracted beams for two cases. One case (Fig. 6.23b) assumes no applied field and no optical activity, the other case (Fig. 6.23a) uses typical values of rotary power and an applied field of 6 kV cm^{-1}. The [001] direction is assumed to correspond to the horizontal. In the first case no diffracted signal is produced for horizontally polarized light and a maximum signal is obtained for vertically polarized light. This is a simple result of the form of the permittivity variation when a field is applied in the [001] direction, as described in eqn (6.32). The presence of optical activity and birefringence

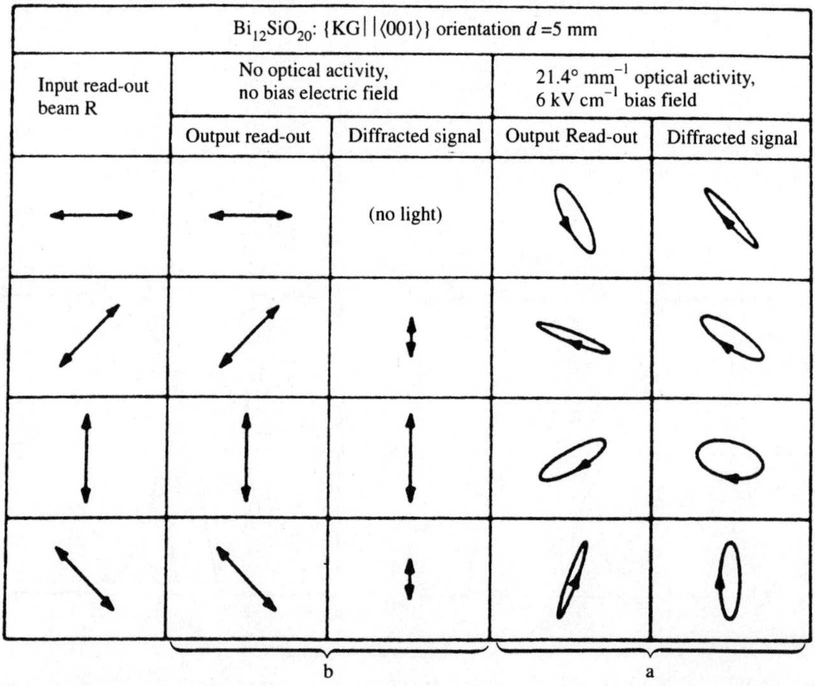

Fig. 6.23 Theoretical calculations of the state of polarization of the transmitted and diffracted beams in the longitudinal geometry with an applied field in BSO and without a field in a hypothetical similar material possessing no optical activity (after Marrakchi et al. 1986b).

6.9 Diffraction efficiency

complicates the situation, with elliptical states being generated for all the linear input polarizations considered[17].

Figure 6.24 shows similar results for the same crystal in the transverse geometry where, in the absence of any applied field, the diffraction efficiency is independent of the input polarization state (Sturman et al. 1994). Note that for horizontal or vertical input states, in the case where optical activity and birefringence are neglected the output states are orthogonal, and that even when they are included the two states are approximately orthogonal. This feature can be used to good effect to separate out diffracted and transmitted beams using a polarizer, in order to obtain good signal to noise ratios (Apostolidis et al. 1985). The dependence of diffraction efficiency on the direction of polarization of a linear read-out beam has been studied in BSO in the transverse geometry by Miteva and Nikolova (1982).

An exact analytical solution to the problem of Bragg diffraction from gratings in sillénites has been obtained by Sturman et al. (1994) using Pauli matrices. The

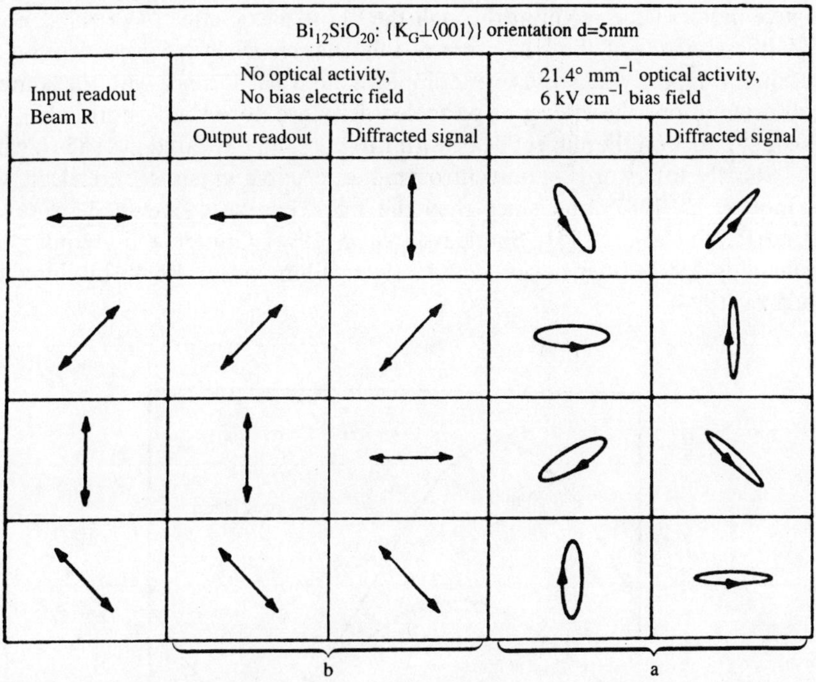

Fig. 6.24 Theoretical calculations of the state of polarization of the transmitted and diffracted beams in the transverse geometry with an applied field in BSO and without a field in a hypothetical similar material possessing no optical activity (after Marrakchi et al. 1986b).

---

[17] It has been shown that energy exchange between two circularly polarized beams is possible, which is dependent on the presence of optical activity and field-induced birefringence (Rouède et al. 1989).

solution is however fairly complex and a simpler approach can be taken when the coupling can be considered weak and the pump beam is therefore undepleted (Mallick *et al.* 1987). For example, Fig. 6.25 shows a comparison of this weak coupling solution with experiments performed using linearly polarized light incident on two crystals of BSO in the longitudinal geometry. Note the strong dependence of the diffraction efficiency on the input polarization orientation; this is again a consequence of the fact that only vertically polarized light interacts with a spatially modulated component of the permittivity tensor, and is in contrast to the transverse geometry where (as has already been stated) the diffraction efficiency is largely independent of the input polarization state.

6.10 The piezoelectric effect

It has recently become clear that any general treatment of the field equations must include the combined effects of piezoelectricity and photoelasticity. Depending on the geometry, the presence of a space charge field can induce a strain in the medium due to the piezoelectric effect, and this strain can in turn affect the refractive index of the medium through the photoelastic effect. We chose not to include these effects at the beginning of this chapter when we were deriving the wave equation, partly to avoid over-complexity and partly since with the common crystal orientations the effects of piezoelectricity are not usually noticeable.

The first paper on the subject dealt with $LiNbO_3$ and appeared in 1986 (Izvanov *et al.*) Shortly afterwards, non-centrosymmetric cubic crystals were dealt with (Stepanov *et al.* 1987) and since then the work has been extended by several authors (Pauliat *et al.* 1991; Shandarov *et al.* 1991; Günter and Zgonik 1991; Anastassakis 1993; Shepelevich *et al.* 1994; Ellin *et al.* 1994; Litvinov and Shandarov 1994).

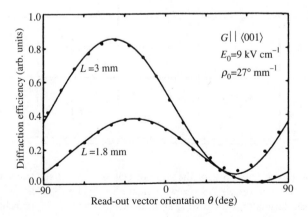

Fig. 6.25 Comparison of experimental and theoretical diffraction efficiency in BSO in the longitudinal geometry as a function of the input angle of the linearly polarized read-out beam (after Mallick *et al.* 1987).

6.10 The piezoelectric effect

Here, by way of an example, we describe one particular situation where piezoelectricity does have a significant role, and must be included in order to obtain any sort of agreement between theory and experiment.

We consider a crystal of BSO in the geometry shown in Fig. 6.26. An external field is applied as in the standard longitudinal configuration, but in this case we consider the situation where the grating written by the two beams lies in the y–z plane at an angle φ to the [001] direction (the z axis). In Appendix B it is shown that when piezoelectricity is to be included, an extra term must be added to the expression for the electric field-induced change in the permittivity (eqn 6.12), giving in the general case

$$(\triangle \varepsilon_\mathrm{r})_{il} = -\varepsilon'_{im} p_{mqkn} n_n \gamma_{kj} e_{rjs} n_s \varepsilon'_{ql} E_r \tag{6.69}$$

where $\underline{\underline{e}}$ is the piezoelectric tensor, $\underline{\underline{p}}$ is the photoelastic tensor, and **n** is a unit vector in the direction of the electric field. In this equation, E_r represents the amplitude of the spatially varying component of the space charge field (the components of E_1 in eqn 6.4) and explicitly does not include any externally applied field (Pauliat *et al.* 1991). In a sense the externally applied field does provide a piezoelectric contribution to the change in the permittivity, in that for a given field, the permittivity change will depend on whether the crystal is clamped (strain equals zero) or unclamped (stress equals zero). For a static field in the [001] direction, the form of the permittivity change due to piezoelectricity mirrors exactly the change in the permittivity due to the electro-optic effect and therefore the two may be combined by appropriate choices of the electro-optic moduli, r_{mqr}.

For the geometry of Fig. 6.26, the magnitude of the normalized space charge field may be obtained in the form (Ellin 1994)

$$E_\mathrm{w} = \frac{-[E_\mathrm{o} \cos \varphi - \mathrm{j} E_\mathrm{D}]}{\left[1 + \dfrac{E_\mathrm{D}}{E_\mathrm{q}} - b \dfrac{E_\mathrm{o}}{E_\mathrm{M}} \cos \varphi \right] + \mathrm{j} \left[b + b \dfrac{E_\mathrm{D}}{E_\mathrm{M}} + \dfrac{E_\mathrm{o}}{E_\mathrm{q}} \cos \varphi \right]} \tag{6.70}$$

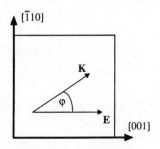

Fig. 6.26 Two-wave mixing in BSO with the grating in the y–z plane at an angle φ to the z axis. In this geometry the effects of piezoelectricity are particularly important.

The space charge field is in the direction of the grating vector, and hence the y and z components of the normalized space charge field are given by

$$E_{w_y} = E_w \sin\varphi \qquad E_{w_z} = E_w \cos\varphi \tag{6.71}$$

Furthermore, it may be shown that when piezoelectricity is included (and the same approximations are made) the equations analogous to (6.36) and (6.37), which apply to the longitudinal geometry in the absence of piezoelectric effects, are

$$\frac{d}{dx}\begin{pmatrix}\mathcal{E}_{1y}\\\mathcal{E}_{1z}\end{pmatrix} = -j\begin{bmatrix}\kappa E_{w_z}\frac{m}{2}+\kappa' E_{p2} & \kappa E_{w_y}\frac{m}{2}+\kappa' E_{p4}\\\kappa E_{w_{yz}}\frac{m}{2}+\kappa' E_{p4} & \kappa' E_{p3}\end{bmatrix}\begin{pmatrix}\mathcal{E}_{2y}\\\mathcal{E}_{2z}\end{pmatrix}$$
$$-\left\{j\kappa\begin{bmatrix}E_0 & 0\\0 & 0\end{bmatrix}+\begin{bmatrix}\frac{\alpha}{2} & -\rho\\\rho & \frac{\alpha}{2}\end{bmatrix}\right\}\begin{pmatrix}\mathcal{E}_{1y}\\\mathcal{E}_{1z}\end{pmatrix} \tag{6.72}$$

$$\frac{d}{dx}\begin{pmatrix}\mathcal{E}_{2y}\\\mathcal{E}_{2z}\end{pmatrix} = -j\begin{bmatrix}\kappa E_{w_z}^*\frac{m^*}{2}+\kappa' E_{p2}^* & \kappa E_{w_y}^*\frac{m^*}{2}+\kappa' E_{p4}^*\\\kappa E_{w_y}^*\frac{m^*}{2}+\kappa' E_{p4}^* & \kappa' E_{p3}^*\end{bmatrix}\begin{pmatrix}\mathcal{E}_{1y}\\\mathcal{E}_{1z}\end{pmatrix}$$
$$-\left\{j\kappa\begin{bmatrix}E_0 & 0\\0 & 0\end{bmatrix}+\begin{bmatrix}\frac{\alpha}{2} & -\rho\\\rho & \frac{\alpha}{2}\end{bmatrix}\right\}\begin{pmatrix}\mathcal{E}_{2y}\\\mathcal{E}_{2z}\end{pmatrix} \tag{6.73}$$

where κ' is given by

$$\kappa' = \frac{\mu_0\varepsilon_0\omega_1^2(\varepsilon')^2}{2k_1\cos\theta} \approx \frac{\mu_0\varepsilon_0\omega_2^2(\varepsilon')^2}{2k_2\cos\theta} \tag{6.74}$$

and where the coefficients describing the piezoelectric contributions are given by

$$E_{p2} = \frac{m}{4}\begin{pmatrix}p_2 R_3 E_{w_z} + p_3 R_3 E_{w_z} + \frac{1}{\sqrt{2}} p_1 R_1 E_{w_y} + \frac{1}{\sqrt{2}} p_3 R_1 E_{w_y} +\\ \sqrt{2} p_4 R_1 E_{w_y} + \frac{1}{\sqrt{2}} p_1 R_2 E_{w_y} + \frac{1}{\sqrt{2}} p_2 R_2 E_{w_y} + \sqrt{2} p_4 R_2 E_{w_y}\end{pmatrix} \tag{6.75}$$

$$E_{p3} = \frac{m}{2}\left(p_1 R_3 E_{w_z} + \frac{1}{\sqrt{2}} p_2 R_1 E_{w_y} + \frac{1}{\sqrt{2}} p_3 R_2 E_{w_y}\right) \tag{6.76}$$

$$E_{p4} = \frac{m}{2}\left(p_4 R_3 E_{w_y} + \frac{1}{\sqrt{2}} p_4 R_1 E_{w_z} + \frac{1}{\sqrt{2}} p_3 R_2 E_{w_z}\right) \tag{6.77}$$

6.10 The piezoelectric effect

$$R_1 = \gamma_{11}Q_1 + \gamma_{12}Q_2 + \gamma_{13}Q_3,$$
$$R_2 = \gamma_{21}Q_1 + \gamma_{22}Q_2 + \gamma_{23}Q_3,$$
$$R_3 = \gamma_{31}Q_1 + \gamma_{32}Q_2 + \gamma_{33}Q_3$$
$$\gamma_{11} = (\Gamma_{22}\Gamma_{33} - \Gamma_{23}^2)/D, \qquad \gamma_{22} = (\Gamma_{11}\Gamma_{33} - \Gamma_{13}^2)/D,$$
$$\gamma_{33} = (\Gamma_{11}\Gamma_{22} - \Gamma_{12}^2)/D, \qquad \gamma_{12} = \gamma_{21} = (\Gamma_{13}\Gamma_{23} - \Gamma_{12}\Gamma_{33})/D,$$
$$\gamma_{13} = \gamma_{31} = (\Gamma_{12}\Gamma_{23} - \Gamma_{13}\Gamma_{22})/D, \qquad \gamma_{23} = \gamma_{32} = (\Gamma_{12}\Gamma_{13} - \Gamma_{11}\Gamma_{23})/D \quad (6.78)$$
$$D = \Gamma_{11}(\Gamma_{22}\Gamma_{33} - \Gamma_{23}^2) - \Gamma_{22}\Gamma_{13}^2 - \Gamma_{33}\Gamma_{12}^2 + 2\Gamma_{12}\Gamma_{23}\Gamma_{13}$$
$$\Gamma_{11} = c_1 m_1^2 + c_3(m_2^2 + m_3^2), \qquad \Gamma_{22} = c_1 m_2^2 + c_3(m_1^2 + m_3^2),$$
$$\Gamma_{33} = c_1 m_3^2 + c_3(m_1^2 + m_2^2), \qquad \Gamma_{12} = \Gamma_{21} = (c_2 + c_3)m_1 m_2,$$
$$\Gamma_{13} = \Gamma_{31} = (c_2 + c_3)m_1 m_3, \qquad \Gamma_{23} = \Gamma_{32} = (c_2 + c_3)m_2 m_3$$
$$Q_1 = 2em_2 m_3, \qquad Q_2 = 2em_1 m_3, \qquad Q_3 = 2em_1 m_2$$

Here Γ_{ik} is Christoffel's tensor (see Appendix B); $p_1 \equiv p_{11}, p_2 \equiv p_{12}, p_3 \equiv p_{13}$, and $p_4 \equiv p_{44}$ are the coefficients of the photoelastic tensor; $c_1 \equiv c_{11}, c_2 \equiv c_{12}$, and $c_3 \equiv c_{44}$ are the coefficients of the rigidity tensor, and e is the value of the nonzero components of the piezoelectric tensor.

The first thing that is apparent from the preceding treatment is the great increase in complexity that results from the inclusion of piezoelectric effects. One might ask whether going to such lengths is ever necessary, or put another way, whether the inclusion of piezoelectricity produces significant changes in the behaviour of a photorefractive crystal? Unfortunately (for those who would prefer the simple life) it is essential to incorporate piezoelectricity to describe even qualitatively the results of two-wave mixing in some geometries. The geometry of

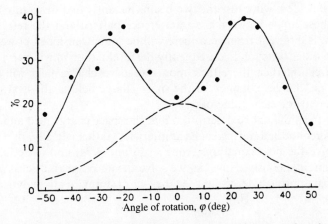

Fig. 6.27 Two-wave mixing gain at optimum detuning as a function of angle between the grating vector and the [001] direction (see Fig. 6.26 for geometry): experimental data (●); theory including piezoelectricity (——); theory excluding piezoelectricity (- - - -).

Fig. 6.26 is a case in point. Figure 6.27 shows plots of two-wave mixing gain at optimum detuning frequency, as a function of angle, φ, defined in Fig. 6.26. The figure includes theoretical curves, both including and ignoring piezoelectricity, along with experimental data obtained using a 1 cm cube of BSO and 514 nm light. The applied field was $7\,\mathrm{kV\,cm^{-1}}$, the input beam ratio was 830, the input polarization was in the [110] direction, and the grating spacing was 15 μm. A fourth-order Runge–Kutta technique was used to solve eqns (6.72) and (6.73). It is clear from the figure that the inclusion of piezoelectricity is essential when the grating vector lies at an angle to the [001] direction.

6.11 The interaction between the space charge field and the optical field

At this point, we should like to address in a little more detail an issue that was briefly discussed in Section 3.9: namely the response when the incident optical beams are time-varying. To recap: eqns (3.73) and (3.74) describe the effect that the space charge field has on the optical waves, while eqn (3.76) represents the other side of the coin, describing the effect that the optical waves have on the material.

Suppose that we restrict ourselves for the moment to the relatively simple case where two beams are turned on and meet inside a photorefractive crystal in which there is initially no space charge field; an accurate picture of what is happening at the beginning may be described as follows. The two beams are switched on at $t = 0$. Since the time taken by the light to traverse the crystal is negligible in comparison with all the relevant time constants describing the response of the material, nothing happens to begin with. The two beams are initially unaffected. The optical fields remain constant along the x direction. A time $\triangle t$ later however, an elementary space charge field will appear which is independent of x but varies periodically in the z direction. What will be the consequence of this elementary space charge field? It will cause the two beams instantly to diffract into each other. If Φ, the phase angle between the interference pattern and the refractive index distribution, is different from zero there will be some transfer of power from one beam to the other beam and consequently the modulation m will vary. If m varies in the x direction then the effect upon the space charge field will also be x-dependent, i.e. the development of the space charge field in the next interval $\triangle t$ will be different for each value of x.

It is clear from the above description that the space charge field and the optical field are interconnected during each $\triangle t$ interval. Mathematically, this means that we must solve the field equations (eqns 3.73 and 3.74) and materials equation (eqn 3.76) simultaneously. How can we solve them? We shall discuss a little later the numerical approach; let us first see the steps leading to an analytical solution.

6.11.1 An analytical solution

It turns out that an exact solution of the above equations is possible under the conditions when the applied field is time-invariant, the input optical fields are

6.11 The interaction between the space charge field and the optical field

unchanging, the beam ratio is high enough, and the pump beam may be assumed to be undepleted. We may then disregard eqn (3.73) and $\mathcal{E}_1$ can be regarded as a real constant $\mathcal{E}_{10}$ in eqn (3.76), leaving us with

$$\frac{\partial \mathcal{E}_2}{\partial x} = j\kappa \frac{E_1^*}{2} \mathcal{E}_1 \tag{6.79}$$

$$\frac{\partial E_1}{\partial t_n} + pE_1 = C\mathcal{E}_{10}\mathcal{E}_2^* \tag{6.80}$$

where we have introduced a new constant, C, replacing $2q/I_0$ in eqn (3.76) and have made use of eqn (6.3).

In order to reduce the above equations to their standard form known to mathematicians, we need to introduce new variables with the transformation

$$F = E_1 e^{t_n} \quad \text{and} \quad G = \mathcal{E}_2^* e^{t_n} \tag{6.81}$$

The remaining two equations then take the forms

$$\frac{\partial G}{\partial x} = j\frac{\kappa F}{2}\mathcal{E}_{10} \tag{6.82}$$

and

$$\frac{\partial F}{\partial x} = C\mathcal{E}_1 G \tag{6.83}$$

Next we need to introduce the new independent variables ξ and η with the relationships

$$\xi = \frac{-\kappa}{2}\mathcal{E}_{10} x \quad \text{and} \quad \eta = |C|\mathcal{E}_{10} t_n \tag{6.84}$$

Differentiating eqn (6.82) by η and substituting it into eqn (6.83) we obtain the following single partial differential equation:

$$\frac{\partial^2 G}{\partial \xi \partial \eta} = -j\frac{C}{|C|} G \tag{6.85}$$

which is a differential equation investigated in detail by mathematicians.

What are the boundary conditions in ξ and η? It may be seen from the previous calculations that the function G is proportional to the optical field, and its derivative by η is proportional to the space charge field. We mean then by boundary conditions that we need to know the optical field at $\eta = 0$ for all values of ξ, and we need to know the space charge field at $\xi = 0$ for all values of η. The optical field at $\eta = 0$ is simply the unperturbed field before interaction starts. The space charge field at $\xi = 0$ is not automatically known but can be easily obtained from eqn (6.80). At the input of the crystal the right-hand side of eqn (6.80) is constant (the input optical fields are not functions of time) so the time variation can be readily obtained.

The rest is mathematics, but far from trivial mathematics. The solution subject to these boundary conditions (known as the Cauchy type when the function and its derivative are known on the $\eta = 0$ and $\xi = 0$ lines) may be found (see for example Courant and Hilbert 1953) as the Riemann solution (Solymar and Heaton 1984). Unfortunately, such a solution does not provide much physical insight into what is happening in the process being studied, but it does allow the evolution of the optical or space charge fields to be calculated relatively easily. This approach has been extended, including detuning of the optical beams, though numerical methods were required (Goltz and Tschudi 1988).

It should be remembered that this solution is subject to a number of stringent assumptions. One of them – that the input optical fields are unvarying – can actually be relaxed whilst still permitting an analytic solution; see for example Cronin-Golomb (1987), Anderson and Feinberg, (1989) and Webb and Solymar (1990a,b). One application of this theory concerns the amplification of high frequency signals (Hamel de Montchenault *et al.* 1987, 1988) as described in Section 3.10. Amplification resulting from a chopped signal beam has also been reported by Kawata and Kawata (1991). Practical applications would of course demand much higher modulation frequencies; the amplification of microwave signals in $BaTiO_3$ has been reported by Dolfi *et al.* (1990).

In all cases where time-varying effects are considered, an analytical solution can only be derived for the undepleted pump case. In the more general case we need to resort to numerical methods.

6.11.2 Numerical solution

The numerical solution of partial differential equations is still a developing branch of numerical analysis. It is not possible as yet to provide the differential equation and the boundary conditions and trust the rest to a library program. We are however lucky in the present case. The solution is simple (Heaton and Solymar 1985) because the accompanying physics is simple. Let us remember what we said earlier of the physical phenomena. At time t there is a certain spatial distribution of the optical fields. In the next Δt interval there will be a change in the space charge field at all points in space caused by the optical fields. Owing to this change in the space charge field there will be an immediate redistribution of the optical fields which causes variation in the space charge field, and so on.

The mathematical equivalent of this physical picture is as follows. We know at $t = 0$ the optical field distribution as a function of x. We substitute this known distribution into the right-hand side of eqn (3.76) and solve for E_1 at all points in space at the end of an interval Δt. Having got this value of E_1 everywhere in space we substitute it into the right-hand side of eqns (3.73) and (3.74), which are then just ordinary differential equations in x. We find the changes in $\mathcal{E}_1$ and $\mathcal{E}_2$ at every point in space. We may then proceed to the next time interval and solve the time-varying equation, i.e. we may make good use of the fact that one of our differential equations varies in time and the other two vary in space. Thus we can neatly separate the time domain and the space domain.

It is fair to ask whether it is worth while to go to the complication of combining the materials and field equations. Wouldn't the materials equation alone give sufficient accuracy? Let us look at a few examples, taking parameters typical of

6.11 The interaction between the space charge field and the optical field

a BSO crystal. For a sufficiently high beam ratio that the gain is always independent of beam ratio, we can plot the gain as a function of time in the diffusion regime for a crystal length of 4 mm and for a range of inter-beam angles. The result is shown in Fig. 6.28, from which it may be seen that the two solutions are very close to each other. There is only a slight difference between the curves, the combined solution lagging slightly. If we now increase artificially the electro-optic coefficient by a factor of 10 in order to increase the gain (see Fig. 6.29) the general shape of the curves is unchanged but the combined solution now lags considerably more behind the 'materials-equations-only' solution.

The situation is fairly similar in the presence of optimum or near-optimum detuning where the temporal evolution resembles that in the diffusion regime, though of course the gain is much higher. Major differences between the two methods will only arise when the applied electric field is sufficiently large to lead to high transient but small stationary gain. For electric fields of 3, 5, 7 and 9 kV cm^{-1} the gain as a function of time is shown in Fig. 6.30. It may be seen that the differences between the two solutions become considerably greater as the applied field increases. The oscillations in the two solutions are quite different but if the time scale is long enough we find that both solutions tend to the same stationary value.

The conclusions are clear. When the space charge field rises smoothly, the combined solution predicts some delay relative to the materials-equations-only solution. But the two solutions can be significantly different when the applied electric field dominates and both solutions display strong oscillatory tendencies.

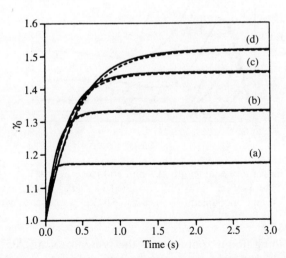

Fig. 6.28 Signal gain as a function of time for a crystal of length 4 mm, no applied electric field, and an effective coefficient $r_{\text{eff}} = 3.4$ pm V^{-1}. The inter-beam angle is (a) 10°, (b) 20°, (c) 30°, (d) 40°. The combined field and material equation solutions (- - - -) are in this case nearly the same as the uncoupled solutions (———) (after Solymar et al. 1994).

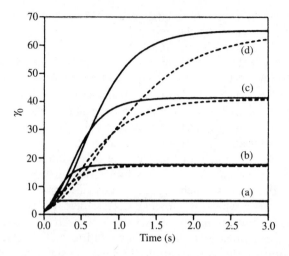

Fig. 6.29 As for Fig. 6.28, but with $r_{eff} = 34\,\text{pm}\,\text{V}^{-1}$. The combined solution now lags significantly behind the uncoupled solution (after Solymar *et al.* 1994).

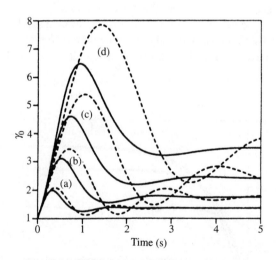

Fig. 6.30 Signal gain for a crystal length of 4 mm and $r_{eff} = 3.4\,\text{pm}\,\text{V}^{-1}$. The electric field $E_0 =$ (a) $3\,\text{kV}\,\text{cm}^{-1}$, (b) $5\,\text{kV}\,\text{cm}^{-1}$, (c) $7\,\text{kV}\,\text{cm}^{-1}$, (d) $9\,\text{kV}\,\text{cm}^{-1}$. The discrepancy between the combined (- - - -) and uncoupled (———) theories is seen to increase with applied field.

Can we say anything about transients in the presence of an AC applied field? Would the combined solution look significantly different from those shown in Figs 6.28, 6.29 and 6.30? This is a subject that has not so far been investigated at all. It seems unlikely that one could find an analytical solution, but perhaps not impossible. The numerical solution is however quite straightforward. We shall

6.11 The interaction between the space charge field and the optical field

give here only one pair of curves, Fig. 6.31, showing the gain as a function of time for an interbeam angle of 2° and a length of 4 mm. The AC field is a square wave with amplitude 7 kV cm^{-1} and a frequency of 7 Hz, the other parameters remaining unchanged from those used above. It may be seen that in the combined solution the ripples are considerably reduced. Note also that for an AC applied field the two solutions no longer tend to the same steady state behaviour, since both the optical fields and the space charge field keep on varying.

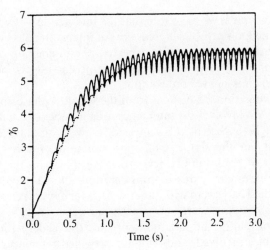

Fig. 6.31 Signal gain as a function of time for an applied square wave field. $E_0 = 7\,\text{kV cm}^{-1}$, crystal length = 1 mm, modulation frequency = 7 Hz, $r_{\text{eff}} = 3.4\,\text{pm V}^{-1}$, and inter-beam angle = 2°. The oscillation of the gain is reduced for the combined solution (dashed lines) in comparison with the uncoupled solution (solid lines) (after Solymar *et al.* 1994).

7
MULTI-WAVE MIXING

7.1 Introduction

In the preceding chapters we have seen that the interaction of two beams within a photorefractive medium involves a great deal of interesting physics and can result in optical amplification. As might be expected, the interaction of three or more waves in such a medium leads to additional surprising effects, perhaps the most remarkable of which is phase conjugation.

True phase conjugation (as distinct from the forward phase conjugation introduced in Section 7.5) involves the interaction of a number of counter-propagating beams. As might be expected, this can lead to some quite complex theoretical models under all but the simplest of approximations. Furthermore, there are already a number of books and review articles dealing with the subject of phase conjugation; therefore partly due to considerations of space and partly due to a desire not to needlessly overlap with other works, we have taken the decision here to provide only an outline of this fascinating subject.

There is hopefully sufficient material to provide the reader with an understanding of the principles involved, and we have provided references to some of the more important recent developments along with the more detailed reviews. We shall concentrate most of our efforts, though, on describing some of the multi-wave mixing interactions that occur when all the beams are travelling in the same direction through the crystal.

7.2 Forward three-wave mixing (I)

We shall consider two versions of the forward three-wave mixing process. A schematic representation of the first kind is shown in Fig. 7.1(a). Three coplanar beams are incident on a crystal such that beam 3, which we call the signal beam, bisects beams 1 and 2 which are the pump beams. The two piezoelectric mirrors serve to detune the pump beams. The corresponding Ewald diagram is shown in Fig. 7.1(b). The off-Bragg vector, ψ, is included, as with this geometry it is possible for beam coupling to occur between, for example, beams 1 and 3 mediated by the grating written by beams 3 and 2. As we shall see later, this off-Bragg coupling is most important in forward three-wave mixing.

We shall derive here the field equations for three-wave mixing based on the analysis of Ringhofer and Solymar (1988b), but before starting the mathematics let us first ask the question whether it is possible to obtain amplification in this configuration. In the two-beam case we have seen that steady-state amplification

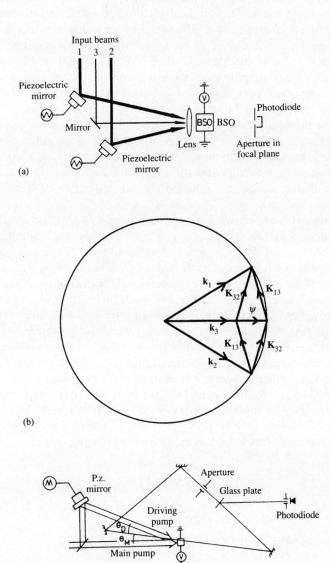

Fig. 7.1 (a) Experimental arrangement for three-wave mixing in geometry I. (b) Ewald diagram of forward three-wave mixing in geometry I. (c) Three-wave mixing using a ring-resonator (after Erbschloe and Solymar 1988a).

goes in one direction, i.e. a pump beam on one side of the signal beam will lead to amplification whereas a pump beam on the other side will produce attenuation of the signal beam. This physical picture suggests that three-wave amplification is unlikely to be possible. The signal beam will gain power from one of the pump beams and will deliver power to the other, the net result being close to zero. The first experiments showing that it is after all possible to obtain a significant amount of amplification were performed by Erbschloe and Solymar (1988a). A schematic representation of their experimental set-up is shown in Fig. 7.1(c). A DC voltage can be applied across the crystal and one of the pump beams can be detuned. There is no input for a third beam, but if the conditions are right, it can be generated from noise with the aid of the ring resonator. The principles are the same as discussed in Section 3.11 (see Fig. 3.14). When the gain is high for a beam in the direction in which the resonator is set up, then, with the aid of the feedback, significant optical power can be coupled into the resonator.

The experiment is as follows. First, the main beam is switched on alone. Then radiation appears in the resonator by means of two-wave amplification. The resonating beam adjusts its frequency in order to maximize the imaginary part of the space charge field. Next, the driving beam is switched on with an intensity equal to that of the main beam. As expected from the qualitative arguments given above, the two pump beams work against each other and the resonating beam vanishes. However, the resonating beam reappears with high intensity when we detune the driving beam. Similar experiments were also performed with a linear resonator (Erbschloe and Solymar 1988b). What is the explanation? The explanation must be that the optimum phase angle between the interference pattern and the refractive index distribution must be different for three-wave mixing than for two-wave mixing. In order to find out what the optimum phase angle is and how it is related to the other parameters, we shall now perform the analysis.

We write the total electric field as the sum of the components of the three waves in the form

$$\mathcal{E}(\mathbf{r}, t) = \frac{1}{2} \begin{bmatrix} \mathcal{E}_1(\mathbf{r}, t) e^{j(\omega_1 t - \mathbf{k}_1 \cdot \mathbf{r})} + \mathcal{E}_2(\mathbf{r}, t) e^{j(\omega_2 t - \mathbf{k}_2 \cdot \mathbf{r})} \\ + \mathcal{E}_3(\mathbf{r}, t) e^{j(\omega_3 t - \mathbf{k}_3 \cdot \mathbf{r})} + \text{c.c.} \end{bmatrix} \quad (7.1)$$

The resulting intensity pattern within the crystal we take as

$$\begin{aligned} I &= I_1 + I_2 + I_3 + \mathcal{E}_1^* \cdot \mathcal{E}_2 e^{j(\Omega_{12} t - \mathbf{K}_{12} \cdot \mathbf{r})} + \mathcal{E}_1^* \cdot \mathcal{E}_3 e^{j(\Omega_{13} t - \mathbf{K}_{13} \cdot \mathbf{r})} \\ &\quad + \mathcal{E}_3^* \cdot \mathcal{E}_2 e^{j(\Omega_{32} t - \mathbf{K}_{32} \cdot \mathbf{r})} + \text{c.c.} \\ &= I_0 \left[1 + \frac{m_{12}}{2} e^{j(\Omega_{12} t - \mathbf{K}_{12} \cdot \mathbf{r})} + \frac{m_{13}}{2} e^{j(\Omega_{13} t - \mathbf{K}_{13} \cdot \mathbf{r})} + \frac{m_{32}}{2} e^{j(\Omega_{32} t - \mathbf{K}_{32} \cdot \mathbf{r})} + \text{c.c.} \right] \end{aligned} \quad (7.2)$$

where $\Omega_{il} = \omega_i - \omega_l$ is the angular frequency difference between beams i and l, $\mathbf{K}_{il} = \mathbf{k}_i - \mathbf{k}_l$ is the grating vector recorded by these two beams, I_1, I_2, and I_3 are the intensities of beams 1, 2 and 3, respectively, I_0 is the total intensity, and m_{il} is

7.2 Forward three-wave mixing (I)

the fringe modulation produced by interference between beams i and l, which we define using complex notation as

$$m_{il} = \frac{2\mathcal{E}_i \cdot \mathcal{E}_l^*}{I_0} \tag{7.3}$$

We assume that in the presence of this intensity pattern the photorefractive material exhibits a linear response, so that we may represent the field in the crystal in the form

$$\mathbf{E} = \mathbf{E}_0 + \frac{1}{2}\left\{\begin{array}{l}\mathbf{E}_{12}e^{j(\Omega_{12}t-\mathbf{K}_{12}\cdot\mathbf{r})} + \mathbf{E}_{13}e^{j(\Omega_{13}t-\mathbf{K}_{13}\cdot\mathbf{r})} \\ +\mathbf{E}_{32}e^{j(\Omega_{32}t-\mathbf{K}_{32}\cdot\mathbf{r})} + \text{c.c.}\end{array}\right\} \tag{7.4}$$

where the space charge field resulting from interference between beams i and l, E_{il}, is given by

$$E_{il} = E_{w_{il}} m_{il} \tag{7.5}$$

where $E_{w_{il}}$ is the appropriate normalized space charge field.

In order to emphasize the physics of the three-wave mixing process, we take the simplest possible approach in deriving coupled equations governing the evolution of the amplitudes of the three waves. We ignore absorption and optical activity and we assume a scalar value for the electro-optic coefficient, r, so that the change in the relative permittivity ε_r in the presence of the field takes the form

$$\Delta\varepsilon_r(\mathbf{E}) = -(\varepsilon_r)^2 rE \tag{7.6}$$

In addition we use the scalar form of the wave equation

$$\nabla^2 \mathcal{E} = \mu_0 \varepsilon_0 \frac{\partial^2 \varepsilon_r \mathcal{E}}{\partial t^2} \tag{7.7}$$

As in Section 3.2, we proceed by substituting eqns (7.1) and (7.3)–(7.6) into eqn (7.7). If we assume the field amplitudes to be time-independent and slowly varying (so we can neglect their second spatial derivative), assume that the beams subtend small angles with the x axis, and neglect the change in the permittivity brought about by the externally applied field, the result is

$$\left\{\sum_{i=1}^{3} 2j\mathbf{k}_i \cdot (\nabla \mathcal{E}_i) e^{j(\omega_i t-\mathbf{k}_i\cdot\mathbf{r})}\right\} + \mu_0 r \frac{\varepsilon_r^2}{\varepsilon_0}\left\{\sum_{\substack{l=1 \\ l\neq m}}^{3}\sum_{m=1}^{3} E_{w_{lm}} \frac{\mathcal{E}_l \mathcal{E}_m^*}{I_0} e^{j(\Omega_{lm}t-\mathbf{K}_{lm}\cdot\mathbf{r})}\right\}$$
$$\times \left\{\sum_{p=1}^{3} \omega_p^2 \mathcal{E}_p e^{j(\omega_p t-\mathbf{k}_p\cdot\mathbf{r})}\right\} = 0 \tag{7.8}$$

We take a coupled wave approach, similar to that used in the treatment of two-wave mixing, and gather together all terms with exponents $j(\omega_1 t - \mathbf{k}_1 \cdot \mathbf{r})$, $j(\omega_2 t - \mathbf{k}_2 \cdot \mathbf{r})$, and $j(\omega_3 t - \mathbf{k}_3 \cdot \mathbf{r})$. However, in this case the situation is more complicated. As stated earlier, it is possible for beams 1 and 3 to couple via grating $\mathbf{K}_{32}$ and in a similar way, beams 3 and 2 may interact via grating $\mathbf{K}_{13}$.

In order to take account of these additional interactions it is necessary to make use of the off-Bragg vector, ψ, defined as in Fig. 7.1 so that

$$\mathbf{k}_1 = \mathbf{k}_3 + \mathbf{K}_{32} - \psi; \quad \mathbf{k}_2 = \mathbf{k}_3 - \mathbf{K}_{13} - \psi \qquad (7.9)$$

If we now gather terms with similar exponents, we obtain the following coupled wave equations:

$$\frac{d\mathcal{E}_1}{dx} = -j\frac{\kappa}{I_o}[E_{w_{13}}I_3\mathcal{E}_1 + E_{w_{32}}\mathcal{E}_3\mathcal{E}_3 e^{-j\psi x}\mathcal{E}_2^* + E_{w_{12}}I_2\mathcal{E}_1] \qquad (7.10)$$

$$\frac{d\mathcal{E}_3}{dx} = -j\frac{\kappa}{I_o}[I_1 E_{w_{13}}^* + I_2 E_{w_{32}}]\mathcal{E}_3 - j\frac{\kappa}{I_o}[E_{w_{13}} + E_{w_{32}}^*]e^{j\psi x}\mathcal{E}_1\mathcal{E}_2\mathcal{E}_3^* \qquad (7.11)$$

$$\frac{d\mathcal{E}_2}{dx} = -j\frac{\kappa}{I_o}[E_{w_{32}}^* I_3 \mathcal{E}_2 + E_{w_{13}}^* \mathcal{E}_3 \mathcal{E}_3 e^{-j\psi x}\mathcal{E}_1^* + E_{w_{12}}^* I_1 \mathcal{E}_2] \qquad (7.12)$$

where κ is given by eqn (6.38).

Solution of these coupled equations may be facilitated by a few simplifying assumptions. Firstly we note that for small inter-beam angles the gratings $\mathbf{K}_{13}$ and $\mathbf{K}_{32}$ lie in approximately the same direction, so we can equate the normalized space charge fields associated with these gratings, writing

$$E_{w_{13}} = E_{w_{32}} = E_w \qquad (7.13)$$

Next we make the assumption that we can ignore interactions between the two pump beams. There are two reasons for making this assumption: firstly, in the sillénites it is found experimentally that two-wave mixing between the pump beams occurs in a different range of detuning frequencies from those exhibiting interesting phenomena due to forward three-wave mixing; secondly, neglecting the interaction between the pump beams considerably simplifies the solution of eqns (7.10) to (7.12) but without affecting the more interesting aspects of forward three-wave mixing. Implementing these simplifications, eqns (7.10) to (7.12) become

$$\frac{d\mathcal{E}_1}{dx} = -j\frac{\kappa}{I_o}[E_w I_3 \mathcal{E}_1 + E_w \mathcal{E}_3 \mathcal{E}_3 e^{-j\psi x}\mathcal{E}_2^*] \qquad (7.14)$$

$$\frac{d\mathcal{E}_3}{dx} = -j\frac{\kappa}{I_o}[I_1 E_w^* + I_2 E_w]\mathcal{E}_3 - j\frac{\kappa}{I_o}[E_w^* + E_w^*]e^{j\psi x}\mathcal{E}_1\mathcal{E}_2\mathcal{E}_3^* \qquad (7.15)$$

$$\frac{d\mathcal{E}_2}{dx} = -j\frac{\kappa}{I_o}[E_w^* I_3 \mathcal{E}_2 + E_w^* \mathcal{E}_3 \mathcal{E}_3 e^{-j\psi x}\mathcal{E}_1^*] \qquad (7.16)$$

In order to solve these equations we make use of the undepleted pump approximation, writing the field amplitudes of the two beams as

$$|\mathcal{E}_{1,2}(x)| = |\mathcal{E}_{1,2}(0)| = \sqrt{I_{1,2}} \qquad (7.17)$$

7.2 Forward three-wave mixing (I)

Equations (7.14) to (7.16) become decoupled and it is only necessary to solve eqn (7.15). After making the substitution

$$\mathcal{E}_3(x) = A_3(x)e^{j\psi x/2} \tag{7.18}$$

this equation becomes

$$\frac{dA_3}{dx} = -j\frac{\kappa}{I_o}[I_1 E_w^* + I_2 E_w]A_3 - j\frac{\kappa}{I_o}[E_w^* + E_w]\sqrt{I_1 I_2}A_3^* \tag{7.19}$$

Solution is now straightforward: eqn (7.19) is differentiated with respect to x to give

$$\frac{d^2 A_3}{dx^2} = -j\frac{\kappa}{I_o}[I_1 E_w^* + I_2 E_w]\frac{dA_3}{dx} - j\frac{\kappa}{I_o}[E_w^* + E_w]\sqrt{I_1 I_2}\frac{dA_3^*}{dx} \tag{7.20}$$

Equation (7.19) and its complex conjugate may then be used to eliminate all terms involving A_3^* in eqn (7.20) to leave a second-order differential equation, the solution of which may be obtained in the usual way, leading to the following expression for the amplitude of beam 3:

$$\mathcal{E}_3 = (a_3 e^{\gamma_+ x/2} + b_3 e^{\gamma_- x/2})e^{j\psi x/2} \tag{7.21}$$

where

$$\gamma_\pm = -2\kappa \mathrm{Im}(E_w)\frac{(I_2 - I_1)}{I_0} \pm \sqrt{-4\kappa^2 \mathrm{Re}(E_w)^2 \frac{(I_2 - I_1)^2}{I_0^2} - 4\psi\kappa \mathrm{Re}(E_w) - \psi^2} \tag{7.22}$$

and a_3 and b_3 are determined by the boundary conditions at the input face of the crystal, $x = 0$.

If $\mathrm{Re}(\gamma_+)x$ is large, implying that there is appreciable gain, the second term in eqn (7.21) may be ignored and the signal wave amplitude is seen to increase exponentially with distance. When one pump beam (beam 2 say) is much stronger than the other, the term under the square root in eqn (7.22) is always negative and so $\mathrm{Re}(\gamma_+)$ simplifies using eqn (6.48) to

$$\mathrm{Re}(\gamma_+) = -2\kappa \mathrm{Im}(E_w) = \Gamma \sin \Phi \tag{7.23}$$

which is equivalent to the expression obtained in Section 3.2 (eqn 3.35) for the exponential gain in two-wave mixing when the phase, Φ, is $\pi/2$. The gain in this case is dependent on the imaginary part of the space charge field.

Let us now consider the case where the two pump beams are of equal intensity. Equation (7.22) then reduces to

$$\gamma_\pm = \pm\sqrt{-4\psi\kappa \mathrm{Re}(E_w) - \psi^2} \tag{7.24}$$

Under such conditions, exponential gain will be obtained so long as the condition $\kappa \mathrm{Re}(E_w) < -\psi/2$ is satisfied. A notable feature of this amplification process is that unlike in two-wave mixing, with this geometry the exponential gain may be seen to depend on the real part of the space charge field. Whilst this difference may not seem to be of great significance for the case of photorefractive media where the argument of the space charge field may be controlled by a number of parameters, it may be significant for other non-linear mechanisms which naturally exhibit a local response[1]. As yet, this interaction geometry does not seem to have been studied in media other than photorefractive crystals.

The gain is also crucially dependent on the magnitude of the off-Bragg parameter. If this is too small the gain is negligible, while if it is too large the ψ^2 term dominates, γ_+ becomes imaginary and there is no exponential gain. Figure 7.2 shows the theoretical dependence of $\text{Re}(\gamma_+)$ as a function of the pump–signal beam angle and the detuning frequency. The parameters used for this calculation, which are typical of BSO, were: $N_A^- = 10^{22}\,\text{m}^{-3}$; $N_D = 10^{25}\,\text{m}^{-3}$; $\varepsilon_s = 56$ $\lambda = 515\,\text{nm}$; $T = 300\,\text{K}$; $n = 2.62$; $r_{\text{eff}} = 3.4\,\text{pm}\,\text{V}^{-1}$; $\mu = 10^{-5}\,\text{m}^2\,\text{V}^{-1}\,\text{s}^{-1}$; $s = 1.06 \times 10^{-5}\,\text{m}^2\,\text{J}^{-1}$; $\gamma_R = 1.65 \times 10^{-17}\,\text{m}^3\,\text{s}^{-1}$; $I_0 = 10\,\text{mW}\,\text{cm}^{-2}$; $E_0 = 10\,\text{kV}\,\text{cm}^{-1}$.

One experimentally observed feature of three-wave mixing that is not explained well by the theory presented here concerns the relative magnitudes of the gains obtained in two-wave and three-wave mixing. Detailed examination of the theory suggests that for a given crystal, the optimum gain available under two-wave mixing is greater than that obtainable with three-wave mixing. Experiments have suggested that this is not the case: in one experiment a maximum gain of less than 150 was observed under two-wave mixing, while over 8000 was obtained with three-wave mixing (D. C. Jones and Solymar 1991b). This discrepancy may arise because the theory presented here assumes a linear dependence of the

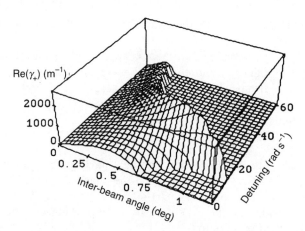

Fig. 7.2 Theoretical plot showing the dependence of the exponential gain coefficient on the detuning frequency and the angle between the pump and signal waves in forward three-wave mixing.

---

[1]Examples include interaction via the third-order non-linear susceptibility, $\chi^{(3)}$ (Boyd 1992), via thermal gratings (Khoo et al. 1988), and via the $\chi^{(3)}$-like response of liquid crystals (Pilipetski et al. 1981; Webb and Solymar 1991b).

7.2 Forward three-wave mixing (I)

amplitude of the space charge field on the intensity modulation, m. As we have seen, this is an approximation that is only strictly valid when m is small, and clearly with equal power pump beams in three-wave mixing this condition is violated. Some of the non-linearities and instabilities that can result when the intensity modulation is large are described in Sections 2.9 and 5.6.1, where we discuss the origin of spatial subharmonic gratings.

Further surprising features of this forward three-wave mixing interaction become apparent if we use the initial conditions

$$\mathcal{E}_{1,2,3}(0) = e^{j\phi_{1,2,3}}\sqrt{I_{1,2,3}(0)} \tag{7.25}$$

to calculate the coefficients a_3 and b_3 in eqn (7.21). The reason for the explicit inclusion of the arbitrary phase terms ϕ_1, ϕ_2, ϕ_3 will become clear shortly. If we assume for simplicity that the pump beams are of equal power, the result is

$$a_3 = \sqrt{2I_3(0)} \left[\frac{1}{2} - j \frac{\frac{\psi}{2} + \kappa \mathrm{Re}(E_\mathrm{w})(1 + e^{j(\phi_1+\phi_2-2\phi_3)})}{2\gamma_+} \right] e^{j\phi_3} \tag{7.26}$$

$$b_3 = \sqrt{I_3(0)} \left[\frac{1}{2} + j \frac{\frac{\psi}{2} + \kappa \mathrm{Re}(E_\mathrm{w})(1 + e^{j(\phi_1+\phi_2-2\phi_3)})}{2\gamma_+} \right] e^{j\phi_3} \tag{7.27}$$

We shall further assume that the exponential gain is large, so that after a short distance we may neglect the second term in eqn (7.21) containing b_3. The amplitude of the signal beam is then given by

$$\mathcal{E}_3(x) = a_3 e^{(j\frac{\psi}{2}x + \gamma_+ x)} \tag{7.28}$$

Substituting eqn (7.26) into eqn (7.28) and rearranging leads to the following expression for the amplitude of the signal beam:

$$\mathcal{E}_3(x) = -j\sqrt{I_3(0)} \frac{\kappa \mathrm{Re}(E_\mathrm{w})}{\gamma_+} e^{j(\phi_1+\phi_1+\phi')/2} \cos[(\phi_1+\phi_2-2\phi_3-\phi')/2] e^{(\gamma_+ + j\frac{\psi}{2})x} \tag{7.29}$$

where

$$\tan(\phi') = \frac{\gamma_+}{\kappa \mathrm{Re}(E_\mathrm{w}) + \frac{\psi}{2}} \tag{7.30}$$

From eqn (7.29), the intensity of the signal beam after traversing a crystal of length L may readily be shown to be

$$I_3(L) = I_3(0) \frac{\kappa^2 \mathrm{Re}(E_\mathrm{w})^2}{2\gamma_+^2} [1 + \cos(\phi_1+\phi_2-2\phi_3-\phi')] e^{2\gamma_+ L} \tag{7.31}$$

244   *Multi-wave mixing*

From eqns (7.29) and (7.31), two surprising conclusions may be drawn. Firstly, the phase of beam 3 when it leaves the crystal (its output phase) is independent of its input phase. That is to say, there is no complex exponential phase term in eqn (7.29) that depends on ϕ_3. Instead the phase is determined by ϕ_1 and ϕ_2, the phases of the pump beams and by ϕ', which depends only on parameters of the material and the experimental geometry. It thus appears as if the forward three-wave mixing process is able to remove phase fluctuations from the signal beam – so long as they occur slowly enough for the material to respond. As yet, to our knowledge, this effect has not been observed. The second conclusion which may be drawn from eqn (7.31) is that the output intensity depends in a cosinusoidal fashion on the phases of the three input waves. The three-wave mixing process is seen to convert phase information to amplitude information, providing phase-dependent amplification.

Experimental evidence for this second phenomenon is shown in Fig. 7.3 (Webb and Solymar 1991a). In the experiment, which was carried out with a crystal of BSO, the input phase of beam 3 was controlled by reflecting the beam from a mirror mounted on a piezoelectric transducer. The transducer was driven with a low frequency (much slower than the response time of the crystal) triangular waveform, while the intensity of the beam was monitored behind the crystal. The figure shows an oscilloscope trace of the intensity of beam 3 (upper trace) and the triangular waveform (lower trace). Initially, there was no electric field applied to the crystal. Point A denotes the time when the electric field was first applied and point B corresponds to the time when the signal beam was unblocked. Thereafter, the amplification is clearly seen to be a sinusoidal function of the drive voltage, with the gain varying from 0 to 160.

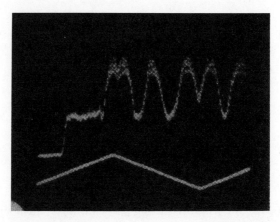

Fig. 7.3 Detected signal intensity (upper trace) and voltage controlling the signal beam phase (lower trace). Point A indicates the time at which the electric field is applied to the crystal. Point B indicates when the signal beam was first incident (after Webb and Solymar 1991a).

7.3 Forward three-wave mixing (II)

The second version of forward three-wave mixing that we shall consider is shown in Fig. 7.4. Only two beams are incident on the crystal: beam 1 (a strong pump beam) and beam 2 (a weak signal beam). The third, higher-order beam arises due to off-Bragg diffraction. This geometry has been studied much longer than that in Fig. 7.1 and in non-linear materials other than photorefractives, such as liquid crystals (Khoo and Liu 1987), sodium vapour (Heer and Griffin 1979), and silicon (Eichler *et al.* 1987).

When the space charge field possesses an imaginary component, amplification of the signal beam is possible via the usual 2WM process. However, as we shall see when the higher-order beam is taken into account, some amplification of the signal beam is possible even when the material exhibits a local response.

We proceed in the same fashion as in the previous section. The difference is that instead of eqn (7.9) we have the following relationships between the wave and grating vectors

$$\mathbf{k}_3 = \mathbf{k}_1 + \mathbf{K}_{12} - \psi; \quad \mathbf{k}_2 = \mathbf{k}_1 - \mathbf{K}_{12} \tag{7.32}$$

Using these relationships, the following coupled equations are obtained, which are analogous to eqns (7.10) to (7.12)

$$\frac{d\mathcal{E}_3}{dx} = -j\frac{\kappa}{I_o}[E_{w31}I_1\mathcal{E}_3 + E_{w12}\mathcal{E}_1\mathcal{E}_1 e^{-j\psi x}\mathcal{E}_2^* + E_{w32}I_2\mathcal{E}_3] \tag{7.33}$$

$$\frac{d\mathcal{E}_1}{dx} = -j\frac{\kappa}{I_o}[I_3 E^*_{w31} + I_2 E_{w12}]\mathcal{E}_1 - j\frac{\kappa}{I_o}[E_{w31} + E^*_{w12}]e^{j\psi x}\mathcal{E}_3\mathcal{E}_2\mathcal{E}_1^* \tag{7.34}$$

$$\frac{d\mathcal{E}_2}{dx} = -j\frac{\kappa}{I_o}[E^*_{w12}I_1\mathcal{E}_2 + E^*_{w31}\mathcal{E}_1\mathcal{E}_1 e^{-j\psi x}\mathcal{E}_3^* + E^*_{w32}I_3\mathcal{E}_2] \tag{7.35}$$

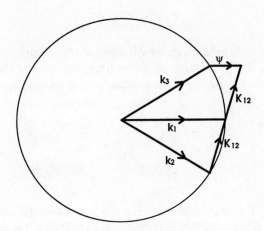

Fig. 7.4 Ewald diagram of forward three-wave mixing in geometry II.

Once again we now make some simplifying assumptions. Firstly, we neglect interactions between the two weak beams, as these will be much smaller than any coupling involving the pump beam. Secondly, we equate the normalized space charge fields:

$$E_{w31} = E_{w32} = E_w \tag{7.36}$$

Equations (7.33) to (7.35) then become

$$\frac{d\mathcal{E}_3}{dx} = -j\frac{\kappa}{I_o}[E_w I_1 \mathcal{E}_3 + E_w \mathcal{E}_1 \mathcal{E}_1 e^{-j\psi x}\mathcal{E}_2^*] \tag{7.37}$$

$$\frac{d\mathcal{E}_1}{dx} = -j\frac{\kappa}{I_o}[I_3 E_w^* + I_2 E_w]\mathcal{E}_1 - j\frac{\kappa}{I_o}[E_w + E_w^*]e^{j\psi x}\mathcal{E}_3\mathcal{E}_2\mathcal{E}_1^* \tag{7.38}$$

$$\frac{d\mathcal{E}_2}{dx} = -j\frac{\kappa}{I_o}[E_w^* I_1 \mathcal{E}_2 + E_w^* \mathcal{E}_1 \mathcal{E}_1 e^{-j\psi x}\mathcal{E}_3^*] \tag{7.39}$$

For the moment, we shall also assume that the pump beam (beam 1) is undepleted. With this approximation we are left with

$$\frac{d\mathcal{E}_3}{dx} = -j\frac{\kappa}{I_o}[E_w I_1 \mathcal{E}_3 + E_w \mathcal{E}_1 \mathcal{E}_1 e^{-j\psi x}\mathcal{E}_2^*] \tag{7.40}$$

$$\frac{d\mathcal{E}_2}{dx} = -j\frac{\kappa}{I_o}[E_w^* I_1 \mathcal{E}_2 + E_w^* \mathcal{E}_1 \mathcal{E}_1 e^{-j\psi x}\mathcal{E}_3^*] \tag{7.41}$$

These are easily decoupled and solved to give

$$\mathcal{E}_3(x) = (a_3 e^{\gamma_+ x/2} + b_3 e^{\gamma_- x/2})e^{-j\psi x/2} \tag{7.42}$$

$$\mathcal{E}_2(x) = (a_2 e^{\gamma_+^* x/2} + b_2 e^{\gamma_-^* x/2})e^{-j\psi x/2} \tag{7.43}$$

where the complex exponential gain coefficients are given by

$$\gamma_\pm = \pm\sqrt{4\psi\kappa E_w - \psi^2} \tag{7.44}$$

and a_3, b_3, a_2, and b_2 are determined by the boundary conditions at the input face of the crystal, $x = 0$. For example, if we assume that initially there is no higher-order beam and that the complex amplitude of the incident signal wave is $\mathcal{E}_{20}$, then the result is

$$a_2 = -j\frac{\kappa E_w^* \mathcal{E}_{20}}{\gamma_+^*} + \frac{\mathcal{E}_{20}}{2} + j\frac{\psi\mathcal{E}_{20}}{2\gamma_+^*} \quad b_2 = j\frac{\kappa E_w^* \mathcal{E}_{20}}{\gamma_+^*} + \frac{\mathcal{E}_{20}}{2} - j\frac{\psi\mathcal{E}_{20}}{2\gamma_+^*} \tag{7.45}$$

$$a_3 = -j\frac{\kappa E_w \mathcal{E}_{20}^*}{\gamma_+} \quad b_3 = j\frac{\kappa E_w \mathcal{E}_{20}^*}{\gamma_+} \tag{7.46}$$

Remember now that we are considering the case when the material exhibits a local response, so E_w is actually a real quantity. In that case it is easy to see that so long as the product κE_w is positive there will be a range of interbeam angles (and therefore values of ψ) for which the expression under the square root in eqn (7.44) is always positive and hence there will be exponential growth of both signal and higher order waves.

What will limit this exponential growth? In amplification by two-wave mixing, the signal beam intensity continues to grow until the pump beam is depleted and the signal beam then saturates. Will that happen here? The answer, obtained by solving eqns (7.37) to (7.39), is no. Unfortunately there appears to be no simple analytic solution. Instead numerical techniques must be used. Figure 7.5 shows the results of such numerical calculations with $\Gamma L = 10$, input beam ratio = 40, and for three different values of the product ψL, where L is the crystal length. It can be seen that rather than saturation there is an oscillatory behaviour, somewhat akin to that observed in diffraction from a static grating as discussed in Section 1.2.

We shall return to forward three-wave mixing in this geometry once again in Section 7.5 when we discuss the imaging properties of forward wave interactions.

7.4 Forward four-wave mixing

Figure 7.6 shows Ewald diagram representations of the forward four-wave mixing geometry first proposed by Ringhofer and Solymar (1988a,b). Here we again have three coplanar beams incident but now beam three does not bisect the two pump beams. Off-Bragg interactions can lead to the build up of a fourth beam shown in the figure: for example, light may be coupled into beam four from beam 1 via off-Bragg diffraction from grating $\mathbf{K}_{32}$.

Proceeding in the same fashion as for forward three-wave mixing we can derive the following coupled equations (analogous to eqns 7.10 to 7.12) describing the evolution of the amplitudes of the four waves within the material:

$$\frac{d\mathcal{E}_1}{dx} = -j\frac{\kappa}{I_o}\begin{bmatrix} E_{w13}I_3\mathcal{E}_1 + E_{w14}I_4\mathcal{E}_1 + E_{w12}I_2\mathcal{E}_1 \\ +E_{w42}\mathcal{E}_3\mathcal{E}_4 e^{-j\psi x}\mathcal{E}_2^* + E_{w32}\mathcal{E}_3\mathcal{E}_4 e^{-j\psi x}\mathcal{E}_2^* \end{bmatrix} \quad (7.47)$$

$$\frac{d\mathcal{E}_3}{dx} = -j\frac{\kappa}{I_o}[I_1 E_{w13}^* + I_2 E_{w32} + I_4 E_{w34}]\mathcal{E}_3$$
$$\qquad - j\frac{\kappa}{I_o}[E_{w42}^* + E_{w14}]e^{j\psi x}\mathcal{E}_1\mathcal{E}_2\mathcal{E}_4^* \quad (7.48)$$

$$\frac{d\mathcal{E}_4}{dx} = -j\frac{\kappa}{I_o}[I_1 E_{w14}^* + I_2 E_{w42} + I_4 E_{w34}^*]\mathcal{E}_4$$
$$\qquad - j\frac{\kappa}{I_o}[E_{w13} + E_{w32}^*]e^{j\psi x}\mathcal{E}_1\mathcal{E}_2\mathcal{E}_3^* \quad (7.49)$$

$$\frac{d\mathcal{E}_2}{dx} = -j\frac{\kappa}{I_o}\begin{bmatrix} E_{w32}^* I_3\mathcal{E}_2 + E_{w42}^* I_4\mathcal{E}_1 + E_{w12}^* I_1\mathcal{E}_2 \\ +E_{w13}^*\mathcal{E}_3\mathcal{E}_2 e^{-j\psi x}\mathcal{E}_1^* + E_{w14}^*\mathcal{E}_3\mathcal{E}_4 e^{-j\psi x}\mathcal{E}_1^* \end{bmatrix} \quad (7.50)$$

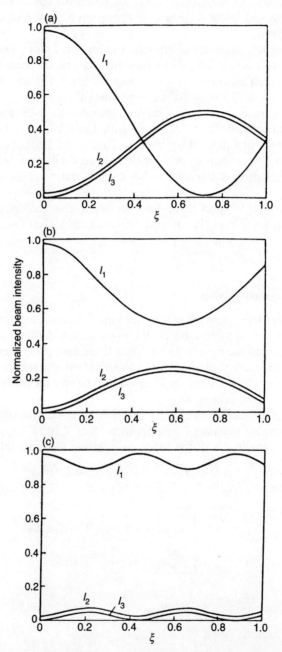

Fig. 7.5 (a)–(c) Plots of normalized beam intensity as a function of normalized crystal length for $\Gamma L = 10$ and $\psi L = 0, 1, 5$, respectively (after Au and Solymar 1988c).

7.4 Forward four-wave mixing

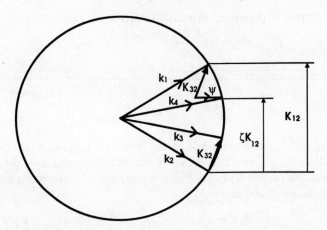

Fig. 7.6 Ewald diagram showing forward four-wave mixing.

In order to simplify things we once again neglect interaction between the pump beams, taking them to be undepleted; we also ignore interactions between beams 3 and 4, which are assumed to be negligible compared to the coupling to the pump beams, and we assume that

$$E_{w13} = E_{w42} \quad \text{and} \quad E_{w14} = E_{w32} \tag{7.51}$$

Applying these simplifications to eqns (7.47) to (7.50) we are then just left with the following two equations:

$$\begin{aligned}\frac{d\mathcal{E}_3}{dx} &= -j\frac{\kappa}{I_o}[I_1 E_{w13}^* + I_1 E_{w14}]\mathcal{E}_3 \\ &\quad - j\frac{\kappa}{I_o}[E_{w13}^* + E_{w14}]e^{j\psi x}\mathcal{E}_1\mathcal{E}_2\mathcal{E}_4^*\end{aligned} \tag{7.52}$$

$$\begin{aligned}\frac{d\mathcal{E}_4}{dx} &= -j\frac{\kappa}{I_o}[I_1 E_{w14}^* + I_2 E_{w13}]\mathcal{E}_4 \\ &\quad - j\frac{\kappa}{I_o}[E_{w13} + E_{w14}^*]e^{j\psi x}\mathcal{E}_1\mathcal{E}_2\mathcal{E}_3^*\end{aligned} \tag{7.53}$$

These equations are easily decoupled into two ordinary second-order differential equations leading to solutions for the amplitudes of beams 3 and 4 of the form

$$\mathcal{E}_3(x) = (a_3 e^{\gamma_+ x/2} + b_3 e^{\gamma_- x/2})e^{j\psi x/2} \tag{7.54}$$

$$\mathcal{E}_4(x) = (a_4 e^{\gamma_+^* x/2} + b_4 e^{\gamma_-^* x/2})e^{j\psi x/2} \tag{7.55}$$

where the complex exponential gain coefficients are given by

$$\gamma_\pm = j\kappa(E_{w14} - E^*_{w13})\frac{(I_1 - I_2)}{I_0}$$
$$\pm \sqrt{-\kappa^2(E_{w14} + E^*_{w13})^2 \frac{(I_2 - I_1)^2}{I_0^2} - 2\psi\kappa \frac{(I_2 + I_1)}{I_0}(E_{w14} + E^*_{w13}) - \psi^2}$$
(7.56)

and $a_3, b_3, a_4,$ and b_4 are determined by the boundary conditions at the input face of the crystal, $x = 0$. If we assume that the complex amplitude of the third beam when it is incident on the crystal is $\mathcal{E}_{30}$ and there is no fourth beam incident, then these constants are found to be

$$a_3 = -j\frac{\Gamma\mathcal{E}_{30}}{\gamma_+ I_0}[I_1 E^*_{w13} + I_2 E_{w14}] + \frac{\mathcal{E}_{30}}{2} \quad b_3 = j\frac{\Gamma\mathcal{E}_{30}}{\gamma_+ I_0}[I_1 E^*_{w13} + I_2 E_{w14}] + \frac{\mathcal{E}_{30}}{2}$$
(7.57)

$$a_4 = -j\frac{\Gamma\mathcal{E}^*_{30}\mathcal{E}_1(0)\mathcal{E}_2(0)}{\gamma_+ I_0}[E^*_{w13} + E_{w14}] \quad b_4 = j\frac{\Gamma\mathcal{E}^*_{30}\mathcal{E}_1(0)\mathcal{E}_2(0)}{\gamma_+ I_0}[E^*_{w13} + E_{w14}]$$
(7.58)

With the exponential gain coefficients given by eqn (7.56), in general both the real and imaginary components of the space charge fields can contribute to gain. As with forward three-wave mixing in the first considered geometry, things become clearer if we assume that the pump beams possess equal intensities. Then we may simplify eqn (7.56) to give

$$\gamma_\pm = \pm\sqrt{-2\psi\Gamma(E_{w14} + E^*_{w13}) - \psi^2} \tag{7.59}$$

Again the gain depends on the magnitude of the off-Bragg parameter, but now the first term under the square root sign will in general be complex, as the two normalized space charge fields will be different. Figure 7.7 shows a plot of the

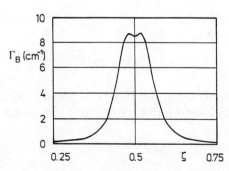

Fig. 7.7 Exponential gain coefficient vs. ζ for forward four-wave mixing in BSO (after Ringhofer and Solymar 1988a).

calculated maximum value of $\gamma_\pm$ (obtained by varying the detuning frequency) as a function of ζ, defined as in Fig. 7.6, which in the paraxial limit may be approximated by

$$\zeta = \frac{\zeta K}{K} \approx \frac{|K_{42}|}{|K_{12}|} \tag{7.60}$$

The following material parameters appropriate for BSO were chosen: $\mu = 10^{-5}\,\mathrm{m^2\,V^{-1}\,s^{-1}}$; $\gamma_n = 1.65 \times 10^{-17}\,\mathrm{m^3\,s^{-1}}$; $N_{A-} = 10^{22}\,\mathrm{m^{-3}}$; $\varepsilon_s = 56\varepsilon_o$; $n = 2.62$; $\lambda = 514\,\mathrm{nm}$; $r = 1.72 \times 10^{-12}\,\mathrm{m\,V^{-1}}$; $E_0 = 8\,\mathrm{kV\,cm^{-1}}$; grating spacing between beams 1 and 2 = $16.8\,\mathrm{\mu m}$. It can be seen that high gain is obtainable for a range of angles about the central position, but interestingly there is a slight local minimum at the centre.

7.5 Forward phase conjugation

We have seen in Section 3.12 (and will discuss further in the latter part of this chapter) that the process of phase conjugation leads to many interesting properties. As we shall now show, the beams that arise as a result of off-Bragg processes in forward four-wave mixing (beam 4) and forward three-wave mixing in the second geometry (beam 2) possess some of the attributes of the true phase conjugate of the signal beam (beam 3) (Kukhtarev and Odulov 1980). Firstly by inspecting eqns (7.45) and (7.46) (forward three-wave mixing) and (7.57) and (7.58) (forward four-wave mixing) it is possible to see in both cases that the amplitudes of the generated beams are proportional to the complex conjugate of the amplitude of the signal beams.

In the case of forward four-wave mixing this fact can be used to explain the strange behaviour of the signal beam observed in forward three-wave mixing in the first geometry, which was reported at the end of Section 7.2. Forward three wave mixing may be considered as a limiting case of forward four-wave mixing when the third and fourth beams overlap. In that case, and assuming the gain is large so that the third and fourth beams have comparable amplitude, the net result is the addition of the amplified signal beam to its complex conjugate. Let us consider the case first of all where the two beams have the same phase on leaving the crystal. They will add constructively giving the maximum possible output. If the phase of the input signal beam is advanced in some way (this could be done by using a piezoelectrically driven mirror, as in the experiment described in Section 7.2, for example), then the phase of beam 4 will be retarded by the same amount. Now the two waves will not add constructively, resulting in a reduced intensity, and it is not difficult to see that this variation will be periodic, repeating every time the input phase of beam 3 varies by π. Furthermore, because the phase of each beam shifts by equal but opposite amounts, the phase of the wave resulting from their superposition actually does not change and this explains the independence of the output phase of the signal beam in forward three-wave mixing on its input phase.

Returning now to forward four-wave mixing and forward three-wave mixing in the second geometry, we have seen that in each case the phase of the generated wave is the conjugate of that of the signal beam. Furthermore, from Figs 7.4 and 7.6 it may be seen that the component of the wave vector of the generated beams in the transverse (y) direction is opposite to that of the signal beam. Some thought will also show that when the signal wave has a component in the z direction (out of the plane of the paper in Figs 7.4 and 7.6), then this component will also be reversed in the generated beams. Only the components in the x direction are shared. Thus in each case the generated wave has the same form as the true phase conjugate of the signal wave, with the exception that the x component of the wave vector is in the same direction as that of the signal wave, instead of in the opposite direction. It is for this reason that the generated waves are known as forward phase conjugates.

The forward phase conjugate beams then share many of the properties of the true phase conjugate; for instance if the signal wave is a diverging spherical wave, the forward phase conjugate will be a converging spherical wave, with a similar radius of curvature. As an extension of this, if the signal wave comes from an object a certain distance in front of the crystal, then the forward phase conjugate will produce an image which is a similar distance behind the crystal (see Fig. 7.8). Thus in this geometry the crystal is able to act as a strange sort of lens where the image distance is always equal to the object distance.

This process is illustrated in Fig. 7.9, which shows several photographs of the images of the four beams leaving the crystal in a forward wave mixing experiment (D. C. Jones *et al.* 1990). Figure 7.9(a) and (b) are concerned with forward four-wave mixing. Figure 7.9(a) shows the case when the signal wave is diverging; then the fourth beam is brought to a focus. Figure 7.9(b) shows the reconstructed image of a vertical raster placed in the signal beam. Figure 7.9(c) and (d) concern forward three-wave mixing in the second geometry. Figure 7.9(c) shows the reconstructed image of a horizontal raster placed in the signal beam. Figure 7.9(d) is similar but with an image of the US Air Force resolution chart.

Forward phase conjugation in BSO was also obtained by Takacs *et al.* (1992b) in the configuration shown in Fig. 7.10(a). An electric field of $7\,\mathrm{kV\,cm^{-1}}$ was

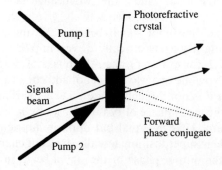

Fig. 7.8 Forward phase conjugate imaging.

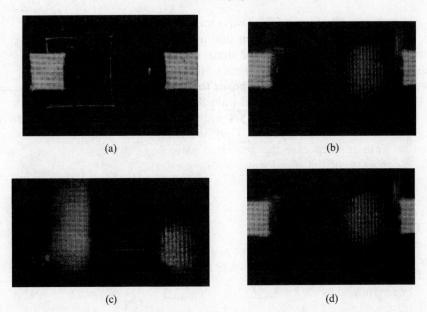

Fig. 7.9 (a,b) Imaging in forward four-wave mixing: (a) diverging beam 3 produces focused beam 4; (b) reconstructed image of vertical raster in the signal beam. (c,d) Imaging in forward three-wave mixing: (c) image of a horizontal raster placed in the signal beam; (d) image of US Air Force resolution chart placed in the signal beam. (After Jones *et al.* 1990.)

applied between the electrodes. The frequencies of the two pump beams could be adjusted by the piezoelectric mirrors PM1 and PM2. The angle between the pump beams was $2\alpha = 2°$. The signal beam was incident at an angle ϕ to the plane of the pump beams. Without detuning, only the three input beams could be detected at the output. In the presence of detuning, the output pattern shown in Fig. 7.10(b) appeared, where PC1, PC2 and PC3 were phase conjugates and S^{-1} and S^{-2} were higher order replicas of the signal beam. It is interesting to note that the maximum phase conjugate output (each time at optimum detuning) was obtained when ϕ was equal to a small fraction of $1°$. The phase conjugate output was considerably smaller when the phase matching condition $\phi = \alpha = 1°$ was satisfied.

7.6 Phase conjugation

In Section 3.12 we introduced phase conjugation in general terms, and we have discussed forward phase conjugation in the previous section. Phase conjugation is a subject to which we could easily devote several chapters. However, as mentioned in the introduction, this time we shall make no attempt at comprehensive

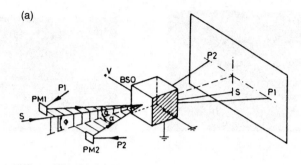

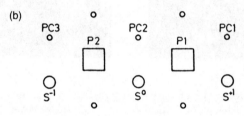

Fig. 7.10 Out-of-plane forward phase conjugation (after Takacs et al. 1992b). (a) Experimental arrangement. (b) Pattern of output beams in the presence of suitable detuning.

coverage[2]. We shall discuss it only briefly, trying to focus attention on the aspects we find more interesting.

When one talks about phase conjugation it is usually tacitly understood that it is done in the reverse (in contrast to forward) configuration, with the two pump beams being antiparallel (or nearly antiparallel), and that is also our assumption in the present section. We could look at a phase conjugator as a 'black box' which generates the phase conjugate replica of any beam which impinges upon its front surface. This process is often described as the 'time reversal' of the incident beam, because the reflected beam behaves as if it were the original beam moving back in time. An advantage of the concept of time-reversal is that it makes it easier to understand the perhaps single most amazing property of phase conjugation, namely aberration correction. When a plane wave is transmitted through a piece of aberrating glass it becomes distorted. Common sense would then suggest that if the distorted beam is reflected back into the aberrating medium it will emerge even more distorted. This is certainly true if an ordinary mirror is used.

---

[2] For a more detailed description of phase conjugation in non-linear media see for example Fisher (1983), Shkunov and Zel'dovich (1985), Zel'dovich et al. (1985), Pepper (1986), White et al. (1989), Feinberg and MacDonald (1989), Odoulov et al. (1990), Hall and Powell (1990), Tikhonchuk and Zozulya (1991), Petrov et al. (1991), Yeh (1993), and Petersen (1994).

7.6 Phase conjugation

If, as was shown in Fig. 3.15(b), a phase conjugate mirror is used, then the beam will emerge undistorted upon passing back though the same distorter! This in itself is very surprising, but what of the light which is reflected (in spurious directions) from the front surface of the distorter? Surely this power is lost. Surprisingly enough, the answer is no. Reformulating the problem as that of a partially reflecting mirror placed in front of a phase conjugate mirror (i.e. a Fabry–Perot interferometer), as shown in Fig. 7.11, it was shown by Drummond and Friberg (1983) and confirmed experimentally by Lindsay and Dainty (1986) that for incident light of intensity I_o, the intensity of the phase conjugate beam, I_c, and the specularly reflected light, I_s, are given by

$$I_c = R_c \left(\frac{1 - R_M}{1 - R_M R_c} \right)^2 I_o, \quad I_s = R_M \left(\frac{1 - R_c}{1 - R_M R_c} \right)^2 I_o \qquad (7.61)$$

where R_M and R_c are the reflectivities of the ordinary mirror and the phase conjugate mirror, respectively. Note that as the phase conjugate reflectivity approaches unity the specular reflection tends to zero. This remarkable result shows that even for a highly reflecting mirror the scattered light can be completely eliminated by placing a perfect phase conjugator behind the mirror. This is possible owing to the multiple reflections between the mirrors. The light which is reflected from the phase conjugate mirror interferes destructively with the specularly reflected beam. In other words, the whole set-up works as if there was no ordinary mirror present at all.

The appeal of phase conjugation should be immediately apparent. Among the many applications, we can mention the prospect of transmitting images undistorted through optical fibres or through the atmosphere, lensless imaging of objects having submicrometre resolution, and tracking of fast-moving satellites (with possibly devastating effect if the phase conjugate reflectivity is much greater than unity). Other applications include using phase conjugation for phase-locking

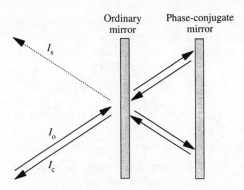

Fig. 7.11 A Fabry–Perot interferometer consisting of a semi-transparent ordinary mirror in parallel with a phase conjugate mirror.

of lasers, image processing, reconfigurable optical interconnects, beam quality improvement, interferometry, and novelty filters, to mention just a few (see Part III).

How are phase conjugate beams generated? In 1949, Dennis Gabor 'invented' holography[3] and in the process showed that if a hologram is written by interfering an image-bearing beam with a reference plane wave, then the phase conjugate of the image beam could be generated by reading the hologram with a plane wave propagating antiparallel to the original reference beam. The underlying principle has been discussed in Section 3.12. In 1965 Kogelnik demonstrated the possibility of aberration correction, although it was exceedingly impractical at the time owing to the long developing time of holographic film and the fact that once written, the hologram is fixed. Furthermore, the hologram had to be replaced back in exactly the same position in order to have good quality phase conjugation. Real-time phase conjugation became a reality in 1972 when Zel'dovich et al. observed phase conjugation due to stimulated Brillouin scattering in a high pressure methane gas. In 1977 Hellwarth proposed that the phase conjugate beam could be generated by four-wave mixing in a non-linear medium, and this has since been demonstrated using a large variety of different non-linear materials.

In the following we shall look at how four optical waves can interact inside a photorefractive material. We shall concentrate on the steady-state phase conjugation, looking at various aspects which influence the reflectivity. Section 7.6.2 examines a number of resonator configurations which allow for self-pumped phase conjugation. We conclude our survey in Section 7.6.3 with a brief look at mutually pumped phase conjugation.

7.6.1 Four-wave mixing

The basic configuration for phase conjugation via four-wave mixing is depicted in Fig. 7.12. With two counterpropagating plane-wave pump beams, beams 1 and 2, illuminating the crystal, any additional beam (i.e. beam 4) which is incident on the

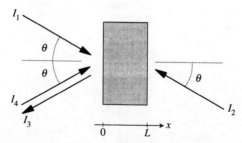

Fig. 7.12 Basic configuration for phase conjugation via four-wave mixing in a photorefractive material.

---

[3]Incidentally, Gabor's research on holography won him the Nobel Prize in physics in 1971.

7.6 Phase conjugation

crystal will be phase conjugated and retroreflected (beam 3). The four beams will interact with each other by setting up four diffraction gratings[4]: one transmission grating, one reflection grating, and two gratings of period $\lambda/2$ created by the pairs of counterpropagating beams. A complete treatment of the wave interactions is now exceedingly complicated and in general it becomes necessary to resort to numerical solutions. Proper attention then has to be paid to the space charge fields set up by the various interference patterns of different spatial frequency and direction with respect to any applied electric field. Furthermore, the propagation direction, polarization, and wavelength of the beams dictate the effective electro-optic coefficient that they experience. The seminal work in understanding phase conjugation in photorefractive materials was carried out by Cronin-Golomb et al. (1984), who presented analytical solutions under various approximations, the most notable being the one-grating approximation. This stipulates that although there are in general four gratings, in many situations one grating is dominant. We shall in the following adopt this approach and assume that only the transmission grating is present. In this case, if absorption and optical activity are neglected then the coupled equations for the amplitudes of the four waves simplify to:

$$\frac{d\mathcal{E}_1}{dx} = j\kappa E_w g \mathcal{E}_4 \tag{7.62a}$$

$$\frac{d\mathcal{E}_2^*}{dx} = j\kappa E_w g \mathcal{E}_3^* \tag{7.62b}$$

$$\frac{d\mathcal{E}_3}{dx} = -j\kappa E_w g \mathcal{E}_2 \tag{7.62c}$$

$$\frac{d\mathcal{E}_4^*}{dx} = -j\kappa E_w g \mathcal{E}_1^* \tag{7.62d}$$

where

$$g = \frac{\mathcal{E}_1 \mathcal{E}_4^* + \mathcal{E}_2^* \mathcal{E}_3}{I_o}, \quad \kappa = \frac{\pi r_{\text{eff}} n^3}{\lambda \cos \theta} \tag{7.62e}$$

We shall proceed by further assuming that the pump beams are undepleted, i.e. I_1, $I_2 \gg I_3, I_4$. This greatly simplifies the mathematics, while still retaining most of the important features. With $\mathcal{E}_1$ and $\mathcal{E}_2$ constant, eqns (7.62c) and (7.62d) are readily integrable. Using the boundary conditions $\mathcal{E}_3(L) = 0$ and $\mathcal{E}_4^*(0)$ known, the solution for the amplitude of beam 3 is:

$$\mathcal{E}_3(x) = \mathcal{E}_4^*(0) \frac{\mathcal{E}_2}{\mathcal{E}_1^*} \left[\frac{\exp(-j\kappa E_w(x-L)) - 1}{\exp(j\kappa E_w L) + r} \right] \tag{7.63}$$

---

[4]There will be six separate gratings if the pump beams are not counterpropagating.

where r is the ratio of the pump intensities, I_1/I_2. Let us consider what this solution means. First, it is seen that the amplitude of beam 3 is directly proportional to the phase conjugate amplitude of the beam 4. Second, Bragg matching conditions require that $\mathbf{k}_2 - \mathbf{k}_3 = \mathbf{k}_4 - \mathbf{k}_1$. Therefore if the pump beams are counterpropagating (i.e. $\mathbf{k}_1 = -\mathbf{k}_2$) then so are beams 3 and 4. Taken together, we have that both the phase and direction of propagation of beam 4 have been reversed. In other word, beam 3 is the phase conjugate of beam 4.

Let us now turn to the phase conjugate reflectivity, R_c, defined as the ratio of the intensities of beams 3 and 4 at $x = 0$. It follows from eqn (7.63) that the reflectivity is given by

$$R_c = \frac{I_3(0)}{I_4(0)} = r \left| \frac{\exp(j\kappa E_w L) - 1}{\exp(j\kappa E_w L) + r} \right|^2 \tag{7.64}$$

Let us concentrate first on the case of purely imaginary E_w (i.e. $\Phi = 90°$), applicable to diffusion recording of holograms in photorefractive materials. For equal pumping (i.e. $r = 1$), the reflectivity simplifies to $R_c = \tanh^2(\kappa |E_w| L/2)$. This cannot exceed unity. In order to obtain reflection with gain it is necessary to have asymmetric pumping. By comparison, if the response of the material is local (i.e. $\Phi = 0°$) then the optimum reflectivity always occurs for equal pumping, and reflectivities exceeding 100% are possible. In Fig. 7.13 we plot the reflectivity versus pump ratio for $|\kappa E_w L| = 3$ and $\Phi = 0°$, $30°$, $60°$, and $90°$. It may be seen from eqn (7.64) that the optimum pumping condition occurs when (Fischer et al. 1981)

$$\kappa \mathrm{Im}(E_w) L = \ln(r) \tag{7.65}$$

giving a maximum reflectivity of

$$R_{c,\max} = \left| \frac{\sinh(-j\kappa E_w L/2)}{\cos(-\kappa \, \mathrm{Re}(E_w) L/2)} \right|^2 \tag{7.66}$$

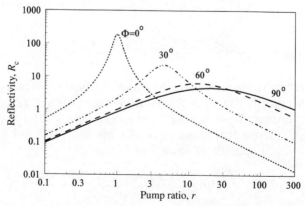

Fig. 7.13 Phase conjugate reflectivity as a function of pump beam ratio for $|\kappa E_w L| = 3$ and $\Phi = 0°$, $30°$, $60°$, and $90°$.

7.6 Phase conjugation

Referring back to Fig. 7.13, this means that for $\Phi = 90°$ a maximum reflectivity of 4.5 is possible at a pump ratio of 20.

It is interesting to note that according to eqn (7.66) it is possible to have infinite reflectivity or so-called self-oscillation (Yariv and Pepper 1977). What this means is that we may generate an output beam without any input beam. The condition for self-oscillation is:

$$\text{Im}(E_w)L = \ln(r) \quad \text{and} \quad k\,\text{Re}(E_w)L = \pm\pi, \pm 3\pi, \ldots \quad (7.67)$$

Note that this rules out self-oscillation for $\Phi = 90°$. This is not to say that self-oscillation is not possible with photorefractive materials, but it does mean that it becomes necessary to change the phase shift by either applying an electric field to the crystal or detuning the pump beams (J. F. Lam, 1983).

At this stage it is worth while to digress slightly and recast the process of phase conjugation in a slightly different form. Figure 7.14 illustrates our starting point. By looking at phase conjugation as two separate processes of two-wave mixing, we may capitalize on our familiarity (obtained in Part I) of the simpler two-wave mixing process and develop a better intuitive understanding of some of the peculiarities of phase conjugation in photorefractive materials. We start by recalling that for two-wave mixing in photorefractive materials the energy transfer is in one specific direction (in the direction of the c axis) irrespective of whether the optical beams enter from the left or right side. Thus if beam 4 gains energy, as depicted in Fig. 7.14, then beam 3 will lose energy. The two processes clearly work against one another. Albeit an extremely over-simplified view of four-wave mixing, this does help, however, to explain qualitatively why it is beneficial at $\Phi = 90°$ to have asymmetric pumping which will imbalance the energy transfer and consequently lead to a net gain of the phase conjugate beam. It was these types of arguments which prompted Stepanov and Petrov (1984a,b), to propose what they called a positive feedback configuration for phase conjugation. This entails exploiting the fact that for certain anisotropic crystals it is possible to control the direction of energy transfer by changing the input polarization. Thus, if beams 1 and 4 are polarized orthogonal to beams 2 and 3, then both two-wave mixing processes will contribute to the amplification of the signal beam and its phase conjugate. Mathematically it means that κ changes sign in eqns (7.62) and (7.63). For

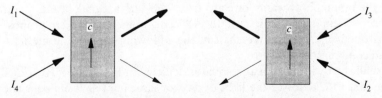

Fig. 7.14 Two-wave mixing interpretation of phase conjugation, showing how the signal beam, I_4, is amplified on traversing from left to right, but then loses energy upon traversing back (from right to left) as the phase conjugate beam, I_3.

unity beam ratio the phase conjugate reflectivity then becomes (Stepanov and Petrov, 1984a)

$$R_c = \left| \frac{-j\kappa E_w L}{2 + j\kappa E_w L} \right|^2 \tag{7.68}$$

which for $\Phi = 90°$ means that the self-oscillation threshold is $\kappa \, \text{Im}(E_w)L = 2$. The reflectivity improvements of the positive feedback configuration have been demonstrated experimentally for BTO (Stepanov and Petrov 1984a,b) and GaAs (Rajbenbach et al. 1989b). Fischer and Weiss (1988) have since derived a more general equation to account for optical activity, asymmetric pumping, and pump depletion.

Our treatment so far has concentrated on the simplest imaginable case where one grating is dominant and the pump beams are undepleted. These approximations are, however, very limiting and are consequently unsuited for many practical applications. In particular, the undepleted pumps approximation is clearly inappropriate when the pumping beams are derived from the signal beam itself, as is the case for the self-phase-conjugate mirrors of the next section. Many different solution techniques incorporating various approximations have been proposed in the past. These include, for the one grating approximation, the exact solution by Cronin-Golomb et al. (1982a, 1984), which accounts for pump-depletion, and the perturbational solution by Petersen (1991), which provides a simple analytical solution for the case of detuning. We shall not here expound on virtues and properties of the many solutions, but shall instead refer to a good recent review on the subject by Yeh (1993).

7.6.2 Self-pumped phase conjugate resonators

So far we have looked at how a phase conjugate mirror can be realized by pumping a non-linear material with two mutually phase conjugate beams. In this way it is possible to 'turn on' the phase conjugate mirror at will and even control the reflectivity by varying the intensity of the pump beams. This method, however, places a high demand on the quality and alignment of the external pumping beams and the uniformity of the crystal itself. A different class of phase conjugate mirrors which have obvious appeal are the passive phase conjugate mirrors (PPCM). These do not require external pumping, but instead derive their pumping beams directly from the input signal beam. Figure 7.15 illustrates schematically a few of the PPCMs which have been proposed and demonstrated in the past. We shall in the following discuss the merits of each of these configurations in turn.

We shall start with the linear resonator depicted in Fig.7.15(a). Historically this was the first PPCM to see the light of day, or laser light as it happens (Feinberg and Hellwarth 1980; White et al. 1982; Cronin-Golomb et al. 1982b,c, 1984, 1986; Odoulov and Soskin 1983; Odoulov and Sukhoverskova 1984; Cronin-Golomb and Yariv 1986; Kalinin and Solymar 1988a,b). To understand how it works we

7.6 Phase conjugation

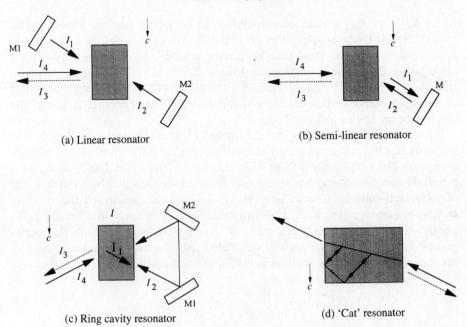

Fig. 7.15 Self-pumped phase conjugate resonators.

need to look at how a resonator beam can be set up inside the Fabry–Perot cavity. When a single beam is incident on the crystal, light will be scattered in various directions due to impurities, crystal imperfections, or random fluctuations of the space charge field (see Chapter 8). The light which is scattered in the direction of the c axis is amplified owing to two-wave mixing in the crystal. However, of all the beams which are amplified only the beam in the mirror cavity undergoes successive amplification due to the optical feedback. As a result, this beam will eventually win out and a steady-state resonator beam will be established. Once created, beams 1 and 2 act as pumping beams which will transform the photorefractive crystal into a phase conjugate mirror.

Two aspects are worth noting at this stage: (1) because the pumping beams are derived from the input signal beam, the reflectivity cannot exceed 100% in steady state; (2) the nature of the build-up of the resonator beam requires that the non-linear medium has a non-local response. Can we say anything about the oscillation conditions of the linear resonator? Conventional resonator arguments would suggest that oscillation can be sustained as soon as the gain exceeds the losses. Assuming that the only significant losses are due to non-perfect mirrors (i.e. neglecting crystal absorption and losses due to scattering and Fresnel reflections), we would expect the threshold coupling strength to be given by

$$R_M = R_{M1} R_{M2} = \exp[2\kappa \operatorname{Im}(E_w) L] \qquad (7.69)$$

where R_{M1} and R_{M2} are the reflectivities of the respective mirrors. This is in fact also the result Cronin-Golomb et al. (1982c), arrived at using the one-grating approximation. In the ideal case of perfect mirrors the coupling threshold is zero, giving the linear resonator the lowest threshold of all the passive phase conjugate mirrors. The bad news is that the alignment of the mirrors is critical to the performance of this device, and furthermore, the coherence length of the incident beam has to be quite large.

It is important to emphasize that the linear PPCM is self-starting, meaning that the resonator beam will appear from 'nothing' provided the threshold coupling is exceeded. The semi-linear PPCM (Fig 7.15b), on the other hand, also has a threshold coupling strength above which it can sustain oscillations, but it is not normally self-starting. Cronin-Golomb et al. (1984) demonstrated theoretically and experimentally that oscillation can build up in this configuration if a sufficiently strong seeding beam is used to initiate the oscillation. Using the single-grating approximation the authors calculated that the oscillation would be self-sustaining if the coupling coefficient was above the coupling threshold of

$$(\kappa E_w L)_{th} = \sqrt{1+R_M} \ln\left(\frac{\sqrt{1+R_M}-1}{\sqrt{1+R_M}+1}\right) \tag{7.70}$$

For an ideal mirror this simplifies to $(\kappa E_w L)_{th} = 2.49$. A clarification is called for here. Namely, the fact of the matter is that the semi-linear configuration *can* be self-starting. This can come about if scattering (discussed in Section 8.2) of the incident optical beam is such as to supply the necessary seed beam.

We turn now to the ring cavity configuration of Fig. 7.15(c) (Odintsov and Rogacheva 1982; Cronin-Golomb et al. 1983) which, with good reason, is one of the most popular PPCM configurations. The set-up is now such that the signal beam which traverses the crystal is redirected back into the crystal to act as a pump beam. Again two-wave amplification is crucial to the setting up of the cavity beam. The coupling threshold for the ring resonator PPCM is

$$(\kappa E_w L)_{th} = \frac{R_M+1}{R_M-1} \ln\left(\frac{R_M+1}{2R_M}\right) \tag{7.71}$$

For $R_M = 1$ this becomes $(\kappa E_w L)_{th} = 1$. The are several other attractive features. For one, it is a self-starting phase conjugate mirror. Secondly, there are essentially no problems associated with alignment of the cavity beam. Furthermore, Cronin-Golomb et al. (1984) and Cronin-Colomb and Lau (1985) noted that the requirements on coherence length are not very severe for this configuration. This is because beam 1 is formed by diffraction of beam 4 and the two beams subsequently travel the same distance before they intersect again in the crystal (as beams 2 and 3). This ensures that both pairs of beams are automatically coherent. This means that phase conjugation should be unaffected by changes in cavity length. However, it should be pointed out that Cronin-Golomb et al. (1983), found that high quality phase conjugation is only observed for signal beams

7.6 Phase conjugation

with a complicated structure. This minor complication can easily be circumvented by placing an aberrator in front of the phase conjugate mirror.

There is one last benefit of the ring resonator which is worth mentioning. Although the threshold of the ring cavity PPCM is higher than that of the linear mirror, once the oscillation threshold is crossed, the reflectivity almost immediately exceeds that of the linear resonator PPCM. This behaviour is clearly seen in Fig. 7.16, which compares the reflectivities of the various configurations in Fig. 7.15 for the case of ideal cavity mirrors (Cronin-Golomb *et al*. 1984).

The last passive phase conjugate mirror we shall consider here is the class of conjugators for which the optical feedback loop is provided by total internal reflection, thus eliminating altogether the need for additional optics. The first device of this kind was demonstrated experimentally by Feinberg (1982b) and later analysed in more detail by MacDonald and Feinberg (1983). The basic configuration is that of Fig. 7.15(d). With a single beam incident on a crystal of $BaTiO_3$, fanning beams were seen to form in such a way that they re-intersected the main beam after being retroreflected from one corner of the crystal. The phase conjugate properties of this device were argued to stem from setting up of two separate interaction regions. This configuration has since come to be known as the 'cat' phase conjugator because Feinberg used a picture of his pet cat to convincingly demonstrate the high quality aberration correction abilities of this device. The theoretical reflectivity of this two-interaction-region phase conjugator has been plotted in Fig. 7.16 as the short-dashed curve. The oscillation threshold is fairly high, being $(\kappa E_w L)_{th} = 4.68$ for perfect internal reflection (Cronin-Golomb

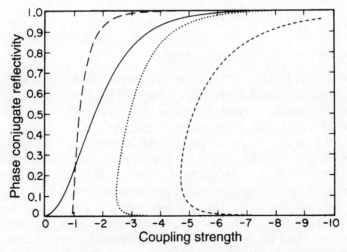

Fig. 7.16 Phase conjugate reflectivity as a function of coupling strength, $\kappa E_w L$, for a linear resonator (———), ring resonator (– – – –), semi-linear resonator(· · · ·), and a two-interaction region 'cat' phase conjugate mirror (- - - -) (after Cronin-Golomb *et al*. 1984).

et al. 1984). It should also be noted that this device is not inherently self-starting, but has to rely on seeding, for example from fanning beams. Despite these disadvantages, total internal reflection PPCMs are in general very popular because they offer the advantage of being compact and relatively insensitive to vibrations. Furthermore the coherence length of the input light only has to be comparable with the size of the crystal. For detailed measurements of the effect of the position of the input beam see Gower and Hribek (1988). As a final comment, it is worth noting that the geometry of Fig. 7.15(d) is not the only, or necessarily the best, configuration for a PCCM based on total internal reflection. There have recently been substantial theoretical and experimental research efforts towards designing and cutting crystals in novel shapes with the aim of reducing build-up time and increasing reflectivity (Cronin-Golomb and Brandle 1989; Medrano *et al.* 1994; H. Y. Zhang *et al.* 1990; Mu *et al.* 1995).

We shall round this section off with a brief look at the performance of some of the passive phase conjugate mirrors which have been demonstrated to date. $BaTiO_3$ has captured most of the limelight by exhibiting reflectivities up to 74% over a large range of wavelengths (Dunning *et al.* 1990; Pepper 1986; Ross and Eason 1992), but appreciable phase conjugation has also been observed in $KNbO_3$ (Rytz and Shen 1989), SBN (Salamo *et al.* 1986), KNSBN (Yue *et al.* 1989), KTN (J. Wang *et al.* 1992), and KLTN (Wei *et al.* 1994), to name just a few. Self-pumped phase conjugation has also been demonstrated with semiconductor and sillénite materials but the reflectivities have typically been much lower, e.g. 11% in InP:Fe (Bylsma *et al.* 1989). Millerd *et al.* (1992a,b) have attributed the relatively poor performance of these materials to the combined effects of large signal effects (which arise because coupling has to be enhanced with one of the enhancement techniques of Section 5.4.) and absorption, with the latter being the biggest problem because it places limits on the length of the sample that can be used.

7.6.3 Mutually pumped phase conjugators

Until now we have looked at phase conjugate mirrors for which all the interacting beams are mutually coherent. In this section we introduce a new class of phase conjugator, the mutually pumped phase-conjugator (MPPC), which has to date only been observed with photorefractive materials. The basic configuration, known as the double phase conjugate mirror (DPCM), is depicted in Fig. 7.17(a) and has two beams incident on a photorefractive material. Sternklar *et al.* (1986) were the first to demonstrate experimentally that this configuration could lead to the phase conjugation of both beams once the coupling strength exceeded a certain threshold. The surprise in their discovery was that the two incident beams did not need to be mutually coherent or, in fact, even of the same colour (Sternklar and Fischer 1987; Fischer and Sternklar 1987; Kaczmarek *et al.* 1994b)[5], as discussed in Section 3.12. We shall briefly repeat here the main

---

[5]In which case it is commonly referred to as a double-colour pumped oscillator (DCPO).

7.6 Phase conjugation

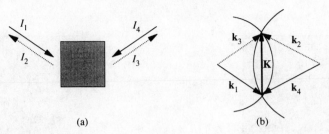

Fig. 7.17 (a) Basic configuration of the double phase conjugate mirror. (b) Ewald diagram showing the 'shared' grating which is responsible for phase conjugation.

arguments. Any single beam which is incident on the crystal will undergo some scattering (see Section 8.2). The scattered light will subsequently interfere with the main beam and so give rise to a multitude of 'noise' gratings. Now, if two beams are incident they may interact with one another via those gratings which are common to both beams (Ewbank 1988). In fact, all those gratings which are not common will be coherently erased (He et al. 1992). The Ewald diagram of Fig. 7.17(b) illustrates how this sharing of a transmission grating will lead to the coupling of the incoherent beams and the subsequent generation of the phase conjugate beams.

Using the undepleted pumps approximation it is possible to show that for $\Phi = 90°$ the oscillation threshold for the double phase conjugate mirror is given by (Petrov et al. 1989)

$$\kappa \, \text{Im}(E_\text{w})L = \frac{r+1}{r-1} \ln r \tag{7.72}$$

Thus, the lowest threshold occurs for unity pump ratio, giving $\kappa \, \text{Im}(E_\text{w})L = 2$. We should note at this stage that if the incident beams are plane waves they will give rise to conical scattering (see Section 8.2.4). So, in order to have good phase conjugation the incident beams must be of a complex (speckle-like) pattern so as to to avoid the degeneracies in the Bragg matching angle on a given volume grating.

The double phase conjugate mirror configuration depicted in Fig. 7.17(a) is but one of many different configurations for which mutually pumped phase conjugation has been observed. In addition to the DPCM (Weiss et al. 1987; Eason and Smout 1987; Smout and Eason 1987), there are the 'bird-wing' conjugator (Ewbank 1987), 'bridge' conjugator (D. Wang et al. 1989; Sharp et al. 1990), 'frog-legs' conjugator (Ewbank et al. 1990), and 'fish-head' conjugator (C. C. Chang and Selviah 1995), all of which are illustrated in Fig. 7.18. These configurations are technically distinguished by the number of total internal reflections and the number of interaction regions. The imaginative names stem from the 'close' resemblance to their namesakes.

Having mentioned some of the main properties and configurations of the mutually pumped phase conjugator, let us consider what makes these devices

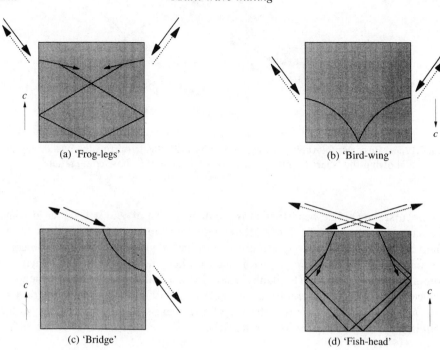

Fig. 7.18 Mutually pumped phase conjugate mirrors.

particularly attractive. Their main advantage is that they are self-aligning. Thus, even when the incident angle or position of the beams is changed dynamically it will not (significantly) affect the phase conjugation, provided the crystal reacts fast enough. This makes the MPPC especially attractive for realizing free-space bidirectional optical links, such as interconnections between optical fibres (Caulfield *et al.* 1987; Weiss *et al.* 1987; Wolffer and Gravey 1994). Furthermore, it means that the MPPC lends itself very well to efficient beam steering, which can be realized by using a variable wavelength source for one of the beams (Fischer and Sternklar 1987). Among the other capabilities which have been demonstrated are locking of lasers (Sternklar *et al.* 1986; MacCormack *et al.* 1995), and image colour conversion (Sternklar and Fischer, 1987; Fischer *et al.* 1989).

Finally, before concluding this section we should note there has recently been some discussion as to whether the double phase conjugate mirror is an oscillator or an amplifier. In other words, the question is whether the DPCM has a gain threshold after which the reflectivity grows exponentially (i.e. an oscillator), or whether there is no well-defined threshold and different seeds are amplified to different levels, i.e. an amplifier. It has been suggested (Zozulya 1991; Belic *et al.* 1994; Korneev and Sochava 1995) that while the DPCM may be an oscillator in the plane-wave approximation, in the more correct two-dimensional treatment the DPCM is in fact a convective amplifier. However, the latest calculations of Engin *et al.* (1994) indicate that the DPCM behaves as an oscillator at and above a certain threshold. The authors showed that there are in fact two thresholds: one for the phase conjugate reflectivity and one for phase conjugate fidelity.

8
SPURIOUS BEAMS

8.1 Introduction

A spurious beam may be defined as a beam which by certain simple considerations should not be there. For example, if we have one or more beams incident upon a piece of ordinary optical material (say, a window pane) the expectation is that apart from a small attenuation the beams will be unchanged. When it comes to photorefractive materials we know that the various beams may interact with each other and consequently their relative output intensities may change. But can the beams wander away from their original paths or, even worse, can new beams arise out of nothing? Yes, we have discussed such phenomena a number of times in this book: higher diffraction orders in Section 3.8, ring resonators in Section 3.11, the double phase conjugate mirror in Section 3.12, and a string of phase conjugate resonators in Section 7.6. And of course self-phase-conjugation of the types shown in Figs 7.15 and 7.18 are prime examples of beams appearing rather unexpectedly and following rather unconventional paths. So why do we need an extra chapter on spurious beams? The reason is that apart from the examples quoted, there are still plenty of configurations in a number of photorefractive materials which lead to the appearance of new beams and for which some classification is desirable.

The simplest classification may be to divide the subject into two parts: scattering and instabilities. This is essentially the plan we have adopted. Scattering is caused by random inhomogeneities in the crystal whereas instability comes about because the system cannot remain in its quiescent state in the presence of any small perturbations. The distinction can be even better emphasized in mathematical terms. The scattering of the input optical beam is described by the field equations whereas the instability of the space charge field can be deduced from the materials equations. Accordingly, we shall discuss scattering in Section 8.2 where all the arguments will be related to optical beams, and we shall discuss in Section 8.3 the best-documented instability, the subharmonic instability, in terms related to the properties of the material. We are well aware of the fact that there are other kinds of instabilities too, completely describable by field equations, as for example the instability of the phase conjugate beam, and there is substantial literature dealing with them. Our compromise solution is to add one more Section, 8.3, under the name of spatio-temporal instabilities, which will briefly summarize recent research.

8.2 Scattering

8.2.1 Introduction

There is no accepted classification for scattering studies, partly because the effects are still only partially understood and because this is not a finished subject: new scattering mechanisms are still reported every year. Is it clear where to start? Well, scattering studies started with the very first report (Ashkin *et al.* 1966) on photorefractive properties. The authors observed a nice clean beam going in and a seriously distorted beam coming out at the other end of the crystal. A number of similar patterns were found afterwards. See for example the later studies by Avakyan *et al.* (1978, 1983).

We shall start in Section 8.2.2 with the case when the transverse intensity distribution of the light beam is non-uniform and the appearing refractive index variation affects the passage of the beam through the crystal. In Section 8.2.3 we shall discuss scattering amplified by two-wave mixing, followed in 8.2.4 by the Bragg diffraction of one beam on 'noise' gratings recorded by another beam and its own scattered radiation. A combination of these effects is discussed in Section 8.2.5 which may also be interpreted as forward four-wave interactions. Scattering due to backward four-wave interactions is described in Section 8.2.6. Anisotropic scattering, i.e. scattering due to the anisotropic nature of some photorefractive materials, is the subject of Section 8.2.7. It is a fairly big subject related strongly to anisotropic Bragg diffraction which has already appeared in Section 6.9 concerned with diffraction efficiency. A few further topics which did not quite fit into the previous sections will be discussed in Section 8.2.8. Having discussed the build-up of scattering, it makes good sense to finish this study with a section on attempts at how to avoid that build-up. That will be Section 8.2.9.

8.2.2 Beam distortion due to non-uniform transverse intensity distribution

This is going back again to early observations of the photorefractive effect as 'optical damage'. F. S. Chen (1967) already calculated the transverse space charge field due to the finite size of the light beam. The cause of beam distortion is even more clearly stated in a later paper by the same author (F. S. Chen 1969, quoted already in the history chapter): 'The effect is attributed to the drifting of the photoexcited electrons out of the illuminated region followed by their retrapping near the beam periphery. The space charge field between these retrapped electrons and the positive ionized centers in the illuminated region causes the observed change of refractive indices via the electro-optic effect of the samples'. Further studies of this mechanism were made by W. D. Johnston (1970), Moharam and Young (1976), Feinberg (1982a), El Guibaly (1983), and El Guibaly and Young (1980, 1983). Perhaps the most important characteristic of this effect is that it needs for its existence no more than an intensity distribution varying in the transverse plane. It can be either coherent or incoherent, in contrast to phenomena we shall discuss later when the scattering beam must be coherent with the beam causing the scattering.

8.2.3 Scattering due to two-wave amplification

This type is mentioned in Section 3.11 where the basic physics is described. A single beam, let's call it a pump beam, is directed at the photorefractive material and it is scattered by the various inhomogeneities. Scattered beams in each direction may then be imagined as plane waves capable of interacting with the pump beam. If the scattering is weak then the process may be regarded as a two-wave interaction between the pump beam and an individual scattered wave. Waves scattered in the direction in which two-wave amplification is sufficiently high will then be observed. These were the conclusions reached by Voronov et al. (1980) and later confirmed by Obukhovskii and Stoyanov (1985) in scatter studies in SBN (mentioned already in Section 3.11). This is the fundamental mechanism which is always present to some degree in all the scattering processes discussed, though we shall have to take into account some other effects as well when we come to the complicated scattering patterns observed e.g. in $LiNbO_3$ and $BaTiO_3$.

The scattering diagrams measured by Voronov et al. (1980) were in the forward direction but, on the same principle, backward scattering should also be possible. T. Y. Chang and Hellwarth (1985) explained the observed phase conjugation in $BaTiO_3$ as originating from backscattering, i.e. from the interaction of the input beam and a backscattered wave. Backscattering from a $LiNbO_3$:Fe crystal was in fact used by Litvinenko and Odoulov (1984) to provide an external mirror for a copper vapour laser. Theoretical calculations by Valley (1987) showed that backscattering will be optimized in a hole-dominated crystal with a large trap density.

Scattering diagrams observed in BSO (Imbert 1986; Rajbenbach et al. 1989a; Ellin 1994) can also be explained with this simple two-wave mixing model although the actual calculations are quite complicated since absorption, optimum detuning, optical activity, and the piezoelectric effect (see Section 6.10) must all be included. The numerical analysis by Ellin (1994) was based on the vector differential equations of Ellin et al. (1994). The piezoelectric effect was also needed (Montemezzani et al. 1995) for explaining the petal type of scattering from $BaTiO_3$.

8.2.4 Bragg diffraction from noise gratings

This phenomenon was discovered by Moran and Kaminow (1973) by observing scattering rings in the diffraction pattern of poly(methyl methacrylate), a holographic material. It was explained in terms of Bragg diffraction by Forshaw (1974). The basic principles are as follows: The material is illuminated by a single beam incident with wave vector k_1. We assume that it scatters in all possible directions. The input beam and a scattered beam produce a so-called noise grating. Since the scattered beams are in all possible directions, it follows that in the

8.2 Scattering

developed material (it may of course be self-developing, i.e. photorefractive) there will be grating vectors originating at point A and terminating at all points on the Ewald sphere: see e.g. the scattered beam with wave vector $\mathbf{k}_{1s}$ in Fig. 8.1 which reaches the Ewald sphere at B and produces the grating vector $\mathbf{K}_s$. Next we shall introduce a second (replay) beam incident with wave vector $\mathbf{k}_2$ (chosen here of different length since we allow the possibility of a change in wavelength at replay). This beam will now be diffracted by the noise gratings. The diffracted beam will however be weak in all directions except in those in which the Bragg conditions are satisifed, i.e. for the diffracted beam $\mathbf{k}_{2d}$ which also ends in point B, and for which

$$\mathbf{k}_{2d} = \mathbf{k}_2 + \mathbf{K}_s \tag{8.1}$$

More generally, the Bragg conditions are satisfied at all points where the sphere of radius k_1 intersects the sphere of radius k_2. The intersection defines a circle leading to a conical scattered beam originating at O_2. Note that depending on the input wave vector of the replay beam (Forshaw 1974) we may observe another type of conical scattering which occurs when the scattered wave vector $\mathbf{k}_2 - \mathbf{K}_s$ satisfies the Bragg conditions.

In a holographic material in which scatter is strong (e.g. silver halide emulsions) the noise gratings may have high amplitudes and may, in fact, entirely deplete the input beam (Solymar and Riddy 1990). Photorefractive materials of good quality do not strongly scatter, hence the noise gratings are weak and the conical diffraction described is not regularly observed. It was first reported in LiNbO$_3$ by Magnusson and Gaylord (1974).

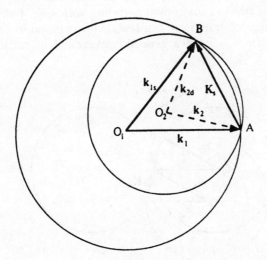

Fig. 8.1 Generation of scattering rings, illustrated by means of an Ewald diagram for the writing and (Bragg-matched) reading of noise gratings.

8.2.5 Combination of Bragg diffraction and two-wave mixing

The noise gratings in the previous section were described as being due to beams scattered in all directions. In photorefractive materials we may however assume that only those scattered beams are of any significance which are in a direction in which two-beam amplification is significant. Such scatter rings were found in BSO by Ellin and Solymar (1994) but their intensity was weak and the rings could only be seen when the input beam causing the scattering was strong ($\approx 3 \text{ W cm}^{-2}$).

Another interesting phenomenon is the appearance of dark rings in a background of fairly uniform scattering as observed first by T. Y. Chang and Yeh (1987). The explanation of this phenomenon is slightly more difficult. It was given by Horowitz and Fischer (1992) who provided further experimental results as well. There are now two beams incident (Fig. 8.2) with wave vectors $\mathbf{k}_1$ and $\mathbf{k}_2$ which are assumed to be of the same wavelength though not coherent with each other. Scattering is assumed again in all possible directions. It needs to be noted that incoherent beams may interact with each other via their scattered beams, provided that beam 1 and its scattered beam produce the same grating as beam 2 and its scattered beam. With reference to Fig. 8.2 this occurs when

$$\mathbf{k}_1 - \mathbf{k}_{1s} = \mathbf{k}_{2s} - \mathbf{k}_2 = \mathbf{K}_s \tag{8.2}$$

We need to remember that two-wave amplification has a preferred direction, the direction of the c axis when the carriers are electrons. Thus if the c axis of the material ($BaTiO_3$ in the experiments of Horowitz and Fischer) is in the direction shown, then the scattered beam with wave vector $\mathbf{k}_{1s}$ will be amplified and the scattered beam with wave vector $\mathbf{k}_{2s}$ will be de-amplified due to the interaction between the four beams via the common grating vector $\mathbf{K}_s$. But this will occur only when the Bragg conditions are satisfied. If the material has high scatter, as $BaTiO_3$ has, then there will be a general background of scattered radiation.

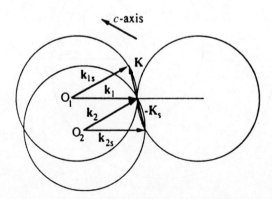

Fig. 8.2 Ewald diagram to understand the appearance of dark scattering rings by combining Bragg diffraction criteria with two-wave mixing.

8.2 Scattering

However, there will be a reduction of scatter in the directions in which the Bragg conditions are satisfied. Consequently, a dark ring will appear.

Equation (8.2) can be interpreted as forward four-wave mixing. A similar mechanism is at play in the experiments of Dolfi *et al.* (1993), who found interaction between two incoherent beams incident on the same side of a $BaTiO_3$ crystal. They showed that the scattering persists for a while when one of the beams is switched off.

Four waves (two incident, two scattered) in the forward direction were involved in the experiments of Salamo *et al.* (1988) as well, with the difference that the interaction occurred between waves of different wavelengths. They used single beams from a multi-line argon laser to illuminate samples of cerium-doped SBN and BSKNN. Owing to material dispersion, blue light refracted more than green light (for simplicity, choosing only one further colour for the discussion), as shown in Fig. 8.3 where their wave vectors are represented by k_B and k_G. Assuming scattering in all directions, there must exist in the same plane a scattered green wave vector and a scattered blue wave vector (denoted by k_{Bs} and k_{Gs}) which give rise to the same grating vector, K_s, i.e.

$$k_B - k_{Bs} = k_G - k_{Gs} = K_s \tag{8.3}$$

The consequence is that the scattered waves for which this relationship is satisfied will be reinforced. Carrying out the analysis for all the wavelengths, Salamo *et al.* showed that the scatter diagram is a rainbow in which all the colours delivered by the multi-line laser add their contributions.

8.2.6 Backward four-wave mixing

The double phase conjugate mirror discussed in Section 3.1.2 was a particular example of two input waves and two scattered waves producing the same grating in a backward interaction. One could obviously have the same effect if the second

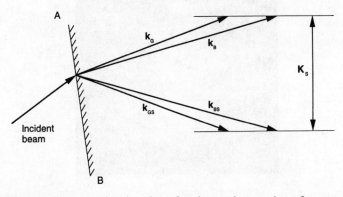

Fig. 8.3 Wave vector diagram showing the refraction and scattering of waves. The input beam is multicoloured. The blue wave is refracted more than the green wave. AB is the crystal surface.

274 *Spurious beams*

beam is obtained by boundary reflection, as shown by Rupp *et al.* (1987). In their experiments a single beam was incident perpendicularly upon the surface of $LiNbO_3$ and $LiTaO_3$ crystals whose *c* axis was in the direction of propagation. One of the scattering patterns they obtained is shown in Fig. 8.4. The evidence that boundary reflection played a significant role was obtained by placing an index-matching liquid behind the crystal. Reduced reflection led to reduced scattering.

8.2.7 Anisotropic scattering

Up to now it has been tacitly assumed that the polarization of the scattered beam is the same as the polarization of the input beam. This is not necessarily so. We have seen in Section 6.9 that an ordinary and an extraordinary beam might be coupled to each other by what we called intermode coupling. Such a case was despicted in Fig. 6.19 where the anisotropic material was $BaTiO_3$ (Kukhtarev *et al.* 1985). The interpretation was that two extraordinary beams recorded a grating and two ordinary beams were produced when the grating was read by the extraordinary beams. The result could also be interpreted as an example of scattering, in which two new beams are produced. We shall now look at some other examples of anisotropic scattering in a little more detail.

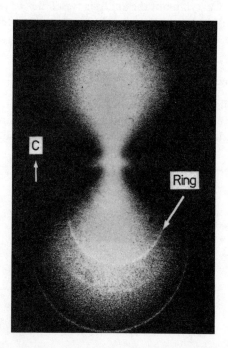

Fig. 8.4 Diffuse scattered light and part of a scattering cone due to a single beam incident upon a $LiTaO_3$ crystal whose *c* axis is in the direction of propagation (after Rupp *et al.* 1987).

8.2 Scattering

Conical scattering due to one incident beam. As discussed in Sections 8.2.4 and 8.2.5, conical scattering patterns may be obtained without relying on anisotropy, but then we need a writing beam and a reading beam. In the presence of anisotropy a single beam will suffice. The c axis should again be perpendicular to the plane of incidence and, for simplicity, we assume a single beam incident perpendicularly on the surface of a material with positive birefringence ($n_e > n_o$). Following the arguments of Belabaev *et al.* (1986) we present graphically the various scattering possibilities in Fig. 8.5. The mechanism is a kind of four-wave forward interaction in which the incident beam has a double role.

Let us first look at Fig. 8.5(a). The wave vector of the incident beam is $\mathbf{k}_o$. Scattered beams will arise from noise provided that the phase matching condition

$$\mathbf{k}_{e1} - \mathbf{k}_o = \mathbf{k}_{e2} - \mathbf{k}_o \tag{8.4}$$

is satisfied. It is clear from the geometry that the condition is satisfied for all extraordinary waves subtending an angle θ_a with the incident ordinary wave. Hence the scattering pattern is a cone of half-angle θ_a. However, this is not the only possible conical scattering for an incident ordinary beam. The phase matching condition is also satisfied when

$$\mathbf{k}_{e1} - \mathbf{k}_o = \mathbf{k}_{o2} - \mathbf{k}_o \tag{8.5}$$

as shown in Fig. 8.5(b), leading now to conical scattering of half-angles θ_{b1} and θ_{b2}.

If the incident beam has a polarization which has both ordinary and extraordinary components, then there is one more variety of phase matching:

$$\mathbf{k}_{e1} - \mathbf{k}_o = \mathbf{k}_{e2} - \mathbf{k}_e \tag{8.6}$$

as shown in Fig. 8.5(c). The cone half-angle is now θ_c.

Can we have forward scattering when only an extraordinary wave is incident? No phase matching is possible, because there is now only one point at which two circles intersect each other as shown in Fig. 8.5(d).

Can we distinguish the various types of conical scattering shown in Fig. 8.5(a)–(c)? For a LiTaO$_3$ crystal the cone angles were determined by Belabaev *et al.* (1986) as $\theta_a = 6.67°$, $\theta_{b1} = \theta_{b2} = \theta_c = 4.71°$. Thus θ_a is sufficiently different from the other cone angles to be observed separately and cases (b) and (c) can of course be distinguished from each other because they correspond to input beams of different polarization.

A clear proof for the above discussed mechanism of four-wave forward scattering was provided by Odoulov *et al.* (1985) in LiTaO$_3$. They found that the cone angle was a function of temperature, and the relationship corresponded to that expected from the temperature variation of birefringence. It needs to be mentioned here that LiNbO$_3$ has negative birefringence ($n_e < n_o$); thus for the arguments presented above to be valid, the ordinary and extraordinary beams need to be interchanged. Conical patterns in LiNbO$_3$:Cu were reported by Kiseleva *et al.* (1987), in KNbO$_3$ by van Olfen *et al.* (1992), and in BaTiO$_3$ by Rupp and Drees

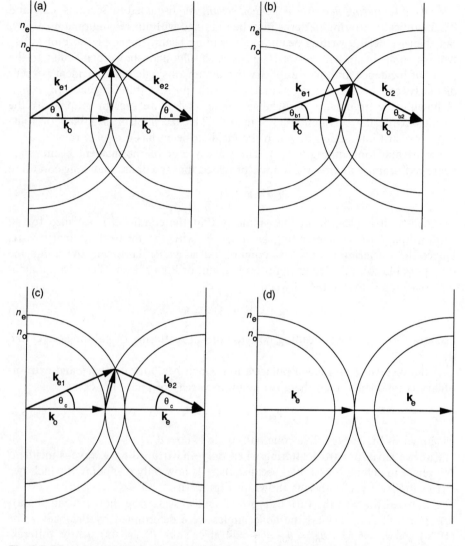

Fig. 8.5 Phase matching conditions satisfying (a) eqn (8.4), (b) eqn (8.5), (c) eqn (8.6). (d) Phase matching condition cannot be satisfied for an incident extraordinary wave.

(1986), Temple and Ward (1986), Ewbank et al. (1986), Z. Zhang et al. (1988), and Hu et al. (1989).

For an analysis of the temporal behaviour of scattering see Marotz et al. (1986).

Conical backward scattering was reported by Grousson et al. (1984a) and Obukhovskii et al. (1993). The latter authors not only identified the scattering mechanism (reflection gratings due to counterpropagating waves are involved) but also modelled the variation of intensity as a function of the azimuth angle around the cone.

8.2 Scattering

Conical scattering due to two incident beams. One ordinary and one extraordinary wave incident upon a $BaTiO_3$ crystal in the plane perpendicular to the c axis is the subject of a series of papers (Odoulov et al. 1991, 1992; Odoulov and Sturman 1992; Sturman et al. 1992e). The incident pump beams (this is for $LiNbO_3$ with negative anisotropy) $\mathbf{k}_{op}$ and $\mathbf{k}_{ep}$ must be at such incident angles that the two scattered beams of ordinary polarization, $\mathbf{k}_{o1}$ and $\mathbf{k}_{o2}$, can satisfy the phase-matching conditions

$$\mathbf{K}_1 = \mathbf{k}_{op} - \mathbf{k}_{o2} = \mathbf{k}_{o1} - \mathbf{k}_{ep} \quad \text{and} \quad \mathbf{K}_2 = \mathbf{k}_{o1} - \mathbf{k}_{op} = \mathbf{k}_{ep} - \mathbf{k}_{o2} \quad (8.7)$$

as may be seen in Fig. 8.6.

A variation on the phase-matching conditions of Fig. 8.5(b) was worked out by Novikov et al. (1992). This time $\mathbf{k}_{e1}$ and $\mathbf{k}_{o2}$ are the wave vectors of the incident beams and $\mathbf{k}_o$ is the wave vector of the scattered beam. Since this is approximately half-way between the two input beams it was described as a 'subharmonic' by Novikov et al. (1992), but its excitation is based on quite different principles from those to be discussed later in this chapter.

Straight lines have been found in scattering patterns in addition to rings both with (Sturman et al. 1993c) and without (Goulkov et al. 1993) change of polarization in the scattered beam. The crystal used was $LiNbO_3$:Fe in forward scattering. A scattering pattern found by Sturman et al. (1993c) is shown in Fig. 8.7. It was successfully described by the theory given in the same paper. Ilyenkov et al. (1992) also found a combination of rings and straight lines in a backward scattering configuration with two mutually incoherent beams incident from the opposite sides of a $LiNbO_3$:Fe crystal. The latter authors also observed a scattered wave in a nearly retropropagating direction which they called a 'forbidden' interaction geometry because existing theories do not permit radiation in those directions.

The exact cause of anisotropic scattering is not always known. It is believed that off-diagonal elements of the photovoltaic tensor (for further details see Section 9.2), known also as the circular photovoltaic effect, play a significant role (Belinicher 1978; Belincher and Sturman 1980; Odoulov 1982). In particular,

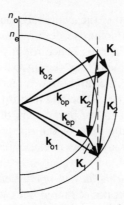

Fig. 8.6 Phase matching conditions satisfying eqn (8.7).

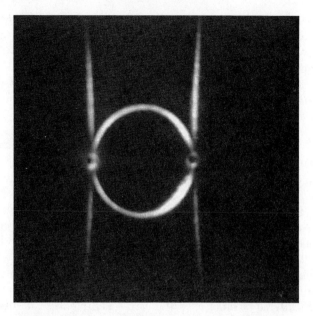

Fig. 8.7 Scattered light from a LiNbO$_3$:Fe crystal. Two extraordinary waves are incident upon the sample with an inter-beam angle (in air) of 22° in the plane normal to the c axis of the crystal (after Sturman *et al.* 1993).

it was shown by Kiseleva *et al.* (1987), who studied both copper-doped and iron-doped LiNbO$_3$, that the magnitude of the off-diagonal elements (in particular the sign of the antisymmetric component) depends on the nature of the activator rather than on the host matrix in which it is located.

8.2.8 Further notes on scattering

Alternative mechanisms. Neither the available experimental results nor the theories are sufficiently refined for a definitive description of all scattering phenomena in photorefractive materials. There is hence room for theories which may describe part of the observed phenomena, e.g. the introduction of kinky paths by Brown and Valley (1993), or point in entirely new directions. Papers by Obukhovskii *et al.* (1987) and Obukhovskii (1989) belong into this category. The proposal is that small scale non-uniformities in the parameters of the crystal (particularly in the photovoltaic constants on the scale of the optical wavelength) may significantly contribute to scattering or may actually cause an instability.

Temporally periodic scattering for c.w. input. Lemeshko and Obukhovskii (1985) reported the appearance of cones scattered backwards from a LiNbO$_3$:Fe crystal. The beam was incident along the c axis of the crystal. The scattered radiation

8.2 Scattering

appeared in pulses with a period of about 2 seconds. The phenomenon did not depend on the polarization of the input wave but did depend on the wavelength of the radiation: present at $\lambda = 0.44\,\mu m$ but absent at the longer wavelengths, $\lambda = 0.488$ and $0.63\,\mu m$.

Explosive instability. It was shown theoretically by Novikov *et al.* (1986a) that in counterpropagating four-wave and six-wave interactions the intensity of the scattered wave varies as $(L - L_{crit})^{-2}$. At the critical interaction length, strong scattering was expected. They called it an explosive instability owing to its similarity to an instability known in plasmas. In the experiments strong transient scattering was obtained which they attributed to the above mechanism.

Resonance without external mirrors. This is obviously not the place to discuss here all the possible methods which might lead to radiating beams in resonators. That is a very big subject to which we have not been able to devote sufficient space. We would just like to mention here one resonator, obtained with a $LiNbO_3$ crystal, when the cavity was formed by the two opposite crystal surfaces (Odoulov and Soskin 1983; Odulov 1984; Novikov *et al.* 1986b).

As a final comment on scattering mechanisms we wish to recommend to the reader's attention a review of scattering phenomena by Odoulov and Soskin (1989).

8.2.9 Noise reduction techniques

There are many ways to reduce noise in signal processing applications. Our aim is here to mention a few which may be employed owing to the specific properties of photorefractive crystals.

Integration of N images. The technique is to amplify an image, first diffused by a screen, by conventional phase conjugation, then repeat the process by moving the diffusing screen. The original image is erased then and a new image appears in the same position. The image is the same but the noise is uncorrelated. Hence by integrating N images, very clear, nearly speckle-free amplified images can be obtained as shown by Huignard *et al.* (1980a).

Tilting the crystal. Part of the noise is caused by multiple reflection of the input beams. A twice-reflected beam may be significantly amplified depending on the angle the reflected beam subtends with the pump beam. By tilting the crystal this particular angle can be increased so as to reduce the available gain. The noise power may be reduced by a factor of 20 as shown by Rajbenbach *et al.* (1991).

Rotation of the crystal. This technique can be used owing to the following two properties of noise in photorefractive crystals: (i) the build-up of noise (being of lower intensity) is slower than the build-up of signal amplification; and (ii) the noise sources are bound to the crystal and move with it. Hence by slowly rotating the crystal (a typical figure is 1–2° per second) the noise build-up is prevented while the signal remains unchanged. Image amplification in the absence ($\Omega = 0$) and presence ($\Omega \neq 0$) of rotation may be seen in Fig. 8.8 as reported by Rajbenbach et al. (1989a). For $\Omega = 0$, the amplified images ($\times 10^3$ and $\times 50$ for BaTiO$_3$ and BSO respectively) are corrupted by strong background amplified scattered noise and multiple-reflection parasitic images for BaTiO$_3$. For $\Omega \neq 0$ no noise can be seen in the amplified images and the parasitic image also disappears. For further discussion see Huignard and Rajbenbach (1993).

Heterodyne detection. This is again amplification by two-wave mixing, but one of the beams is amplitude- or phase-modulated at a frequency higher than the inverse of the material response time. If, in addition, the detector aperture is much larger than the grain of the random noise pattern, then part of the noise is integrated out as shown by Breugnot et al. (1993).

Incoherent erasure. Let us assume that the image amplified by two-wave mixing is rather small and recall (Section 8.2.5) that due to the interaction of two incoherent beams a dark ring may appear in the scattering pattern. Thus by a judicious choice of a third beam that is incoherent with the pump beam, the dark ring

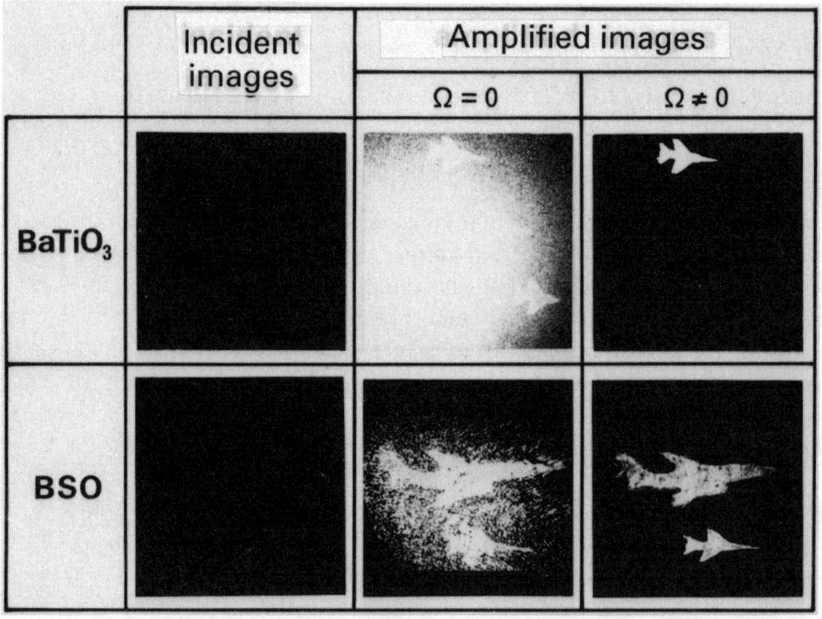

Fig. 8.8 High gain, low noise image amplification. Left: low intensity incident images. Middle: amplified images. Right: noise-free amplified images obtained by rotating the crystal.

will appear just in the region where the image is, i.e. the scattered noise is erased. Noise reduction in this manner was demonstrated by Breugnot *et al.* (1994). A method based on incoherent erasure has also been presented by He and Yeh (1994).

8.3 Subharmonic instabilities

8.3.1 Introduction

In Part I of this book we have seen how two coherent optical beams may interact with each other inside a photorefractive crystal by setting up a refractive index grating of spatial frequency K, such that both incident beams are automatically Bragg-matched to the grating. In addition, we saw how this could lead to an exchange of energy or phase between the two beams. However, in our considerations we made the implicit assumption that the grating thus created is stable. Following the experiments of Mallick *et al.* (1988) it became apparent that this assumption does not necessarily hold under all conditions. With two beams intersecting inside a crystal of BSO, Mallick *et al.* demonstrated that for certain experimental parameters one or more 'subharmonic' beams could be observed to exit the output face of the crystal. The beams were so named because they subtended angles between the pump beams which were consistent with the existence inside the crystal of gratings with submultiple grating vectors, i.e. K/n where n is an integer.

How are subharmonics generated? A typical experimental set-up is depicted schematically in Fig. 8.9(a), while Fig. 8.9(b) illustrates the corresponding Ewald diagram for how the $K/2$ gratings appear from an instability of the K grating. It has been shown that both the use of the moving grating technique (Mallick *et al.* 1988) and the application of an AC electric field (Takacs and Solymar 1992) may

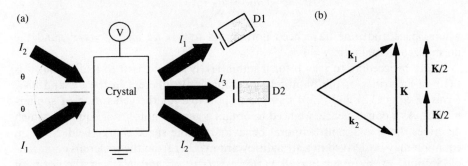

Fig. 8.9 (a) Schematic representation of an experiment showing the generation of the second subharmonic. D_1 and D_2 are detectors. (b) Ewald diagram demonstration of subharmonic splitting.

give rise to the appearance of spatial subharmonic gratings. So far, subharmonics have been reported in crystals of BSO (Mallick et al. 1988), BGO (Grunnet-Jepsen et al. 1993) and BTO (Takacs et al. 1992a), with theory suggesting that they should also be observable in GaAs, InP, and CdTe:V (see e.g. Sturman et al. 1992a,c; Grunnet-Jepsen et al. 1994c). This brings us to the theory. The first attempt at explaining the spontaneous appearance of subharmonic beams attributed the phenomenon to the preferential amplification of noise (Ringhofer and Solymar 1988a), which seems plausible when one recalls from Chapter 7 the large gains possible for forward three-wave mixing. However, when this approach was unsuccessful in predicting the threshold conditions for the appearance of subharmonic beams (i.e. the range of detuning frequencies, applied electric fields, inter-beam angles, and beam ratios for which subharmonic beams were observed), the attention turned to the materials equations. Could the grating become unstable? The emergence of the subharmonic beam is now widely accepted to be caused by an instability of the space charge field against spatial period doubling (or tripling etc.) (Au et al. 1990). Here, the non-linearity of the material is believed to initiate the instability process, after which three-wave interaction may further amplify the subharmonic beam.

The aim of this section is to give a brief insight into the main concepts involved in a theoretical investigation of the subharmonic instability and to address the problem of mapping the threshold conditions for subharmonic generation.

8.3.2 Theory and experiment

We shall start our investigation by examining in this section the subharmonic instabilities which may arise during two-wave mixing in a photorefractive material when using the moving grating technique (see Sections 2.8 and 5.4.2). In this case the optical excitation is in the form of a travelling wave of wave vector K and frequency Ω:

$$\frac{I_p}{I_o} = m\cos(bt_n - \zeta) \tag{8.8}$$

where space and time have been normalized to $\zeta = Kz$ and $t_n = t/\tau_d$, and the normalized detuning is $b = \Omega\tau_d$.

Before proceeding to search for instabilities, it is important to realize that it is the non-linearities of the photorefractive material (inherent in the nE and nN_D^+ terms in eqns (2.3) and (2.8)) that are responsible for the generation of subharmonics. As a consequence, we need to obtain a general non-linear equation which describes the spatial and temporal behaviour of the space charge field. Such an equation may be derived in a straightfoward manner from the materials equations of Section 2.1 by assuming all variables (J, n, E, and N_{D1}^+) in the form of $X(z,t) = X_o(t) + X_p(z,t)$ and isolating the space charge field, E_p (Sturman et al. 1992a; Kwak et al. 1992, 1993a). If we for simplicity assume that diffusion and saturation can be neglected (i.e. small inter-beam angles) so that

8.3 Subharmonic instabilities

E_D, $E_M \ll E_o \ll E_q$, we obtain our starting equation in the form (Kwak et al. 1993a)

$$(E_o + E_p) \frac{\partial^2 E_p}{\partial \zeta \partial t_n} - E_M \frac{\partial E_p}{\partial t_n} - E_M E_p [1 + m\cos(bt_n - \zeta)] = mE_M E_o \cos(bt_n - \zeta) \tag{8.9}$$

Let us recall what this differential equation means. The excitation by the interference pattern is represented by the term $m\cos(bt_n - \zeta)$. What do we expect to happen? We expect, and it is borne out by the linearized model (see Section 2.2), that excitation at the fundamental spatial frequency will cause the appearance of a space charge field at the same frequency. Going from a linear to a non-linear model it is easy to see that the product of two quantities, both varying at the fundamental frequency, will generate a second harmonic component. But how can one generate a subharmonic? We shall try to make this process plausible first by a proper mathematical solution showing the onset of instability, and later, in Section 8.3.3, by an analogy with an electrical circuit and a parametrically excited pendulum. In all cases we shall make the following simplifications to eqn (8.9):

(i) assume that $E_p \ll E_o$ and so neglect the non-linearity in the first term of eqn (8.9);
(ii) disregard the excitation term on the right-hand side of eqn (8.9), which is equivalent to disregarding the response at the fundamental spatial frequency, K.

Under the above assumptions eqn (8.9) reduces to

$$E_o \frac{\partial^2 E_p}{\partial \zeta \partial t_n} - E_M \frac{\partial E_p}{\partial t_n} - E_M E_p [1 + m\cos(bt_n - \zeta)] = 0 \tag{8.10}$$

We shall now restrict ourselves to an examination of the experimentally predominant $K/2$ subharmonic instability. This means that we shall assume the solution to eqn (8.10) in the form of a $K/2$ subharmonic space charge field travelling at the same speed as the interference pattern:

$$E_p = \frac{1}{2} E_{1/2} \exp\left(\frac{j}{2}(bt_n - \zeta)\right) + \text{c.c.} \tag{8.11}$$

Equating now the coefficients of $\exp[j(bt_n - \zeta)/2]$ and $\exp[-j(bt_n - \zeta)/2]$ separately to zero and neglecting higher exponential orders, this trial solution leads to the following ordinary differential equation:

$$\frac{dE_{1/2}}{dt_n} + \alpha E_{1/2} - \gamma E_{1/2} = 0 \tag{8.12}$$

where

$$\alpha = \frac{1}{D}\left[E_M - \frac{b}{4}(E_o - 2jE_M)\right], \quad \gamma = -\frac{mE_M}{2D}, \quad D = E_M + \frac{j}{2} E_o \tag{8.13}$$

This is a linear differential equation with constant coefficients for the complex function $E_{1/2}(t)$. It is well known that solutions exist in the form $\exp(\lambda t_n)$. The criterion for the instability is now that the real part of λ is positive, i.e. the function grows exponentially. The corresponding mathematical conditions may be obtained from the coefficients by applying the Routh–Hurwitz criterion well known in control theory (see e.g. Hagedorn 1981):

$$(\alpha\alpha^* - \gamma\gamma^*) < 0, \quad (\alpha + \alpha^*) < 0 \tag{8.14}$$

If either of these conditions is valid, eqn (8.10) becomes unstable and as a consequence the $K/2$ subharmonic beam will emerge. Performing the calculations indicated in the inequality eqn (8.14), the threshold curve for the instability may be obtained in the simple analytical form

$$\frac{E_o}{E_M} = \frac{2}{b} [2 \pm \sqrt{m^2 - b^2}]. \tag{8.15}$$

In Fig. 8.10 the experimentally obtained instability boundary has been plotted as a function of detuning frequency and applied field for a $1\,\text{cm}^3$ cubic crystal of BSO for an incident beam ratio of unity, an interbeam angle of 2 degrees and a total intensity of $7.6\,\text{mW}\,\text{cm}^{-2}$. For the theoretical threshold curve (solid line) the parameters chosen were $\tau_d = 0.325\,\text{ms}$ and $E_M = 0.365\,\text{kV}\,\text{cm}^{-1}$. The region inside the bounded area represents the parameters for which a $K/2$ subharmonic would exist. The agreement between theory and experiment is sufficiently good

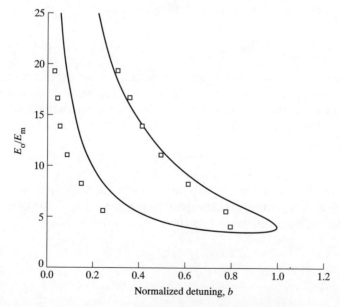

Fig. 8.10 Experimental ($\square$) and theoretical (———) instability region for the second subharmonic in BSO (after Solymar et al. 1994).

8.3 Subharmonic instabilities

for us to be confident that we have pinpointed the basic mechanism for the generation of subharmonic beams.

At this stage it is worthwhile to attempt to make some predictions about which photorefractive materials are most likely to become unstable to the generation of subharmonic beams. By looking at eqn (8.15) we see that the lowest threshold occurs for $m = 1$. If we set $b = 1$ the threshold simplifies to $E_o/E_M = 4$. By inserting the expression for E_M we find that a subharmonic instability may arise when

$$\mu\tau_e > \frac{2\Lambda}{\pi E_o} \tag{8.16}$$

Thus, a requirement for subharmonic instability is that the mobility–lifetime product of a material is large, as was first realized by Sturman et al. (1992c). Let us take an example. Setting $\Lambda = 15\,\mu\text{m}$ and $E_o = 10\,\text{kV cm}^{-1}$ means that the material requirement for subharmonic instability is that $\mu\tau_e > 10^{-7}\,\text{cm}^2\,\text{V}^{-1}$. Sillénites (BSO, BGO, and BTO) are likely to have high mobility–lifetime products. GaAs, CdTe, and InP are also good candidates from amongst the semiconductors, whereas subharmonics are unlikely to be seen in $BaTiO_3$ and $LiNbO_3$.

The above treatment has been instructive in conveying the main ideas associated with a subharmonic instability analysis. However, it does not contain the whole truth. In our analysis we made several assumptions which do not hold in practice. By neglecting E_p in the first term of eqn (8.9) we conveniently removed the need to know anything about the fundamental space charge field. This approach was adopted by Nestiorkin (1991), Kwak et al. (1992, 1993a), and Shershakov and Nestiorkin (1993) and is useful in that it leads to simple expressions for the instability threshold. Unfortunately, the inclusion of this term does affect the instability threshold. As a first approximation one could subsequently use the linear-in-modulation approach and assume E_p in a form containing only the K and $K/2$ gratings. This method has been used with fairly good results for predicting subharmonics for the moving grating technique (Richter et al. 1994; Nestiorkin and Shershakov 1993) as well as the AC electric field technique (Kwak et al. 1993b; Grunnet-Jepsen et al. 1994b).

However, in Section 2.9 we noted that this linear-in-m approximation is only valid for $m \ll 1$. Subharmonics, in contrast, have only been observed experimentally for beam ratios close to unity (i.e. m. close to 1) where we expect large signal effects to play a significant role. As a consequence, the correct method would be to replace eqn (8.11) by a Fourier series which includes the K and $K/2$ space charge fields and several of their higher harmonics (Sturman et al. 1992a; Serrano et al. 1994c). Unfortunately, analytical results then become inaccessible and extensive numerical modelling is required.

So far we have blissfully ignored the further complication of wave interactions, having argued that they are unimportant when concerned only with threshold values. Obviously wave interactions would need to be taken into consideration if we wished to say anything concrete about the intensities of the subharmonic beams. This has not yet been successfully tackled. However, Sturman et al. (1995a) suggested an ingenious experimental set-up for studying subharmonic instabilities which would eliminate optical interactions altogether. It consists of having the two pump beams incident on the (001) face of a crystal, with an electric field applied in the $\langle 110 \rangle$ direction. For this configuration the electro-optic properties of crystals of point group 23 (e.g. BSO, BTO, BGO) or 43m (e.g. GaAs or CdTe) are such that the pump beams do not experience any electric field-induced changes in the optical dielectric tensor, and hence no ensuing wave mixing. Any gratings in the crystal may however easily be monitored by having a weak Bragg-matched probe beam incident on the (110) face. Using this set-up McClelland et al. (1994) and Sturman et al. (1995a) have shown unambiguously that subharmonics are the result of a material instability and not dependent on any optical nonlinearity. The added benefit of this configuration is that any gratings set up by the pump beams can be probed at a specific depth into the crystal. Thus, even though the intensity of the pump waves decays exponentially into the crystal due to absorption, at any given cross-sectional depth the total intensity will be uniform. This is an important advantage when we consider that the subharmonic instability is dependent on the intensity, through τ_d.

Before concluding this section it should be mentioned that although most attention within this field has been given to the $K/2$ subharmonic, there has been some work on explaining the appearance of the $K/3$ and $K/4$ subharmonics (Nestiorkin and Shershakov 1993; Shershakov and Nestiorkin 1993; Liberman and Zel'dovich 1993; Sheng et al. 1995). Shershakov and Nestiorkin (1993) extended the above analysis to consider the non-degenerate case where the subharmonics of wave vector $a\mathbf{K}$ and $(1-a)\mathbf{K}$ are generated, where a lies between 0 and 0.5. They suggested that although the $K/2$ subharmonic was seen to have the lowest threshold for generation, the discrete value $a = 1/3$ was also favourable for reasons of symmetry. The $K/4$ was argued to stem from the doubling process of the $K/2$.

8.3.3 Analogies

We shall now digress slightly and attempt to present the subharmonic instability in a more familiar light. In 1883 Lord Rayleigh turned his attention to parasitic resonances in pipe organs, and found that they could be modelled by what has come to be known as the damped Mathieu equation (Mathieu 1868), an equation which describes a damped oscillator in which the energy storage parameter is modulated at a certain frequency (see e.g. Pipes 1965). In the world of electronics the damped Mathieu equation describes the voltage across the RCL circuit in Fig. 8.11(a) if the capacitance is modulated in time (see e.g. Yariv 1991):

$$\frac{d^2 V}{dt^2} - \kappa \frac{dV}{dt} - \frac{V}{LC_o}\left[1 + \frac{\Delta C}{C_o} \cos(\omega_p t)\right] = 0 \qquad (8.17)$$

where V is the voltage, L is the inductance, κ is the loss, and C_o and ΔC are respectively the average value and modulated amplitude of the capacitance. The most intriguing feature of the Mathieu equation is that for certain parameters the oscillator becomes unstable and it becomes possible to sustain oscillations at a

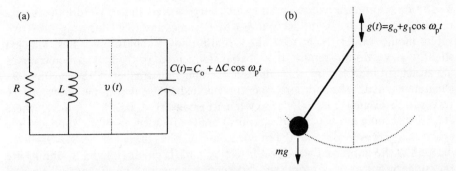

Fig. 8.11 (a) A resonant non-linear circuit capable of exciting temporal subharmonics. (b) A parametrically excited pendulum.

frequency which is half the value of the applied frequency, ω_p. The relevance to subharmonic instability should become clear when we note that if in eqn (8.10) we change to a moving coordinate system by the transformation $\xi = bt_n - \zeta$, then the equation for the space charge field takes the form of a damped Mathieu equation, as was realized by Kwak et al. (1992):

$$E_o \frac{d^2 E_p}{d\xi^2} - E_M \frac{dE_p}{d\xi} - E_M E_p[1 + m\cos\xi] = 0 \qquad (8.18)$$

So in analogy with the known calculations for the above mentioned circuit, we could derive the threshold conditions for the subharmonics as was done in Kwak et al. (1992). In this way we may gain an insight into the nature of the instability by looking at a conceptually simpler analogy.

We shall conclude this section with one last analogy which provides perhaps the best intuitive means of visualizing the subharmonic instability, namely the parametrically excited pendulum shown schematically in Fig. 8.11(b), which also obeys the Mathieu equation. In this case, if 'gravity' is modulated at a frequency ω_p, we find that for low frequencies the pendulum will merely move up and down in synchronism with the excitation. However, above a certain threshold frequency the motion of the pendulum becomes unstable and it will start to swing back and forth with a frequency of $\omega_p/2$. Moreover, for even higher frequencies or amplitudes of oscillation yet other frequencies will arise and we may eventually observe a transition to chaos (see e.g. Mullin 1994).

8.3.4 Subharmonic space charge waves

Sturman et al. (1992b,d, 1993a,b, 1995b) have adopted a somewhat different approach to explaining subharmonics. By regarding photorefractive gratings in the context of space charge waves, they look upon the subharmonics as being the result of a parametric excitation of space charge waves. In this picture the fundamental space charge wave is believed to become unstable, decaying into two subharmonic waves. Note that the definition of 'subharmonic' has changed slightly, as we do not require that spatial frequencies of the subharmonic waves are exact submultiples of the imposed fundamental frequency. Sturman et al. suggested that three conditions must be fulfilled for subharmonic space charge waves to be excited. First, the energy must be conserved, i.e. $\Omega = \omega_1 + \omega_2$, where Ω is the detuning frequency and ω_1 and ω_2 are the frequencies of the subharmonic space charge waves. Second, the momentum must be conserved, i.e. $\mathbf{K} = \mathbf{k}_1 + \mathbf{k}_2$, where $\mathbf{K}$ is the impressed wave vector (set by inter-beam angle) and $\mathbf{k}_1$ and $\mathbf{k}_2$ are the subharmonic wave vectors. The third condition is that the subharmonic waves which are excited parametrically must obey the dispersion relation for weakly damped waves in the photorefractive medium. In Section 5.3 we show that this relation is given by

$$\omega(k) = \frac{1}{\mu k E_o \tau_e \tau_d} \tag{8.19}$$

It immediately follows that the resonance condition for the parametric excitation of $K/2$ subharmonics occurs when the detuning frequency $\Omega = 2\omega(K/2) = 4\omega(K)$. Thus we note that the optimum driving frequency for the $K/2$ subharmonic is four times the resonance frequency for two-wave mixing gain. This is the same result as was obtained in eqn (8.15), when we recall that the optimum detuning frequency for two-wave mixing gain was shown in Section 2.8 (eqn 2.109) to be E_M/E_o. On other counts, however, the space charge wave approach leads to quite different predictions about the instability. For instance, the space charge wave theory suggests that a continuous spectrum of subharmonic space charge waves should be excited. This prediction seems to be in direct conflict with the experimental observations of discrete K/n subharmonics. On the other hand, the three-dimensional treatment introduced by Sturman et al. (1993a) probably stands the best chance of explaining the splitting of subharmonics which has been observed experimentally by Pedersen and Johansen (1994, 1995).

8.3.5 Other mechanisms for subharmonic generation

So far we have concentrated on subharmonic instabilities brought on by the moving grating technique. Are there other ways of generating subharmonics? Several methods have been proposed. Nestiorkin and Shershakov (1993) examined the possibility of generating subharmonics by using the so-called phase-locked detection scheme discussed in Section 5.4.7 which consists of applying an AC square wave field in synchronism with a 'hopping' (as opposed to a 'running') interference pattern. Liberman and Zel'dovich (1993) suggested using the moving grating technique in conjunction with an AC sine wave field, calculating instability thresholds using space charge wave concepts. Finally, Sturman et al. (1993a) proposed that subharmonics could be excited by a weakly oscillating external field of the form $E_o(t) = E_{oo}[1 + s\cos(\Omega t)]$, where $s \ll 1$. However, it

8.3 Subharmonic instabilities

should be stressed that, apart from the moving grating technique, spatial subharmonics have as yet only been observed experimentally for one other method, namely the AC field enhancement technique described in Section 5.4.6.

The $K/2$ subharmonic instability threshold for the AC field technique is very different from that of the moving grating technique. Figure 8.12 depicts for BSO the experimental electric field amplitude threshold as a function of AC frequency for (a) an applied AC square wave and (b) an AC sine wave. The $K/2$ subharmonics are spontaneously generated for parameter space above the threshold curves. Three important observations may be made. First, we no longer have the characteristic 'banana' shape of Fig. 8.10. Instead, the threshold decreases monotonically and exhibits no upper frequency limit. Second, the instability

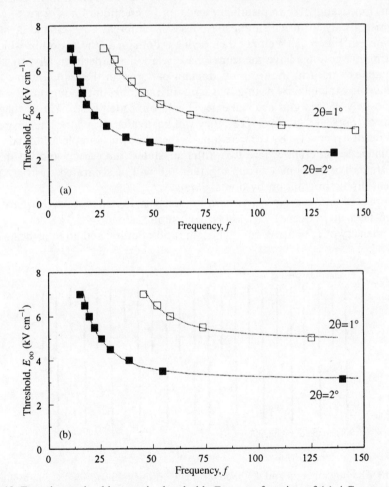

Fig. 8.12 Experimental subharmonic threshold, E_{oo}, as a function of (a) AC square wave electric field frequency and (b) AC sine wave electric field frequency of inter-beam angles $2\theta = 1°$ and $2\theta = 2°$, $m_1 = 1$, $I_0 = 6\,\text{mW cm}^{-2}$. Dotted lines are visual guides.

threshold is seen to decrease as the inter-beam angle is increased from 1 to 2 degrees. Finally, comparing the instabilities for the sine and square wave applied fields, we note that the threshold is consistently lower for the AC square wave. All these observations were successfully predicted by theory (Grunnet-Jepsen et al. 1994b).

8.3.6 Subharmonic domains

There is a further very interesting aspect of subharmonic gratings which deserves mention. From a geometric point of view, it is obvious that when a K/n subharmonic grating forms, it may form in one of n different phases relative to the fundamental grating, which are identical unless viewed from a fixed coordinate system. For example, if an instability leads to the creation of a $K/2$ grating, then this grating may form with opposite phases in different areas of the crystal, as illustrated in Fig. 8.13. We may then define a domain as the region in which the subharmonic gratings have the same phase relations. Experimentally, this property manifests itself in the form of 'domain patterns' in the subharmonic beam. This may be explained by noting that light diffracted from two regions consisting of oppositely phased gratings will emerge with opposite phases. Thus, where two regions of 'opposite' gratings meet, they will be divided by a dark line where light interferes destructively. By comparison, the best known example of domains is in ferromagnetic materials where two different states, dipoles up and down, may exist. Here, like states may group together and will be separated from groups of different dipole orientation by domain walls.

Figure 8.14 shows a typical pattern obtained in BGO for an AC square wave applied field of frequency $f = 144\,\text{Hz}$ and amplitude $E_o = 6.7\,\text{kV}\,\text{cm}^{-1}$, a total input intensity of $I_o = 5\,\text{mW}\,\text{cm}^{-2}$, a beam ratio of unity, and an inter-beam angle

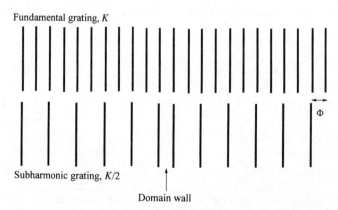

Fig. 8.13 The fundamental and $K/2$ subharmonic gratings shown side by side to illustrate the two degenerate subharmonic states, which differ only in that they are shifted a half period with respect to the fixed coordinate system. Where two different subharmonic states meet, there exists a region of competition that constitutes a domain wall.

8.3 Subharmonic instabilities

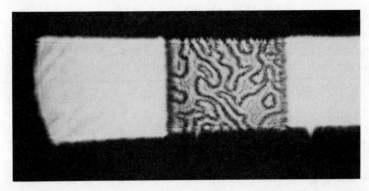

Fig. 8.14 Photograph of the domain structure in a BGO crystal (after Grunnet-Jepsen *et al.* 1993).

of 4 degrees. This photograph of a screen placed about 30 cm behind the crystal shows the $K/2$ subharmonic beam situated half-way between the two pump beams. An interferometric measurement has confirmed that as one traverses a domain boundary the phase of the diffracted light shifts by 180 degrees (Grunnet-Jepsen *et al.* 1993).

8.3.7 Competition effects

Before concluding our investigation of subharmonic instabilities we may wonder why there is an interest in subharmonics at all. Are there any apparent applications? Well, Mallick *et al.* (1988) suggested that they may be useful in beam steering, and one could envision other applications where the control of diffraction gratings afforded by the instabilities may be beneficial. However, perhaps more importantly, subharmonic instabilities could be detrimental to other already established applications: they will increase noise or cross-talk by the presence of unwanted beams and of course they will represent a power loss. The latter point begs the question: do subharmonic instabilities have a detrimental effect on two-wave mixing gain? Intuitively, we would expect yes, for the reason that (1) the creation of any new gratings is bound to diffract energy in corresponding directions, thus detracting from the energy transferred to the signal beam, and (2) the strength of the fundamental space charge field responsible for the refractive index grating will most likely decrease when it becomes unstable to the parametric generation of subharmonic spatial frequencies.

Unfortunately, a theoretically investigation of the effect of subharmonics on two-wave mixing would be extremely difficult, as it would require the inclusion in the theory of multi-wave optical interactions, higher diffracted orders, and higher orders of the space charge field. However, a recent experimental study by Grunnet-Jepsen *et al.* (1994c) has confirmed unambiguously that subharmonic instabilities can indeed have a significant detrimental effect on two-wave mixing gain. The authors correlated a decrease in two-wave mixing gain with the emergence of the $K/2$ subharmonic. Figure 8.15 shows the evidence. When an AC square wave of amplitude $E_o = 7\,\text{kV}\,\text{cm}^{-1}$ and frequency 10 Hz (i.e. just below the subharmonic threshold) is applied to a crystal of BSO, as shown in Fig. 8.15(a), the gain is seen to increase significantly, as would be expected. However, if the experiment is repeated, as shown in Fig. 8.15(b), for a frequency of 60 Hz (i.e. above the subharmonic threshold) the gain rises steadily until the slower subharmonic beam starts to emerge, whereupon the gain falls markedly. Figure 8.16 shows the corresponding steady-state two-wave mixing gain and subharmonic intensity versus applied frequency. We should mention that the spread of the data points in Fig. 8.16 does not represent noise, but is a reflection of the oscillation amplitude of the steady-state space charge field, as described in Section 5.4.6 (Fig. 5.22). Note that the peak in gain, or more correctly, the subsequent sudden drop in gain is seen to correlate exactly with the appearance of the subharmonic

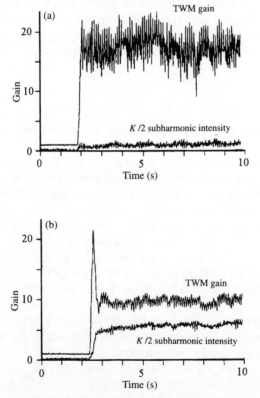

Fig. 8.15 Transient two-wave mixing gain and subharmonic intensity for an applied field AC square wave electric field of amplitude $E_o = 7\,\text{kV}\,\text{cm}^{-1}$ and a frequency of (a) 10 Hz, which is below the instability threshold, and (b) 60 Hz, which is above the threshold. $\beta = 350$, $I_o = 8\,\text{mW}\,\text{cm}^{-2}$.

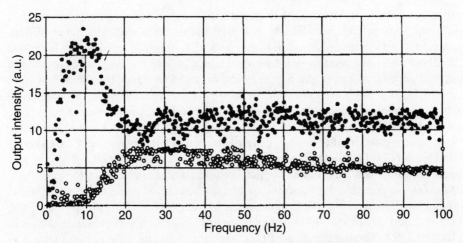

Fig. 8.16 Upper trace: two-wave mixing output as a function of frequency for an input beam ratio of $\beta = 315$, a total intensity of $I_o = 9.2\,\text{mW}\,\text{cm}^{-2}$, and an applied square wave electric field of amplitude $E_o = 7\,\text{kV}\,\text{cm}^{-1}$. Lower trace: output intensity of the $K/2$ subharmonic beam. (After Grunnet-Jepsen et al. 1994c.)

beam. The reason is simply, as stated previously, that the subharmonic beam when it appears draws its power from the pump beam, so less power is available for the signal beam. In fact, the amount of power in the subharmonic beam that was measured could be more than the amount lost from the signal beam. Thus it is not a simple case of power going to the subharmonic instead of going to the signal beam. It would be more correct to say that the two beams compete for power from the pump beam. This is actually quite similar to the phenomenon observed some time ago in which the subharmonic beam competed for power with (i) a signal beam in three-wave amplification (D. C. Jones et al. 1991), and (ii) a resonating beam (D. C. Jones and Solymar 1989).

It is interesting to note that there have been some recent articles reporting peaks in gain similar to that in Fig. 8.16 in CdTe:V by Belaud et al. (1994) and by Moisan et al. (1994), and in InP:Fe by Turki (1993). This would seem to suggest that these materials also are becoming unstable to subharmonic instabilities, a possibility which has not yet been investigated.

In conclusion, this section has shown that in choosing or designing a crystal for a particular application there is an obvious need to map out for which parameters, if any, the diffraction gratings will become unstable.

8.4 Spatio-temporal instabilities

Until now, we have discussed how gratings written in photorefractive materials may become unstable to the generation of new gratings which will diffract light in spurious directions. Instabilities in photorefractive wave mixing are, however, not restricted to material instabilities. It is well known that spatial patterns of light may form spontaneously in non-linear optical systems in which there is optical feedback. To illustrate this point we shall consider here the simplest example of a system which consists of a non-linear medium and a feedback mirror, as shown in Fig. 8.17. If a single beam is incident such that the reflected beam propagates in the exact opposite direction, you would be in good company if you thought that the only effect would be that the counterpropagating beams could exchange energy or phase. Reality is surprisingly different. Firth (1990) and d'Alessandro and Firth (1991) predicted theoretically that above a certain threshold of intensity and coupling such a non-linear optical system could become unstable and spontaneously reorganize the beam into a regular hexagonal pattern. This transverse instability has been observed experimentally using sodium (Grynberg et al. 1988; Pender and Hesselink 1989, 1990), nematic liquid crystals (Macdonald and Eichler 1992; Tamburrini et al. 1993), and liquid crystal light valves (Türing et al. 1993; Pampaloni et al. 1993) as the non-linear material. Photorefractive materials, with their large non-linearities for low optical powers, are particularly well suited for displaying hexagonal pattern formation, as was demonstrated using $KNbO_3$ (Honda 1991; Banerjee et al. 1995; Honda and Matsumoto 1995a) and $BaTiO_3$ (Honda and Matsumoto 1995b). Figure 8.18 shows an example of the rearrangement of a single beam into a hexagonal spot array as obtained with $KNbO_3$ and viewed in the (a) far field and (b) near field regimes. It is interesting to note that Honda and Matsumoto (1995a) recently showed that the motion of the patterns can be controlled. For instance, a slight misalignment (a fraction of a degree) of the counterpropagating beams was shown to lead to a movement or 'flow' of the near field pattern in one direction. Moreover, if the coupling constant was changed spatially by illuminating the crystal non-uniformly with an additional incoherent beam, it was possible to induce a velocity gradient perpendicular to the direction of flow. The upshot of this is that the far field pattern can be made

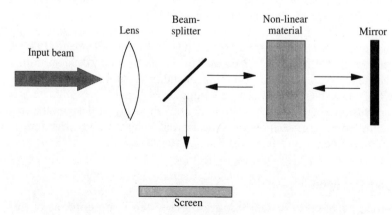

Fig. 8.17 Schematic representation of the experiment leading to spatio-temporal instability.

8.4 Spatio-temporal instabilities

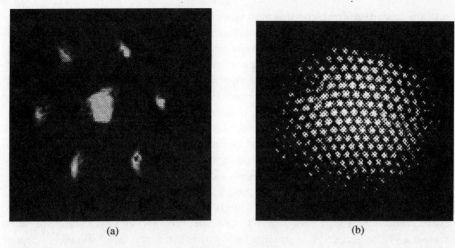

Fig. 8.18 (a) Far-field transmission pattern showing central spot and hexagonal spot array. (b) Near-field pattern showing hundreds of spots arranged in a hexagonal array. (After Banerjee *et al.* 1995.)

to rotate at an angular frequency $\omega = 0.5 \times \mathbf{v}(r)$, where $\mathbf{v}(r)$ is the velocity of the near field pattern.

We shall not here go into any further detail concerning the origin of the self-organization in this system. Suffice it to say that the formation of hexagonal patterns occurs widely in non-linear systems, having been seen in such fields as biology, chemistry, hydrodynamics, and optics. The common factor which pervades the theoretical investigations seems to be that expensive computers are needed in order to make a theoretical study of the complex spatio-temporal instabilities feasible.

Let us now turn to instabilities arising in optical phase conjugators. Such instabilities are perhaps of more immediate and direct concern due to the great potential applicability of phase conjugation. It should come as no surprise that phase conjugate resonators may exhibit instabilities and chaos. After all, resonators represent textbook examples of systems apt to become chaotic because they incorporate all the necessary ingredients for instability: they are non-linear, they are driven (by pump beams), and there is feedback (Zozulya and Tikhonchuk 1989; Bélic *et al.* 1991; Schuster 1988). We may loosely refer to two types of instability: the temporal and the spatio-temporal instability. Looking first at the former, in the present context we mean that the phase conjugate reflectivity may become unstable leading to oscillations (Günter *et al.* 1985; Gower and Hribek 1988; Horowitz *et al.* 1991) or even (deterministic) chaotic fluctuations (Gauthier *et al.* 1987; Valley and Dunning 1984; Reiner *et al.* 1988) in time. Theoretically this behaviour has been attributed separately to such key factors as the existence of multiple interaction regions (Gauthier *et al.* 1987), the presence of a DC field which shifts the frequency of the phase conjugate wave (Królikowski *et al.* 1990), and similar strength transmission and reflection gratings (Bélic et al. 1991). While a plane wave approximation was sufficient to model this behaviour, a more complex three-dimensional treatment is called for when it comes to explaining spatio-temporal instabilities and the transverse dynamics of phase conjugating systems, i.e. how the beams behave in space as well as time. Any theoretical work on the subject will undoubtedly be an extremely arduous exercise in computational modelling. Although there is no abundance of theoretical work, several examples of spatio-temporal instabilities have been studied experimentally. S. R. Liu and Indebetouw (1992) examined the emergence of spatio-temporal instabilities in an externally pumped phase conjugate resonator based on $BaTiO_3$, showing that complex dynamic structures would appear in the self-oscillating beam. Kobialka et al. (1989) showed that the self-oscillating beam in a ring resonator was spatially chaotic. Hussain et al. (1990) noted that fluctuations in phase conjugate reflectivity could stem from a competition between two different self-phase-conjugate reflection geometries set up inside a crystal of $BaTiO_3$. Zheng et al. (1995) recently showed that competition effects were also the likely cause of instability in a Cu:KNSBN self-pumped phase conjugate mirror. However, in this case, the competition was between fanning and self-phase conjugation.

It is likely that a future study of the spatio-temporal dynamics of phase conjugate mirrors will throw some light on the nature of the instabilities, but more importantly it might suggest ways of eliminating or avoiding the instabilities altogether. Meanwhile, some methods for suppressing the fluctuations in reflectivity have been found experimentally. Zheng et al. (1995) demonstrated that by rotating the polarization of the incident light, the extraordinary component of the light would create the self-phase-conjugate beam while the ordinary component served as an erase beam (Kwong et al. 1984; Orlov et al. 1994) to suppress the fanning. In this way, they were able to practically eliminate the chaotic fluctuations in reflectivity by rotating the incident beam by 35 dgrees from the extraordinary axis. Note that the fanning only has to be reduced enough for the self-pumped phase conjugation to 'win' over the fanning effects. It is worth mentioning one other method for eliminating instabilities. While Ross et al. (1993) were investigating the self-pumped phase conjugation characteristics of a 'blue' $BaTiO_3$ crystal they noted that the reflectivity was stabilized when a vibration source, in their case a nearby cooling fan, was turned on. The authors suggested that the substantial improvement in the stability resulted from a washing-out of competing, parasitic gratings.

9

SIX TOPICS IN SEARCH OF A CHAPTER[1]

9.1 Introduction

In a well-organized book every topic finds a chapter. Unfortunately, this is not the case in the present one. However hard we have tried, some of the topics consistently refused to fit into any of the chapters written so far. The natural solution was to put these 'anomalous' topics into one chapter in spite of their diversity. Their common feature is that in some sense they depart from the norm.

The usual mechanism of grating formation is diffusion or drift in an applied field. The photovoltaic effect is the odd man out. The effect can of course be used for recording gratings and indeed we did include a phenomenological description of it in Section 2.4. Nevertheless the photovoltaic field relies on a different mechanism which deserves to be summarized. We shall do that in Section 9.2. The material in which the photovoltaic effect has been most often investigated is $LiNbO_3$, the very first photorefractive material. We shall then jump from the first material to the latest ones, which are of course polymers. Their properties cannot be discussed in great detail because not a great deal is known about them. Recent achievements though have been spectacular and the potential for applications is great. We shall give a brief summary of their origin, their promise and their state of the art in Section 9.3.

Another deviation from the norm is band-edge photorefractivity, to be discussed in Section 9.4. It occurs only in a particular wavelength region, it is based on some complex physical process, and the change in the refractive index depends quadratically on the electric field, a departure from the well known linear electro-optic effect. Further complication in the mechanics of grating formation comes with quantum wells, which are available only in the form of very thin films, usually below 1 μm. So that is certainly a new departure. Even photopolymers would have thicknesses of the order of 100 μm, which is at least large relative to the wavelength of the light. There is also a new effect in quantum wells which has a very impressive name (the quantum confined Stark effect) and may significantly enhance the achievable change in the index of refraction. A brief discussion of these effects will be given in Section 9.5. Stratified media also rely on thin films. They consist of sandwich-like structures of photorefractive materials and pure

[1](with apologies to Pirandello)

dielectrics which may give rise to new properties and new applications. We shall mention them in Section 9.6. Finally, in Section 9.7 we shall talk about a topic that is getting fashionable, solitons. There are two types. One of them we have already mentioned in Section 3.9. Under some conditions erasure can be regarded as the grating envelope moving out of the material without changing its shape. The other type of soliton, called a spatial soliton, means trapping of the input light in a narrow corridor.

9.2 The photovoltaic effect

We have already mentioned the photovoltaic effect and its role in recording gratings in Section 2.4. We shall go here into a little more detail. The basic experimental result is that a current flows when certain materials are illuminated by light. It might be caused by the Dember effect or a macroscopic inhomogeneity like a p–n junction. It may even be possible to postulate a series of p–n junctions which could lead to voltages (for open-circuited samples) well exceeding that corresponding to the band gap in the material. These explanations were more or less satisfactory up to the time when F. S. Chen (1969) made his measurement of photocurrent in $LiNbO_3$. From the field associated with this current he realized that the success of recording holograms in $LiNbO_3$ was due to very large internal fields. He postulated a value of $70\,\text{kV}\,\text{cm}^{-1}$. It was clear that a new effect of unknown origin was at play. The explanation was provided by Glass et al. (1974) in terms of the polar character of the crystal, i.e. the asymmetry of the lattice: 'The Nb–Fe^{2+} distances in the $\pm c$ directions are different so that the probabilities p_+, p_- of intervalence charge transfer in the $\pm c$ directions will differ'. The photovoltaic current may then be written in the form

$$J_{\text{ph}} = e\,\frac{\alpha I}{\hbar\omega}\,(p_+ l_+ - p_- l_-) \qquad (9.1)$$

where l_+ and l_- are the appropriate mean free paths. When the carrier is scattered it loses its preferential momentum, after which it contributes to photoconductivity but not to the photovoltaic current.

The simplest model for explaining the effect is an asymmetric potential well which serves as a model of an impurity (von der Linde and Glass 1975; Glass 1978). There are then three contributions to the preferential current as discussed by Belinicher and Sturman (1980). The first contribution comes from asymmetric emission. As stated before, the chance of a carrier emerging to the right is higher than that of it emerging to the left. A similar contribution comes from recombination. Owing to invariance under time inversion, recombination of carriers arriving from the right is more likely. This means carriers going towards the left disappear, which is a positive contribution to the current flowing to the right. The third contribution comes from scattering. Carriers moving from left to right have a higher probability of scattering by the impurity; hence they have a negative contribution, they reduce the current. Summarizing the three contributions, Belinicher and Sturman (1980) obtained the photovoltaic current in the form

$$J_{\text{ph}} = e \frac{\alpha I}{\hbar \omega} \frac{1}{\tau_{\text{iso}}} (v_{\text{o}} - v_{\text{t}}) \xi \qquad (9.2)$$

where v_{o} is the velocity with which the carriers are emitted to the right, v_{t} is the thermal velocity, t_{iso} is the time needed for thermalization, and ξ is a measure of the asymmetry of the potential well that needs to be determined by quantum mechanical calculations. If the medium is in equilibrium with the radiation, then the emission of the carriers by thermal photons occurs with a velocity v_{t} and the photovoltaic current is zero, as it should be.

An alternative impurity model consisting of two neighbouring point defects was considered by Obukhovsky and Stoyanov (1982). A number of other mechanisms like interband and intraband transitions (which are dependent on the properties of the crystal rather than on those of the impurities) were discussed by Belinicher and Sturman (1980). Their general conclusion is that the photovoltaic effect is a manifestation of the breakdown of the principle of detailed balancing. An attempt was also made (Chanussot and Glass 1976) to explain the photovoltaic effect in terms of the Franck–Condon relaxation of the ionic subsystem during ionization and recombination but, as shown by Levanyuk and Osipov (1977), that mechanism cannot lead to a finite constant current.

The photovoltaic current happens to depend also on the polarization of the input light. The most general relationship may be described with the aid of the third rank photovoltaic tensor, β_{ijk}, as

$$(J_{\text{ph}})_i = \beta_{ijk} \mathcal{E}_j \mathcal{E}_k^* \qquad (9.3)$$

Since the photovoltaic current is real, it cannot change sign if the complex conjugate is taken, whence it follows that

$$\beta_{ijk} = \beta_{ikj}^* \qquad (9.4)$$

Thus the real part of β_{ijk} is symmetric in respect of the last two indices and the imaginary part is antisymmetric. Owing to the above relationship it is again possible to represent the photovoltaic tensor in the reduced notation as introduced by eqn (B.6), and some people do so, but for clarity we shall stick to the full notation here.

Let us now look at some of the elements of the tensor. In the experiment most often done, the light beam is incident with extraordinary polarization and the current flows in the direction of the c axis. The tensor element responsible for this effect is β_{333}. For input light with ordinary polarization and for current flowing perpendicular to the c axis it is the β_{222} element that plays a role. What about the off-diagonal elements? Looking at eqn (9.3), an off-diagonal element implies that the optical electric field must be present both with polarization i and j. Hence the input light beam must be circularly or elliptically polarized. Adopting this terminology, the current due to diagonal elements is described as a linear current, and that due to a circularly polarized light as a circular current.

How could we determine the photovoltaic constants? If, for simplicity, we describe the photovoltaic current with the relationship $J = pI$ and apply a voltage with an opposite polarity so as to stop the current flowing, then the equation

$$pI + \sigma E = 0 \tag{9.5}$$

(where σ is the sum of dark conductivity and photoconductivity for the particular incident light intensity) must be valid. For a sample of $KNbO_3$, which has a rather small photovoltaic effect, the result of such experiments (Günter and Micheron 1978) is shown in Fig. 9.1, where the density of photocurrent is plotted against applied voltage.

We have so far been concerned with single beams incident. The photovoltaic effect can of course be used for recording gratings, in which case we need two beams incident. The corresponding space charge field may be determined by the techniques described in Section 2.4. There has been a lot of experimental work on gratings of that kind, which have no special properties so they need not be reviewed here. We would just like to mention that transients can be used for determining the photovoltaic constant, as shown by Kondilenko et al. (1982).

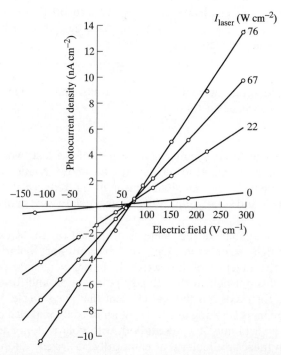

Fig. 9.1 Photocurrent density as a function of applied voltage for $KNbO_3:Fe^{3+}$ at a wavelength of 488 nm (after Günter and Micheron 1978).

9.2 The photovoltaic effect

For a theoretical treatment of transients based on the 'Kiev' model see Carrascosa et al. (1988).

A grating can also be recorded by waves with orthogonal polarization, as first proposed by Belinicher (1978). He showed that the resulting 'circular' photovoltaic current can actually exceed the 'linear' current calculated by Belinicher et al. (1977). The effect was found experimentally by Odoulov (1982) and used for four-wave mixing by Novikov et al. (1986b). For its role in anisotropic diffraction see Odoulov and Soskin (1989).

What is the mechanism of recording gratings by two orthogonally polarized beams? The argument we have used throughout this book is that the sinusoidally varying intensity pattern causes the appearance of photoexcited electrons which diffuse, drift, etc. However, when the two input beams have orthogonal polarizations the excited carrier density does not vary as a function of space. So how does the electric field come about that is needed to change the index of refraction? Very simply. Owing to the antisymmetric terms of the photovoltaic tensor the two input beams create a current which will cause an electric field due to the simple relationship $E = J_{ph}/\sigma$. But will the photovoltaic current vary periodically as a function of space? It will, as shown below.

The element responsible for the effect is $\beta_{113} = \beta^s{}_{113} + j\beta^a{}_{113}$. The two incident waves may be written as

$$\mathcal{E}_e \mathbf{i}_e \exp(-j\mathbf{k}_e \cdot r) \quad \text{and} \quad \mathcal{E}_o \mathbf{i}_o \exp(-j\mathbf{k}_o \cdot r) \tag{9.6}$$

where $\mathcal{E}_e$ and $\mathcal{E}_o$ are the amplitudes (for simplicity we have taken them real) and $\mathbf{i}_e$ and $\mathbf{i}_o$ are the unit vectors of the extraordinary and ordinary waves respectively. The geometry is shown in Fig. 9.2(a). The c axis of the crystal is

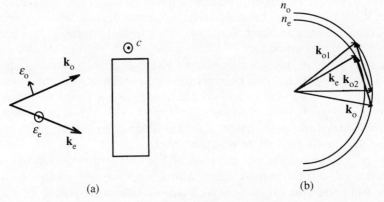

Fig. 9.2 An ordinary and an extraordinary wave recording a grating via the circular photovoltaic effect. (a) Schematic representation. (b) Ewald curves perpendicular to the c axis showing the anisotropy, the recording wave vectors, the resulting grating vector, and replay by an ordinary wave.

perpendicular to the input plane of the waves. The complete expression for the current is then

$$\begin{aligned}J_{ph} &= \tfrac{1}{2}[\beta_{113}\mathcal{E}_1\mathcal{E}_3^* + \text{c.c.}] \\ &= \tfrac{1}{2}[(\beta_{113}^s + j\beta_{113}^a)\mathcal{E}_e\mathcal{E}_o \exp[-j(\mathbf{k}_o - \mathbf{k}_e)\cdot\mathbf{r}] + \text{c.c.}] \\ &= \beta_{113}^s \cos(\mathbf{K}\cdot\mathbf{r}) + \beta_{113}^a \sin(\mathbf{K}\cdot\mathbf{r})\end{aligned} \quad (9.7)$$

i.e. we have a grating with grating vector

$$\mathbf{K} = \mathbf{k}_o - \mathbf{k}_e \quad (9.8)$$

It may be seen from Fig. 9.2(b) that although the grating is recorded by two orthogonally polarized waves it is also suitable for isotropic interaction. An ordinary wave incident with wave vector $\mathbf{k}_{o1}$ will give rise to a diffracted ordinary wave of wave vector $\mathbf{k}_{o2}$.

9.3 Photorefractive polymers

All our photorefractive materials so far have been inorganic crystals. It is not easy to grow them, it is not easy to control their properties, and they cost a lot. Will photorefractive polymers take their place in the near future? As with everything in this field, it is difficult to tell. It is certain though that polymers have made great advances since the report of Ducharme *et al.* (1991) when they first appeared on the scene.

What do we need for producing a photorefractive polymer? We need to get charged carriers from somewhere, the carriers must be able to move in the material for a space charge to appear, we need traps to sustain the space charge, and, finally, the electric field obtained should induce a change in the index of refraction. The beauty of polymers is that these functions could all be separately controlled. Remembering that the photorefractive sensitivity is inversely proportional to the static dielectric constant (eqn 2.94), it is a further advantage that polymers have low dielectric constants.

Some of the basic ideas for producing photorefractive polymers have been around for a long time. Their photosensitivity has been known, their photoconductivity has been known, and of course they have been used for decades for recording gratings. In spite of all that, progress was rather slow at the beginning. The stumbling block, presumably, was the difficulty to put all these elements together in order to find photorefractive properties.

How do photorefractive polymers work? The charge-transporting property is provided by so called transport agents which, if present in high enough concentration, will permit the charge to move from one transport agent to another one. Sensitization of photoconductivity at a particular wavelength can be achieved by having a suitably chosen dye in a low concentration. The molecules responsible for the non-linear optical effects are called chromophores. These are conjugated organic molecules with an asymmetric charge distribution caused, for example, by

9.3 Photorefractive polymers

donor and acceptor substituents placed at opposite ends of the chromophore. In order to have a macroscopic non-linearity, the chromophores must be partially aligned by a fairly high electric field. In fact, charge generation and charge movement are also favourably influenced by an applied field. A low glass transition temperature is also a requirement because that makes poling at room temperature possible.

A schematic diagram of a possible realization, the one chosen by Ducharme *et al.* (1991), is shown in Fig. 9.3, which is taken from the review article of Moerner and Peyghambarian (1995). Here CT stands for charge transport and CG for charge generation, NLO is the non-linear optical chromophore, and 'trap' means of course traps. Note that only the chromophores are attached to the polymer backbone. There are other 'paradigms' as well in which some other, or all of the constituents, are attached to the polymer backbone, but it is too early to say which one of them will lead to the best realization.

A diffraction efficiency of 10^{-4} was reported by Ducharme *et al.* (1991) in the partially cross-linked epoxy polymer bisA–NPDA, where bisA is bisphenol-A diglycidyl ether and the chromophore NPDA is 4-nitro-1,2-phenylenediamine. NPDA, besides being the non-linear chromophore, also provides optical absorption for charge generation. The rise time of the grating was however rather slow, of the order of minutes. Rise time below 1 second was reported by Silence *et al.* (1992a) using the photorefractive polymer PMMA–PNA:DEH, where PNA provides the chromophore and DEH is the charge transport agent. The maximum diffraction efficiency measured was however only 10^{-6}. Two-wave mixing experiments were performed by C. A. Walsh and Moerner (1992), showing that the phase angle is close to $\pi/2$ in the presence of a high enough electric field. Is this what we should expect? Well, the values of our characteristic fields are not known in photorefractive polymers, but one would hope that our previous derivations are still meaningful. If we look at eqn (2.58) we may see that the phase angle is $\pi/2$ in the absence of an applied electric field, that it decreases for increasing E_o but

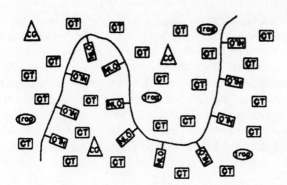

Fig. 9.3 The elements of a photorefractive polymer shown in one of the possible realizations (after Meerholz *et al.* 1994). See text for key.

for high enough E_o it is $\pi/2$ again[2]. We should also note here that in contrast to most experimental configurations the mean angle of the two input beams does not coincide with the surface normal, i.e. we are concerned here with slanted gratings. The reason is the same as in the case of $BaTiO_3$ where the grating is usually slanted in order to ensure the contribution of the large r_{42} electro-optic coefficient. In the present case a slanted grating means that there is a component of the grating vector in the direction of the applied electric field which enhances efficiency.

A new photopolymer, poly(N-vinylcarbazole) (PVK) was introduced by Y. Zhang *et al.* (1992), who doped it with C_{60} (on C_{60} sensitization see also Silence *et al.* 1992b) and with a second-order non-linear optical molecule, DEANST. Using a very similar polymer, net two-wave mixing gain was found by Donckers *et al.* (1993). The gain coefficient measured was quite high, reaching $10\,cm^{-1}$ at a field of $50\,V\,\mu m^{-1}$, but the net gain was still negligible, considering that the thickness of the sample was only $125\,\mu m$. The maximum diffraction efficiency measured was 1%. It was noted by Moerner *et al.* (1994) that there is a further enhancement mechanism due to the effect of the space charge field on the chromophores. For PVK:F–DEANST:TNF (the polymer used by Donckers *et al.* (1993) and Silence *et al.* (1993)) they estimated the enhancement factor to be about 20. It should be emphasized that this is an enhancement with no equivalent in inorganic crystals. Kippelen *et al.* (1993) reported photorefractivity in carbazole–tricyanovinylcarbazole (PVK–TCVK).

Yet another material, PMMA to which C_{60} and DTNBI are added, was reported by Silence *et al.* (1994a). They found an enhancement effect associated with trap activation when the sample was uniformly irradiated. Using the same material, Silence *et al.* (1994b) reported another interesting effect. The erasure rate at low intensity was much smaller than expected and, in fact, the grating strength increased during reading (see also Section 3.9). Improved photoconductive sensitivities were reported by B. E. Jones *et al.* (1994). They obtained recording times of the order of 100 ms at an intensity of $1\,W\,cm^{-2}$.

The best material to date, reported by Meerholz *et al.* (1994), was also based on PVK (for further comments see Moerner 1994). The chromophore was 2.5-dimethyl-4-(p-nitrophenylazo)anisole (DMNPAA), the sensitizer was 2,4,7-trinitro-9-fluorenone (TNF), and N-ethylcarbazole (ECZ) was added to decrease the glass transition temperature. The diffraction efficiency as a function of the applied electric field is shown in Fig. 9.4. This is a curve we have seen before. For a transmission grating, diffraction efficiency is a periodic function of index modulation as shown in Fig. 1.2. We have also seen a similar curve (Fig. 6.18) when demonstrating the erasure of a grating in Section 6.9. The possibility of overmodulation indicates a large change in the index of refraction which can indeed be worked out from the thickness ($105\,\mu m$) and the slant angle (60°) as coming to

---

[2]This follows actually quite simply from the equivalent circuit discussed in Section 5.6.2. High enough E_o means that the equivalent resistance in Fig. 5.29 tends to infinity, therefore that branch of the circuit can be disregarded. Thus only the capacitance remains, which leads to a phase angle of $\pi/2$.

9.4 Band-edge photorefractivity

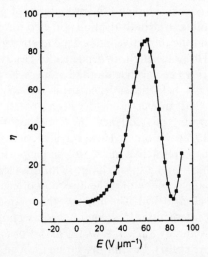

Fig. 9.4 Diffraction efficiency as derived from a degenerate four-wave mixing experiment: recording with waves polarized perpendicular to the plane of incidence; replay with a wave with parallel polarization (after Meerholz *et al.* 1994).

5.5×10^{-3}, a very large value indeed. The net gain coefficient obtained from four-wave mixing experiments was $207 \, \text{cm}^{-1}$ in the presence of an absorption of $13 \, \text{cm}^{-1}$.

In summary, one has to acknowledge the spectacular success of photorefractive polymers, which have come far in a span of no more than four years. There are, though, many questions still unanswered; the most important ones among them are stability, repeatability, fatigue (can they withstand numerous writing–reading cycles), and temperature sensitivity. There is not much known about scattering, noise, and optical quality either. The potential is there and hopes are high.

9.4 Band-edge photorefractivity

Up to now we have been free to choose the wavelength at which we operated (one of the assets of photorefractive interactions) and we have relied on the linear electro-optic effect for providing the grating. When we talk about band-edge photorefractivity, our choice of wavelength is limited, the mechanism at play is the quadratic electro-optic effect, the underlying physics is a lot more complicated, but the outcome is more or less the same. We can record a grating and we can have two-wave mixing gain. Perhaps the only major difference is that the direction of gain depends on the sign of the electric field.

Band-edge photorefractivity follows from the Franz–Keldysh effect (Franz 1958; Keldysh 1958). The wavelength has to be just above that causing band-to-band excitations. The excitation may still occur using the good offices of the photons themselves. This is called photon-assisted tunnelling. If we now apply an electric field, the band edge is tilted and the probability of photon-assisted tunnelling increases. The result is electro-absorption and electrorefraction, meaning that both the absorption and the refractive index depend strongly both on wavelength and on electric field. The wavelength dependence of electro-absorption and electrorefraction are related to each other by the Kramers–Kronig relations (see e.g. Yariv 1991). The electric field dependence turns out to be a nearly quadratic function. This means that an electric field with a spatial frequency K will lead to a refractive index grating of frequency $2K$. Consequently, the two beams will not be Bragg-matched to each other. However if a constant electric field, E_o, is applied, the change in the refractive index will be proportional to the cross-product $2E_o E_s$, which is linear in the space charge field. It is clear now why the sign of the refractive index change (and with it the direction of energy transfer) is dependent on the sign of the electric field. The origin of the space charge field is the same as in the conventional interaction. It is due to carriers photoexcited out of traps.

The effect was first found by Partovi et al. (1990a). For a more detailed account see Partovi and Garmire (1991). In semi-insulating GaAs a two-wave mixing gain of $2.8\,cm^{-1}$ was found which was higher than the absorption coefficient of $2.0\,cm^{-1}$. By applying a moving grating technique and combining the band-edge grating with the conventional photorefractive grating they succeeded in obtaining a gain coefficient of $16.3\,cm^{-1}$ in the presence of an absorption coefficient of $3.0\,cm^{-1}$ in a crystal of 4 mm interaction length. The band-edge effect was also shown to exist in InP:Fe (Millerd et al. 1990), where a gain of $19\,cm^{-1}$ was achieved by using both band-edge resonance and intensity resonance (Picoli et al. 1989a), which we discussed briefly in Section 5.4.3. Using the same technique Millerd et al. (1992a) investigated in InP:Fe both the ring self-pumped phase conjugate mirror and double phase conjugation at a wavelength of 970 nm. At a wavelength of 1064 nm (where attenuation is lower) Eichler et al. (1992) found that the refractive index change due to the Franz–Keldysh effect was still larger than that due to the linear electro-optic effect at an applied field of $12\,kV\,cm^{-1}$.

9.5 Quantum well structures

A quantum well means a potential well to which electrons are confined. A low-band-gap material between two high-band-gap materials will provide such a quantum well. If we keep on doing this, i.e. join together materials in the high–low–high–low fashion, and if each layer is rather thin (say between 7 and 30 nm) then we obtain a multiple quantum well structure. It is operated near the band edge, so the effects are similar to those obtained by band-edge refractivity discussed in the previous section. There are, though, two major differences: (i) the interaction length is bound to be a lot smaller in a quantum well structure than in a bulk material since 100 layers, each of them being 30 nm, still amount to no more than 3 μm; (ii) the sensitivity of the index of refraction to changes in the electric field is much higher in the quantum well structure than in the bulk material.

9.5 Quantum well structures

Where does the enhanced sensitivity come from? From the properties of excitons in quantum wells. A detailed discussion of the properties of excitons is obviously beyond the scope of this book, and it is not a topic that can be easily dispensed with in a few sentences; nevertheless we shall make the attempt here. Excitons are electron–hole pairs bound together by Coulomb forces. Obviously less energy is needed to create an exciton than to create an electron–hole pair. The difference is the binding energy. Hence photons with just-below-bandgap energies can create excitons. In bulk materials however, the exciton lifetime is so short (a few hundred femtoseconds) that the transition is not easily observable. In a quantum well however, the excitons have much longer lifetimes. The reason is that when an exciton is confined to a well comparable with its orbit, then it is much more difficult to tear the electrons and holes apart. In a sense they are protected by the walls. The exciton resonance (i.e. the photon wavelength which optimally creates excitons) can then be observed. Since the absorption depends on the wavelength, so will the refractive index according to the Kramers–Kronig relations. For our purpose all that matters is that a change in electric field will influence the absorption spectrum and will lead to significant variation in the index of refraction. The relation between refractive index and electric field is still given by the quadratic relation but there are now two possible variations. If the applied electric field is along the wells (which may be called the parallel geometry) then we are still relying on the Franz–Keldysh effect. If the applied field is across the wells (which may be called the perpendicular geometry) then the effect is called the quantum confined Stark effect, which actually gives a greater change in the index of refraction.

The effect was first observed in the parallel geometry by Glass et al. (1990), who called it the photodiffractive effect. The experiments were done in semi-insulating GaAs/AlGaAs quantum wells at a wavelength of 830 nm in the four-wave mixing configuration (defects were introduced by proton implantation). The interbeam angle was 20° corresponding to a grating spacing of 2.4 μm. In a more detailed account by Nolte et al. (1990) using the same material with the same parameters, the diffraction efficiency was measured as a function of the applied field and was found to follow a quartic relationship for low electric fields. The full curve is shown in Fig. 9.5.

The two-wave mixing gain was measured by Q. N. Wang et al. (1991), also in GaAs/AlGaAs quantum wells. In the experiments of Nolte et al. (1991) the grating is written by He–Ne laser light at 633 nm which creates electron–hole pairs across the gap, so it is greatly attenuated. The probe beam is at 840 nm and is absorbed only in the GaAs quantum wells. They showed that the grating was rather robust. There was no significant erasure of the grating even for probe beam intensities higher than the pump intensities. More detailed experiments and analyses were carried out by Q. N. Wang et al. (1992). They measured both diffraction efficiency and the gain coefficient (for variation of gain with applied field see Fig. 9.6). The phase shift deduced from these experiments came close to $\pi/2$ whereas the theoretical calculations using conventional bipolar analysis yielded much smaller phase shifts. The discrepancy was resolved later by Q. N. Wang et

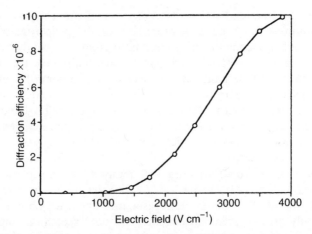

Fig. 9.5 Diffraction efficiency as a function of applied electric field at a wavelength of 830 nm and a grating space of 2.4 μm (after Nolte et al. 1990).

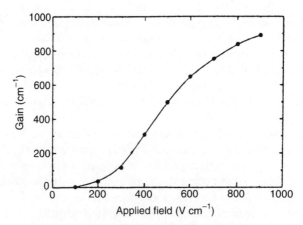

Fig. 9.6 Gain coefficient as a function of applied electric field for an AlGaAs/GaAs quantum well structure for a grating spacing of 14.5 μm at an intensity of 1.6 mW cm^{-2} for a modulation of $m = 0.67$ (after Wang et al. 1992).

al. (1994a,b). They showed that the theoretical value of the phase angle can be close to $\pi/2$ if hot electron effects (carrier heating by the applied field) are taken into account (see also Section 5.2.8). The space charge electric field was calculated on the basis of a simplified bipolar model by Carrascosa et al. (1992). Some of the expressions there are actually incorrect owing to some printing mistakes. For the corrected formulae and curves see Magana et al. (1994). Diffraction with femtosecond pulses was measured by Brubaker et al. (1994), obtaining efficiencies close to 10^{-4}.

Experiments in the perpendicular geometry were also conducted. Partovi *et al.* (1991) obtained diffraction efficiencies up to 0.25% in CdZnTe/ZnTe quantum well structures using visible light in the range 570–613 nm. In later work by the same authors (Partovi *et al.* 1992), in the same kind of quantum well both c.w. and single pulse experiments were conducted yielding diffraction efficiencies slightly in excess of 1%. Using Cr-doped GaAs/AlGaAs quantum wells, Partovi *et al.* (1993) reported a diffraction efficiency of 3% with a rise time of 3 µs. The perpendicular geometry for the reflection case was investigated theoretically by Shkunov and Zolotarev (1995). They showed that gain coefficients up to $10^4\,\mathrm{cm}^{-1}$ were in principle possible.

9.6 Stratified holographic optical elements

The idea is to put layers of dielectrics between holograms producing a sandwich-like structure. The technique was pioneered by Yakimovich (1980), Zel'dovich and Yakovleva (1984), and Zel'dovich *et al.* (1984), who worked out the diffraction efficiency and selectivity properties just for two elements. Generalization to more elements was done by Tanguay and Johnson (1986), Johnson and Tanguay (1988), and Nordin *et al.* (1992), who proposed several applications. One of the applications demonstrated was to divide an input beam into seven beams of equal intensity. Experiments on two holograms whose distance from each other could be finely adjusted with the aid of a piezoelectric actuator were performed by Sheridan (1990).

All the above mentioned structures used static holograms. The extension to photorefractive holograms was performed by de Vré and Hesselink (1994), who obtained the first-order diffraction efficiency in closed form. They also proposed multiple storage of holograms, utilizing the fact that the photorefractive phase in each layer can be separately controlled by an applied voltage.

It is not feasible to use thick photorefractive polymers, because large electric fields are needed. However, this disadvantage does not appear when they are used as elements in stratified holograms. Such a combination was reported by Stankus *et al.* (1994). They produced both two-layer and four-layer structures. The diffraction efficiency measured during a 60 s recording period is shown in Fig. 9.7. A certain diffraction efficiency is obtained when an electric field is applied only to layer 1, and a similar diffraction efficiency when the electric field is applied only to layer 2. It is gratifying to notice that when an electric field was applied to both layers, the optical electric fields added coherently and the diffraction efficiency quadrupled.

9.7 Solitons

Solitary waves were discovered by the naval architect J. Scott Russell (1844) on the Edinburgh to Glasgow canal in 1834. He observed 'a rounded, smooth and well-defined heap of water which continued its course along the channel apparently without change of form or diminution of speed', and subsequently he wrote:

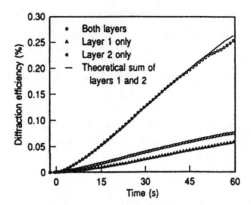

Fig. 9.7 Diffraction efficiency as a function of time when an electric field is applied to one of the layers or to both layers simultaneously (after Stankus *et al.* 1994).

'I followed it on horseback, and overtook it still rolling on at a rate of some eight or nine miles an hour, preserving its original figure some thirty feet long and a foot to a foot and a half in height'. Such solitary waves have been observed since, under a large variety of conditions. The name 'soliton' was coined by Zabusky and Kruskal (1965), the 'on' at the end emphasizing that it is a bit like a particle which may keep its identity after an interaction.

Solitons in optical fibres are also well known. They represent little bundles of light energy which travel along the fibre without changing their shape. One might say that that bundle of energy is confined to a narrow temporal range and a narrow one-dimensional spatial range. By analogy one may define a 'spatial soliton' when a light beam is confined in two spatial dimensions. This is a possible definition, but it seems unlikely that J. Scott Russell would have approved of it. A spatial soliton is not a single heap rolling along. It would make a lot more sense to call the phenomenon 'self-trapping', a term which is also in use although it sounds less glamorous. Since in photorefractive literature the term 'spatial soliton' has now been widely accepted, we shall reluctantly follow the usage.

Photorefractive spatial solitons were predicted by Segev *et al.* (1992) and by Crosignani *et al.* (1993). The phenomenon is due to the balance of two effects: ordinary diffraction, and Bragg diffraction between a large number of Fourier components making up the light beam. The effect was first observed by Duree *et al.* (1993) in an SBN crystal with a 10 mW laser beam at a wavelength of 457 nm. The solitons could be erased by incoherent light from a flashbulb. For a brief qualitative description, see Segev *et al.* (1993).

The stability of solitons against small perturbations was investigated by Segev *et al.* (1994a). Both light and dark solitons were observed in BTO by Iturbe Castillo *et al.* (1994) using the drift mechanism of non-linearity, whereas Valley *et al.* (1995) predicted similar effects in $LiNbO_3$ due to the photovoltaic effect. Finally, we wish to mention the prediction and observation of screening solitons by Segev *et al.* (1994b) and Shih *et al.* (1995) respectively. The trapping is caused

9.7 Solitons

by an applied electric field. Where the optical intensity is higher, the electric field is lower (due to the higher conductivity), which modifies the refractive index and traps the beam. The spatial solitons observed by Shih *et al.* (1995) had a diameter as small as 9.6 µm at microwatt power level.

Solitons which travel without changing their shape may also exist in photorefractive materials. A solution based on the non-linear materials equations was obtained by Shamonin (1993b). Another solution by Jeganathan *et al.* (1994,1995) has already been described in Section 3.9 and demonstrated in Fig. 3.11. Their analysis uses the field equations (eqns 3.31 and 3.32) and the temporal variation of the space charge field (eqn (2.33) as the starting point. By introducing a new variable in the form of $R = j\mathcal{E}_2/\mathcal{E}_1$ they end up with a single partial differential equation in terms of a variable which is the spatial integral of the index of refraction distribution. They show then that an approximate solution of that differential equation may be obtained in the form of solitons.

PART III
Applications

GENERAL INTRODUCTION

In this third section of the book, we address as best we can in the space available the enormous range of applications that have been proposed for photorefractive materials. We should start by saying exactly what we mean by applications. At the time of writing, whilst a few prototype devices have been produced that make use of some form of photorefractive signal processing element, these have almost all been constructed as technology demonstrators, and there is to our knowledge only one commercial product on the market (Rakuljic and Leyva 1993). Nevertheless, there is hope that with the rapid progress being made in photorefractive materials research it will not be long before other photorefractive devices do achieve commercial success.

So by applications here we refer to systems that make use of dynamic gratings recorded in photorefractive crystals to perform some (potentially useful) function. The system may consist of just the photorefractive crystal itself or a complex arrangement of many different optical and optoelectronic devices. It is not possible here to do full justice to all the ingenious uses which have been proposed for photorefractives; we have had to be selective in choosing topics for in-depth treatment and have tried to make choices based on current research trends. We certainly do not intend this to imply that these areas represent the 'best science' in any respect, we are simply attempting to maximize the appeal of the book. We hope that sufficient references have been supplied to satisfy the reader who is interested in one of the areas treated less comprehensively.

We treated the amplification of plane waves in great detail in the previous two parts. In Chapter 10 we shall look at the more general case where images need to be amplified. We shall also treat in the same chapter other image processing applications: thresholding, edge enhancement, the ability to identify changes in an image (novelty filtering), and spatial light modulation.

Perhaps the most attractive application of photorefractive materials is pattern recognition, for which the most important operation is correlation. This will be the subject of Chapter 11, along with the related task of recalling a stored image from a noisy or incomplete copy of that image – associative recall.

In Chapter 12 we look at the various issues surrounding the storage and subsequent replay of many images within a single photorefractive crystal: such as multiplexing techniques, storage capacity and recording schedules. Finally, in Chapter 13 we hedge our bets as to which applications are liable to be commercially successful in the future; there we briefly survey a representative cross section of the many other uses that have been suggested for photorefractive devices.

10
IMAGE AMPLIFICATION AND IMAGE PROCESSING

10.1 Introduction

First of all it is important to examine in general terms why there is interest in optical signal processing. The capacity of our electronic processing systems continues to increase at such a rate that the average scientist can have on their desk today a small PC with a processing power comparable to the departmental mainframe of scarcely ten years ago. To date, this rapid increase in performance shows few signs of abating; nevertheless there are many signal processing applications where the speed of current electronic devices still leaves much to be desired. Typical are those applications where the data is in a parallel format such as an image and those where an operation is to be carried out simultaneously on a whole set of data.

As a more concrete example, sometimes it is necessary to look at the difference between two images – so that a computer can recognize that an intruder has entered an area under surveillance, for example. Electronically, this would usually be done by sequentially subtracting the values stored in each pair of corresponding pixels in the two images, whereas it is potentially far more efficient if all pixel pairs could be processed at the same time. This is most important when the maximum speed is required. In our example, if the intruder were to be a human thief, then serial processing would be more than fast enough. On the other hand, if the 'intruder' is in the form of a missile approaching a military craft at several times the speed of sound, then any increase in speed could make a real difference between life and death.

Optical techniques are in many cases extremely well suited to the parallel processing of information. One- and two-dimensional arrays of data may be easily constructed by spatially modulating the intensity, phase, or polarization of a light beam, and complex interconnections between many sources and detectors can be readily set up in three dimensions, since light beams are able to pass unaffected through each other (needless to say, such complex interconnections are not possible using wires carrying conventional electrical signals). Unfortunately, the fact that light beams are able to traverse one another with no effect poses a problem when it comes to the actual processing of information in the optical domain.

The problem is that once we have chosen to encode our sets of data in the optical domain, then in many cases we ideally want to be able to process the data by having one set of data operate in some way directly on another. So one of the features of light is at the same time an advantage, when we wish to make interconnections, and a disadvantage, when we wish to process entirely in the optical domain.

The reason that in our everyday experience light beams do usually pass unhindered through each other lies in the fact that in most ordinary situations, the wave equation describing the propagation of light is linear. If we consider two intersecting beams of light, each beam may be described by a separate solution of that wave equation. It is well known from calculus that if two solutions to a linear equation are added, then the result is also a solution, so in our example, when the two beams are both present, each behaves in exactly the same fashion as if it were present alone, i.e. it travels on unaffected by the presence of the other.

In order for beams of light to interact, then, it is clear that they must meet in a medium with a non-linear response; only in materials in which the wave equation is non-linear can we hope to carry out processing entirely in the optical domain. This is the reason behind the interest in the use of non-linear optics in general – and photorefractive crystals in particular – for signal processing.

Photorefractive crystals have a number of general attributes that are attractive for signal processing applications. The first and perhaps most important is that the 'strength' of the photorefractive response (which may be characterized by the amplitude of the refractive index grating recorded in the material) is independent of the total incident intensity. This is not the case in most other forms of non-linear optical interaction. For example, when dealing with the third-order non-linear susceptibility $\chi^{(3)}$ (Boyd 1992), the refractive index $n(\mathbf{r})$ associated with a given position, $\mathbf{r}$, is given by

$$n(\mathbf{r}) = n_0 + n_2 I(\mathbf{r}) \tag{10.1}$$

where n_0 is the refractive index in the limit of small intensity, $I(\mathbf{r})$ is the local optical intensity, and n_2 is a constant of the material. In most such materials, very high intensities are required to achieve changes in the refractive index comparable with those obtainable in photorefractive media. This often necessitates the use of high power, short pulse (sub-microsecond) lasers, which are both bulky and expensive – neither of which is a trait that is likely to endear such systems to a potential customer. Furthermore, it is usually necessary to focus the light into the interaction region to further increase the intensity. This places severe limitations on the fidelity of any optical image processing that is carried out with the aid of the non-linear interaction. By contrast, with photorefractive materials there is the potential for devices to operate using low power beams of light, which in the case of certain materials (particularly the semiconductors) may be derived from low cost, rugged, and compact semiconductor diode laser sources.

Another feature of the photorefractive effect is its inherently non-local response, which, as we have seen, means that in two-wave mixing it is possible to amplify one beam of light at the expense of another. This amplification process is made use of in a number of the applications to be described[1]. The final feature

[1] In contrast, as may be clearly seen from eqn (10.1), when dealing with materials displaying a $\chi^{(3)}$-based non-linearity, the response is always local, i.e. maxima in the refractive index always coincide with maxima in the intensity.

which is again important in many applications is the fact that there is a time constant associated with the writing or erasure of the index grating that may be set anywhere within a wide range by a suitable choice of material, incident intensity, and experimental geometry.

We begin our tour of the applications of photorefractive materials by looking at the amplification of image-bearing beams.

10.2 Image amplification

We should say at the start that whilst it may be interesting from a purely scientific viewpoint, it is difficult to identify applications in which the photorefractive effect would simply be used as a means of amplifying an image. Instead it is more likely that the amplifier would be simply a small part of a larger and more complex optical or hybrid optoelectronic system.

When it was realized that the non-local nature of the photorefractive response could lead to optical amplification and it became apparent that amplification was possible with expanded and collimated beams, it was perhaps only natural that the amplification of image-bearing beams should be considered. Indeed, in one of the seminal papers (Kukhtarev et al. 1979b) there is an image of a test chart that has been amplified ten times by passage through a crystal of $LiNbO_3$. With the development of grating enhancement techniques using running gratings or AC fields (Huignard and Marrakchi 1981a; Stepanov and Petrov 1985) and the use of materials possessing a larger electro-optic coefficient (such as $BaTiO_3$ or SBN), much higher gains became available: for example, a gain of about 4000 was reported in 1983 for $BaTiO_3$ (Laeri et al. 1983; Tschudi et al. 1986), and this figure was arrived at after losses due to absorption and Fresnel reflections had been taken into account. More recently, Brignon et al. (1995) reported image amplification in $BaTiO_3$ by a factor of 1.5×10^5.

In general, image amplification is carried out in either the image or the Fourier transform domains. In both cases a collimated beam is directly modulated with the desired spatial information using, for example, a transparency or better a spatial light modulator. As shown in Fig. 10.1(a), for image domain amplification, the object is imaged into the crystal using one or more lenses. Figure 10.1(b) shows the approach taken for Fourier domain amplification (Hong et al. 1990); a convex lens is used to focus the image-bearing beam into the crystal, where the transverse amplitude distribution of the signal beam is the Fourier transform of the amplitude distribution of the object. The latter approach allows the amplification of large images using small crystals, which is no small advantage given the current price of some photorefractive materials and the difficulty of producing large, high quality, single domain crystals! This is the most common approach when using $BaTiO_3$ for example. It also facilitates a number of optical image processing schemes (described later), which rely on operations carried out in the Fourier transform domain.

Several departures from ideal behaviour in the photorefractive gain process have become apparent. The main ones are a limited dynamic range in the output

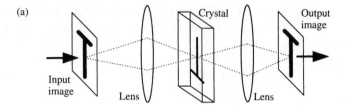

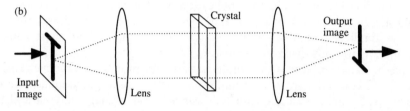

Fig. 10.1 Image amplification in (a) the image domain and (b) the Fourier domain.

image (or put another way, a reduced signal to noise ratio) due to amplified scattered light (beam fanning), a non-uniform modulation transfer function, and non-uniform amplification of the image due to pump beam depletion. We shall look at the effects of these limitations and some of the techniques put forward for getting around them.

10.2.1 Scattered light and beam fanning

In addition to performing the useful task of amplifying a signal beam, the photorefractive effect may also act so as to amplify any light scattered within the crystal. This amplified scattered light provides one source of noise and is most severe in those directions that provide the maximum gain, as discussed in Chapter 8. Unfortunately, this usually means that the noise is at its greatest in a direction close to that in which the signal beam is travelling, resulting in a poor signal-to-noise ratio (SNR)[2]. Furthermore, the situation is worsened by the fact that there are inelastic scattering mechanisms that broaden the spectral width of the scattered light, compared to the incident beams. In particular, this means that there is always some component of the scattered light that has the correct frequency detuning from the pump beam for optimum gain. The SNR may be improved by reducing the gain (e.g. the signal beam may be incident in a direction providing

---

[2] The signal and noise are those components of the intensity leaving the crystal in the direction of the signal beam that arise from the amplified image and amplified scattered light, respectively.

less than optimum gain (Joseph *et al.* 1991b)), but this approach is often not very attractive.

In photorefractive materials that are capable of very high gain, such as SBN, the scattered light may be amplified to such an extent that significant depletion of the pump beam occurs, even in the absence of a signal beam, and the scattered light forms an intense beam in the direction of optimum gain (Voronov *et al.* 1980). Figure 10.2 illustrates this effect in a crystal of $BaTiO_3$. This process is known as beam fanning and is a serious limitation to the use of photorefractive crystals as image amplifiers. On the other hand, as we shall see later, it may be constructively exploited in a number of optical signal processing schemes.

There are a number of approaches that may be taken to reduce the deleterious effects of amplified scattered light. All essentially rely on exploiting different characteristics in the behaviour of the amplified signal and scattered light. Firstly, if it is not necessary to use the maximum possible gain obtainable with a particular crystal, some considerable improvement in SNR may be obtained by careful choice of the orientation of the signal beam with respect to the crystal c axis. In many cases, however, it is desirable to have as much gain as possible, which means orienting the signal beam in the same direction as the amplified scattered light.

If a particular application does require the maximum possible gain, use can be made of the fact that in general the dependencies of the amplified signal and noise on the pump beam power are rather different. In the undepleted pump regime, the amplified signal intensity is independent of pump intensity. On the other hand, as shown in Fig. 10.3, the amplified noise depends strongly on the pump intensity. In particular, at low pump intensities there is very little amplified noise.

It is possible to exploit this dependence to obtain high gain with low noise using a pulsed read-out scheme (Joseph *et al.* 1990, 1991a). The idea is firstly to record a

Fig. 10.2 Example of beam fanning in $BaTiO_3$ (after Feinberg 1982a). The crystal is the small rectangle which is inside the larger square (an oil-filled cuvette). The lower beam emerging towards the right is the pump beam. Fanned light is visible above this. The beams are made visible due to fluorescence in iodine gas.

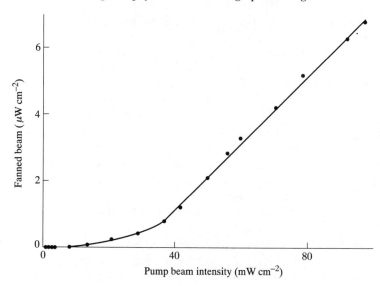

Fig. 10.3 Dependence of noise (fanned beam) intensity on pump intensity for beam fanning in BaTiO$_3$ (after Joseph *et al.* 1990).

photorefractive grating using the image-bearing signal beam along with a pump beam with an intensity below the threshold for the onset of beam fanning. Once this grating has been recorded, it is replayed using a much higher pump intensity to give a greatly amplified signal beam. As long as this read-out process is carried out on a time scale short in comparison with the characteristic formation time for the beam fanning process (typically of the order of 3 s for a pump beam intensity of 3.7 W cm^{-2} in a crystal of BaTiO$_3$), a good SNR may be obtained. Figure 10.4 shows images of a test chart amplified in a crystal of BaTiO$_3$ both with and without the pulsed read-out technique. The benefits of the noise reduction are clearly apparent, with gains of around 11 000 being obtained with an SNR of 1300.

The characteristic formation time of the gratings produced by the signal beam and the scattered noise are also rather different, the former being in general much shorter than the latter. This difference in the recording times may be exploited to provide low noise amplification in BaTiO$_3$ by rotating the crystal during the recording process (Rajbenbach *et al.* 1989a; Huignard and Rajbenbach 1993) as has been described in Section 8.2.9. Some other techniques for reducing scattered noise are also described in that section.

Low noise image amplification in LiNbO$_3$ has also been reported (J. Xu *et al.* 1994). These authors obtained a transient gain of over 1000 in a 1 mm thick sample.

10.2.2 Pump beam depletion

Another departure from ideal behaviour occurs when the amplification is sufficient to cause significant depletion of the pump beam intensity in those regions

10.2 Image amplification

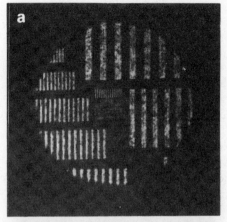

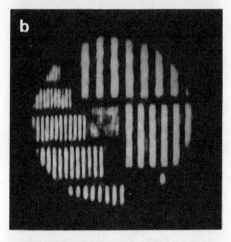

Fig. 10.4 Amplification using pulsed read-out (after Joseph *et al.* 1990). (a) Unamplified image without pump beam. (b) Image amplified with pulsed read-out. (c) Image amplified with c.w. read-out.

that overlap with the brighter parts of the image (Vachss and Yeh 1989). Figure 10.5 illustrates the process. The darker areas of the image do not cause appreciable pump depletion and experience the maximum gain, while the brighter areas deplete the pump beam and experience a smaller gain. As a result, the ratio of the maximum to minimum intensities at the input to the crystal is greater than that when the signal beam leaves the crystal. This ratio is known as the dynamic range of the image and the net result then is a decrease in the dynamic range, a process generally referred to as contrast reduction or compression.

If this were to be all there was to it, then the treatment of pump beam depletion would be straightforward. Given the exponential gain factor, γ, relevant to the particular crystal and experimental geometry, and the pump and signal intensities at the input to the crystal, the evolution of the signal intensity can be found from eqns (3.42) and (3.43).

However, the situation is not quite as straightforward as this. The pump and signal beams must subtend an appreciable angle with each other and this means that, as one part of the pump beam propagates through the crystal, it will overlap with different parts of the image. The effect of this can be most easily seen if one considers the case of an image in which there is a sharp transition between a region of low intensity and a region of high intensity, the intensity being in the form of a step function as shown in Fig. 10.6.

All signal rays above the ray marked C in the figure will experience maximum gain, as none of the pump light with which they interact has been significantly depleted. Conversely, all the signal rays below the ray marked D will experience minimum gain, as the pump light with which they interact *is* appreciably depleted. The complication arrives for all those signal rays between C and D. Although they originate from the low intensity part of the image, for some distance within the crystal they interact with pump light which has been partially depleted through

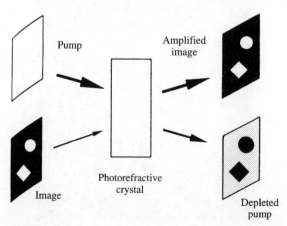

Fig. 10.5 Image amplification by two-wave mixing, showing the resultant non-uniform depletion of the pump (after Vachss and Yeh 1989).

10.2 Image amplification

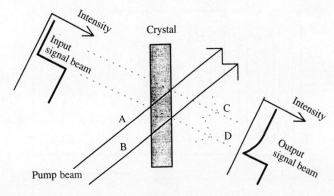

Fig. 10.6 The effect of the varying pump depletion occurring due to a non-uniform image intensity.

interaction with light from the bright part of the image. As a result the gain in this region is less than that experienced by light from the rest of the low intensity part of the image. In terms of the pump light, all rays below ray B suffer the most depletion, while those above ray A suffer minimal depletion. The amount of depletion decreases from B to A.

Figure 10.7 shows typical plots of pump and signal intensities at the input to the crystal and at depths of 5 mm and 1 cm within the crystal. The data were obtained numerically for a crystal with $\Gamma = 10\,\text{cm}^{-1}$. Severe distortions are plainly visible in the amplified signal and depleted pump beams. There is no way of avoiding this kind of distortion other than by ensuring that the input beam ratio is sufficiently high that the pump beam is not significantly depleted.

10.2.3 Modulation transfer function

As we have seen earlier in Chapter 2, the amplitude of an index grating recorded in a photorefractive crystal by the interference of two plane waves is dependent on the spatial period of the grating, or put another way, on the angle subtended by the pump and signal beams. As a result of this, the gain is also dependent on the inter-beam angle. Whilst this effect can be significant in BaTiO$_3$ (Zhou *et al.* 1993), it is a particular problem in the sillénites and related materials, such as GaAs. These materials have the advantage of possessing some of the fastest response times for a given incident intensity currently available, but suffer from having rather low electro-optic coefficients when compared to BaTiO$_3$ and SBN. Gain enhancement is possible using the moving grating or AC field techniques described in Chapter 3, but high gains are only obtained over a narrow range of inter-beam angles.

As a typical example Fig. 10.8 shows a theoretical plot of the gain as a function of inter-beam angle for a 1 cm long crystal of BSO with typical values chosen for the material parameters. The gain has been optimized by applying a field of

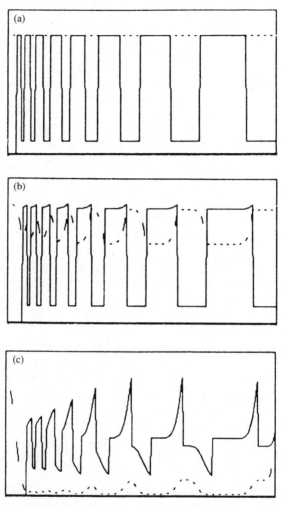

Fig. 10.7 Numerical calculation of pump beam (····) and image beam (——) at three positions within a photorefractive crystal (after Vachss and Yeh 1989): (a) input image; (b) amplified image, 5 mm deep; (c) amplified image, 10 mm deep. Pump scale = 200.0; image scale = (a) 1.000, (b) 70.98, (c) 415.2.

$8\,\text{kV}\,\text{cm}^{-1}$ and detuning one of the beams. Clearly, appreciable gain is only available for a narrow range of input angles. When the signal beam contains an image, the beam may be Fourier-decomposed into a distribution of plane waves, subtending different angles with the pump beam. Only those components which fall within the peak of the gain curve in Fig. 10.8 will experience appreciable amplification. Images with greater detail (i.e. a greater space–bandwidth product) have a greater spread of angles in their Fourier distribution, and therefore with these images more spatial information is not amplified by the photorefractive crystal.

10.2 Image amplification

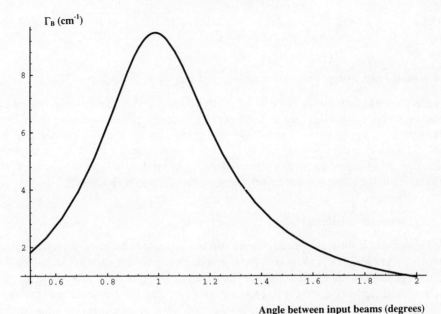

Angle between input beams (degrees)

Fig. 10.8 Theoretical plot of gain as a function of inter-beam angle for BSO. Detuning frequency is held constant.

An approach to overcoming this problem was suggested by Vainos and Gower (1991) and is illustrated in Fig. 10.9. The pump beam takes the form of a divergent spherical wave, which itself contains Fourier components covering a range of angles. The idea is that each component of the signal beam should be able to experience amplification due to an interaction with a component of the pump beam that is at the optimum inter-beam angle. The net result is a much broader, flatter spatial frequency response. The technique was also used to improve the fidelity of phase conjugation by four-wave mixing; in that case, one of the pump

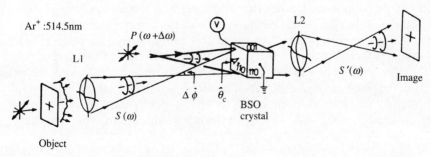

Fig. 10.9 Scheme for high fidelity image amplification (after Vainos and Gower 1991).

waves was a divergent wave, as in Fig. 10.9, and the other pump was its conjugate – i.e. a converging spherical wave.

10.3 Image processing

In the previous section we discussed the amplification of image-bearing beams by photorefractive crystals. This is an interesting enough phenomenon in itself, but there is a range of more remarkable image processing operations that can be carried out using the photorefractive effect, which will be discussed in this section. One operation, correlation, has attracted such a great deal of interest that we shall not cover it here, but shall make it the subject of its own chapter.

10.3.1 Image thresholding

An important optical processing operation is that of thresholding or determining which parts of a grey scale image have an intensity above a certain threshold. As with edge enhancement, discussed in the next section, this form of processing can help a machine vision system to pick out an object from a cluttered background. The key to this operation is the introduction of a strong non-linearity in the transfer function of the system so that parts of the image with an intensity greater than the threshold are enhanced, while those parts below the threshold are dimmed, ideally resulting in a binary image.

One approach making use of a semi-linear self-pumped phase conjugate mirror is shown in Fig. 10.10(a) (Sayano et al. 1988). The image to be thresholded is incident from the right, pumping the cavity formed between the photorefractive crystal ($BaTiO_3$) and the mirror and generating a phase conjugate image. Thresholding is provided by the response of the phase conjugate mirror as illustrated in Fig. 10.10(b). The threshold intensity could be adjusted by translating the mirror towards or away from the crystal. This sets up a moving fringe pattern within the crystal, and because the time constant associated with the build-up of the refractive index grating within the crystal is inversely proportional to the intensity of the incident light, for a given fringe velocity there will be a threshold intensity below which the grating is not able to 'keep up' with the moving fringe pattern.

A different approach, but one that also involves phase conjugation, is shown in Fig. 10.11 (M. B. Klein et al. 1986). Again a semi-linear phase conjugate mirror is involved, this time an externally pumped one. In the absence of the erase beam, a resonator is formed between the photorefractive crystal and the output coupler (95% reflectivity). The effect of the incoherent erase beam is to modify the reflectivity of the phase conjugate mirror, as it can be shown (Marrakchi et al. 1984) that the interference of two coherent beams (of amplitudes $\mathcal{E}_1$ and $\mathcal{E}_2$) in the presence of an incoherent beam (amplitude $\mathcal{E}_3$) results in an index grating with amplitude

$$\Delta n \propto \frac{\mathcal{E}_1 \mathcal{E}_2^*}{(|\mathcal{E}_1|^2 + |\mathcal{E}_2|^2 + |\mathcal{E}_3|^2)} \tag{10.2}$$

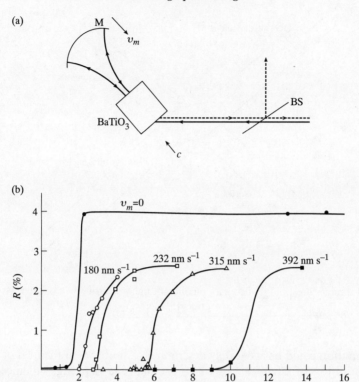

Fig. 10.10 Image thresholding using a semilinear phase conjugate mirror (after Sayano et al. 1988). (a) Experimental arrangement: M = mirror; BS = beamsplitter. (b) Dependence of reflectivity on incident intensity for a number of mirror velocities.

So when beams 1 and 2 are spatially uniform, the index modulation and therefore the diffraction efficiency depend critically on the intensity of the incoherent beam relative to the other two. The non-linearity inherent in eqn (10.2) is responsible for the thresholding behaviour of this device. In contrast to the approach shown in Fig. 10.10, this device employs an incoherent image-bearing beam. It therefore also acts as a spatial light modulator for incoherent-to-coherent image conversion (see Section 10.3.4). Also the fact that bright parts of the input image produce a decrease in the diffraction efficiency of the resonator means that the image is inverted as well as being thresholded.

10.3.2 Edge enhancement

There are some instances when it is desirable to enhance, or locate, the edges of some object. One example would be in a machine vision system, where it might be necessary for a computer to locate the boundary of an object so that its position

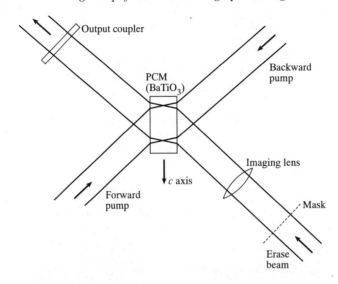

Fig. 10.11 Image thresholding using an incoherent image bearing beam (after M. B. Klein et al. 1986).

and orientation could be accurately determined. Another is in the field of correlation (the subject of Chapter 12), where we shall see that edge enhancement of images aids the ability to discriminate between objects. Edge enhancement may be readily carried out using photorefractive crystals.

One approach involves writing a grating in a photorefractive crystal using one beam bearing an image of the desired object and a second plane reference beam which has an intensity intermediate between those corresponding to the darkest and lightest parts of the image. The image is projected into the crystal (image domain processing – see Fig. 10.1) and with this arrangement the greatest fringe modulation will occur in those regions of the image in which the image intensity and reference intensity are equal. Because we have chosen the reference intensity to be between the brightest and dimmest parts of the image, the highest fringe modulation will occur at the edges of the object where the intensity is changing from bright to dim. If the grating is now read out using a separate plane wave, the brightest parts of the diffracted image will correspond to the edges of the original image.

A second approach involves processing the image in the Fourier domain. The image bearing beam is focused into the crystal, where its Fourier transform produces a grating through interference with a plane reference wave. In the Fourier plane, the brightest parts of most images correspond to the lowest spatial frequencies, whereas the higher spatial frequencies would usually possess comparatively low intensities. If the reference beam intensity is chosen to be comparable to the intensity of the high spatial frequency components of the image, then these components will produce a high index modulation, whereas the much stronger low spatial frequency components will interfere with the reference wave with a low visibility, and hence produce a low index modulation. Again, all that is necessary is to read out the grating with a separate beam, though with this technique it is of course necessary to inverse-Fourier-transform the diffracted beam using a lens in order to recover the edge-enhanced image. Figure 10.12 shows the result of this latter edge enhancement technique performed on the image of a comb (Feinberg 1980).

A related approach is shown in Fig. 10.13 (Joseph *et al.* 1992). Two-wave mixing is carried out between an image-bearing beam and a strong plane pump beam, both derived from the same HeNe beam. In the absence of the beam from the HeCd laser, the result is simply an amplified version of the image contained on the transparency O1. The effect of the HeCd laser beam is to partially erase the grating in the region of the crystal corresponding to low spatial frequencies in the

 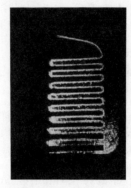

Fig. 10.12 Fourier domain edge enhancement (after Feinberg 1980). (a) Original image of comb. (b) Edge-enhanced image.

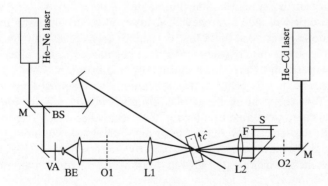

Fig. 10.13 Fourier domain processing (after Joseph *et al.* 1992). M = mirror, BS = beamsplitter, L1, L2 = lenses, VA = variable attenuator, O1 = image transparency, O2 = transparency controlling erasure, F = filter, S = screen.

image (as discussed in Section 10.3.1 in relation to eqn (10.2)), thus reducing the gain in those regions. The net result is that the higher spatial frequencies, relating to the edges of objects in the image, are amplified.

With this approach, more sophisticated Fourier plane filtering is possible. If a transparency is inserted in the HeCd beam at position O2, those spatial frequencies that are present in this transparency will be erased in the two-wave mixing process (Joseph *et al.* 1991c).

10.3.3 Novelty filter

The novelty filter is a device which only transmits those parts of an image that have recently changed in some way. Figure 10.14(a) shows one way of carrying out this process in photorefractive media, as described by Ford *et al.* (1988)[3]. Their approach involves imposing the scene to be monitored as a modulation in the phase of a coherent beam, using a spatial light modulator in the form of a cathode ray tube and a Hughes liquid crystal light valve. This beam is then focused into a crystal of $BaTiO_3$ that is arranged so that when the input is unchanging, the intensity of the transmitted beam is heavily depleted by beam fanning (see Chapter 8). In the steady state, very little light therefore reaches the output plane.

When there is a change in the input scene, this appears as a change in the phase of that portion of the coherent beam. In beam fanning, the depletion of the input beam occurs through the amplification by the beam of scattered light, mediated by noise gratings recorded between the beam and the scattered light. Immediately after the change has occurred, that portion of the beam that has changed will in general no longer have the appropriate phase to amplify the scattered light via the noise gratings, and so for a short time this portion of the beam will be transmitted to the output plane. As the phase of the noise gratings readjust, the output signal will decrease until the beam is depleted once again.

Sample results are shown in Fig. 10.14(b): the upper photograph shows the input scene when the system is first illuminated. Thereafter, so long as nothing changes, the scene will fade from view. The lower photograph shows the result of the subject moving her eyes. Note that care needs to be taken in interpreting the output from a novelty filter, when something has moved as the filter will reveal both the new position of the object and its original position.

A development of this beam fanning technique was reported by Suzuki and Sato (1992). By incorporating a controlled shutter sequence to interrupt the signal bearing beam, they were able to alter the time scale over which the novelty filter responded. In particular they were able to register very slowly moving objects.

Figure 10.15 shows the arrangement used by Anderson *et al.* (1987) to implement novelty filtering using a $BaTiO_3$ phase conjugate mirror[4]. In the steady state,

---

[3]This approach was proposed in a paper by Cronin-Golomb *et al.* (1987).
[4]A similar system has also been reported using phase conjugation in BSO, which for a given intensity provides a far more rapid response time than $BaTiO_3$ and therefore does not respond to slow variations in the image (Khoury *et al.* 1989).

10.3 Image processing

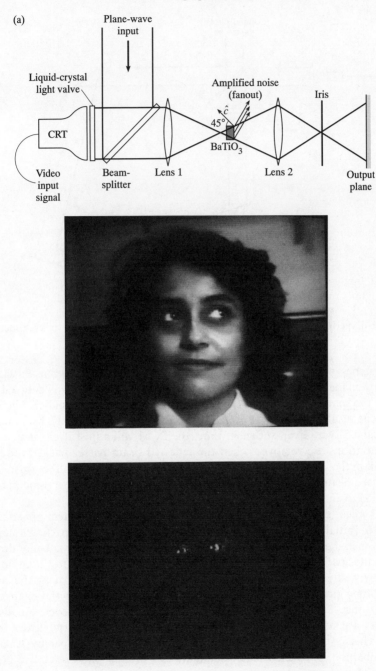

Fig. 10.14 Novelty filter using beam fanning in BaTiO$_3$ (after Ford *et al.* 1988). (a) Experimental arrangement. (b) Results – see text for detail.

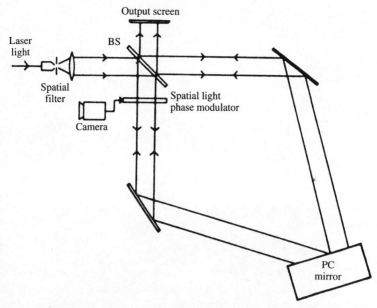

Fig. 10.15 Experimental arrangement for novelty filtering using a PCM (after Anderson *et al.* 1987).

the properties of the PCM ensure that on returning to the beamsplitter both beams are plane waves and their phases are such that all power is coupled back towards the spatial filter, none reaching the output screen. A rapid change in the input scene results in a change in the phases of some regions of the beam passing through the SLM. Given sufficient time, the PCM will adjust itself so that these phase variations are removed when the reflected beam passes back through the SLM, but in the short term there will be some spatial variation in the phase of this beam at the beamsplitter, and hence in these regions of the image some light will be coupled to the output screen.

Some typical results taken from the output screen of this system are shown in Fig. 10.16. Initially, there is no input to the system, the SLM imparting a uniform phase to the whole beam, and the output is blank as all power is being directed towards the spatial filter. Next, some characters are presented to the input, and immediately after this has occurred, their presence is registered at the output screen. Some time later, after the PCM has compensated for the change, the output is once again blank, as the characters are no longer novel. Finally, the characters are removed from the input and their absence is recognized by the system. This system and others are described in the review paper by Anderson and Feinberg (1989) which nicely summarizes work on photorefractive novelty filters up to that date.

Cronin-Golomb *et al.* (1987) described an approach to novelty filtering that made use of two-wave mixing. Their idea was to have an image-bearing signal beam (the image being encoded as a spatial modulation of the phase of that beam)

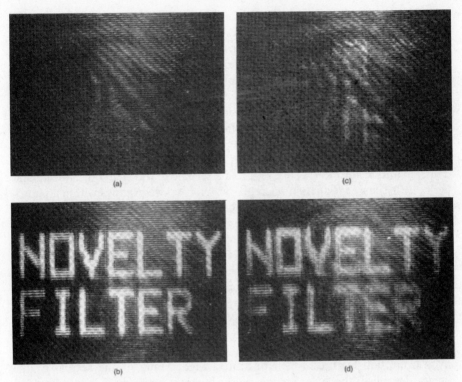

Fig. 10.16 Novelty filtering using a PCM: results (after Anderson *et al.* 1987). (a) Output when input image is uniform. (b) Output immediately after characters are presented at input. (c) Output some time later when the PCM has compensated for the new input. (d) Output immediately after characters are removed from the input.

and a uniform reference beam and to arrange things so that power was coupled strongly from the signal to the reference beam, resulting in the steady state in a heavily depleted signal beam. In this situation, we know that the grating recorded in the photorefractive crystal will be $\pi/2$ out of phase with the intensity fringes. Any change in a region of the image would result in a local variation in that phase angle, thus reducing the beam coupling and allowing some of the signal beam to pass through the crystal, until such time as the crystal was able to readjust and restore the beam coupling.

A different technique has been described by N. S.-K. Kwong *et al.* (1988) and Soutar *et al.* (1991) which relies on transient enhancement of the grating amplitude under two-wave mixing (see Section 6.13) when one beam carries the image to be filtered and the other is a plane reference beam. A change in the intensity of a region of the image causes a transient rise in the space charge field amplitude in the part of the crystal corresponding to that region. In the experiment reported by Soutar *et al.*, the concomitant increase in the diffraction efficiency of that region was monitored using a HeNe read-out beam. The drawback with this approach was that it was necessary to threshold the image on the HeNe beam using a PC (though this could also be done using photorefractives) and the threshold could only be easily set when the image was binary in nature.

Khoury et al. (1991a) made use of the different time constants associated with phase conjugate beams arising from transmission and reflection gratings. The two phase conjugates were arranged to overlap and interfere destructively in the steady state, thus cancelling out. Transients resulted in the faster responding beam dominating for a period of time, thus producing a detectable intensity at the output. The same group also reported another scheme for novelty filtering involving four-wave mixing (Khoury et al. 1991b).

Hussain and Eason (1991) have reported a scheme in which an object beam and a periodically phase modulated reference beam write two gratings within a photorefractive crystal that are phase-shifted by 180°. Under steady state conditions, diffraction of a read-out beam from these gratings leads to destructive interference. However, change in the object beam results in the complementary gratings being of unequal amplitude, and a diffracted beam is observed.

A photorefractive novelty filter has been applied in a microscope system. The novelty filter was used to make moving protozoa visible while largely eliminating the background of stationary algae from the image (Cudney et al. 1988).

Finally, Anderson and Feinberg (1989) point out that there may be useful applications for the opposite of a novelty filter, which they christen a monotony filter. This is a device that rejects moving or changing objects in an input image, only transmitting features which do not change. Many of the several of the schemes described above (excluding those that rely on beam fanning) have an output that provides a monotony filtered image. For example, in Fig. 10.15, if a second beamsplitter is inserted between the spatial filter and the original beamsplitter, the monotony filtered image may be accessed, which is directed back towards the spatial filter.

10.3.4 Spatial light modulation

A spatial light modulator (SLM) is a device which is able to impose spatial information (amplitude and/or phase variations) on a beam of light. Some SLMs are addressed electrically and such devices are found in projection TV systems, for example. Other SLMs receive an optical input. At first sight the ability to take an image on one beam and transfer that image to a second beam may seem of little consequence, but there are many situations where this is useful: for example, the input may be a normal image with a spatially varying intensity and we may require that this information is transferred to another beam in the form of variations in phase; alternatively, the input may possess low coherence – perhaps arising from a video display – and we may wish to impose the information on a laser beam to facilitate the processing of the image in a coherent system.

The latter approach, dubbed incoherent-to-coherent conversion, may be performed using the arrangement shown in Fig. 10.17 (Marrakchi 1988a,b). The technique is centred around the amplification of a weak uniform beam from

10.3 Image processing

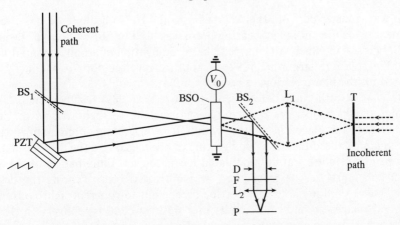

Fig. 10.17 Arrangement for performing incoherent-to-coherent conversion (after Marrakchi 1988b). BS1, BS2 = beamsplitters, L1, L2 = lenses, T = transparency, D = iris diaphragm, F = green filter.

BS1 at the expense of a strong uniform pump beam, which has been given a detuning by the PZT mounted mirror so as to maximize the available gain. Both these beams are derived from the 514 nm (green) line of an Ar^+ laser. The amplified signal beam passes through a diaphragm, a green filter, lens L2, and a polarizer on its way to be viewed by a CCD array (not shown in figure). The incoherent beam comes from an HeCd laser at 442 nm and illuminates the image held on the transparency, which is then imaged into the BSO crystal via lens L2.

Those regions of the image which are the brightest cause the largest reduction in the gain experienced by the signal beam, this reduction occurring due to the decrease in the index modulation according to eqn (10.2). Thus the intensity variations in the input image are transferred to variations in the gain and hence the intensity of the signal beam.

This argument would suggest that the system should produce a contrast-reversed coherent version of the input image. In fact the situation is more complicated: the BSO was used in the transverse geometry (field parallel to the [110] direction). With this orientation, the effect of the optically active and birefringent BSO on the polarization of the emerging beam is that those regions which have experienced maximum gain and those where the gain has been removed by the incoherent beam have very different polarization states. By rotating the polarizer, it is possible to select between the two and obtain a signal beam possessing either a contrast-reversed or a same-contrast image.

The principle of operation of the system just described – namely the erasure by an incoherent beam of a grating written by coherent beams – typifies the operation of most photorefractive SLMs (Shi *et al.* 1983). In perhaps the first photorefractive SLM (Kamshilin and Petrov 1980), a grating written using uniform beams from an HeCd laser was erased using image-bearing light from an Ar^+ laser, the resulting grating being read out using an HeNe laser beam. Thus the image was transferred from the Ar^+ beam to the HeNe beam. In a more recent scheme, the incoherent image-bearing beam was used to cause selective erasure in an SBN crystal in which self-pumped phase conjugation of a uniform input beam was taking place (Sharp et al. 1992). A contrast-reversed copy of the image was imposed on the phase conjugate beam.

SBN was also the material chosen by Ma et al. (1989) for their spatial light modulator, which operates on a different principle. In their scheme the input transparency was placed in contact with a Ronchi grating. Illumination with a halogen lamp caused the object and grating to be imaged into the SBN crystal, the result being an index grating modulated by the object. Illumination of the SBN with a coherent laser beam resulted in a diffracted beam bearing the desired image. This technique is capable of the incoherent-to-coherent light conversion of several images at once and results in a positive contrast conversion. It also possesses a high dynamic range; in particular, those portions of the image which are black do not write a grating in the photorefractive crystal and therefore there is no output from the SLM corresponding to those regions (Vachss et al. 1991). This is not true of some of the other approaches. One feature of this technique, as with other photorefractive SLMs, is that its intensity transfer function is non-linear; in fact the output intensity is proportional to the square of the incoherent input intensity. Using this technique with BSO, improved diffraction efficiency has been obtained by translating the Ronchi grating to produce a moving fringe pattern in the crystal (Vachss et al. 1992). With the appropriate velocity, applying an external electric field then produces an enhanced space charge field analogous to that which occurs when the moving grating technique is used in two-wave mixing experiments (see Section 2.8).

Another approach, shown in Fig. 10.18, makes use of beam fanning in $BaTiO_3$ (Notni et al. 1992). In the absence of the incoherent beam, the coherent beam is heavily depleted due to beam fanning. Incoherent illumination acts to erase the gratings recorded by the scattered light in the photorefractive crystal, thus allowing the passage of the coherent beam through that region of the crystal. Anisotropic self-diffraction (see Section 6.9) in $KNbO_3$ has also been used as a means for realizing an SLM (Voit and Günter 1987).

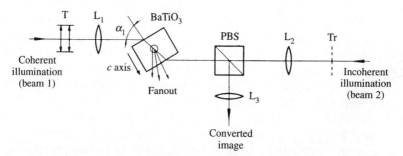

Fig. 10.18 Incoherent-to-coherent conversion using beam fanning in $BATiO_3$ (after Notni et al. 1992). T = beam-expanding telescope, PBS = polarizing beamsplitter, L1, L2, L3 = lenses, Tr = transparency holding input image.

10.3.5 Holographic interferometry

Holographic interferometry is a technique for observing the amplitude of vibrations or static deformation in a structure. In the latter case double exposure holographic interferometry is used, where two holograms are recorded of the same object in between which the deformation of the object is supposed to have occurred. For those parts of the object which have not moved (or have moved so as to change the optical path of the reflected light by a whole number of wavelengths) the two holograms will superimpose in phase. Conversely, for those parts of the object which have moved so as to change the optical path length by an odd number of half wavelengths, the two holograms will be 180° out of phase. On replay, the former regions will produce a bright diffracted beam, while the latter will produce no diffracted beam. Thus the replayed image of the object is seen to be composed of bright and dark fringes mapping out contours of the displacement that has occurred between the two recordings.

Double exposure holographic interferometry may be carried out using conventional holographic recording techniques, but by using photorefractive media no development process is required. The first report of double exposure holography was by Huignard and Herriau (1977), who used BSO. Figure 10.19(a) illustrates the set-up that they used, with some typical results being shown in Fig. 10.19(b). In this case the object was a transistor with heat sink. The procedure in this case was to record a hologram with the transistor turned off, then record a hologram with the transistor on and then replay both holograms with only the reference beam incident. The replay quickly erased the two stored holograms but long term storage of the interferogram was possible on the Vidicon memory tube. BGO has also been used in this kind of application (Ja 1980).

Double exposure holographic interferometry has also been reported using $LiNbO_3$ to visualize the phase changes imparted to an optical beam traversing a wind tunnel (Magnusson *et al.* 1987). With this system, the flow field of the air in the tunnel could be studied. More recent papers by Troth *et al.* (1991) and Sochava *et al.* (1992) report on a detailed study of holographic interferometry using sillénite crystals. In one arrangement, using a simple two-wave mixing approach, they observed the desired interference pattern in a higher-order diffracted beam. This prevents the pattern from being swamped by the image-bearing or reference beams. Another arrangement used a crystal of BSO in the transverse geometry (see Section 6.4). In this geometry, it is possible to arrange things so that the polarization of the diffracted light is orthogonal to that of the incident light, in which case the interference pattern can be observed by using an analyser to block out the transmitted reference and object beams.

Time-average holographic interferometry allows one to visualize the amplitude of vibrations of parts of an object, and by using a photorefractive crystal this may be carried out in real time. Figure 10.20(a) shows the approach taken by Huignard *et al.* (1977) (see also Marrakchi *et al.* 1980). A hologram is recorded between a plane reference beam and light reflected from the object to be studied.

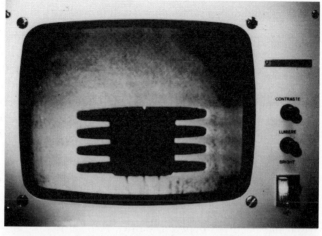
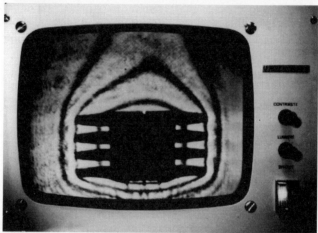

Fig. 10.19 Double-exposure holographic interferometry (after Huignard and Herriau 1977). (a) Experimental arrangement. (b) Sample results: upper trace = initial image of transistor with heat sink; lower trace = interferogram showing induced thermal index gradient.

The hologram is replayed by the retroreflected reference beam, and when the object is stationary the result is an image of the object.

When a given point on the object is vibrating, and we are assuming that the vibration frequency is much greater than the inverse of the response time of the photorefractive crystal, the effect is to partially erase the grating. The degree of erasure depends on the amplitude of the vibration and in fact it can be shown that the dependence of the diffracted beam intensity I_d on the vibration amplitude δ is of the form

$$I_d(\delta) = I_d(0) \left| J_o\left(\frac{4\pi\delta}{\lambda}\right) \right|^2 \tag{10.3}$$

10.3 Image processing

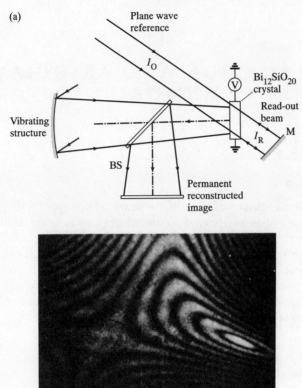

Fig. 10.20 Time-average holographic interferometry (after Huignard *et al.* 1977). (a) Experimental arrangement. (b) Results obtained from vibrating membrane.

where λ is the wavelength of the light and J_o is the zero-order Bessel function. A sample result is shown in Fig. 10.20(b), where the test object was a vibrating membrane.

Time-average holographic interferometry is also possible with two-wave mixing. In this case the gain experienced by a signal beam reflected from the structure of interest is modified by the imposed phase modulation due to the vibration of the structure (Huignard and Marrakchi 1981b; Xie *et al.* 1991).

Finally, while we are on the subject of interferometry, there is considerable interest in the behaviour of otherwise conventional interferometers when one or more of their mirrors is replaced by a phase conjugate mirror (Feinberg 1983). Examples of applications for such interferometers include optical sensing, image subtraction (Chiou and Yeh 1986; Magnusson *et al.* 1992), and wavefront visualization for optical component testing (Erdmann *et al.* 1987; Zurita *et al.* 1991).

11
CORRELATION AND ASSOCIATIVE MEMORIES

11.1 Introduction

As we have already discussed, one of the main advantages of signal processing using non-linear optics as opposed to digital electronics is the ability to handle potentially large quantities of information in parallel. Image processing is an area that lends itself naturally to optical techniques, and in this field two of the most important operations that must be performed are those of convolution and especially correlation. In this chapter we shall see how these operations may be performed using non-linear optics and how the process of correlation may be used to realize an associative memory.

11.2 Correlation and convolution

We begin with a few definitions. Firstly, the function $g(u,v)$ that is the convolution of two functions $f(x,y)$ and $h(x,y)$ is defined by

$$g(u,v) = f(x,y) \otimes h(x,y) = \int_{-\infty}^{\infty} \int_{-\infty}^{\infty} f(x,y)h(u-x,v-y)\mathrm{d}x\mathrm{d}y \quad (11.1)$$

where we denote the process of convolution by the operator $\otimes$. Secondly, when $g(u,v)$ is the cross-correlation of $f(x,y)$ and $h(x,y)$ (which we shall just call the correlation from now on), it is defined by

$$g(u,v) = f(x,y) * h(x,y) = \int_{-\infty}^{\infty} \int_{-\infty}^{\infty} f(x,y)h^*(x-u,y-u)\mathrm{d}x\mathrm{d}y \quad (11.2)$$

where $*$ denotes the operation of correlation. In short-hand notation we denote the two-dimensional Fourier transform of a function with the operator $\mathbf{F}\{\ \}$ so that

$$F(k_x,k_y) = \mathbf{F}\{f(x,y)\} = \int_{-\infty}^{\infty} \int_{-\infty}^{\infty} f(x,y)\mathrm{e}^{\mathrm{j}k_x x}\mathrm{e}^{\mathrm{j}k_y y}\mathrm{d}x\mathrm{d}y \quad (11.3)$$

11.2 Correlation and convolution

and the inverse Fourier transform is denoted by $\mathbf{F}^{-1}\{\ \}$ so that

$$f(x,y) = \mathbf{F}^{-1}\{F(k_x,k_y)\} = \frac{1}{2\pi}\int_{-\infty}^{\infty}\int_{-\infty}^{\infty} F(k_x,k_y)\mathrm{e}^{-\mathrm{j}k_x x}\mathrm{e}^{-\mathrm{j}k_y y}\mathrm{d}k_x\mathrm{d}k_y \qquad (11.4)$$

Finally, we shall need to make use of the convolution theorem (see e.g. Goodman 1968), which states

$$\mathbf{F}\{f \otimes g\} = \mathbf{F}\{f\}\mathbf{F}\{g\}$$
$$\mathbf{F}\{f * g\} = \mathbf{F}\{f\}\mathbf{F}^*\{g\} \qquad (11.5)$$

We first look in general terms at how the correlation and convolution of images may be performed with the aid of non-linear optics. This is intended to be an overview with no mathematical rigour; we shall look at the process in more detail once the general principles have been explained.

Figure 11.1 shows the arrangement needed to carry out correlation and convolution. Three waves are incident on the non-linear medium with electric fields $\mathcal{E}_1$, $\mathcal{E}_2$, and $\mathcal{E}_3$, and we assume that the interaction of these three waves leads to the creation of a fourth wave with electric field, $\mathcal{E}_4$, of the form

$$\mathcal{E}_4 \propto \mathcal{E}_1 \mathcal{E}_2^* \mathcal{E}_3 \qquad (11.6)$$

For the moment we will just take this as a property of the non-linear medium; we shall see later that photorefractive materials can indeed provide a response of this kind. Because of the transforming properties of the imaging lenses in the system, the fields within the medium are the Fourier transforms of the input fields, U_1, U_2, U_3. Therefore, the field $\mathcal{E}_4$ produced within the medium has the form

$$\mathcal{E}_4 \propto \mathcal{E}_1 \mathcal{E}_2^* \mathcal{E}_3 = \mathbf{F}\{U_1\}\mathbf{F}^*\{U_2\}\mathbf{F}\{U_3\} \qquad (11.7)$$

which with the aid of the convolution theorem becomes

$$\mathcal{E}_4 = \mathbf{F}\{U_1\}\mathbf{F}^*\{U_2\}\mathbf{F}\{U_3\} = \mathbf{F}\{U_1 * U_2 \otimes U_3\} \qquad (11.8)$$

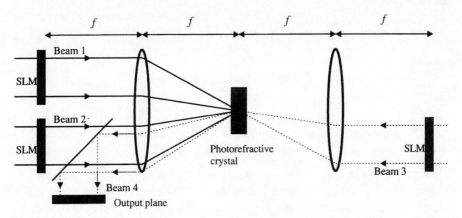

Fig. 11.1 Schematic arrangement used to carry out image correlation and convolution.

or
$$\mathcal{E}_4 = \mathbf{F}\{U_1\}\mathbf{F}^*\{U_2\}\mathbf{F}\{U_3\} = \mathbf{F}\{U_3 * U_2 \otimes U_1\} \quad (11.9)$$

But the output of wave 4, U_4, is again transformed by the action of a lens so that

$$U_4 = \mathbf{F}^{-1}\{\mathcal{E}_4\} \propto U_1 * U_2 \otimes U_3 \quad (11.10)$$

which again may be rewritten as

$$U_4 = \mathbf{F}^{-1}\{\mathcal{E}_4\} \propto U_3 * U_2 \otimes U_1 \quad (11.11)$$

Equations (11.10) and (11.11) show that correlation may be obtained between beams 1 and 2 or between beams 2 and 3. This point will be elaborated on in Section 11.4.

It is clear then that with the set-up of Fig. 11.1, the operations of convolution and correlation may both be carried out. In fact it is simple to arrange to have either operation alone. For example, if just the correlation of two images is required, U_3 or U_1 should take the form of a delta function, which may be implemented using a pin-hole, or by ensuring that $\mathcal{E}_4$ is a uniform plane wave. As we shall see from the literature, most of the interest in this form of image processing is in fact associated with the correlation of two images in this way.

11.3 Material response

We shall now look in more detail at the behaviour of the non-linear medium – in our case a photorefractive crystal – to see how the desired response given by eqn. (11.6) comes about. Let us look first of all at the response of the material to the interference pattern formed between the fields $\mathcal{E}_1$ and $\mathcal{E}_2$ and assume that beam 3 is not present. We shall also assume that the photorefractive crystal is thin, in the sense that the two fields are constant across the thickness of the crystal and we can neglect any coupling between the two waves. In this case the amplitude of the refractive index variation induced in the material is obtainable from eqns (3.3) and (3.5) as

$$\Delta n = -\frac{n^3 r E_w m}{2} \quad (11.12)$$

where

$$m = \frac{2\mathcal{E}_1 \mathcal{E}_2^*}{I_0} \quad (11.13)$$

and I_0 is the total intensity of the two beams, given by

$$I_0 = |\mathcal{E}_1|^2 + |\mathcal{E}_2|^2 \quad (11.14)$$

We next assume that beam 3 is Bragg-matched to the grating produced by beams 1 and 2 and further that it is arranged that the presence of beam 3 does not perturb this grating. This may be achieved by, for example, ensuring that beam 3 is everywhere much weaker than beams 1 and 2, or by deriving beam 3

11.3 Material response

from a separate source, chosen to be of a wavelength where the excitation of electrons from the donor levels is very inefficient. In the latter case, whilst beam 3 cannot readily participate in the recording of a grating, it can still experience fully the grating produced by beams 1 and 2. Beam 3 is diffracted from the grating to produce beam 4.

Assuming that the diffraction efficiency is low and the crystal lossless, the amplitude diffraction efficiency, η, is given by (Kogelnik 1969)

$$\eta = \frac{\pi \Delta n \, d}{\lambda \cos \theta} \quad (11.15)$$

and therefore the amplitude of beam 4 may be obtained using eqns (11.12) to (11.15) as

$$\mathcal{E}_4 = -\frac{\pi \Delta n \, d}{\lambda \cos \theta} \mathcal{E}_3 = -\frac{\pi d n^3 r E_w}{\lambda I_o \cos \theta} \mathcal{E}_1 \mathcal{E}_2^* \mathcal{E}_3 \quad (11.16)$$

which is exactly of the form required in eqn (11.6). So photorefractive materials can be used as the active media for real-time correlation and convolution[1].

This technique was demonstrated for the first time by White and Yariv (1980). Their experimental arrangement was identical to Fig. 11.1 and a crystal of BSO was used as the active element, an applied field being used to enhance the amplitude of the refractive index grating. The three incident beams were all derived from the same Ar^+ laser emitting at 514 nm. Images took the form of transparencies inserted in the beams. Two-image processing was carried out by ensuring that the third image was of the form of a delta function. In practice this was carried out by inserting another lens in the relevant beam so that a focus was formed at the plane where the appropriate image transparency would have been sited. This generated a collimated beam (plane wave) at the crystal, which is the desired form.

Figure 11.2 presents results from the experiment. The form of the three input transparencies is shown along with a photograph of the output of the system. The upper three sets successfully show the correlation of various pairs of test images, while the lower set illustrates the convolution of two images.

At around the same time, Pichon and Huignard (1981) carried out experiments on correlation using the geometry shown in Fig. 11.3, in which all three beams are incident from the same side of the crystal. Lens L_1 produces a grating within the BSO crystal, modulated by the Fourier transforms of the two transparencies A and B. The grating writing occurs at a wavelength of 488 nm. The grating is then read out by a plane wave from an HeNe laser, and the Fourier transform of the read-out beam, produced by lens L_2, yields the desired correlation in the output plane. At 633 nm, the photorefractive response of BSO is very low and so the read-out beam has little influence on the grating produced by the 488 nm light. In

---

[1]Of course other non-linear mechanisms and different media may also be used, such as thermoplastic film (T. C. Lee et al. 1980). All that is required is that beams 1 and 2 are able somehow to produce a variation in the material's refractive index (or absorption) that follows their interference pattern; in this case the phase difference between that interference pattern and the grating is not important.

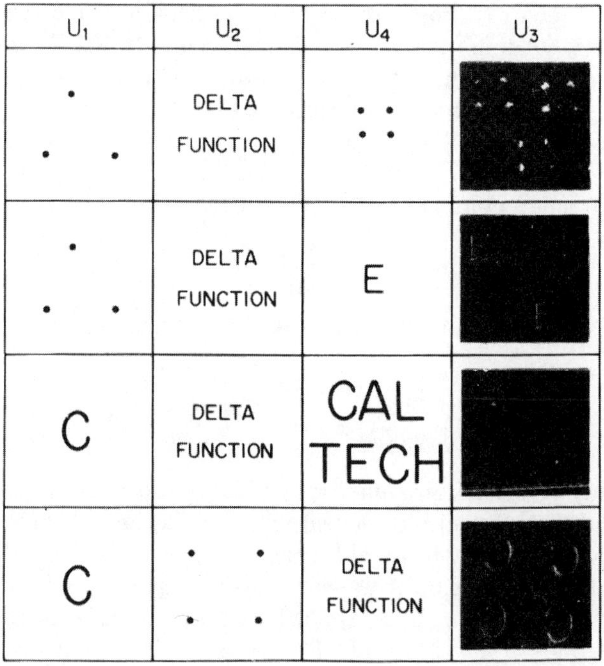

U1
U2
U4
U3
DELTA FUNCTION
DELTA FUNCTION
E
C
DELTA FUNCTION
CAL
TECH
C
DELTA FUNCTION

Fig. 11.2 Image correlation and convolution. First three columns show the input transparencies, fourth column shows the output of the system (after White and Yariv 1980).

(x_o,y_o)
(u,v)
(X,Y)
A
B
λ_1
488 nm
$2b$
L_1
BS
RF
$\delta\theta$
$|A\odot B|^2$
Polarizer
V
L_2
λ_2
633 nm

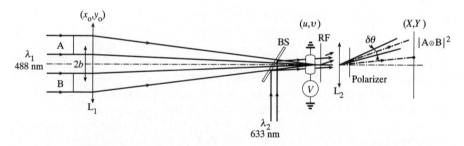

Fig. 11.3 Geometry for correlating the images of two transparencies, A + B, used by Pichon and Huignard (1981).

this arrangement, the signal-to-noise ratio was improved by the inclusion of a polarizer set to reject the majority of the scattered light, which otherwise forms a noise background upon which the desired signal is superimposed (Herriau *et al.* 1978; Apostolidis *et al.* 1985).

More recently, an engineered prototype demonstrator has been described by Rajbenbach *et al.* (1992b) based on the system just described. A schematic diagram of the system is shown in Fig. 11.4, while a photograph of the unit appears in Fig. 11.5. Holograms were written in BSO using a frequency-doubled Nd:YAG

11.3 Material response

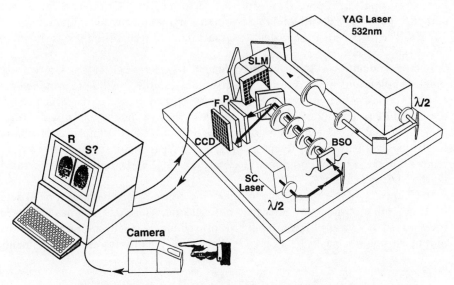

Fig. 11.4 Photorefractive correlator, schematic (after Rajbenbach *et al.* 1992b).

Fig. 11.5 Photorefractive correlator demonstrator (after Rajbenbach *et al.* 1992b).

laser operating at 532 nm. Read-out was performed at 633 nm. The device was compact (60 × 30 × 30 cm) and able to operate in real time with an update rate of 100 ms. The response of the system to a machine tool recognition test is shown in Fig. 11.6.

A face recognition system has been reported by Li *et al.* (1993). Up to forty separate holograms were recorded in a crystal of $LiNbO_3$, each hologram being a combination of several images. By storing a number of images of the same person, under different lighting conditions, a certain degree of invariance to scale, rotation, and facial expression was obtained (see Section 11.6). To improve the ability to distinguish between different people, all images were spatially filtered to remove the DC Fourier component (see Section 11.5) and components generated by the SLM pixelation.

In Chapter 3 we saw that the use of a moving grating along with an applied field is able to enhance the gain under two-wave mixing. This approach has also been investigated as a means for improving the diffraction efficiency and stability of the output correlation peak (Z. Q. Wang *et al.* 1994). Other recent developments

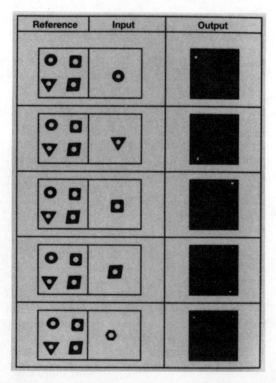

Fig. 11.6 Demonstration of machine tool recognition using the photorefractive correlator of Fig. 11.4. Output shows the correlation peak as viewed by the CCD camera. The lowest pictures show the response to an object that is not in the reference set (after Rajbenbach *et al.* 1992b).

include an optical correlator used as a key element in a residue look-up table processor (K.-Y. Lee et al. 1993) – a system designed to facilitate the parallel arithmetic processing of large amounts of data – and a Fresnel transform used by Soutar et al. (1992) to produce a so-called intensity correlator that was claimed to have superior tolerance to the placement of the input image transparencies than the usual correlator, which is based on the Fourier transform.

Correlation has been demonstrated in materials other than BSO: GaAs has been used with an Nd:YAG laser source emitting at 1.06 μm (Gheen and Cheng 1988; D. T. H. Liu and Cheng 1992). This material has the potential advantage of offering a higher update rate for a given incident intensity, due to the material's fast response. In addition, GaAs is sensitive at wavelengths compatible with potentially cheap and compact semiconductor light sources (Delaye et al. 1994). Faria et al. (1991) have reported correlation in $Bi_{12}TiO_{20}$ (BTO). This material has reasonable sensitivity at 633 nm allowing the use of a comparatively cheap HeNe laser to provide the writing and read-out beams.

11.4 Historical background

The technique of optical correlation using the photorefractive effect owes much to the many innovations in optical signal processing that occurred before photorefractive materials were first investigated. Here we briefly report the key developments that led up to the invention of the photorefractive correlator.

The Fourier transforming properties of a lens have been known for some time and early experiments on the manipulation of images in the Fourier domain were carried out by Abbe (1893) and Porter (1906)[2]. The basic idea is illustrated in Fig. 11.7(a). Two lenses are used to form the image of an illuminated object transparency. The first lens may be considered as generating the Fourier transform of the object in the Fourier plane. The second lens then performs the inverse transform of the resulting field distribution. In the absence of any filter in the Fourier plane, the output image will be similar to the input object, though there will be some differences due to the finite size of the entrance pupil, which will act as a low pass filter to the higher spatial harmonics in the object. However, such an arrangement gives easy access to the Fourier plane and permits the image to be processed. As the field distribution in the Fourier plane reflects the spatial frequencies that were present in the original object, processing in the Fourier domain is known as spatial filtering. Figure 11.7(b) shows the effect of inserting a filter to block out the components in the Fourier plane corresponding to horizontal features in the object.

The techniques of spatial filtering found useful application in the field of phase contrast microscopy. Here, the problem is one of somehow making visible the distortion in a wavefront that results when light passes through a material (often biological) which is approximately uniformly transparent, but in which there is

---

[2] For a somewhat more up-to-date review of Fourier plane processing see Flannery and Horner (1989).

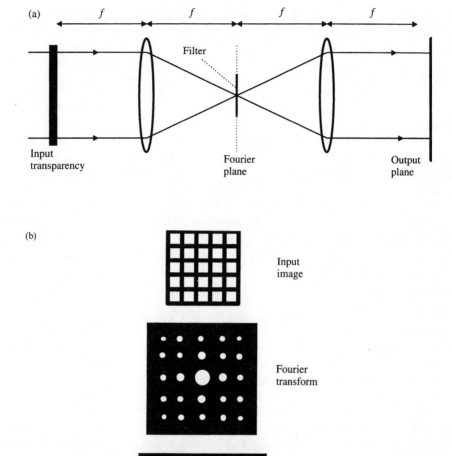
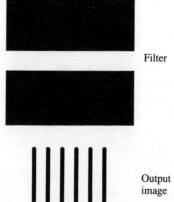

Fig. 11.7 (a) Arrangement to permit Fourier plane processing. (b) Examples of object transparency, filter, and image.

11.4 Historical background

detail of interest in the form of variations in the optical thickness. Solutions to this problem involving spatial filtering include the central dark ground method, in which the central (DC) Fourier component is blocked, the Schlieren technique, in which all spectral components to one side of the central component are blocked, and phase contrast microscopy, introduced by Zernicke (1935), where the phase of the central Fourier component is shifted by 90° with respect to the higher spectral components. The latter technique is often favoured, as it leads to an intensity in the final image that is proportional to the phase delay introduced by the specimen, in the limit when such phase delays are small.

In the 1950s, spatial filtering was applied to the correction of photographic images obtained with defective imaging systems and links were established between spatial filtering and communications theory, where signals in the time domain were often processed by filtering techniques in the corresponding Fourier frequency domain. This work led on to the realization by the end of the 1950s that optical spatial filtering techniques could be used to process electronic signals once those signals had been represented as an image on a photographic film (Cutrona et al. 1960, 1966). Also in the 1950s, incoherent optical systems had been proposed that were able to provide the correlation of two images in the form of transparencies (Kovasnay and Arman 1957). Unfortunately, these incoherent systems suffered from only being able to handle non-negative data (as they were based on intensity) and having a limited space–bandwidth product (low resolution images are required in order to minimize diffraction effects, as the operation of these systems was based on geometrical optics).

Around this time arose the concept of the matched filter, which crops up often in the field of optical correlation. The concept originally arose when the following problem was considered: given a signal $s(t)$ buried in white noise, what filter should be applied to the combination of signal and noise in order to maximize the ratio of instantaneous signal power to average noise power? Analysis shows (Turin 1960) that the impulse response of the desired filter should be

$$h(t) = s^*(-t) \tag{11.17}$$

which implies that in the frequency domain the transfer function is

$$H(\omega) = F\{h(t)\} = F\{s^*(-t)\} = F^*\{s(t)\} \tag{11.18}$$

So in the frequency domain, the desired filter is just the complex conjugate of the frequency response of the signal itself. Because the filter response must be matched to the signal in this way, it is said to be a matched filter. When an unknown signal $u(t)$ is applied to the filter matched to $s(t)$, the resulting output is the convolution of $u(t)$ with the impulse response $h(t)$:

$$\int_{-\infty}^{\infty} u(\tau)h(t-\tau)d\tau = \int_{-\infty}^{\infty} u(\tau)s^*(\tau-t)d\tau \tag{11.19}$$

which may be recognized as the correlation of the two signals $u(t)$ and $s(t)$.

In a coherent optical system, matched filtering of images may be readily carried out by placing the desired filter in the Fourier plane in Fig. 11.7. There are several difficulties with this approach: from eqn (11.18) we see that for correlation, the filter must be the complex conjugate of the Fourier transform of the reference image, and this may be difficult to calculate; the filter must contain amplitude and phase information, and in practice it is difficult to control accurately the phase of the transmitted wave; finally, when the images have a high space–bandwidth product, the positional alignment of the filter may be very difficult.

These problems severely limited the usefulness of optical filtering, but in 1963, holography was able to come to the rescue in the form of the vander Lugt filter (Vander Lugt 1964)[3]. His technique, which at that time had to use photographic emulsion needing development, is essentially a two-stage version of the scheme shown in Fig. 11.1. Initially, the hologram was recorded by exposing the film to the interference pattern produced by beam 1 (a plane wave at the crystal) and beam 2 (being the Fourier transform of the object). After development, the hologram was replaced and replayed by beam 3 (the Fourier transform of the reference image) to generate via beam 4 the correlation of the object and reference images.

This scheme overcomes two of the problems with Fourier plane filtering as described above. Firstly, the generation of the filter is carried out automatically by the optical system utilizing the transform properties of the lens; secondly, even when a pure absorption hologram is used, the filter still contains both amplitude and phase information carried on the 'carrier wave' which is the recorded periodic grating produced by the interference between beams 1 and 2.

Unfortunately, the system is still sensitive to the positioning of the filter within the system after development. In Section 11.2 we saw that with the arrangement of Fig. 11.1, correlation could be obtained between beams 1 and 2 (eqn 11.10) or between beams 2 and 3 (eqn 11.11). Correlation in the Vander Lugt approach involves beams 2 and 3. In such a system, when the hologram recording the interference pattern between beams 1 and 2 has been developed, it must be carefully positioned in the Fourier plane so that it is aligned with the Fourier transform of the reference image.

The problem with positional sensitivity was removed with the development of the joint transform correlator (Weaver and Goodman 1966)[4]. This approach relies on correlation between beams 1 and 2 in Fig. 11.1. The hologram is recorded with the object and reference images in these beams. After development, the hologram is replaced and replayed by beam 3. In this case beam 3 must be a plane wave and therefore alignment of the hologram is not critical.

With this history in mind, the advantages of using photorefractive materials to record the hologram should be apparent. Because the recording and read-out of the hologram occur simultaneously, without the need for development, the

---

[3]Vander Lugt authored a technical report on his filtering technique in 1963, but this is not widely available.

[4]In this paper, Weaver and Goodman actually describe *convolution*. This was essentially done by inverting the reference image, the simple change in coordinate systems resulting in the process of convolution rather than correlation.

correlation process occurs in real time. If the images are transferred to the optical beams using SLMs, the system can be updateable. If the Vander Lugt approach is taken, there is no problem with alignment of the hologram. In fact the Vander Lugt correlator has an important advantage over the joint transform correlator: the speed of the recording of the interference pattern between beams 1 and 2 is limited by the response time of the material, whereas the read-out process using beam 3 is practically instantaneous. Therefore, it is possible even with a slow photorefractive material to carry out correlation with a fast update rate. The fixed set of reference images is imposed on beam 2, and the potentially rapidly changing object is carried by beam 3.

11.5 Departures from ideality in the correlation process

There are several features of the photorefractive correlator that cause the output of the system to differ from the true correlation of the input images, as defined by eqn (11.2). Some of these features have been mentioned earlier in passing, whilst others have so far been ignored. In this section we shall examine them and see that some of the departures from ideality can actually be advantageous.

The first point to make is that eqn (11.14) assumes that the read-out beam is Bragg-matched to the grating written by beams 1 and 2. Strictly speaking, this can only be achieved when beams 1 and 3 are phase conjugates of each other (e.g. they could be two plane waves or alternatively diverging and converging spherical waves with the same radius of curvature). Any spatial variations imposed on beam 3 by the spatial light modulator will in general result in a spread of angles in the Fourier transform domain, when the beam is at the crystal. In general this spread will not be Bragg-matched to the grating in the crystal, especially if there is also structure being imposed on beams 1 and 2 by their respective spatial light modulators.

It turns out that the system is most sensitive to angular variations in the read-out beam (beam 3). To see why qualitatively, consider Fig. 11.8, which shows the recording and Bragg-matched replay of a grating with plane waves, where for generality we have allowed the read-out beam to be at a different wavelength to the writing beams. We shall ignore considerations of angular variations in beam 1, as when correlation is performed this will often be a plane wave at the crystal.

Let us first consider variations in the angle of incidence of beam 2. In this case, should the read-out beam be at the same wavelength as the writing beams, then the read-out process will remain Bragg-matched (this is just the phase conjugation process discussed in Sections 3.12 and 7.6). Essentially what happens is that as the direction of beam 2 changes, there is a rotation of the grating vector, which tends to shift the process off-Bragg. However, concomitant with the rotation is a change in the length of the grating vector which exactly cancels out the effects of the rotation, leaving the process Bragg-matched. When the read-out beam is at a different wavelength, the two effects do not exactly cancel out, though there is still some compensation.

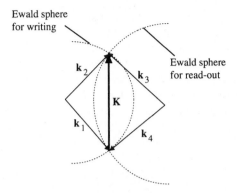

Fig. 11.8 Grating writing with Bragg-matched read-out. The grating, **K**, is recorded by beams 1 and 2. Replay is by beam 3, which in the general case is at a different wavelength, giving beam 4.

If we now consider a variation in the angle of beam 3, this causes the system to be no longer Bragg-matched, and furthermore there is no compensating change in the length of the grating vector. Angular variations in the read-out beam are therefore more serious than angular variations in beam 2. We discuss techniques for overcoming this angular sensitivity in the next section.

The second feature of the simple description of photorefractive correlation presented in Section 11.2 that deserves further discussion concerns eqns (11.6), describing the diffraction efficiency of the read-out process, and (11.7), which relates the fields at the crystal to the input fields. Both equations are strictly only satisfied if the crystal is infinitesimally thin and sited in the focal plane of the Fourier transforming lenses. To have any useful diffraction efficiency the crystal must clearly have a finite width (a few millimetres would be typical for most photorefractive materials). This means that in order that the electric field amplitudes should not vary appreciably through the crystal, it is essential that the focal lengths of the Fourier transform lenses should be sufficiently long that the focused spot size does not vary much over the crystal width. Only in this case will eqn (11.7) be well satisfied.

Furthermore, when the beams intersect in the crystal it is essential that they do so paraxially. Figure 11.9 illustrates the problem. In Fig. 11.9(a) we have a paraxial system in which a light ray in one beam interacts with essentially only one ray in the other beam. In this situation eqn (11.6) is well satisfied. Conversely, in Fig. 11.9(b) the geometry is decidedly off-axis, with the interaction region of the two beams contained entirely within the crystal. Here a ray in one beam interacts with all the rays making up the second beam and eqn (11.6) is definitely not satisfied (Connors *et al.* 1984).

The next point of discussion concerns the expression for the amplitude of beam 4, eqn (11.16). This strictly only matches eqn (11.6) when the intensity, I_0, is constant. However, it is clear from eqn (11.14) that I_0 is not in fact a constant,

11.5 Departures from ideality in the correlation process

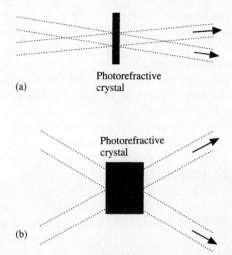

Fig. 11.9 True correlation demands a paraxial system (a) rather than one that is significantly off-axis (b).

but depends on the local intensities of the writing beams. In eqn (11.14) we were only considering the case of two beams incident on the photorefractive medium; in actual fact we have four beams interacting and therefore the correct expression for the intensity at any point is

$$I_0 = |\mathcal{E}_1|^2 + |\mathcal{E}_2|^2 + |\mathcal{E}_3|^2 + |\mathcal{E}_4|^2 \tag{11.20}$$

Equation (11.20) must be modified if the read-out beam is at a different wavelength to the writing beams. This is because in deriving expressions for the space charge field amplitude, I_o is introduced into the model via eqn (2.3) describing the rate of change of ionized donor atoms, which we repeat here for convenience:

$$\frac{\partial N_D^+}{\partial t} = sI(N_D - N_D^+) - \gamma_R n N_D^+ \tag{11.21}$$

If the read-out beam is at a different wavelength from the writing beams, then both it and the diffracted beam must be included with the appropriate value for the photoionization constant, s', so that

$$\begin{aligned}\frac{\partial N_D^+}{\partial t} &= (sI_{\lambda_1} + s'I_{\lambda_2})(N_D - N_D^+) - \gamma_R n N_D^+ \\ &= s(I_{\lambda_1} + aI_{\lambda_2})(N_D - N_D^+) - \gamma_R n N_D^+\end{aligned} \tag{11.22}$$

where the constant a represents the ratio of the photoionization constants at the two wavelengths.

As a result of this, eqn (11.21) should be modified to read

$$I_0 = |\mathcal{E}_1|^2 + |\mathcal{E}_2|^2 + a|\mathcal{E}_3|^2 + a|\mathcal{E}_4|^2 \tag{11.23}$$

to allow for writing and read-out beams of different wavelengths. If true correlation is desired, and eqn (11.16) is to match eqn (11.6), then the intensity of one of the beams should be spatially uniform and should be much greater than the intensity of any of the other beams.

The condition of spatial uniformity implies that the beam must be a plane wave at the crystal (corresponding to a delta function in the image plane). This means that it is not possible to carry out true correlation and convolution at the same time, since only two objects can be imaged on to the crystal. In practice this is no great disadvantage, as usually only one operation is desired. In the majority of experiments this is correlation and so beam 2 might be a strong plane wave and beams 1 and 3 would contain the images to be correlated.

In addition to one of the beams being a plane wave with an intensity large compared to the other beams, it is also important that the coupling between beams should be small if true correlation or convolution is to be obtained. This condition is necessary so that the modulation $2\mathcal{E}_1\mathcal{E}_2^*/I_0$ does not vary with distance through the crystal and so that the intensity of beam 4 is small enough that appreciable depletion of beam 3 does not occur[5]. More detailed discussion of the conditions required to obtain true correlation may be found in Connors et al. (1984).

What now if the intensity, eqn (11.20), is not uniform? It turns out that in some applications this situation can be beneficial, resulting in a 'weighted' correlation that can improve the rejection of unwanted images (Connors et al. 1984; Cooper et al. 1986). Consider the correlation of two images borne on beams 1 and 3. In this case beam 2 is a plane wave and as we have just seen, true correlation will be obtained when $I_2 \gg I_1$, in which case the modulation, m, is proportional to $\mathcal{E}_1\mathcal{E}_2^*$. When the intensity of beam 2 is weaker than the brightest parts of beam 1 the situation is more complicated. At the crystal, beam 1 will have its greatest intensity where the zero spatial frequencies in the object are imaged. In this region the modulation will be of the form

$$m = \frac{2\mathcal{E}_1\mathcal{E}_2^*}{I_0} = \frac{2\mathcal{E}_1\mathcal{E}_2^*}{|\mathcal{E}_1|^2 + |\mathcal{E}_2|^2 + a|\mathcal{E}_3|^2 + a|\mathcal{E}_4|^2} \approx \frac{2\mathcal{E}_1\mathcal{E}_2^*}{|\mathcal{E}_1|^2} \qquad (11.24)$$

which is a quantity much smaller than unity. As one moves to positions corresponding to higher spatial frequencies, the intensity of beam 1 will decrease, whereas that of beam 2 will remain constant (it being a uniform plane wave at the crystal). A point will therefore be reached where the intensities of the two beams are comparable and the modulation is close to unity. It may be seen then that weighted correlation results in a maximum diffraction efficiency at points in the crystal that correspond to higher spatial frequencies in the object in beam 1. The consequence of this is that when the grating is read out by beam 3, correlation

---

[5]It should be borne in mind that with the usual experimental geometries coupling is generally only significant in transmission and therefore only coupling between beams 1 and 2 and between beams 3 and 4 is usually important. Reflection gratings may be ignored to good approximation.

11.5 Departures from ideality in the correlation process

of the higher spatial frequencies results in a greater diffracted signal than correlation of low spatial frequencies (White and Yariv 1982).

A demonstration of the benefits of weighted correlation has been provided by Cooper *et al.* (1986). Figure 11.10 shows the two images that were correlated. The reference image (Fig. 11.10a) is of an edged square, while the test images (Fig. 11.10b) consist of an edged square and a full square. True correlation of the reference image with each test image should result in equal height correlation peaks. Figure 11.11(a) shows linescans though the centre of computer-generated true correlations of the two images: the peak correlation values are the same but the correlation with the full square (on the right) has a much broader structure.

Figure 11.11(b) shows a similar linescan obtained from the correlation plane in an experiment in which correlation was carried out using BSO, and in which the maximum intensities of the reference and test images and the plane wave were in the ratio 60:1:100, respectively. Furthermore, while the plane wave and test image were generated using light of wavelength 514 nm, the reference image was incident at a wavelength of 633 nm where the photorefractive sensitivity is reduced by a factor of $a = 0.01$ (Cooper *et al.* 1986). As a result of this, the effective ratio of the intensities of the three incident beams was more like 0.6:1:100. This was thought to provide a reasonable approximation to the condition summarized in eqn (11.24), where the intensity of the plane wave dominates[6]. The experimental traces are fairly similar to those in Fig. 11.11(a), though there is clearly some distortion in the traces. In practice we would like to be able to distinguish between pairs of images such as these (though images of friendly and hostile aircraft might be of more interest to some). Using these correlation results such discrimination, which is easy to the human observer, would be difficult.

Fig. 11.10 Input images for correlation experiment (after Cooper *et al.* 1986). (a) Reference image. (b) Test images.

---

[6]One must bear in mind though that at the position at the crystal corresponding to the zero spatial frequencies of the test and reference images, the ratio would not be so favourable, as most of the power in the images would be concentrated here, whereas the power in the plane wave is of course uniformly distributed.

(a)

(b)

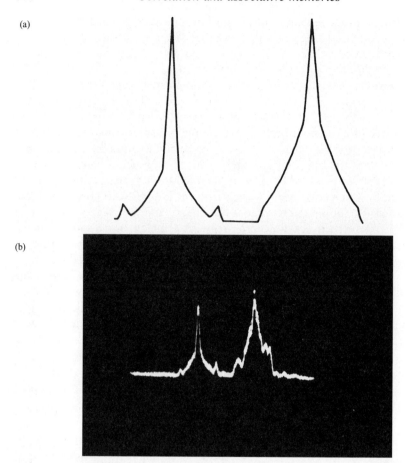

Fig. 11.11 Results of true correlation of images in Fig. 11.10 (after Cooper *et al.* 1986). (a) Computer-generated linescan across centre of true correlation function. (b) Experimentally obtained linescan in correlation plane.

Weighted correlation can do a better job. Figure 11.12 shows a linescan through an experimental correlation of the two images obtained when the ratio of the maximum intensities in the reference, test, and plane waves was 60:50:100, which does not include an effective reduction of a factor of 0.01 in the reference intensity due to its different wavelength. Now the test and plane wave intensities may be comparable, and weighted correlation is obtained which pays greater regard to higher spatial frequencies (corresponding to edges) in the images. Good discrimination between the two test images is obtained, and the system may be said to be able to recognise the edged square as being similar to the test image. Weighted correlation may also be obtained with a system arranged for true correlation by focusing a fourth plane input beam into the region of the crystal where zero spatial frequencies in the test image are found (Connors *et al.* 1984; McCall and Petts 1985). This has the effect of reducing the diffraction

11.6 Shift, scale, and rotation invariance

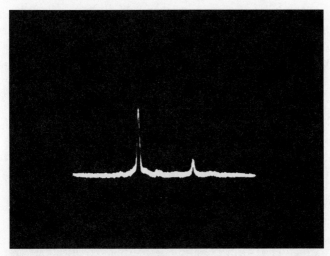

Fig. 11.12 Experimental results of weighted correlation of images in Fig. 11.10 (after Cooper *et al.* 1986).

efficiency in the low spatial frequency region, producing a similar effect to that occurring when the test image and plane beams have similar intensities, as described above.

This same group also demonstrated the advantages of using pulsed correlation. In one experiment with BSO (Nicholson *et al.* 1986), read-out with the test image was still obtained using 633 nm HeNe light, but the reference image and plane wave were derived from a frequency doubled Q-switched Nd:YAG laser emitting at 532 nm. Correlation signals were obtained from single 10 ns laser pulses, the correlated signal having a rise time of just 200 ns. In a second system (Cooper *et al.* 1986), pulses were formed using a c.w. Ar^+ laser along with an acousto-optic modulator. By having the plane wave (and the HeNe read-out beam) both c.w. but the test image pulsed, it was possible to ensure that the grating was written and read-out during a 5 ms pulse of the test image and then erased in less than 1 ms due to the effect of the c.w. plane wave. Pulsed systems of this kind are far less sensitive to vibration than c.w. systems, and by imposing the images on the beams using spatial light modulators, dynamic correlation is easily obtainable with update rates compatible with TV frame rates.

The system shown in Fig. 11.5 in Section 11.3 was the basis of further study by Daniel *et al.* (1995), who analysed the differences between 'true' correlation and the non-linear response of the photorefractive correlator using the whole-beam method (Cronin-Golomb 1992). Detailed modelling of the spatial fidelity of correlation was also undertaken recently by Meigs and Saleh (1994a,b).

11.6 Shift, scale, and rotation invariance

In the previous section we discussed the ways in which the operation of a photorefractive correlator can differ from the mathematical definition given in eqn

(11.2). For many applications, however, true correlation is not exactly what is required. Often we want to recognize whether a particular object is present in a scene. Two examples would be an intelligent missile trying to home in on a particular kind of military vehicle, or an automated building security system trying to recognize if the person trying to gain entry is a resident or not. For such applications we would like some flexibility in the correlation process; ideally the recognition process should not depend on exactly where the image is in the field of view, the scale, or the orientation of the image. We shall see how these features may be built into a photorefractive correlator.

Taking shift invariance first of all, the process of correlation is partially shift-invariant in that a movement of the object does cause a translation of the correlation peak, but the height of the peak remains the same. The position of the correlation peak can of course be used to infer the position of the object in the scene of view. Unfortunately, as we saw in the previous section, the volume holographic nature of the photorefractive correlator limits the shift invariance. Note that in Fig. 11.1 we are really considering the effects of translating the images in the plane of the paper. Translation out of the page does not destroy the Bragg condition and generally there is shift invariance in that direction.

The general problem of angular sensitivity due to off-Bragg replay can be reduced with a beam compression technique involving the use of a Galilean telescope (Yu et al. 1992). It may also be overcome by mechanically shifting the input image to the correlator using a rotating mirror or Bragg cell (Gu et al. 1992). It is also possible to use a thin photorefractive crystal operating in the Raman–Nath regime; for example, He et al. (1993) have used a 48 μm thick plate of $LiNbO_3$ as the non-linear medium. This was heavily doped with Fe to obtain a decent diffraction efficiency. They compared the shift invariance of the reference image with that obtained when the $LiNbO_3$ was replaced with a 1 mm thick SBN crystal. The correlation was carried out using letters of height 1.5 mm as the images; these were transformed using 15 cm focal length lenses and the angle between the beams was 17°. The results, shown in Fig. 11.13, show that the use of the SBN rather than $LiNbO_3$ worsens the shift invariance from 7 mm to 400 μm.

A shift-invariant optical correlator has also been constructed using an optical disk-based storage system, as shown in Fig. 11.14 (Psaltis et al. 1990; Neifeld and Psaltis 1993). A large number of reference images may be stored on the disk, which is able to hold in excess of 10^{10} bits of information. To correlate these images with an unknown input: first of all the disk illumination is turned off and a Vander Lugt style hologram is recorded in the photorefractive crystal by the input image and the reference beam. Then the input is turned off and the disk illuminated, after which the stored images are presented sequentially to the photorefractive crystal and the correlation signals read-out as the disk rotates. Shift invariance is obtained by making use of the rotation of the disk to translate the reference images past the beamsplitter.

If it is desired to make the system scale- and rotation-invariant, a 'brute force' method may be used, where multiple reference images of the same object may be used with different sizes/angles. More detailed information about achieving scale

11.6 Shift, scale, and rotation invariance

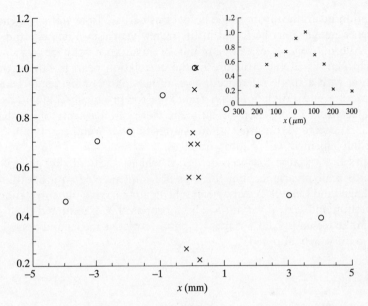

Fig. 11.13 Diffraction efficiency (arbitrary units) of autocorrelation peak of image of letter C as a function of displacement of reference: ($\times$), 1 mm thick SBN crystal; ($\bigcirc$), 48 μm thick LiNbO$_3$ crystal; inset, detail for SBN (after He *et al.* 1993).

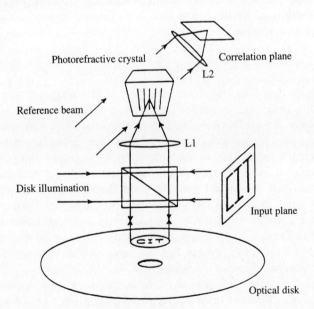

Fig. 11.14 Optical-disk-based photorefractive image correlator (after Neifeld and Psaltis 1993).

and rotation invariance from single holograms derived from statistical ensembles of reference images may be found in the review article by Flannery and Horner (1989). Alternatively, a certain amount of invariance may be obtained if a statistical analysis of the structure of the correlation peak is carried out after correlation with a single reference image, rather than just its height being measured (Merkle and Lörch 1984). A technique for producing a single frequency plane filter from a collection of different views of the same object has been reported (Hester and Casasent 1980; Caulfield and Weinberg 1982), but this was not implemented with photorefractives.

An optical correlator has been described which sacrificed shift invariance for rotation or scale invariance, but this was not implemented with photorefractive media (Leger and Lee 1982). An optical scheme for carrying out correlation that is shift, rotation and scale invariant has been proposed (Casasent and Psaltis 1976), but again the technique did not involve photorefractive media and does not seem to have been widely adopted.

11.7 Efficiency

We have chosen for this section a somewhat ambiguous title, as we wish to discuss two different but closely related issues: namely, how to obtain the highest diffraction efficiency from the correlator, and how to efficiently discriminate between different objects. For correlation purposes, the difffraction efficiency of beam 3 from the hologram recorded in the photorefractive medium by beams 1 and 2 is usually described in terms of the Horner efficiency (Horner 1982). When the hologram is addressed by the image that it is designed to recognize, the Horner efficiency, η_H, is defined as

$$\eta_H = \frac{\text{Power in the central correlation peak}}{\text{Power in the incident optical beam}} \quad (11.25)$$

Clearly for operation with low power lasers in a practical system, we would like to have this value as large as possible.

Unfortunately, a high Horner efficiency is not necessarily associated with good discrimination between objects (Caulfield 1982). This is because most of the diffraction efficiency arises due to the low spatial frequency components of the reference image in beam 2, and as we discussed briefly in Section 11.5, these components do not in general provide good discrimination between different objects. Far better discrimination is provided by the higher spatial frequencies, which in a true correlator contribute comparatively little to the Horner efficiency. As discussed in Section 11.5, discrimination may therefore be improved by suppressing the low spatial frequencies, but this can be at the expense of significantly lowering the resulting Horner efficiency.

True matched filtering then does not necessarily provide the best discrimination between different objects. Furthermore, the matched filter also does not give a good Horner efficiency. For example, Horner (1982) showed that if the object and reference image both took the form of a rectangular aperture, the Horner efficiency could not be greater than 4/9, and therefore for this simple image over half the incident power was wasted. A considerable amount of effort has been devoted towards developing techniques to increase the efficiencies of filters and to create filters that provide better discrimination between desired and undesired objects. Whilst much of this work has relevance to the photorefractive correlator, not all of the schemes are easily implemented in that format; often the desired filter response must be synthesized by computer, and in many cases the performance of the filter has been evaluated by simulation. We shall briefly look at some of these developments.

At the beginning of the 1980s it was realized that when an image was transmitted using a lens system as shown in Fig. 11.7(a), in the Fourier domain the phase information was far more important than the amplitude information as a description of the original image (Oppenheim and Lim 1981). Building on this work, Horner and Gianino (1984) suggested that this idea might also apply to matched filtering and that by using a phase-only filter, a high Horner efficiency might be obtained. In fact it turned out that not only is the phase-only filter more efficient, it also produces a sharper correlation peak.

Just to clarify what we mean by a phase-only filter, if the grating recorded by the interference of beams 1 and 2 in Fig. 11.1 has a complex amplitude $F(u,v)$, where u and v are the two spatial dimensions in the Fourier plane, and $F(u,v)$ is given by

$$F(u,v) = A(u,v)e^{j\phi(u,v)} \tag{11.26}$$

where A is the real amplitude and ϕ the phase of the grating, then a phase-only filter would be realized by a grating with complex amplitude

$$F(u,v) = Ce^{j\phi(u,v)} \tag{11.27}$$

where C is a real constant. Similarly an amplitude-only filter would have the form

$$F(u,v) = A(u,v) \tag{11.28}$$

The efficacy of phase-only filtering is brought out by Fig. 11.15 which shows conventional matched filtering, phase-only filtering, and amplitude-only filtering used to perform the autocorrelation of the capital letter G. The difference between the three approaches is striking and needs no comment.

The improvements brought about by the adoption of the phase-only filter are closely related to those discussed in Section 11.5 relating to weighted correlation. In that case we stated that improved discrimination was brought about by reducing the diffraction efficiency corresponding to the low spatial frequencies in the Fourier plane as compared to the higher spatial frequencies. Since the low spatial frequencies dominate in most images, this weighted correlation is actually acting to even out the amplitude in the Fourier domain, leaving us with something that approximates to the true phase-only filter.

It is clear from eqns (11.12) and (11.15) that the amplitude of beam 4 on leaving the photorefractive crystal is proportional to the local intensity fringe modulation. A device operating in this way is usually said to be acting as a linear correlator. Research has shown that there are generally advantages to be gained when this

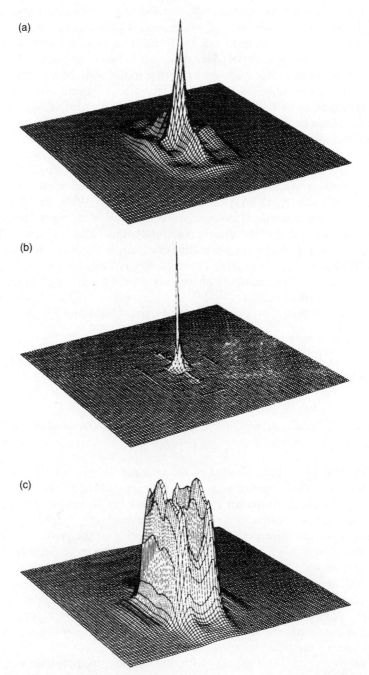

Fig. 11.15 Autocorrelation of image of letter G performed using (a) matched filtering, (b) phase-only filtering, (c) amplitude-only filtering (after Horner and Gianino 1984).

response is non-linear (Javidi 1989; Khoury et al. 1994a), the phase-only filter just discussed being one example of this. Non-linear correlation has been implemented in an optical system incorporating photorefractive devices using the clever arrangement shown in Fig. 11.16 (Khoury et al. 1994b). The object and reference images are contained on a slide S and their Fourier transforms are imaged by beam A_1 on to a photorefractive crystal via lens L_1. This part of the system is acting as a joint transform correlator. A grating is recorded in the crystal as a result of the interference of the resulting field with beam A_4, which is a plane wave. This grating is in turn read by the phase conjugate of the field A_1 at the crystal produced by the self-pumped phase conjugate mirror SP, the result being viewed with the aid of the beamsplitter BS_1.

The system as just described is rather complex, depending critically on the relative intensities of beams 1 and 4. When beam 4 is much greater than beam 1, the system responds as a linear correlator, whereas when beam 1 is much greater than beam 4, the output saturates, becoming constant. By adjusting the beam ratio it is possible to alter the behaviour, either operating the system as a joint transform correlator, or introducing steadily more non-linearity to optimize the discrimination between images.

11.8 Associative memories

Conventional electronic and optical memory systems, typified by those found in the personal computer (eg. RAM, hard disk and CD-ROM) operate in the following fashion. Each location in memory has a unique identifier: its address. By means of that address the central processing unit is able to request the retrieval of the information stored at that memory location. Note that in general, the binary codes forming the address and its associated information will be completely unrelated.

This storage method works very well in modern computers, up to a point, but it is not the only approach to the storage and retrieval of information. For example, current thought is that human memory may operate in quite a different way; our

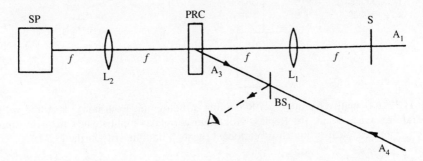

Fig. 11.16 Non-linear correlator. SP = self pumped phase conjugate mirror, PRC = photorefractive crystal, BS_1 = beamsplitter, S = slide containing object and reference images. See text for further details (after Khoury et al. 1994b).

memories often occur as a result of association, as when the sight of an old school friend brings back memories of events in the classroom many years before. This form of associative memory is of great interest to those trying to unravel the behaviour of the human brain; it is also of interest to those trying to get computers to perform image recognition – a task that humans are able to carry out extremely well but which the most complex and powerful machines perform worse than a two-year-old child. The desire is to be able to recall a complete stored image when the system is given a corrupted version of that image, or perhaps just a small part of the image[7].

Associative memories can be constructed using photorefractive crystals. The basis for such a system is usually a correlator operating according to the principles outlined earlier in this chapter. A simple scheme is illustrated in Fig. 11.17, which utilizes a static hologram to hold the reference images. The first phase of

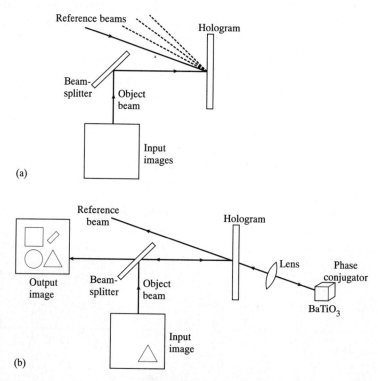

Fig. 11.17 Associative memory. (a) Recording of hologram. Each image is stored with a separate reference beam. (b) Replay. Hologram correlates input images with the stored images. Nearest match is reinforced through non-linearity in the PCM.

---

[7] The fact that the input is not the address of stored information (as with a conventional computer) but a fragment – possibly corrupted – of that information has led to the alternative name of *content addressable memory* for this type of system.

11.8 Associative memories

operation (Fig. 11.17a) is the recording of these images; for each image this is carried out with a reference beam propagating in a different direction.

The associative memory can then be operated as shown in Fig. 11.17(b). An image is input to the system; in this case we have chosen that image to be part of one of the stored references. Correlation occurs between the input image and each of the stored reference images and the reference beams are reconstructed, each with a strength determined by the degree of correlation between the corresponding reference image and the input image.

The next phase in the operation is to somehow select the strongest reference beam, which should be associated with the stored image matching the input image most closely, and return that particular reference beam back to the hologram where it will replay its associated stored image. The selection of the strongest reference beam obviously necessitates the introduction of some non-linearity which can greatly attenuate the other, undesired reference beams. In Fig. 11.17(b) this process is carried out using a self-pumped phase conjugate mirror, which also acts to return the chosen reference beam back towards the hologram. In summary then, the two key elements to this approach towards associative recall are a correlator and the introduction of non-linearity in the processing to facilitate a winner-takes-all competition amongst the outputs from the correlator.

The arrangement of Fig. 11.17(b) with $BaTiO_3$ as the phase conjugator was used in one of the first demonstrations of an associative memory by Soffer *et al.* (1986) working at Hughes Research Labs. They stored just one image, shown in Fig. 11.18(a); replay using a partial image (Fig. 11.18b) resulted in the recovery of the original image (Fig. 11.18c). In two subsequent papers (Dunning *et al.* 1987; Owechko *et al.* 1987), the Hughes group demonstrated associative recall of a grey-scale image and a system in which two separate images were stored, either of which could be recalled by the input of a partial version of itself.

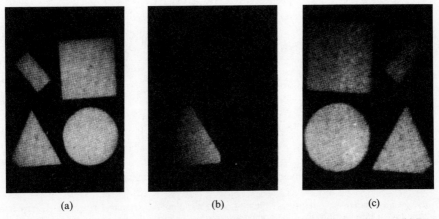

Fig. 11.18 Results of associative replay (after Soffer *et al.* 1986). (a) Stored image. (b) Input image. (c) Replayed image.

The Hughes group were not the only ones working on photorefractive implementations of associative memories at that time. The technique of associative recall described above was also proposed by Yariv and Kwong (1986), and they demonstrated a system working on the same principle as that shown in Fig. 11.17 and containing two stored images (Yariv et al. 1986). A similar technique was also described by Anderson (1986).

More recently an associative memory system has been described by H. Xu et al. (1990) that uses photorefractive KNSBN:Co as the storage element, permitting real time storage and recall. In this system, the thresholding non-linearity was provided in the electrical domain, with liquid crystal electro-optic switches being used to control the reference beams.

The arrangement shown in Fig. 11.19 has been proposed by Wunsch et al. (1993). This is a development of the scheme shown in Fig. 11.17(b). The system has the following features: Firstly, the memory is a photorefractive crystal, so once again images can be stored in real time. Secondly, this system was designed to recognize images by detecting the associated reference beam at the output imager, rather than by replaying the correct stored image[8]. Thirdly, a second phase conjugate mirror (PCM2) has been incorporated, as suggested by Soffer et al. (1986). Because of this, instead of there just being two passes through the storage hologram (as was the case for Fig. 11.17b), a resonator is set up between the two PCMs. The resonator is driven by the input image and filtered by the

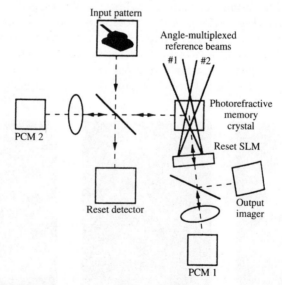

Fig. 11.19 Associative memory (after Wunsch et al. 1993)

[8] An associative memory that operates in this way, recalling the reference beam which may itself contain spatial information, is called a *heteroassociative* memory. If the output of the system is the complete version of the recognized image itself, the system is said to behave in an *autoassociative* fashion.

11.8 Associative memories

storage hologram. When this is combined with thresholding in one of the PCMs, the net result is a system that exhibits strong rejection of undesired stored images. A final feature of this system is that the input image can be compared with the recalled image at the reset detector. If the match is not close enough, that image can be inhibited by blocking the pixel on the reset SLM corresponding to its reference beam; the system would then switch to another stored image. This feature builds-in robustness to the system, preventing misrecognition.

An associative memory based on a correlator implemented in $LiNbO_3$ doped with Ce and Fe was reported by K. Xu *et al.* (1990). In this device, the thresholding non-linearity was provided by two-wave mixing in Fe-doped $LiNbO_3$. The optical disk-based correlator described earlier and shown in Fig. 11.14 has also been used in an associative memory. Non-linearity was introduced into the system in the electronic domain in order to get the system to retrieve the desired image. An associative memory able to recognize full colour images has been described by Yu *et al.* (1994).

A different approach to correlation was described by Ingold *et al.* (1992), and is illustrated schematically by Fig. 11.20. Three objects are stored on a slide which is illuminated with a laser beam, generating three signal beams (the slide could of course be replaced by an SLM). The signal beams pass through a nematic liquid crystal cell and the objects are imaged into a photorefractive crystal. Note that in this correlator the images themselves are present in the photorefractive crystal, not their Fourier transforms. The signal beams are pumped via two-wave mixing by a beam carrying the input image. Because all the images are binary in nature, the gain experienced by each signal beam will depend on the overlap between its associated image and the pump input image.

In this way, the signal that most matches the input image receives the most gain. Thresholding now comes in to play as the signal beams leaving the photorefractive crystal are incident on the nematic liquid crystal. Each beam is arranged to overlap with itself and the liquid crystal is oriented so that the beams incident from the photorefractive crystal are able to control the loss of the beams coming from the slide. This is possible because the birefringence of the liquid crystal

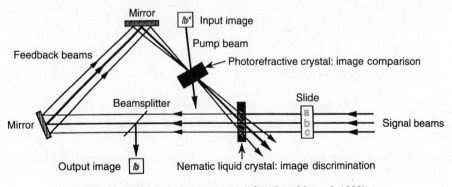

Fig. 11.20 Associative memory (after Ingold *et al.* 1992).

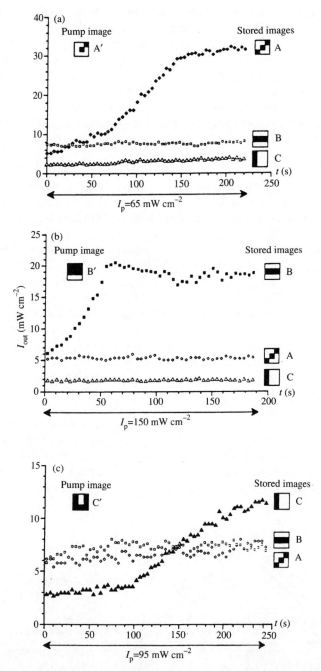

Fig. 11.21 Operation of the system shown in Fig. 11.20 (after Ingold *et al.* 1992): output intensities of the signal beams as a function of time after pump beam is turned on. The plots illustrate recall of the stored images by partial input images.

11.8 Associative memories

depends on the intensity of the light coming from the photorefractive crystal, due to an optical field-induced reorientation of the liquid crystal molecules, and the signal beams pass through crossed polarizers placed either side of the liquid crystal. Higher intensities in the beams from the photorefractive crystal result in higher transmittance through the crystal and the system displays hysteresis (Ingold *et al.* 1989).

The system operates as follows. Initially the liquid crystal is heavily attenuating for all signal beams and so the signal strengths in the photorefractive crystal are low. When the pump input image is applied, the signal beams receive amplification dependent on their overlap with the input image. The signal with the greatest overlap receives the most gain and therefore reduces the absorption of its path through the liquid crystal by the greatest amount. This further increases that signal level, and the two processes of photorefractive gain and reduced attenuation in the liquid crystal act together to raise the signal power until pump depletion in the photorefractive begins to set in. In addition to limiting the growth in the signal with closest match to the input image, pump depletion also reduces the gain possible for the other signal beams and stops their intensities from increasing very much. Examples of this process in action are illustrated in Fig. 11.21. Note that this system is only sensitive to the overlap of the bright regions of the input and signal images; it is not able to compare the similarity of dark portions of the images.

12
STORAGE

12.1 Introduction

Storage was the first application to be suggested for the photorefractive effect and continues to be a promising area for research to this day, as witnessed by the continual steady stream of papers on the subject (see for example the review articles by Pauliat and Roosen (1991) and Hesselink and Bashaw (1993)). The interest is due to the potentially huge amount of information that could be stored holographically in the volume of a photorefractive crystal. There seems to be a real possibility that storage devices based on this technology could be commercialized in the not too distant future.

12.2 Early work

It was not long after the discovery of the photorefractive effect in the guise of optical damage in $LiNbO_3$ and $LiTaO_3$ (Ashkin *et al.* 1966) that people began to realise that this phenomenon might actually be useful. Soon after, F. S. Chen and co-workers (1968b) at Bell Labs suggested that the effect might provide a means for recording information holographically without the need for the development of the stored image, as is required when using photographic film. To demonstrate their concept they used a 1 cm thick crystal of $LiNbO_3$ in which the grating was written at 488 nm using two beams of light from an Ar^+ laser. The hologram was read out at either 488 nm or 633 nm, using an HeNe laser. They found that at a writing intensity of $1\,W\,cm^{-2}$, 100 seconds was required for the grating to reach its maximum diffraction efficiency, and the refractive index variation in the steady state was found to be proportional to the square root of the incident intensity[1]. They succeeded in writing up to 1600 lines mm^{-1} and demonstrated the recording and reconstruction of pictorial information.

$LiNbO_3$ continued to be of interest to researchers, but following the initial work other materials were investigated, such as $BaTiO_3$ (Townsend and La Macchia 1970), $Ba_2NaNb_5O_{15}$ (Amodei *et al.* 1971; Thaxter 1969) and $Sr_xBa_{(1-x)}Nb_2O_6$ (Thaxter and Kestigian 1974; Micheron *et al.* 1974). $BaTiO_3$, whilst providing a reasonable diffraction efficiency of 8% for a 5 mm thick crystal, was shown to possess a very short read-out time constant[2] of 0.25 seconds at $4\,W\,cm^{-2}$ incident

---

[1] The intensity dependence of the refractive index is due to the presence of the photovoltaic effect in $LiNbO_3$ and would not be present in those materials that do not exhibit this effect.
[2] The time taken for the diffraction efficiency to fall to 1/e of its initial value due to erasure of the grating by the read-out beam. See Section 2.3 for more details.

12.2 Early work

intensity (Townsend and La Macchia 1970). As far as long term storage goes, this is of course rather unsatisfactory. $Ba_2NaNb_5O_{15}$ in a nominally pure form also possessed characteristics far from ideal for storage applications. The diffraction efficiency was only around 1% and the read-out time constant was of the order of 1 second with 10 $W\,cm^{-2}$ illumination. However, by doping the crystal with molybdenum and iron it was possible to achieve 67% diffraction efficiency in a 3.2 mm thick crystal and extend the read-out time constant to around 10 seconds under 0.25 $W\,cm^{-2}$ illumination (Amodei *et al.* 1971). In the dark, about 1 hour was needed for the diffraction efficiency to decay to 10% of its original value.

The length of time over which the grating remains recorded in the material is of fundamental importance if photorefractive crystals are to be used for storage applications. Because the photorefractive recording process is a dynamic one that does not require any development, in the normal scheme of things the recorded grating is not stable and will decay with time. Two factors contribute to this decay, which is associated with the excitation and equalization of the spatial distribution of trapped charges: firstly, the grating is erased when the crystal is illuminated with light – during read-out, for example – and secondly, even in the dark there is a very slow decay in the grating amplitude due to thermal excitation of electrons.

As an example, F. S. Chen *et al.* (1968b) reported that under uniform illumination with 0.12 $W\,cm^{-2}$ at 488 nm the grating recorded in their sample of $LiNbO_3$ decayed with a time constant of 10 minutes, whilst under room light illumination the time constant was several hundred hours. In addition to erasure by exposure to a uniform light beam, the grating could be removed quickly by enhancing the thermal decay process by heating the sample up to 170° C.

For the same incident intensities, the time constants associated with the recording and optical erasure of a photorefractive grating are comparable. This symmetry is highly undesirable from the point of view of optical storage. By choosing a crystal where the thermally induced grating decay is very slow and keeping the crystal in the dark after recording a hologram, it is possible to retrieve the hologram after a considerable period of time. However, the main interest in holographic optical storage is the possibility of storing and retrieving at will a huge number of different holograms multiplexed into the crystal in a number of possible ways (to be described later). If the recording and erasure processes are symmetrical, then the recording of a second hologram in the same crystal will partially erase the first hologram.

Clearly, for holographic storage to be practical it is highly desirable to be able to break the record–erase symmetry so that the diffraction efficiency of one hologram is not appreciably affected by the recording of subsequent information. Ideally, it would also be attractive if it were possible in some way to permanently store information, so that no reduction in diffraction efficiency occurs with time. The process of producing a permanently recorded hologram is usually referred to as fixing, though this term is somewhat more loosely often applied to techniques which merely extend the erasure time constant, breaking the record–erase symmetry, rather than producing a truly permanent grating.

12.3 Hologram fixing

We review here several schemes for hologram fixing. The first two – thermal fixing and electrical fixing – involve creating permanent changes in the photorefractive material which mirror the recorded grating, by translating ions in the former case and altering the polarization of the crystal in the latter. The second pair of techniques – two-photon and two-wavelength storage – involve ensuring that the writing and read-out processes have different effects on the material. The final pair – refreshed memories and storage/amplification – accept that erasure of the stored holograms will occur and attempt to compensate for this effect.

12.3.1 Thermal fixing

The first fixing technique to be discovered was a thermal mechanism, first reported by workers at the RCA laboratories (Amodei and Staebler 1971; Staebler and Amodei 1972b). After recording a grating in the usual manner, the fixing process involves heating the crystal of $LiNbO_3$ or $Ba_2NaNb_5O_{15}$ to 100°C in air for 30 minutes. This enables ions present within the crystal to move; these ions drift under the influence of the existing space charge field and take up a distribution so as to cancel out that field. If the sample is cooled to room temperature at this point, there are then two complementary gratings recorded in the crystal: one due to the redistribution of ions (the ionic grating) and one due to the redistribution of electrons amongst the donors (called here the electronic grating). If now the crystal is illuminated with light, the electronic charges are redistributed, revealing the fixed space charge field due to the ions. In $LiNbO_3$, the evidence points to hydrogen ions as being responsible for the fixing process (Vormann et al. 1981). For a mathematical model see Section 5.2.5.

These effects are illustrated in Fig. 12.1 which shows plots of diffraction efficiency as a function of time for three situations. All curves refer to holograms recorded at 488 nm in $LiNbO_3$. Curve A shows the decay in diffraction efficiency that occurs during read-out of a normal, unfixed hologram. The time constant is around 225 seconds. Curve B shows the diffraction efficiency following the thermal fixing of the hologram. Initially the diffraction efficiency is virtually zero because the fields produced by the ionic and electronic space charges cancel out. Under illumination, there is a redistribution of the electronic space charge which reveals the field due to the ions.

Once light begins to be diffracted from the ionic grating, the diffraction efficiency is enhanced because there is a further redistribution of electronic charges under the influence of the interference pattern produced by the transmitted and diffracted light. The conventional photorefractive grating thus produced now adds constructively to the fixed ionic grating to produce a high diffraction efficiency.

The effect of the ionic grating alone may be judged from curve C which shows the diffraction efficiency after the electronic grating has been erased by off-Bragg illumination. The initial diffraction efficiency is therefore solely due to the ionic

12.3 Hologram fixing

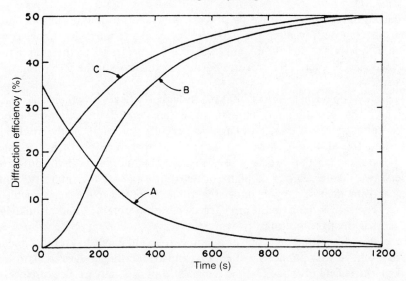

Fig. 12.1 Plot of diffraction efficiency vs. time for a normal hologram (A), a fixed hologram immediately after heat treatment (B), and a fixed hologram with initially uniform electronic charge distribution (C) (after Amodei and Staebler 1971).

grating. Thereafter, the diffraction efficiency is once again enhanced as a result of the usual photorefractive response.

One point that requires a little more explanation is the following apparent discrepancy. At the elevated temperature when the electronic grating may be considered to be constant and the ions are free to move, the redistribution of the ions results in the almost complete cancellation of the space charge field. In contrast, at room temperature when the ionic grating is fixed, uniform illumination of the crystal permits redistribution of electrons and in this case the electron distribution becomes almost homogeneous, revealing the ionic grating.

The two cases are clearly analogous; in one the electronic grating is fixed and the ions mobile, in the other the ions are fixed and the electrons free to move. Why then in the former case is the electronic space charge field cancelled out and in the latter case the ionic space charge field revealed? The physical explanation for this effect rests on the fact that in the first case the space charge field is not quite completely cancelled out, merely reduced, whereas in the second case the electronic distribution is not quite uniform: there is still a periodic component. In both cases, once equilibrium has been reached, at a given point in the crystal there is a balance between the drift of charge carriers under the influence of the remaining space charge field and the diffusion of carriers driven by a concentration gradient.

Whilst the diffusion process is proportional to the carrier concentration gradient, the drift is proportional to the total number of carriers available at that point. So an increase in the relevant carrier concentration tends to favour the drift process, resulting in a reduction in space charge field, whilst a reduction in carrier concentration favours the diffusion process, which acts to homogenize the mobile carrier distribution revealing the fixed grating. For the ions, the carrier concentration is simply the density of ions, whereas for electrons, in the simple model introduced in Chapter 1, the relevant parameter turns out to be the concentration of acceptors, N_A^-.

Experiments have also shown that much higher diffraction efficiencies may be obtained if the recording and fixing processes are carried out simultaneously: i.e. the recording process is carried out at elevated temperature. If the recording of the photorefractive grating is carried out at room temperature, then when the steady state has been reached, as we have seen, there is a balance between the diffusion of carriers away from regions of high concentration (where the intensity is maximum) and the drift of carriers under the influence of the space charge field. This balance determines the amount of charge separation obtained and ultimately the efficiency of the fixed grating.

If the recording is instead done at a temperature sufficient to mobilize the ions, then as we have seen, the ions will tend to redistribute themselves so as to cancel out the electric field produced by the redistribution of electrons. Because of this, a much greater movement of electrons is required to produce a given net space charge field at the elevated temperature than at room temperature and this results in a correspondingly larger redistribution of ions and higher diffraction efficiency of the fixed grating.

By writing and fixing simultaneously, Staebler et al. (1975) demonstrated that over 500 separate holograms could be written in a 1 cm thick crystal of iron-doped $LiNbO_3$. The holograms were written while the crystal was kept at a temperature of 160°C. The crystal was rotated by 0.1° after each recording, this being sufficient to ensure that each hologram could be individually accessed as a result of the Bragg selectivity. Figure 12.2 shows the read-out efficiency of 100 holograms recorded with the same exposure. In a later experiment, 511 holograms were recorded with diffraction efficiencies ranging from 2.5% to 25%.

The authors also measured the time constant associated with the decay of a fixed grating as a function of temperature. This enabled them to predict that at room temperature, the fixed holograms should be stable for 10^5 years. They reported that fixed holograms that had been kept for 2 years showed no measurable degradation of either the diffraction efficiency or the image quality.

Thermal fixing has also been observed in the sillénites. Herriau and Huignard (1986) reported fixing in BSO that occurred at room temperature. Gratings were recorded by the interference of two $1\ mW\,cm^{-2}$ Ar^+ laser beams at 514 nm using an applied electric field of $6\ kV\,cm^{-1}$. The diffraction efficiency was monitored with a weak HeNe beam at 633 nm. Figure 12.3 illustrates the time evolution of the diffraction efficiency. Region **a** corresponds to the initial writing of the grating with both Ar^+ beams present. In region **b** the two beams were blocked and the diffraction efficiency decreases to zero. This effect was attributed to the redistribution of positive charge carriers which were mobile at room temperature. As in the case of $LiNbO_3$ at elevated temperatures, the positive charge redistribution causes an almost complete compensation of the electronic photorefractive space

12.3 Hologram fixing

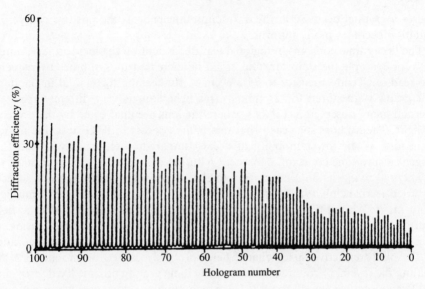

Fig. 12.2 Diffraction efficiency of 100 holograms stored in LiNbO$_3$ (after Staebler *et al.* 1975).

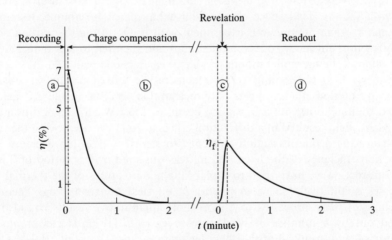

Fig. 12.3 Room temperature fixing in BSO. Plot shows diffraction efficiency as measured with an HeNe probe beam. (a) Recording using AR$^+$ beams. (b) Ar$^+$ beams blocked. (c) Illumination by one Ar$^+$ beam – initially efficiency rises. (d) Continued illumination by Ar$^+$ beam then causes decrease in efficiency. See text for more detail.

charge field. In region **c** one of the Ar$^+$ beams is used to illuminate the sample. This photoexcitation mobilizes the electrons, the distribution of which tends to become more homogeneous thus revealing the presence of the grating due to the positive charges. Finally in region **d**, continuous read-out by the single Ar$^+$ beam causes a gradual decrease in the diffraction efficiency as the positive charge distribution becomes more uniform.

The decay time constant in region **d** was dependent on the incident Ar+ intensity; for example the time constant could be increased to 9 minutes by reducing the read-out beam intensity to 75 $\mu W\,cm^{-2}$. By keeping the crystal in the dark, holograms were stored for 24 hours. This behaviour is very different from the thermal fixing observed in $LiNbO_3$ where the ions are unaffected by the presence of light. The authors suggested that this fixing process in BSO[3] was due to hole transport, as the lower mobility of holes compared to electrons will lead to a larger erasure time constant, consistent with the experimental observations.

Thermal fixing in BSO more akin to that observed in $LiNbO_3$ has also been reported (Arizmendi 1989). A grating was written at 300°C using 514 nm light from an Ar^+ laser. There was in this case no externally applied electric field. During the writing process, the diffraction efficiency as measured using an HeNe probe beam rose to a maximum and then decayed due to compensation of the photorefractive space charge field, probably caused by mobile ions. On cooling the crystal to room temperature, the ionic grating became fixed, providing a diffraction efficiency of about 0.1% which remained constant over a period of several days.

Thermal fixing has also been described in BTO (McCahon et al. 1989). A grating was recorded at 90°C using 633 nm light from an HeNe laser in conjunction with the AC field space charge field enhancement technique described in Section 2.8. As in the previous experiment with BSO, during the writing process the diffraction efficiency rose to a maximum and then fell to less than 5% of its peak value as the thermally mobilized charge carriers compensated the electronic space charge field. On cooling to room temperature, a fixed grating was obtained, though diffraction efficiency was only measurable (3%) when the AC field was applied. Illumination with a 'revealing beam' of $250\,mW\,cm^{-2}$ in addition to the applied AC field resulted in a diffraction efficiency of 13%. Removing the revealing beam caused the diffraction efficiency to return to 3%. No decrease in the diffraction efficiency of the fixed grating was observed over a period of 2 hours.

In this section we have tried to explain the main features of the thermal fixing processes qualitatively so as to provide a physical understanding. Theoretical modelling of these processes to obtain more quantitative results has also been carried out by a number of workers (Meyer et al. 1979; Hertel et al. 1987; Carrascosa and Agulló-López 1990; Jeganathan and Hesselink 1994). One of the models was presented earlier in Section 5.2.5.

12.3.2 Electrical fixing

Electrical fixing of stored holograms is possible in some crystals which, in addition to displaying photorefractivity, are also ferroelectric materials (a term often shortened to ferroelectrics). In order to understand the process of electrical fixation it

---

[3]They also reported having seen the effect in BGO.

is necessary to briefly review some aspects of ferroelectricity. Ferroelectrics display electrical behaviour similar in many ways to the magnetic behaviour of the perhaps more familiar ferromagnetics. In fact the name ferroelectric was coined because of this similarity and not because of any particular dependence on the properties of iron[4]. Below a certain temperature – the Curie temperature – ferroelectrics display a spontaneous electrical polarization, attributable to the lining up of electric dipoles within the material, and the polarization depends non-linearly on any externally applied electric field, displaying hysteresis. The behaviour is illustrated schematically in Fig. 12.4 and this curve may be recognized as being very similar to that of magnetization vs. magnetic field for ferromagnetic materials.

Figure 12.5 illustrates a typical recording geometry for producing electrically fixed holograms. A single domain crystal is poled by applying an electric field, E_1, greater than the coercive field so that the c axis is in the direction shown. Next, both shutters are open so that the hologram is recorded by means of the

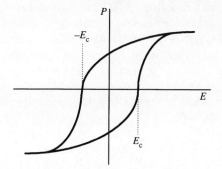

Fig. 12.4 Electrical polarization, P, of a ferroelectric as a function of applied electric field, E. E_c is the coercive field.

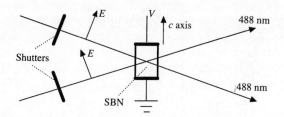

Fig. 12.5 Experimental geometry for electrical fixing in SBN.

[4] Ferroelectric behaviour was first observed in Rochelle salt by Vaselek in 1921 whereas ferromagnetics had been studied since the seventeenth century and the theory of ferromagnetic domains was well established by the early part of the twentieth century. The discovery of ferroelectricity in Rochelle salt led to an even more curious name for these materials: 'Seignette-electrics'. This term, which was more commonly used in continental Europe, referred to P. de la Seignette who made use of the laxative properties of Rochelle salt at La Rochelle in the seventeenth century!

photorefractive effect. Then the shutters are closed and the hologram is fixed by applying a field, $-E_2$, opposite to the original poling field and slightly smaller than the coercive field. This field on its own is insufficient to switch the direction of polarization of the ferroelectric, but it should be remembered that within the crystal there also exists the periodically varying space charge field. When the combined effect of the two fields is considered, there are regions (where the space charge field is most negative) where the total field has a magnitude greater than the coercive field and in these regions the polarization of the crystal is switched in direction.

The regions where the polarization is switched occur once every cycle of the space charge field and therefore accurately mirror the recorded hologram. Furthermore, any incident light beam will be sensitive to these regions since the dielectric constant of the material depends on the polarization. The result is that a fixed grating is now stored within the medium. Under uniform illumination of the material, the photorefractive grating will be erased but the switched polarization grating will still remain.

Electrical fixing was first reported by Micheron and Bismuth (1972) using a $BaTiO_3$ crystal. Soon after, the same authors published more detailed experimental results using $(Sr_{0.75}Ba_{0.25})Nb_2O_6$ (SBN) (Micheron and Bismuth 1973). Figure 12.6 shows the results of a typical fixing experiment using SBN. The curve shows the diffraction efficiency of a grating recorded using 488 nm light from an Ar^+ laser as a function of time, as measured using an HeNe probe beam which did not itself cause appreciable erasure of the grating. The first stage shows the writing of

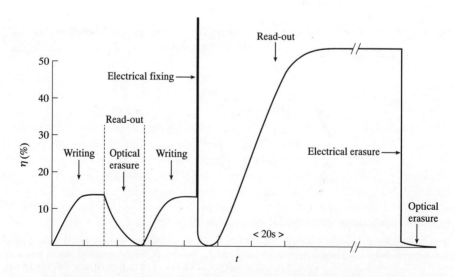

Fig. 12.6 Time history of the diffraction efficiency during electrical fixing in SBN – see text for details (after Micheron and Bismuth 1973).

12.3 Hologram fixing

the grating with two Ar$^+$ beams incident, the diffraction efficiency rising to a maximum. The second stage illustrates the replay of the grating using one Ar+ beam and the consequent erasure of the grating.

Following that erasure, a grating is again written using both Ar$^+$ beams but this time the grating is fixed by applying a negative voltage to the crystal. During this time the diffraction efficiency increases greatly; this is a result of the dependence of the linear electro-optic coefficient on the applied field (the electro-optic coefficient measured under a field equal to the coercive field was five times that measured under zero field). Immediately after fixing, the diffraction efficiency is considerably reduced from its value immediately before. This is because the grating produced by polarization switching is out of phase with the photorefractive grating and the two almost cancel, the photorefractive grating being slightly stronger. Now, when the hologram is replayed by a single Ar$^+$ beam, the diffraction efficiency initially decreases to zero as the photorefractive grating is gradually erased, becoming comparable in strength to the switched polarization grating. Thereafter, the diffraction efficiency rises as firstly the photorefractive grating is completely erased and then a photorefractive grating forms in-phase with the switched polarization grating, a process of enhancement similar to that occurring with thermally fixed gratings. The final diffraction efficiency in this regime is much greater than that of a photorefractive grating alone.

The hologram can be erased by applying a field greater than the coercive field, which has the effect of poling the crystal, all regions now having the same polarization. Then, all that remains is a small photorefractive grating which is erased by uniform illumination of the crystal.

Horowitz et al. (1993) have reported a novel form of electrical fixing in SBN, which is carried out by inducing ferroelectric domains in the crystal using a screening mechanism.

12.3.3 Two-photon storage

A rather different technique for non-destructive read-out of stored holograms has been suggested by von der Linde et al. (1974). Copper-doped LiNbO$_3$ was used as the storage medium in their first experiment. As an optical source, a mode-locked Nd-glass laser emitting at 1064 nm was used and from this laser the second harmonic at 532 nm was also available. The fundamental frequency corresponds closely to an electronic transition from the valence band to a bound state within the band gap, while the second harmonic has sufficient energy to excite electrons from the bound state to the conduction band, thus facilitating charge transport.

The authors showed that it was possible to write a hologram produced by interference between two beams of light at 1064 nm, but only when the crystal was uniformly illuminated at 532 nm. Non-destructive read-out of the stored hologram was obtained by replaying with light at 1064 nm, which possessed insufficient energy to excite electrons into the conduction band. Holograms were also recorded in nominally pure LiNbO$_3$ using only 532 nm radiation, a two-photon absorption process with no intermediate state being responsible for generating the mobile carriers. Good sensitivity was reported, with a diffraction efficiency of 25% being obtained with less than 0.4 J cm^{-2} exposure.

In a later publication (von der Linde *et al.* 1975a), the same authors also reported the recording of holograms by two-photon absorption in potassium tantalate niobate (KTN). This material possessed a sensitivity around 500 times better than LiNbO$_3$. Furthermore, recording and reconstruction of holograms in KTN is only possible in the presence of an electric field and therefore the material offers the possibility of electrical control of the recording or replay processes. Unfortunately, however, storage times were limited to around 10 hours due to the relatively high dark conductivity of the material.

12.3.4 Two-wavelength storage

The symmetry between the writing and erasure processes can be destroyed if the read-out of the hologram is performed using light of a different wavelength to that used for the recording process. Typically, a longer wavelength would be used where the photons have insufficient energy to re-excite the trapped electrons, thus ensuring that the stored grating would only decay by thermal excitation (McRuer *et al.* 1989). Unfortunately, changing the wavelength at replay changes the angle required for Bragg matching. For a single grating stored in a photorefractive crystal this is no great disadvantage, since the Bragg condition can be satisfied by angular tuning of the incident replay beam. However, the storage of appreciable amounts of information requires the recording of many gratings and in general, when replay occurs at a different wavelength, at any given time it is only possible to be Bragg-matched to a subset of these gratings.

This problem is illustrated in two dimensions by Fig. 12.7. In Fig. 12.7(a), the recording process is depicted on the Ewald diagram. Three gratings forming a hologram are produced by interference between the reference beam, with wave vector $\mathbf{k}_r$, and the three signal beams with wave vectors $\mathbf{k}_1$, $\mathbf{k}_2$ and $\mathbf{k}_3$. Figure 12.7(b) shows the replay procedure at a longer wavelength, where the wave vector of the replay beam is now $\mathbf{k}_r'$. Only one grating, $\mathbf{K}_2$, can be replayed on-Bragg to generate a beam with wave vector $\mathbf{k}_2'$.

In fact the situation is complicated somewhat by the fact that if other holograms have been stored in the crystal by using reference beams incident at different angles, then it is possible that when replay is carried out at a different wavelength, parts of different holograms may be read out at the same time. For example, in Fig. 12.7(c) a second hologram consisting of three gratings is recorded in the photorefractive crystal using a different reference wave $\mathbf{k}_{2r}$. When replay is now carried out using the same arrangement as in Fig. 12.7(b), one grating from each hologram is reconstructed on-Bragg. The new replay process is shown in Fig. 12.7(d).

One technique for overcoming this limitation has been suggested by Külich (1991), who showed that spherical signal waves recorded using a spherical reference wave may be replayed at a different wavelength. To do so, it is necessary to use a spherical replay wave with a radius of curvature at the hologram different

12.3 Hologram fixing

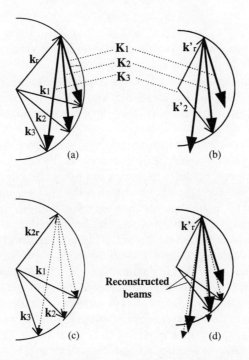

Fig. 12.7 Hologram recording and replay at different wavelengths – see text for details.

from that of the original reference wave. In a practical storage system, it is likely that information would need to be superimposed on the signal wave (and perhaps on the reference wave) in the form of amplitude or phase variations. Unfortunately, in this case when the replay is carried out at a wavelength other than that originally used to record the hologram, the technique shows an inferior resolution in the reconstructed image.

Another technique which facilitates the use of the dual wavelength approach accepts that the read-out of portions of different holograms (as illustrated in Fig. 12.7) will occur (Psaltis et al. 1994). The idea then is that the holograms to be stored would take the form of arrays of point sources representing binary data, being either on or off, and a number of holograms would be stored with reference beams incident at different angles. Furthermore, the recording geometry would be specifically chosen so that for a given replay beam, now at the longer wavelength, the reconstructed hologram takes the form of an array of points, successive lines of which come from successive images stored in the photorefractive crystal. To access a given array of data, either the crystal must be replayed several times and the appropriate lines of data extracted, or the arrays of data could be preprocessed before recording so that on replay the desired array of data is reconstructed from one replay beam.

12.3.5 Refreshed memories

Another general approach to storage in photorefractive crystals allows for the fact that grating erasure is going to occur and accepts that it will be necessary to update the stored images periodically. This is somewhat analogous to electronic storage in the dynamic random access memory (DRAM) that provides the fast access memory (compared with disk and tape based storage) in most current computer systems.

One approach that comes under this category involves copying the stored holograms from one holographic medium to another. By ensuring that the recording process in the second medium is carried out using comparable intensities for the reference beam and the signal beam (which is the image replayed from the first hologram), the second hologram may be recorded with an appreciably higher diffraction efficiency than that of the first. This approach has been demonstrated using a thermoplastic hologram as the second hologram (Brady *et al.* 1990).

Refreshing the holographic memory by simply copying the stored holograms in this way may lead to stored image degradation due to the growth of noise gratings. Better immunity to noise can be achieved by storing digital images and thresholding the images during the copying process. This idea has been demonstrated in a system, illustrated conceptually in Fig. 12.8, in which two SBN crystals were used as the storage media (Sasaki *et al.* 1992). In an experiment, a liquid crystal light valve (a form of spatial light modulator – see Section 10.3.4) was used as the optical amplifier and a 3×4 pixel image was transferred between the crystals 15 times with no visible degradation in image quality.

More recently, workers have reported developments of this technique which still make use of a thresholding device but incorporate just one photorefractive storage element (Boj *et al.* 1992; Qiao and Psaltis 1992). During the refresh procedure, the stored images are read out, thresholded and re-presented to the photorefractive crystal.

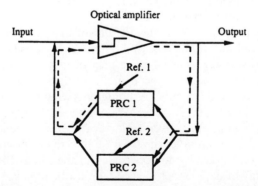

Fig. 12.8 Conceptual dynamic photorefractive memory incorporating two photorefractive crystals (after Sasaki *et al.* 1992). The dashed lines show the path for image transfer between crystals 1 and 2.

12.4 Hologram multiplexing

Another recent approach carries out the refresh procedure entirely optically (Qiao et al. 1991). As Fig. 12.9 shows, the memory refresh is done with the aid of two phase conjugate mirrors operating by means of four-wave mixing. When a given reference beam, r_j, is applied to the system the transmitted beam, t_j, and reconstructed image, i_j, are both phase-conjugated (and perhaps amplified) and fed back to the storage crystal as beams t'_j and i'_j, respectively. Because of the phase conjugation process, the interference of t'_j and i'_j produces a hologram exactly phase-matched to the stored hologram being read out and therefore that hologram is reinforced.

12.3.6 Storage–amplification scheme

One final technique for non-destructive read-out recognizes the fact that the time constant associated with the erasure process is inversely proportional to the intensity of the read-out beam (at least so long as the optical erasure process dominates the thermal one). So by replaying the hologram with a very weak reference beam, it is possible to break the record–erase symmetry. Of course, replaying with a weak reference wave may result in a very weak image with obvious consequences for the signal-to-noise ratio of the process.

To overcome this drawback, Rajbenbach et al. (1992a) have suggested using a second photorefractive crystal as an image amplifier to ensure that the recovered image possesses a good signal-to-noise ratio. In a demonstration experiment, $BaTiO_3$ and $LiNbO_3$ were used as the storage media with an amplifier based around a BSO crystal. With $BaTiO_3$, continuous replay of a stored image was possible for 1 hour, while with the $LiNbO_3$ over 10 hours of continuous replay was obtained.

12.4 Hologram multiplexing

In the preceding section we have said little about the details of recording multiple holograms in a photorefractive crystal. In one or two cases we have made

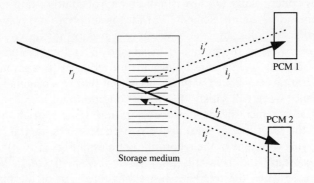

Fig. 12.9 Dynamic photorefractive memory incorporating optical feedback using phase conjugate mirros: r_j = incoming jth reference wave; t_j = transmitted beam; i_j = image beam; t'_j and i'_j = phase conjugates.

reference to recording different holograms using reference beams incident on the crystal at different angles; however, we have not discussed just how many different holograms can be stored in a given crystal. Furthermore, this angular multiplexing technique is not the only option available and indeed there are certain advantages to be gained by using other schemes. Here we give some consideration to those issues.

12.4.1 Spatial multiplexing

Spatial multiplexing involves the recording of a number of holograms in different volume elements of the crystal (Carson 1974). In this simplest case it may be argued that the technique does not really represent true multiplexing, since the holograms do not overlap spatially; indeed it could be viewed as a technique for composing piece-by-piece a single hologram occupying the whole crystal volume. In practice, spatial multiplexing is usually combined with another form of multiplexing – for example angular multiplexing as discussed in the next section (d'Auria et al. 1974).

Spatial multiplexing offers the advantage that it allows pages of data with a relatively small number of bits to be used with a crystal that is able to store a much greater amount of data. As an example, a crystal with a 1 cm^2 input aperture could in principle store a single hologram recorded using visible light containing over 1000×1000 bits. Current input devices are not able to impose this much information on a single beam. However, it would be possible to use a 128×128 pixel SLM to store a page of 16K bits and to spatially multiplex 64 such pages within the crystal. There are other issues too: for a given power of read-out beam, by only reading out part of the crystal the intensity and therefore the signal-to-noise ratio can be made higher, or the read-out time may be reduced.

12.4.2 Angular multiplexing

This is perhaps conceptually the next-simplest way of multiplexing many holograms within the same volume element. The idea is simply to use plain reference waves to record each hologram, these reference waves being chosen so as to be incident on the crystal at different angles, as shown in Fig. 12.10(a)[5]. As the photorefractive crystal is a volume hologram, on replay with a given reference wave only the gratings associated with the hologram that was originally recorded with that reference wave will satisfy the Bragg condition and be reconstructed.

In fact this last statement needs a little elaboration. If we restrict ourselves for the moment to the recording of gratings by two plane waves, then it is certainly true that by varying the angle of the recording beam as in Fig. 12.10(a) sufficiently between each recording, the cross-talk can be made very small. On the other hand, this is not the case when the reference beam is rotated out of the plane of the

---

[5]Alternatively, the reference wave can be held fixed and the crystal rotated between each recording.

12.4 Hologram multiplexing

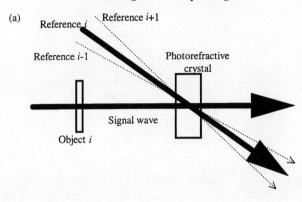

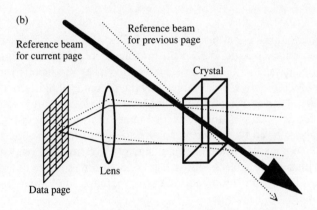

Fig. 12.10 (a) Angular multiplexing of holograms. The figure illustrates the recording of the hologram formed between the ith object and reference waves. The object may be a transparency inserted in the beam or some form of electrically controlled spatial light modulator. (b) Recording a page of digital data using angular multiplexing. Signal waves from two adjacent pixels are shown, along with current and previous reference waves.

paper. Figure 12.11 illustrates the problem, which is due to the degeneracy of the Bragg condition (H. Lee *et al.* 1989).

How many holograms may be stored using angular multiplexing? Or put another way, how much information can be stored? This question was first tackled by van Heerden (1963). In fact there are several issues that need to be taken into account to obtain a reasonable answer to this question, some of which are discussed later. However, the following back-of-an-envelope calculation can be used to obtain an upper limit.

Suppose that to keep things simple we decide to store digital information by interfering the reference beam with plane signal waves which travel in various directions. During the recording process, the presence of a given signal wave

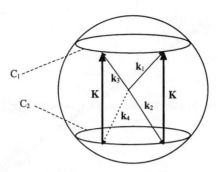

Fig. 12.11 Bragg degeneracy illustrated on the Ewald sphere. A grating **K** is recorded by beams $\mathbf{k}_1$ and $\mathbf{k}_2$. Due to Bragg degeneracy, any other incident beam with its wavevector on circles C_1 and C_2 will also replay the grating. Here $\mathbf{k}_3$ is incident and replays $\mathbf{k}_4$.

denotes a logic '1' and a grating is produced. The absence of a signal beam in a given direction denotes logic '0' and no grating is recorded. On replay with the same reference beam, the presence of a grating causes the reconstruction of the signal wave and a logic state '1' is deduced by the presence of light in that direction. Conversely, the absence of light in that direction would indicate a logic '0'.

A single reference beam is able to record many gratings corresponding to a set of data, usually referred to as a *page*. Figure 12.10(b) illustrates conceptually the recording of one page of data. The plain signal waves may be conveniently generated using quasi-point sources and a lens. These point sources could, for example, be pixels of a spatial light modulator or perhaps individual (but mutually coherent) elements in a laser array.

Two issues determine the volume density of information that may be stored. Firstly we must identify what the angular divergence must be between adjacent signal waves in order for them both to be resolved. Secondly we must decide what angular deviation must separate adjacent reference waves in order that a given page of data is not corrupted by the partial reconstruction of the next page.

The separation of the signal waves is fundamentally limited by diffraction. The reconstructed wave leaving the crystal is limited in diameter to the crystal dimension l. Basic diffraction theory then tells us that this wave will be subjected to an angular spread of the order of λ/l, so each pixel must be allowed to occupy a solid angle of about λ^2/l^2, in order to be resolved from its neighbours. A page of information occupying 4π steradians of solid angle can therefore contain a maximum number of pixels given by $4\pi/(\lambda^2/l^2)$, which is of the order of l^2/λ^2.

The angular separation required of the reference beams may be deduced with reference to Kogelnik (1969), in which it is shown that when an unslanted grating is replayed off-Bragg, the half-power point corresponds to a value of about 1.5 for the parameter ξ, given by

$$\xi = \frac{\delta\theta K d}{2} \tag{12.1}$$

12.4 Hologram multiplexing

where $\delta\theta$ is the deviation from the Bragg condition and d is the crystal thickness[6]. From eqn (12.1) and remembering that $K = 4\pi n/\lambda \sin\theta$, where n is the refractive index of the crystal and λ the wavelength of the light source, we can approximate the angular range over which the grating may be read out as

$$\delta\theta = \frac{3\lambda}{4\pi n d \sin\theta} \tag{12.2}$$

Adjacent reference beams must be separated by this angle, so the number of possible reference beams will be of the order of d/λ.

The total number of bits that may be stored is obtained by simply multiplying the number of bits per page by the number of possible reference beam angles to give a result of magnitude $l^2 d/\lambda^3 = V/\lambda^3$, where V is the crystal volume. Note that this is also the result that we would expect were we to store information by writing bits point by point in the volume using a focused beam of light, where the spot size would be of the order of one wavelength. We can place an upper limit then on the storage capacity of a photorefractive crystal of about 10^{13} bits cm^{-3} at a wavelength of around 500 nm. This is equivalent to being able to store over 1000 CD-ROMs of information in 1cm^3 of material!

How do systems perform in practice? Two research groups have reported using angular multiplexing for the storage of over 500 holograms in iron-doped LiNbO$_3$. In the original experiments of Staebler et al. (1975) the holograms were formed using plane waves. In more recent work (Mok et al. 1991), 500 holograms of images produced using a 320×220 pixel SLM were successfully stored, with diffraction efficiencies of around 0.01%. The reference beam was rotated by 0.01° (measured inside the crystal) between each exposure, no crosstalk being visible on reconstruction.

More impressively, in a later paper Mok (1993) describes the storage of 5000 holograms, corresponding to over 350 million pixels. This was achieved by using a large interbeam angle ($\sim 90°$), which minimizes the angular separation needed between adjacent reference beams (as follows from eqn (12.2)), and by using angular multiplexing both as indicated in Fig. 12.10 and also out of the plane of the paper in that figure, care being taken to avoid Bragg degeneracy. Figure 12.12 shows original and reconstructed images obtained from the system (the images were edge-enhanced prior to storage). The diffraction efficiency was about 4×10^{-6}.

Systems that involve both angular and spatial multiplexing have been developed. Tao et al. (1993) reported the storage of 750 high resolution holograms in a crystal of LiNbO$_3$. Heanue et al. (1994) have described a system with a raw storage capacity of 2.6×10^8 pixels (corresponding to 3×10^9 pixels per cubic centimetre). An impressive feature of this system was that images could be read-out at a rate of over 6 million pixels per second with a bit error rate of 10^{-6}.

---

[6]Note that due to Bragg degeneracy we are only allowing for the reference beam angle to vary in one plane here, as shown in Fig. 12.10(a).

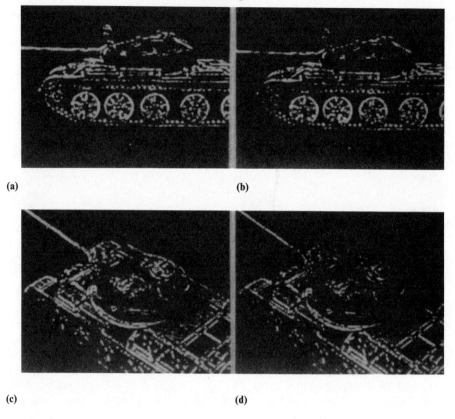

Fig. 12.12 Original (a and c) and retrieved (b and d) images from a holographic storage system holding 5000 images (after Mok 1993).

There is also growing interest in using multimode photorefractive fibres as a storage medium (Hesselink and Redfield 1988; Ito *et al.* 1992). Due to the optical field confinement within the fibre, the recording time can be short even with a source of a modest power level, and the angular sensitivity is very high.

12.4.3 Wavelength multiplexing

At replay, the Bragg condition may be violated by changing the read-out wavelength as well as by altering the propagation direction of the reference wave. This is shown schematically in Fig. 12.13 using the Ewald sphere construction. As a result of this, it is clear that holograms may be recorded and individually accessed by altering the wavelength of the light source, without the need for mechanical rotation or translation of the reference beam. This technique has been demonstrated by Rakuljic *et al.* (1992) who used a wavelength-tunable dye laser to write reflection gratings at different wavelengths in a 2 mm thick crystal of $LiNbO_3$. Two holograms of transparencies depicting integrated circuit masks were

12.4 Hologram multiplexing

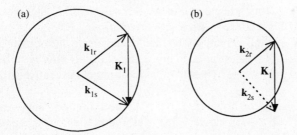

Fig. 12.13 Wavelength multiplexing of stored hologram. (a) Recording or replay of one hologram with grating vector $\mathbf{K}_1$ at a wavelength λ_1, corresponding to a propagation constant $\mathbf{k}_1$. (b) Replay at a longer wavelength, λ_2, violates the Bragg condition for replay from grating $\mathbf{K}_1$; however, any hologram previously recorded at λ_2 will be reconstructed.

recorded at 641 nm and 643 nm. At 642 nm (where no hologram was recorded), the cross-talk from the other images was about −43 dB.

12.4.4 Phase encoding

The implementation of the two previously described techniques for hologram multiplexing is subject to certain technical difficulties. Angular multiplexing may be carried out either by rotating the reference beam (or crystal) or by using some form of SLM to define particular directions. If the former approach is chosen, then access to the data is liable to be slow and there may well be problems in ensuring that the mechanical system is able to operate repeatably to a sufficient degree of accuracy; remember that adjacent reference beams are liable to be differentiated by something of the order of 0.01° if full advantage is to be taken of the storage capacity of the medium (in some circumstances this problem can be avoided by using an acousto-optic modulator to deflect the beam (Mok 1993)). If an SLM is used, with each pixel defining a reference beam direction, then there is considerable inefficiency. This arises because if only one out of n pixels is active (for a system able to store n holograms) only n^{-1} of the total power available will be reaching the crystal, the rest being blocked by the other $(n-1)$ pixels in the SLM.

On the other hand, wavelength multiplexing is currently limited by the technical difficulties involved in producing a laser source with an accurately controllable lasing wavelength and broad tuning range[7]. These deficiencies have prompted the development of another multiplexing technique, known as phase encoding (Krile et al. 1979).

---

[7]Recently, semiconductor lasers have been developed that exhibit single mode operation and are tunable, via an external cavity, over ∼40 nm in the near infrared. At the moment these sources are limited to a few milliwatts of power but it does not seem unreasonable to assume that in the near future such devices will be available with much higher powers and will eventually cover the visible region of the spectrum.

The operation of this scheme is illustrated in Fig. 12.14. The idea is that each image to be stored is recorded with a set of overlapping reference beams that have the same intensities and directions for each image but different relative phases. The particular sets of phases for each image are carefully chosen to be orthogonal so that on replay with a given set, only the image associated with that set is reconstructed.

To get a better feel for how this is possible, note that on replay each reference beam will reproduce a copy of every image stored in the crystal. The set of phases for each reference wave (known as a phase code) is chosen so that all the copies of the desired image have the same phase and add together to produce constructive interference, whereas the copies of any other image cancel out, some copies having one phase and others being shifted in phase by $180°$[8].

More rigorously, we assume that there are M images to be stored and the mth image has an amplitude A^m, which in general will be complex and spatially dependent, and this image is stored with a set of N reference beams each of (real) amplitude S and phase $\phi_1^m \ldots \phi_N^m$. After all holograms have been recorded, the amplitude of the permittivity variation within the crystal takes the form

$$\triangle \varepsilon = \sum_{m=1}^{M} \kappa^m S A^m \sum_{n=1}^{N} \exp[j\phi_n^m] \qquad (12.3)$$

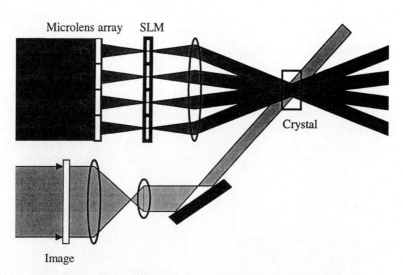

Fig. 12.14 Hologram multiplexing by means of phase encoding.

[8]Phase encoding in this way is a development from an earlier holographic multiplexing technique where a ground glass screen was placed in the reference beam (LaMacchia and White 1968). Translation of the screen between the recording of each hologram ensured that holograms were stored with reference waves possessing different – essentially random – wavefronts. With this technique, a high degree of positional accuracy is required to read out a chosen hologram and it would be difficult to duplicate the ground glass screens so that different users might access the stored data.

12.4 Hologram multiplexing

where κ^m depends on the recording sequence used and also on material parameters. On replay with a given set of reference beams with the same amplitudes but with different phases, $\phi_1^p \ldots \phi_N^p$, the amplitude of the reconstructed wave, R, as a function of distance, z, through the crystal satisfies the following differential equation:

$$\frac{dR}{dz} = K \sum_{m=1}^{M} \kappa^m SA^m \sum_{n=1}^{N} \exp[j(\phi_n^m - \phi_n^p)] \qquad (12.4)$$

where K also depends on material parameters and on the read-out beam amplitude.

The essence of the phase encoding technique is to choose sets of phases for each image so that

$$\sum_{n=1}^{n} \exp[j(\phi_n^m - \phi_n^p)] \begin{cases} = N \ (m = p) \\ = 0 \ (m \neq p) \end{cases} \qquad (12.5)$$

In this case, the phases associated with each image form orthogonal sets. If we consider binary phase encoding ($\phi = 0$ or π) and N is a power of 2, then it turns out that we can store $M = N$ images. For example, with four reference beams we could store four images using the orthogonal codes $(0,0,0,0)$, $(\pi, \pi, 0, 0)$, $(\pi, 0, \pi, 0)$, and $(0, \pi, \pi, 0)$. Figure 12.15 shows the recall of four images stored in this fashion in a crystal of $BaTiO_3$ (Denz et al. 1991).

The interference fringes visible in the reconstructed images in Fig. 12.15 were attributed to cross-talk: i.e. to the simultaneous reconstruction of unwanted

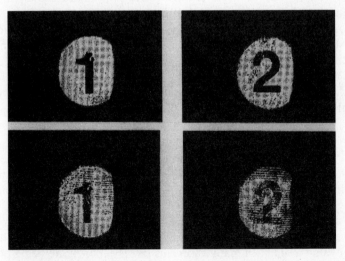

Fig. 12.15 Input images (upper photos) and recalled images (lower photos) from an experimental storage system based on phase encoding (after Denz et al. 1991).

images. In an experiment where two images were stored with two reference beams, the intensity of the unwanted image was around 20% of the desired image. This cross-talk arises from a number of sources. Firstly, in a real system, the phases of the reference beams will not be exactly 0 or π and indeed the exact values of the phases may fluctuate both in time and spatially across a given reference beam. These errors destroy the orthogonality of the phase encoding technique. Secondly, even when the phase codes are all exactly orthogonal, there will still be cross-talk if the copies of a given image replayed by each reference beam have different amplitudes, so that the superposition of the phase-shifted unwanted images do not exactly cancel out. This might occur because the reference beam amplitudes were different or because the amplitudes of the stored hologram of each copy were different. Analysis by Denz et al. (1992) has shown that in a real system, the phase modulator is the critical component producing most of the cross-talk.

12.4.5 Selective erasure

We have seen that it is possible to erase stored holograms by uniform illumination of the crystal (for thermally fixed holograms this must be carried out at high temperatures). Unfortunately, this process erases indiscriminately all holograms stored within the crystal. A much more useful feature in a system where many holograms have been stored using some kind of multiplexing scheme is the ability to erase one particular hologram – in order to update that particular memory location, for example. This is possible by re-recording a copy of the original image, but with a π phase shift in the reference beam. The original and the new holograms take the form of variations in refractive index of the opposite sense, the two superimposing to give no net refractive index variation (Huignard et al. 1975a,b, 1976a,b). The selective erasure process has been shown to be faster than incoherent erasure (Sasaki et al. 1992).

It is possible to use this process to erase part of a given hologram. Figure 12.16(a) shows the reconstruction of a hologram of a page of binary data. Figure 12.16(b) shows a reconstruction of the same page; however, before reconstruction part of the data page (contained within the shape of the letter L) had been re-recorded with a π phase shift, and this portion has been clearly erased. In this way it is possible to erase a single pixel if that is required.

12.4.6 Recording schedules

Assuming that many holograms are to be stored in the crystal using some form of multiplexing scheme, it is obviously important to maximize the diffraction efficiency of each hologram to ensure the greatest possible signal strength. On exposure to a signal and reference beam (assumed here to be equal in intensity) the refractive index modulation, $\triangle n$, of the grating being recorded rises towards a saturation value, $\triangle n_{max}$. In many cases this rise is exponential, being of the form

$$\triangle n(t) = \triangle n_{max}(1 - e^{-t/\tau_1}) \qquad (12.6)$$

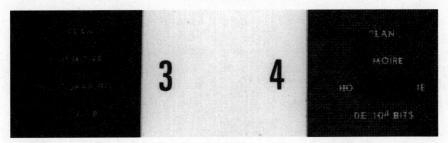

Fig. 12.16 Reconstruction of a page of data (a) before and (b) after partial erasure of the data.

where the time constant τ_1 depends on the incident intensity and the material being used as the storage medium and $\Delta n_{\max}$ depends on the material and the grating spacing, a value of around 10^{-3} being fairly typical for $LiNbO_3$. Unfortunately it is not possible to store many holograms all with diffraction efficiencies approaching $\Delta n_{\max}$, as the recording of one hologram partially erases any previously stored holograms. Usually it would be desirable for all the stored holograms to have the same diffraction efficiency and this means that the exposure times of earlier holograms must be longer than those of later ones. The exposure times needed in order to end up with equal diffraction efficiencies may be calculated as follows.

Firstly, we note from eqn (12.6) that if the ith hologram is exposed for a time T_i, the grating amplitude immediately after recording is given by

$$\Delta n_i^r = \Delta n_{\max}(1 - e^{-T_i/\tau_1}) \quad (12.7)$$

Secondly, we assume that when subsequent holograms are exposed, the grating amplitude of the ith hologram decays exponentially[9] as

$$\Delta n_i(t) = \Delta n_i^r e^{-t/\tau_2} \quad (12.8)$$

so that when N holograms have been recorded, the final amplitude of the ith hologram is given by

$$\Delta n_i = \Delta n_i^r \exp\left(\frac{-1}{\tau_2} \sum_{j=i+1}^{N} T_j\right) = \Delta n_{\max}(1 - e^{-T_i/\tau_1}) \exp\left(\frac{-1}{\tau_2} \sum_{j=i+1}^{N} T_j\right) \quad (12.9)$$

where T_j is the exposure time of the jth hologram. Since we want the final amplitudes of all the gratings to be equal, we can equate Δn_i and Δn_{i-1} to give the following recursive equation:

$$e^{T_i/\tau_2} - e^{T_i(1/\tau_2 - 1/\tau_1)} = 1 - e^{-T_{i-1}/\tau_1} \quad (12.10)$$

It is advantageous if the time constant associated with erasure, τ_2, is much longer than that associated with recording, τ_1. In that case the recording of subsequent

---

[9] Note that we are assuming here that each page of data that is stored has the same number of zeros and ones in it. The total exposure associated with a page of zeros (when only the reference beam is present) is half that associated with a page of ones and would cause appreciably less erasure.

holograms may have little effect on those already recorded. In the section on fixing holograms we have looked at ways of breaking the read–write symmetry to achieve this effect. However, in many cases the two time constants are approximately equal and since this allows us to reach a simple analytic solution for the set of recording times, we shall make this assumption. By setting $\tau_1 = \tau_2 = \tau$, eqn (12.10) may be simplified to give

$$e^{T_i/\tau} - 1 = 1 - e^{-T_{i-1}/\tau} \tag{12.11}$$

which has the simple solution

$$e^{-T_i/\tau} = 1 - \frac{1}{i+c} \tag{12.12}$$

where c is a positive constant.

With this recording schedule the amplitudes of all previous gratings will be equal to the amplitude of the final grating which may be found from eqns (12.12) and (12.7) to be

$$\Delta n = \frac{\Delta n_{\max}}{N+c} \tag{12.13}$$

The maximum amplitude is obtained for $c = 0$, but this is an impractical choice, implying as it does from eqn (12.12) an infinite recording time for the first grating. A more realistic choice is $c = 1$; this has little effect on the grating amplitude given by eqn (12.13) (assuming we wish to store many holograms) but results in only a factor of 2 difference in recording times between the first and last holograms. For small diffraction efficiencies, the diffraction efficiency, η, is proportional to Δn^2 (see eqn 2.92) and so for a large number of stored holograms varies as $1/N^2$.[10]

The recording schedule described above is known as sequential (or scheduled) recording. There is an alternative scheme possessing some advantages known as incremental recording (Anderson and Lininger 1987; Taketomi et al. 1991). The process is illustrated in Fig. 12.17. The idea is to expose each hologram in turn for the same period, but to choose a period much shorter than the characteristic time constant for recording. After exposing each hologram once, the index modulation associated with each hologram will have a small but finite value. This procedure is then repeated so that each hologram is exposed many times. After each round of exposures, although each hologram is erased for a longer time than it is recorded, the index modulation of each hologram will have increased slightly. This is due to an asymmetry in the rate of change of index modulation during recording and erasure that occurs even if the time constants for recording and erasure are the same – see Fig. 12.18.

It may be shown that given a certain total recording time, the sequential and incremental schedules result in approximately the same final index modulation

---

[10] In reality, the writing of a hologram is a dynamic process with the recording grating affecting in its turn the writing beams. The simple theory presented here has been extended by Maniloff and Johnson (1991) using a coupled wave approach, which shows that in fact when many holograms are being stored it is possible to exceed the $1/N^2$ limit derived above.

12.4 Hologram multiplexing

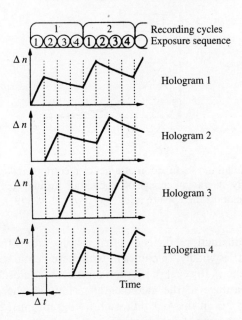

Fig. 12.17 Incremental recording. After each cycle, the amplitude of each grating has risen slightly towards the saturation value (after Taketomi *et al.* 1991).

(Maniloff and Johnson 1992). Theoretically, in contrast to the sequential approach the incremental technique will produce holograms of slightly different diffraction efficiencies, the last to be recorded being the highest. The spread in efficiencies may be arbitrarily reduced by shortening the individual exposure time for each hologram and increasing the total number of exposures. In practice this is not an issue, as even with the sequential schedule, experimental factors prohibit an exact matching of diffraction efficiencies.

The advantages of the incremental approach are twofold. Firstly it is unnecessary to determine the recording and erasure time constants of the material – any error in which will result in non-uniformities of the diffraction efficiencies of the various holograms when the recording schedule is implemented. Secondly, a recording schedule is calculated for particular values of response time and saturation index modulation. This presents no problem when large pages of binary information with similar numbers of zeros and ones are to be recorded. However, problems occur if some images are on average much darker than others or if grey scales are to be recorded (Sasaki *et al.* 1993; Taketomi *et al.* 1992).

The problem is illustrated in Fig. 12.19 which concerns the recording of five holograms consisting of a bright region of intensity 2 and a darker region of intensity 1 in conjunction with a reference beam of intensity 2. The recording time constants for the two regions will differ by a ratio of 4:3 and therefore a schedule cannot be devised that will suit both regions. Figure 12.19(b) shows the resulting diffraction efficiency when the schedule is calculated for the bright region.

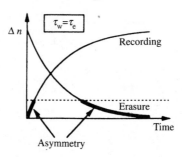

Fig. 12.18 Index modulation as a function of time for recording and erasure with the same time constants. Initially, the rate of change of modulation during recording is much greater than during erasure (after Taketomi *et al.* 1991).

For this region the diffraction efficiency of each hologram is the same but that of the darker region varies widely. In contrast, with incremental recording, the ratios of diffraction efficiencies for each hologram are far more closely matched.

The main disadvantage of the incremental technique when compared to the sequential approach lies in the fact that each hologram must be exposed several times and for each exposure the interference pattern must closely match the index grating already recorded. This places severe constraints on the multiplexing system – for example, current technology does not possess sufficient precision to allow the reliable use of an angular multiplexing system making use of a mechanical translation/rotation of the reference beam. Techniques which require no moving parts such as phase coding have been used successfully.

12.4.7 Storage capacity

We have seen earlier that diffraction places an upper limit of V/λ^3 on the number of gratings that may be stored in a photorefractive crystal. For example, this limit implies that in a $1\,\text{cm}^3$ crystal something like 10^{13} gratings could in principle be stored. However, there are other features of the recording and replay processes that act to reduce this figure still further.

The total number of holograms that can be stored depends on the desired bit error rate on read-out, which in turn is a function of the signal-to-noise ratio with which a particular grating can be replayed. The signal strength is determined by the product of the diffraction efficiency and the incident replay beam intensity, and therefore both of these should be as high as possible. The noise can arise from a number of sources: it may be due to cross-talk from off-Bragg replay of other pages of data or replay of noise gratings recorded by scattered light (Burke and Sheng 1977); it may also be due to electrical noise in the detector circuitry. In this latter case the signal-to-noise ratio will depend on the rate at which information is to be read out of the crystal (Bløtekjaer 1979) as the noise will increase with the electrical bandwidth of the system.

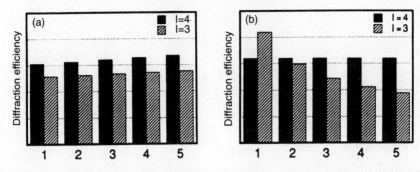

Fig. 12.19 Comparison of (a) incremental and (b) sequential recording. The sequential recording was calculated with an initial exposure equal to the recording time constant associated with the bright region. The incremental exposure time was 1% of this value, the total exposure time in each case being the same (after Taketomi *et al.* 1992).

Taking into account these various factors and assuming that a suitable optimized recording schedule has been used, Bløtekjaer (1979) has estimated that a $1\,\text{cm}^3$ crystal of $LiNbO_3$ could store over 10^{11} bits of information, all of which could be read out in about one second.

13
OTHER APPLICATIONS

13.1 Introduction

For reasons of space we have had to limit our coverage of the applications of photorefractives, and in the previous chapters we have tried to choose areas of activity that we feel are of continuing interest today. Nevertheless, there are a host of other applications that we have not yet touched upon and we hope to make amends for that here with a brief summary of other applications. The fact that we have relegated a subject to this section does not mean that it is somehow less worthy than the preceding chapters, or that these are applications that are less likely to be commercialized. Far from it, because the first topic concerns what is to our knowledge so far the only commercialized photorefractive product (excluding the crystals themselves, which can sometimes change hands for quite large sums of money!).

13.2 Narrow band interference filter

The first commercialized application of volume holography in photorefractive crystals is a holographic filter produced by Accuwave Corporation of Santa Monica, California (Rakuljic and Leyva 1993). This device contains a reflection grating recorded in $LiNbO_3$, and the first application is as a narrow band filter for solar astronomy. Figure 13.1 shows the recording geometry of the filter and a schematic of it in place in a telescope system. After recording, the grating is fixed (see Section 8.2) and possesses the following characteristics: length 8 mm, peak reflectivity at 656.28 nm (Hα line) >50%, FWHM 0.012 nm. The device was shown to compare favourably with a Lyot filter, which is the current filter of choice for many astronomical applications. There are of course many other likely applications for a narrow band filter of this kind, such as in spectroscopy, lidar, remote sensing, optical sensing, and optical communications. See also Mills and Paige (1985), Petersen *et al.* (1991), and Müller *et al.* (1994*a,b)*.

13.3 Reconfigurable array interconnection

One approach to overcoming the fundamental speed limitations of electronic processors that has been gaining ground in recent years is parallel processing, where several processing units simultaneously work on the same problem. For this process to operate efficiently, there has to be some communication between the various processing units. When comparatively small numbers of processors are

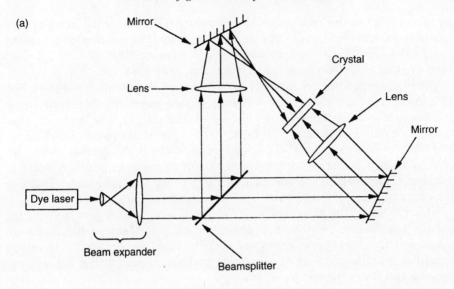
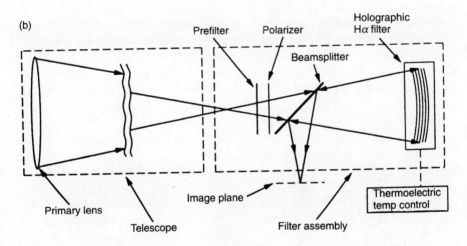

Fig. 13.1 Holographic filter (after Rakuljic and Leyva 1993). (a) Recording geometry. (b) Arrangement of filter with telescope.

involved, this communication can be carried out electronically along wires. However, if the speed of the fastest computers is to keep pace with the growing demand, it is likely that in the future massively parallel machines will be required and implementing reconfigurable connections between all the processors using electronics will become increasingly difficult. Therefore there is considerable interest in carrying out this task optically, and around 1987–88 the use of holograms recorded in photorefractive crystals was suggested as offering a promising approach to realizing reconfigurable arrays of large numbers of such interconnects (Wilde *et al.* 1987; Psaltis *et al.* 1988; Yeh *et al.* 1988).

The idea which forms the basis of most interconnect systems is shown in Fig. 13.2. An array of mutually coherent sources, S, each associated with an electronic processor, is to be connected with an array of detectors, D. The lenses are positioned to ensure that the crystal is in the Fourier transform plane of each array, and the second source array, S′, is the image of the detector array. With this configuration, a hologram recorded between any two sources in arrays S and S′ will then act to direct light from the source in S to the detector in array D that is the image of the source used in array S′ to record the hologram.

Since the late 1980s, there has been a continual output of papers studying the various ramifications of this interconnection scheme. Many of the issues are shared with the topic of storage covered in the previous chapter, such as recording schedules, selective updating of connection patterns, erasure during the read-out process, and Bragg degeneracy of gratings.

If it is necessary to reconfigure the array of interconnections, this may be carried out by erasing the old interconnections and recording new ones. Marrakchi (1989) has shown how the interconnection weight (i.e. the diffraction efficiency) can be controlled by recording two equal-amplitude gratings with a controllable phase shift between them. A phase shift of zero causes the gratings to reinforce one another, leading to maximum interconnection weight; conversely, a phase shift of 180° causes the gratings to cancel each other out. The double exposure technique was extended by Marrakchi *et al.* (1990). They showed that by keeping control over the time-varying phase of the writing beams, the inter-

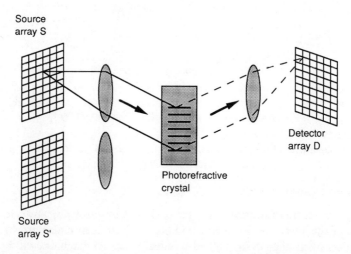

Fig. 13.2 Reconfigurable holographic interconnects implemented in a photorefractive crystal.

connection weight can be adjusted. However, if such phase modulation is applied to a number of superimposed gratings there is some cross-talk between them, as shown by Andersen *et al.* (1994).

In order not to erase the stored hologram containing the interconnection pattern, it may be desirable that the sources carrying the signal should operate at a wavelength where the sensitivity of the photorefractive material is low. This means that the recording of the interconnects and their replay may occur at different wavelengths and in general, because of the angular dependence of the Bragg condition on wavelength, each interconnection would have to be written by two separate sources. However, McRuer *et al.* (1989) have shown that by adding the restriction that the writing beams should propagate in the crystal on the surface of a cone, the number of beams needed to generate n^2 interconnections could be reduced from order n^2 to n. Another approach to the problem of grating erasure was suggested by Rastani and Hubbard (1992), who recorded gratings in separate volumes of the photorefractive crystal so that the beams writing one interconnection would not overlap with previously written gratings.

An alternative approach to avoiding the problem of erasing the stored gratings during replay has all the desired interconnection patterns stored in advance in a fixed hologram (see Section 12.3). A given pattern may then be selected using wavelength or spatial multiplexing (S. Wu *et al.* 1990).

An interconnection scheme using self-pumped phase conjugate mirrors has been reported by Owechko and Soffer (1991). This system has the advantage that because any input beam gets diffracted from several gratings, the Bragg degeneracy associated with a single grating (see Section 12.4.2) can be removed, thereby increasing the number of interconnections that can be made.

A different approach to writing the interconnection hologram was suggested by Ford *et al.* (1990, 1994). They showed how to produce arbitrary interconnections by correlating in a photorefractive crystal a phase-coded version of the input array with a control image constructed from the desired interconnection weights. In a later paper, they demonstrated a compact, packaged system able to connect elements of 5×5 arrays (Takahashi *et al.* 1994).

Finally, Weiss *et al.* (1988) described a novel system to produce a set of dynamic interconnects based on the double phase conjugate mirror (see Section 3.12), Brady and Psaltis (1991) have discussed the implementation of interconnects in photorefractive planar waveguides and Chiou and Yeh (1992) have reported a detailed study based on an experimental 2×8 (two lasers and eight detectors) reconfigurable interconnection system.

13.4 Adaptive interferometry

The dynamic nature of the recording process in photorefractive materials is put to good use in adaptive interferometry (Hall *et al.* 1980; Kamshilin and Mokrushina 1986). In the field of interferometric sensing, the problem is one of converting a measurand induced phase shift in an optical signal beam into a measurable change in intensity at a detector. Conventionally this is carried out using a refer-

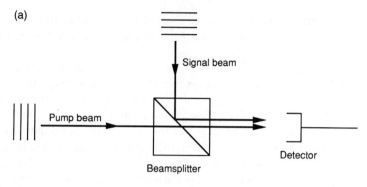

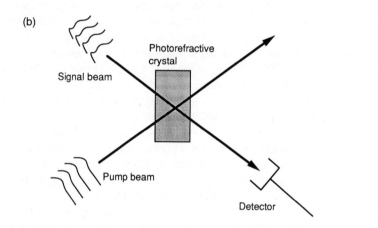

Fig. 13.3 Interferometry using (a) a beamsplitter, (b) a photorefractive crystal.

ence beam and a beamsplitter as shown in Fig 13.3(a). The signal reaching the detector then has the familiar sinusoidal dependence on the signal phase, and by choosing the phase difference between the two beams to be 90° (the so-called quadrature condition), an approximately linear dependence of intensity on phase is obtained, for small phase excursions.

There are two problems with this approach. Firstly, slow environmentally induced drifts in the phases of both beams tend to move the operating point out of quadrature, and secondly, both beams must have similar wavefronts (usually plane) if they are to interfere. Both problems may be overcome using a dynamic grating recorded in a photorefractive crystal as an adaptive holographic beamsplitter, as shown in Fig 13.3(b). All that is necessary is to arrange for the system to be at the quadrature point, which corresponds to a grating that is in phase with the interference fringes (i.e. optimized for phase coupling between the beams – see Section 3.4). This is well satisfied in the drift regime where an external electric field is applied to the crystal (Stepanov 1989).

Adaptive interferometers have been used to create a beam combiner in a 50 Mbit s^{-1} homodyne optical communications receiver (Davidson et al. 1992) and for the detection of broadband acoustic signals (Ing and Monchalin 1991).

Rossomakhin and Stepanov (1991) and Ing and Monchalin (1991) have shown how the anisotropic properties of photorefractive crystals may be used to obtain a linear dependence of output intensity on phase in the diffusion regime, for those applications where high electric fields are undesirable. Hofmeister and Yariv (1992) were also able to obtain a linear dependence without an applied field by using paraelectric KTN and KLTN crystals, materials which lack a linear electro-optic coefficient but exhibit the zero external field photorefractive (Zefpr) effect (Agranat et al. 1989).

13.5 Self-organizing optical circuits

A number of interesting developments in the area of photorefractive applications have come from Anderson's group at the University of Colorado. Particularly intriguing is their work on self-organizing optical circuits, based on competition for gain amongst overlapping resonator modes (Benkert and Anderson 1991). They have developed a bistable flip-flop (Anderson et al. 1991) and incorporated this key element in several complex self-organizing circuits. For example, the

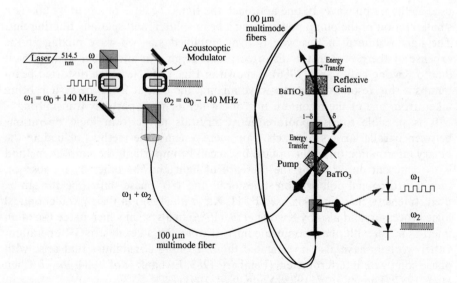

Fig. 13.4 Self-organizing circuit able to demultiplex two optical communication channels (after Saffman et al. 1991).

arrangement shown in Fig. 13.4 is able to demultiplex two optical communication channels (Saffman et al. 1991).

The input to the system consists of two optical carriers separated by 280 MHz, each modulated with a signal in the kilohertz region[1] and each carried in different mode distributions of the same multimode fibre. Light emerging from this fibre acts as the pump to the photorefractive oscillator circuit. When the system has reached steady state, which takes around 10 s with 10 mW incident on the $BaTiO_3$, the two carriers will be carried by separate fibres in the oscillator and be available from spatially separate outputs. The system is entirely passive, requiring no power other than the input signals, and is able to operate without any prior knowledge of the carrier frequencies, the only requirement being that their frequencies should differ by more than the inverse of the response time of the photorefractive media.

13.6 Some further applications

The dynamic nature of holograms recorded in photorefractives allows them to be used to 'clean up' a beam – to take a beam with a distorted, unknown and possibly changing wavefront and generate from it a beam with a well-defined and stable wavefront. Such a device can be used to enable a badly distorted laser beam to be focused into a diffraction-limited spot, for example, or allow light to be coupled from a multimode optical fibre to a single-mode fibre efficiently.

This process is readily achieved using two-wave mixing. The distorted beam is used as the pump wave. In one approach the signal beam is obtained by taking a small fraction of the pump power using a beamsplitter, and spatially filtering this. The light obtained in this way is then amplified via two-wave mixing at the expense of the power in the distorted pump beam. An alternative arrangement due to Kwong and Yariv (1986) is shown in Fig. 13.5. Here, the distorted beam pumps a ring resonator via two-wave mixing, the output from the system being taken from one of the resonator mirrors, which is made partially transmitting.

It is possible to use photorefractive crystals to perform logic operations between parallel arrays of pixels. For such systems, the method of coding the binary information on to the optical beam can be important; the simplest method is to represent one state by the presence of light and the other by its absence, though orthogonal polarization states or 0° and 180° phase-shifted beams can be used. Intensity encoding was used by H. Xu et al. (1992) in their parallel optical logic processor, and also by Kwong et al. (1986) in a system which made use of an interferometer with phase conjugate mirrors to realize an exclusive OR operation. Other workers have also investigated the interesting possibilities that arise with phase conjugate interferometers (Feinberg 1983; Ewbank et al. 1984; W. H. Chen et al. 1987; Tomita et al. 1988; Yang et al. 1991).

---

[1] The signal frequency could be much higher than this.

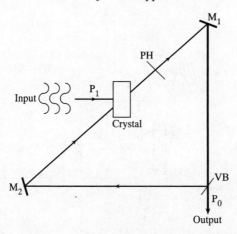

Fig. 13.5 Wavefront conversion using a unidirectional ring resonator (after Kwong and Yariv 1986). P_1 = pump beam, P_o = output beam, M_1, M_2 = mirrors, PH = pinhole, VB = partially transmitting mirror.

A number of workers have suggested techniques for deflecting an optical beam using a photorefractive crystal (Rak *et al.* 1984; Pauliat *et al.* 1986b; Henshaw and Todtenkopf 1986). Everard *et al.* (1992) have suggested a technique for the self-routing of an optical beam: the beam is incident on a self-pumped phase conjugate mirror which is able to pump several semi-linear resonators (see Fig. 7.15b) of different lengths. A particular resonator may be selected by arranging that the autocorrelation of the optical beam should have a peak at a delay corresponding to the length of that resonator.

RF frequency electronic signals may be correlated optically by using the signal to modulate the carrier driving an acousto-optic deflector. The diffracted optical beams from two such deflectors can be combined with the aid of a photorefractive crystal to provide a correlation signal (Psaltis *et al.* 1985; Krainak *et al.* 1988).

Wavelength multiplexing has been used by Ma *et al.* (1994) in order to produce an adjustable focal length lens. The idea is to record many holographic lenses of different focal lengths at different wavelengths; any particular focal length can then be selected by choosing the appropriate wavelength.

Optical phase conjugation has been used for image projection with high numerical aperture and micrometre resolution. Figure 13.6 shows the arrangement used by Levenson *et al.* (1981) for their experiments on photolithography

Holograms recorded in photorefractive media have been used to compress optical pulses in the picosecond range (Roblin *et al.* 1987). Rajbenbach *et al.* (1987) have used a photorefractive crystal in an optical system to perform matrix inversion. A novel pattern recognition system called the bifurcating optical pattern recognizer has been described by H. K. Liu (1993).

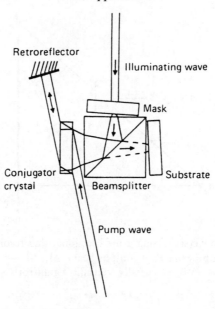

Fig. 13.6 Schematic showing the process of photolithographic image projection using phase conjugation by four-wave mixing in LiNbO$_3$ (after Levenson et al. 1981).

Finally, the photorefractive effect has been used in applications as diverse as fuzzy logic (W. Wu et al. 1995), imaging through scattering media (Hyde et al. 1995), ambiguity processing (T. C. Lee et al. 1980), laser frequency bandwidth narrowing (Chomsky et al. 1992), creating a delay line for time-dependent signals (Zhou and Anderson 1993), the storage and replay of 200 Mbit s^{-1} bit sequences (Ito and Kitayama 1992), particle image velocimetry (Petersen et al. 1992), velocity filtering (Vachss and Hesselink 1988c), and neural networks (Wagner and Psaltis 1987).

APPENDIX A
WAVE PROPAGATION IN ANISOTROPIC MATERIALS

The aim of this appendix is to introduce the properties of wave propagation in anisotropic materials. For the majority of readers this is probably a superfluous exercise. Those who entered the field of optics would have very likely come across problems in which anisotropy played a role. There is however a substantial minority who are thoroughly familiar with wave propagation in isotropic materials, know all the tricks of the trade employed in solving Maxwell's equations in isotropic media, and show no fear in tackling tensors. Nevertheless, when it comes to anisotropy their physical intuition grinds to a halt. They have heard of birefringence, of ordinary and extraordinary waves, of index ellipsoids, etc. but they have not had a chance manipulating them. This appendix should serve as a gentle introduction to them.

Let's start with Maxwell's equations in an isotropic medium. We shall write them in the form

$$\nabla \times \mathcal{E} = -j\omega\mu\mathcal{H} \tag{A.1}$$

$$\nabla \times \mathcal{H} = j\omega\varepsilon\mathcal{E} \tag{A.2}$$

where we have assumed time variation in the form $\exp j\omega t$ (ω—frequency, t—time), $\mathcal{E}$ and $\mathcal{H}$ are the electric and magnetic fields, μ is the magnetic permeability, and ε is the dielectric permittivity, a scalar for the time being. Taking the curl of eqn (A.1) we obtain

$$\nabla \times (\nabla \times \mathcal{E}) = -j\omega\mu(\nabla \times \mathcal{H}) = \omega^2\mu\varepsilon\mathcal{E} \tag{A.3}$$

Using the relationship

$$-\mathbf{k} \times (\mathbf{k} \times \mathcal{E}_o) = -(\mathbf{k} \cdot \mathcal{E}_o)\mathbf{k} + k^2\mathcal{E}_o \tag{A.4}$$

it may be clearly seen that a solution exists in the form

$$\mathcal{E} = \mathcal{E}_o \exp(-j\mathbf{k} \cdot \mathbf{r}) \tag{A.5}$$

provided $\mathcal{E}$ and $\mathbf{k}$ are perpendicular to each other and

$$|\mathbf{k}| = k = \omega\sqrt{\mu\varepsilon} = \frac{\omega}{v} \tag{A.6}$$

where $v = (\mu\varepsilon)^{1/2}$ is the velocity of light in the medium. The vector $\mathbf{k}$ is known as the wave vector.

What is the difference when the wave propagates in an anisotropic dielectric? The dielectric constant becomes a tensor. Equation (A.3) modifies to

$$\nabla \times (\nabla \times \mathcal{E}) = \omega^2\mu\underline{\underline{\varepsilon}}\mathcal{E} \tag{A.7}$$

(in our notation a quantity twice underlined is a second-rank tensor). All the properties of wave propagation follow from eqn (A.7). We shall not take the most general case when the permittivity may be different in all three directions, known as the biaxial case. We shall be content to concentrate upon the uniaxial case when two of the permittivities are equal, i.e. $\varepsilon_\perp = \varepsilon_{xx} = \varepsilon_{yy}$ and only $\varepsilon_\parallel = \varepsilon_{zz}$ is different. The corresponding matrix is

$$\underline{\underline{\varepsilon}} = \varepsilon_o \begin{pmatrix} \varepsilon_\perp & 0 & 0 \\ 0 & \varepsilon_\perp & 0 \\ 0 & 0 & \varepsilon_\parallel \end{pmatrix} \quad (A.8)$$

in the representation when the z axis is the optical axis of the uniaxial crystal. In the usual terminology it is called the c axis. It is also worth mentioning at this point that depending on the sign of $\varepsilon_\parallel - \varepsilon_\perp$ we talk about positive or negative birefringence or about positive and negative uniaxial crystals.

Let us solve now the problem for the case when the wave travels in the z direction, i.e. $\mathbf{k} = k\mathbf{i}_z$. Can we still have a solution for $\boldsymbol{\mathcal{E}}_o$ in the transverse x–y plane? Equation (A.4) will yield $-k^2 \boldsymbol{\mathcal{E}}_o$ and the right-hand side of eqn (A.7) will be

$$\underline{\underline{\varepsilon}} \boldsymbol{\mathcal{E}}_o = \varepsilon_o \begin{pmatrix} \varepsilon_\perp & 0 & 0 \\ 0 & \varepsilon_\perp & 0 \\ 0 & 0 & \varepsilon_\parallel \end{pmatrix} \begin{pmatrix} \mathcal{E}_{ox} \\ \mathcal{E}_{oy} \\ 0 \end{pmatrix} = \varepsilon_o \varepsilon_\perp \underline{\underline{\delta}} \boldsymbol{\mathcal{E}}_o \quad (A.9)$$

where

$$\underline{\underline{\delta}} = \begin{pmatrix} 1 & 0 & 0 \\ 0 & 1 & 0 \\ 0 & 0 & 0 \end{pmatrix} \quad (A.10)$$

Quite obviously, we have again a solution provided that

$$k^2 = \omega^2 \mu \varepsilon_o \varepsilon_\perp \quad (A.11)$$

We find that propagation in the direction of the optical axis will not be affected by the $\varepsilon_\parallel$ element, and the wave vector is still perpendicular to the electric field vector, $\boldsymbol{\mathcal{E}}_o$.

Let us next look at the situation when wave propagation is perpendicular to the optical axis, say, in the y direction. We shall consider two cases, namely when the electric field vector is first in the x direction and secondly when it is in the z direction.

When the electric field is in the form

$$\boldsymbol{\mathcal{E}} = \mathcal{E}_{ox}\mathbf{i}_x \quad \text{and} \quad \mathbf{k} = k\mathbf{i}_y \quad (A.12)$$

(where $\mathbf{i}_x$ and $\mathbf{i}_y$ are unit vectors in the direction of the x and y axes respectively) then eqn (A.4) comes to $k^2 \mathcal{E}_{ox}\mathbf{i}_x$ and the right-hand side of eqn (A.7) to $\omega^2 \mu \varepsilon_o \varepsilon_\perp \mathcal{E}_{ox}\mathbf{i}_x$, whence

$$k^2 = \omega^2 \mu \varepsilon_o \varepsilon_\perp \quad (A.13)$$

once more. Now propagation is not in the direction of the optical axis but perpendicular to it in the y direction and still the wave propagates with the same velocity determined by the transverse dielectric constant, $\varepsilon_\perp$. The provisional conclusion we may draw is that what matters is not so much the direction of propagation but the fact that in both cases the electric field vector has been perpendicular to the optical axis.

Is anything going to change when the electric field vector is in the direction of the optical axis? Let's have the direction of propagation again in the y direction, i.e. assume

$$\mathcal{E} = \mathcal{E}_{oz}\mathbf{i}_z \quad \text{and} \quad \mathbf{k} = k\mathbf{i}_y \tag{A.14}$$

Equations (A.4) and (A.7) yield now $k^2 \mathcal{E}_{oz}\mathbf{i}_z$ and $\omega^2 \mu \varepsilon_o \varepsilon_\| \mathcal{E}_{oz}\mathbf{i}_z$ respectively, leading to

$$k^2 = \omega^2 \mu \varepsilon_o \varepsilon_\| \tag{A.15}$$

There *is* now a difference. The velocity of the wave in the y direction depends on its polarization. If polarized in the x direction it propagates with a velocity $v = (\mu \varepsilon_o \varepsilon_\|)^{-1/2}$. If polarized in the z direction it propagates with a velocity $v = (\mu \varepsilon_o \varepsilon_\perp)^{-1/2}$. The wave with polarization in the direction of the optical axis is called an *extraordinary* wave, and the one with polarization perpendicular to the optical axis is called an *ordinary* wave. The terms extraordinary and ordinary ray are also often used.

The mathematics has so far been trivial. We have had some very simple equations to solve and very simple formulae for the modulus of the wave vector and for the velocity of the wave. Let us try now something a little more complicated. Let us assume that the wave vector, the electric field vector and the z axis are all in the same plane and there is an angle θ between k and the z axis as shown in Fig. A.1. Taking k perpendicular to $\mathcal{E}_o$ we may write them in the form

$$\mathcal{E} = \mathcal{E}_o(\cos\theta \mathbf{i}_y - \sin\theta \mathbf{i}_z) \tag{A.16}$$

and

$$\mathbf{k} = k(\sin\theta \mathbf{i}_y + \cos\theta \mathbf{i}_z) \tag{A.17}$$

Equation (A.7) reduces then to

$$\omega^2 \mu \varepsilon_o \mathcal{E}_o(\varepsilon_\perp \cos\theta \mathbf{i}_y - \varepsilon_\| \sin\theta \mathbf{i}_z) = k^2 \mathcal{E}_o(\cos\theta \mathbf{i}_y - \sin\theta \mathbf{i}_z) \tag{A.18}$$

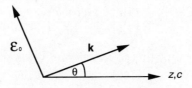

Fig. A.1 The wave vector, the polarization vector, and the optical axis are all in the same plane.

leading to the simple scalar equations

$$k^2 = \omega^2 \mu \varepsilon_o \varepsilon_\| \quad \text{and} \quad k^2 = \omega^2 \mu \varepsilon_o \varepsilon_\perp \tag{A.19}$$

There is something obviously wrong! When $\varepsilon_\perp \neq \varepsilon_\|$ it is not possible simultaneously to satisfy both equations. What can be wrong? Surely, we are entitled to assume that a wave travels in a certain specific direction determined by the angle θ. What's wrong is the assumption that the electric field vector is perpendicular to the direction of propagation. So let us start again, dropping that assumption and taking the electric field and the wave vector in the more general form

$$\mathcal{E} = \mathcal{E}_{oy}\mathbf{i}_y + \mathcal{E}_{oz}\mathbf{i}_z \quad \text{and} \quad \mathbf{k} = k_y \mathbf{i}_y + k_z \mathbf{i}_z \tag{A.20}$$

Since $\mathbf{k}$ and $\mathcal{E}_o$ are no longer perpendicular to each other, eqn (A.4) looks now a little more complicated. The complete expression for eqn (A.7) is now

$$\begin{aligned} &- (k_y \mathbf{i}_y + k_z \mathbf{i}_z)(k_y \mathcal{E}_{oy} + k_z \mathcal{E}_{oz}) + (k_y^2 + k_z^2)(\mathcal{E}_{oy}\mathbf{i}_y + \mathcal{E}_{oz}\mathbf{i}_z) \\ &= \omega^2 \mu \varepsilon_o (\varepsilon_\perp \mathcal{E}_{oy}\mathbf{i}_y + \varepsilon_\| \mathcal{E}_{oz}\mathbf{i}_z) \end{aligned} \tag{A.21}$$

yielding the scalar equations

$$(k_z^2 - \omega^2 \mu \varepsilon_o \varepsilon_\perp)\mathcal{E}_{oy} - k_y k_z \mathcal{E}_{oz} = 0 \tag{A.22}$$

and

$$-k_y k_z \mathcal{E}_{oy} + (k_y^2 - \omega^2 \mu \varepsilon_o \varepsilon_\|)\mathcal{E}_{oz} = 0 \tag{A.23}$$

This is now a linear, homogeneous algebraic equation in $\mathcal{E}_{oy}$ and $\mathcal{E}_{oz}$. A nontrivial solution exists if the determinant of the coefficients vanishes. With a little algebra we obtain the condition as

$$\frac{k_y^2}{k_\|^2} + \frac{k_z^2}{k_\perp^2} = 1 \tag{A.24}$$

where

$$k_\|^2 = \omega^2 \mu \varepsilon_o \varepsilon_\| \quad \text{and} \quad k_\perp^2 = \omega^2 \mu \varepsilon_o \varepsilon_\perp \tag{A.25}$$

This is clearly an ellipse in the k_y, k_z plane. If the wave vector subtends an angle θ with the z axis (eqn A.17) it may be easily shown that the modulus of the wave vector is equal to

$$k = \left(\frac{\sin^2 \theta}{k_\|^2} + \frac{\cos^2 \theta}{k_\perp^2} \right)^{-1/2} \tag{A.26}$$

the corresponding wave velocity is

$$v = \frac{\omega}{k} = \left[\frac{1}{\mu \varepsilon_o} \left(\frac{\sin^2 \theta}{\varepsilon_\|} + \frac{\cos^2 \theta}{\varepsilon_\perp} \right) \right]^{1/2} \tag{A.27}$$

and the equivalent value of the relative dielectric constant comes to

$$\varepsilon_{eq} = \left(\frac{\sin^2\theta}{\varepsilon_\parallel} + \frac{\cos^2\theta}{\varepsilon_\perp}\right)^{-1} = \frac{\varepsilon_\parallel \varepsilon_\perp}{\varepsilon_\perp \sin^2\theta + \varepsilon_\parallel \cos^2\theta} \qquad (A.28)$$

Clearly, ε_{eq} varies monotonically from $\varepsilon_\perp$ at $\theta = 0$ to $\varepsilon_\parallel$ at $\theta = 90°$. Since in practice the index of refraction is used more often than the dielectric permittivity, we shall introduce n_o and n_e, the indices of refraction for the ordinary and extraordinary wave respectively with

$$n_o^2 = \varepsilon_\perp \quad \text{and} \quad n_e^2 = \varepsilon_\parallel \qquad (A.29)$$

whence

$$n_{eq} = \frac{n_e^2 n_o^2}{n_o^2 \sin^2\theta + n_e^2 \cos^2\theta} \qquad (A.30)$$

As we have proven above, the electric field vector is not perpendicular to the wave vector in the general case. The actual angle can be determined from eqns (A.22) and (A.23). It turns out that taking practical values for n_o and n_e (for LiNbO$_3$ at $\lambda = 633$ nm for example the respective values are 2.286 and 2.200) the angle is very close to $90°$. In fact, none of the phenomena discussed in this book depends on that angle, so it will not be mentioned again, but just as a matter of interest we note here that the angle between **D**, the dielectric displacement vector, and the direction of propagation is always $90°$, as follows from Maxwell's $\nabla \cdot \mathbf{D} = 0$.

Our most significant conclusion so far is that the speed of propagation (or the modulus of the wave vector) depends on the polarization, i.e. on the direction of the electric field vector. If the polarization is perpendicular to the y–z plane we have an ordinary wave with the same value of **k** in all directions. The geometrical locus of **k** is then given by a circle as shown in Fig. A.2. For polarization in the y, z–plane the geometrical locus of **k** is given by the ellipse of eqn (A.24), shown also in Fig. A.2 for a positive uniaxial crystal, i.e. $n_e > n_o$. What happens if the electric field vector has components both in the y–z plane and perpendicular to it? Then the two components must be treated separately; they will propagate with different speeds.

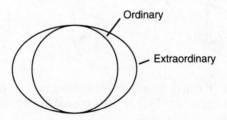

Fig. A.2 The geometrical locus of the wave vector for an ordinary (circle) and an extraordinary (ellipse) wave for a positive uniaxial crystal, $n_e > n_o$.

We have illustrated the effect of anisotropy by choosing the y–z plane but, obviously, the waves would behave in an identical manner in any plane containing the optical axis. Hence the curves of Fig. A.2 could be turned into surfaces by rotating them around the z axis. These surfaces are known as dispersion surfaces, momentum surfaces, or normal surfaces. Even more popularly, when one draws a few wave vectors stretching from the origin to one of these surfaces, the name of wave vector diagram conveys what the whole thing is about. In X-ray diffraction and in static or ordinary holography this surface is known as the Ewald sphere. When one wishes to determine the efficiency of the diffracted wave, one performs certain operations upon the Ewald sphere (see Section 1.2). Since the concept of a wave diffracted by a grating is one of the most important ones in the study of wave interactions with photorefractive materials, it follows that arguments relating to the Ewald sphere appear often in this book. In the more general anisotropic case we have to deal, of course, with both a sphere and an ellipsoid as discussed above. It makes then good sense to emphasize the analogy by referring to both surfaces as Ewald surfaces.

In a plane containing the optical axis the Ewald curves would look the same as shown in Fig. A.2. The wave vectors of the extraordinary waves lie on an ellipse and those of the ordinary waves on a circle. Recording of gratings has now a wider variety. It can be done by two extraordinary waves, by two ordinary waves, and by one of each[1] as shown in Fig. A.3 (a), (b), and (c) respectively. In a plane perpendicular to the optical axis the Ewald curves are concentric circles.

Optical activity. This is one more important property photorefractive crystals (e.g. those of the sillénite family: BSO, BGO, and BTO) may have. Independently of the direction of propagation, the optical activity will cause a linearly polarized wave to rotate its polarization as it propagates in the crystal. Mathematically, it may be described by off-diagonal elements in the dielectric tensor. For an isotropic material the tensor may be written in the form

$$\underline{\underline{\varepsilon}} = \varepsilon_o \begin{pmatrix} \varepsilon_r & 0 & 0 \\ 0 & \varepsilon_r & -jG \\ 0 & jG & \varepsilon_r \end{pmatrix} \tag{A.31}$$

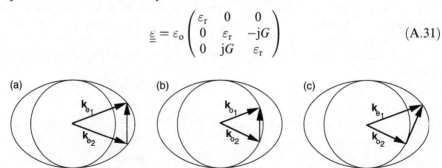

Fig. A.3 Recording of a grating in an anisotropic medium by (a) two extraordinary waves, (b) two ordinary waves, and (c) an ordinary and an extraordinary wave.

---

[1]This is a rather special case. It may occur due to the off-diagonal element of the photovoltaic tensor as explained in Section 9.2.

where ε_r is the relative dielectric constant and the electric field vector is supposed to be in the y–z plane. The differential equation to solve is once more eqn (A.7) and the solution is still to be found in the form of eqn (A.5) with $\mathbf{k}$ and $\boldsymbol{\mathcal{E}}_o$ perpendicular to each other. Taking the direction of propagation in the x direction, the differential equation reduces to the algebraic equations

$$k^2 \begin{pmatrix} 0 \\ \mathcal{E}_{oy} \\ \mathcal{E}_{oz} \end{pmatrix} = \omega^2 \mu \varepsilon_o \begin{pmatrix} \varepsilon_r & 0 & 0 \\ 0 & \varepsilon_r & -jG \\ 0 & jG & \varepsilon_r \end{pmatrix} \begin{pmatrix} 0 \\ \mathcal{E}_{oy} \\ \mathcal{E}_{oz} \end{pmatrix} \quad (A.32)$$

The arguments are now the same again as those leading from eqn (A.21) to the characteristic equation of (A.24). Following the same steps we obtain

$$k_\pm = \omega\sqrt{\mu\varepsilon_o}(\varepsilon_r \pm G)^{1/2} \cong \omega\sqrt{\mu\varepsilon_o}\left(n_r \pm \frac{G}{2n_r}\right) \quad (A.33)$$

where $n_r = (\varepsilon_r)^{1/2}$ and we have assumed $G \ll n_r$, a relationship that always holds in practice. The corresponding values of the electric field vector are

$$\boldsymbol{\mathcal{E}}_\pm = \mathcal{E}_o(\bar{\mathbf{i}}_y \pm j\bar{\mathbf{i}}_z) \quad (A.34)$$

where the indices $+$ and $-$ refer to left and right circularly polarized waves. It may be shown that the rotation per unit length for a linearly polarized wave is equal to

$$\rho = \frac{1}{2}(k_+ - k_-) = \frac{\pi G}{\lambda n_r} \quad (A.35)$$

APPENDIX B

THE ELECTRO-OPTIC EFFECT, INCLUDING THE PIEZOELECTRIC CONTRIBUTION

In Appendix A we have discussed the propagation of waves in anisotropic materials in which the dielectric constant is a second-rank tensor. When the components of this tensor are dependent on the electric field we describe it as the electro-optic effect. For some obscure historical reason the electro-optic coefficients were defined as affecting the components of the impermeability tensor, which is the reciprocal of the dielectric constant tensor. By definition,

$$\underline{\underline{\eta}} = \varepsilon_0 \underline{\underline{\varepsilon}}^{-1} \tag{B.1}$$

When the change in the components of the impermeability tensor is linearly dependent on the electric field, we talk of the linear electro-optic effect. We write it in the form

$$\underline{\underline{\Delta\eta}} = \underline{\underline{r}}\mathbf{E} \tag{B.2}$$

where r is the third-rank (underlined three times) electro-optic tensor containing in general 27 components. More usually the above relationship is written in coordinate form as

$$\Delta\eta_{ij} = r_{ijk} E_k \tag{B.3}$$

where the convention is used that there is summation over repeated subscripts. The subscripts range from 1 to 3 corresponding to three-dimensional space and, of course, 1, 2, and 3 refer to the coordinates x, y, and z. Thus, for example, the 12 component is the same as the xy component, and can be written either as

$$\Delta\eta_{xy} = r_{xyx} E_x + r_{xyy} E_y + r_{xyz} E_z \tag{B.4a}$$

or as

$$\Delta\eta_{12} = r_{121} E_1 + r_{122} E_2 + r_{123} E_3 \tag{B.4b}$$

We are going to use mostly the numerical notation, but not always in the above form. The reason is that a simplification in notation is possible owing to the symmetry relationship

$$r_{ijk} = r_{jik} \tag{B.5}$$

valid for all electro-optic crystals. The so called contracted subscript notation may then be introduced with the relations

$$1 = (11),\ 2 = (22),\ 3 = (33),\ 4 = (23) = (32),\ 5 = (31) = (13),\ 6 = (12) = (21)$$
$$\tag{B.6}$$

The electro-optic effect, including the piezoelectric contribution 417

which may be best remembered by the diagram

$$\begin{pmatrix} 1 & 6 & 5 \\ 6 & 2 & 4 \\ 5 & 4 & 3 \end{pmatrix} \tag{B.7}$$

Returning to our previous example, eqn (B.4) may now be written as

$$\Delta \eta_{12} = r_{61} E_1 + r_{62} E_2 + r_{63} E_3 \tag{B.8}$$

This looks a little odd, since the subscript on the l.h.s. is 12 whereas on the r.h.s. we must now use subscript 6. One can of course get used to this notation, but the advantages are rather dubious. The artificially created second-rank tensor does not obey the usual transformation properties of second-rank tensors and a great chunk of physical intuition is also lost in the process. Unfortunately, this notation is so widely used in the literature that we have no option but to introduce it. We shall, however, use it sparingly. We shall return to the three-index form whenever we deem it necessary.

In practice, of course, we use the permittivity tensor and not the impermeability tensor; hence it is the change in the permittivity tensor that needs to be related to the electric field. This can be done relatively easily, considering that the electro-optic effect causes only a minute change in the values of any of the components. When the electro-optic effect causes a change in the tensors it is still true that

$$(\underline{\underline{\varepsilon}} + \underline{\underline{\Delta \varepsilon}})(\underline{\underline{\eta}} + \underline{\underline{\Delta \eta}}) = \varepsilon_0 \underline{\underline{I}} \tag{B.9}$$

Since the changes are small we may neglect the product $\underline{\underline{\Delta \varepsilon}} \, \underline{\underline{\Delta \eta}}$ and obtain from eqns (B.1) and (B.9) the relationship

$$\underline{\underline{\Delta \varepsilon}} \, \underline{\underline{\eta}} = -\underline{\underline{\varepsilon}} \, \underline{\underline{\Delta \eta}} \tag{B.10}$$

Post-multiplying both sides of eqn (B.10) with the inverse of the impermeability tensor leads to

$$\underline{\underline{\Delta \varepsilon}} = -\frac{1}{\varepsilon_0} \underline{\underline{\varepsilon}} \, \underline{\underline{\Delta \eta}} \, \underline{\underline{\varepsilon}} \tag{B.11}$$

We shall use the above expression in some specific cases only. It will however save some space and time if we do not do each derivation separately but we do it here in some generality (not in complete generality because only the case of a uniaxial crystal will be considered). The relative dielectric constants in the transverse and longitudinal directions are then $\varepsilon_\perp$ and $\varepsilon_\parallel$ as already described in eqn (A.7). We may then evaluate $\underline{\underline{\Delta \varepsilon}}$ by performing the required operations:

$$\underline{\underline{\Delta \varepsilon}} = -\varepsilon_0 \begin{pmatrix} \varepsilon_\perp & 0 & 0 \\ 0 & \varepsilon_\perp & 0 \\ 0 & 0 & \varepsilon_\parallel \end{pmatrix} \begin{pmatrix} \Delta \eta_{11} & \Delta \eta_{12} & \Delta \eta_{13} \\ \Delta \eta_{21} & \Delta \eta_{22} & \Delta \eta_{23} \\ \Delta \eta_{31} & \Delta \eta_{32} & \Delta \eta_{33} \end{pmatrix} \begin{pmatrix} \varepsilon_\perp & 0 & 0 \\ 0 & \varepsilon_\perp & 0 \\ 0 & 0 & \varepsilon_\parallel \end{pmatrix} \tag{B.12}$$

yielding

$$\underline{\underline{\Delta\varepsilon}} = -\varepsilon_o \begin{pmatrix} \varepsilon_\perp^2 \Delta\eta_{11} & \varepsilon_\perp^2 \Delta\eta_{12} & \varepsilon_\perp \varepsilon_\parallel \Delta\eta_{13} \\ \varepsilon_\perp^2 \Delta\eta_{21} & \varepsilon_\perp^2 \Delta\eta_{22} & \varepsilon_\perp \varepsilon_\parallel \Delta\eta_{23} \\ \varepsilon_\perp \varepsilon_\parallel \Delta\eta_{31} & \varepsilon_\perp \varepsilon_\parallel \Delta\eta_{32} & \varepsilon_\parallel^2 \Delta\eta_{33} \end{pmatrix} \quad (\text{B.13})$$

where

$$\begin{aligned}
\Delta\eta_{11} &= r_{11i}E_i = r_{1i}E_i, & \Delta\eta_{12} &= r_{12i}E_i = r_{6i}E_i, \\
\Delta\eta_{13} &= r_{13i}E_i = r_{5i}E_i, & \Delta\eta_{22} &= r_{22i}E_i = r_{2i}E_i, \\
\Delta\eta_{23} &= r_{23i}E_i = r_{4i}E_i, & \Delta\eta_{33} &= r_{33i}E_i = r_{3i}E_i
\end{aligned} \quad (\text{B.14})$$

Having got the general formula, we need to know how to evaluate it. The electro-optic coefficients are available in the standard coordinate system with the X, Y, Z coordinates in the [1,0,0], [0,1,0], and [0,0,1] crystallographic directions, where the numbers are Miller indices[1].

There are, in principle, 18 independent electro-optic coefficients but many fewer when crystal symmetry is taken into account. For the more often used photorefractive materials we shall give below the non-zero values and the relationships between the others.

BGO, BSO, BTO, GaAs, InP, CdTe $\quad r_{41} = r_{52} = r_{63}$
LiNbO$_3$, LiTaO$_3$ $\quad r_{12} = -r_{22} = -r_{61}, \ r_{13} = r_{23} = r_{33}$,
$\quad r_{42} = r_{51}$
BaTiO$_3$, KTN, SBN, KNbO$_3$ $\quad r_{13} = r_{23} = r_{33}, \ r_{42} = r_{51}$

The actual values of the electro-optic coefficients have been measured only for specific frequencies and not with great accuracy. For BSO, for example, values have been quoted for the r_{41} coefficient ranging from 3.2 pm V^{-1} to 5 pm V^{-1}. We shall discuss these difficulties in somewhat more detail in Appendix C. For the time being our aim is to determine $\underline{\underline{\Delta\varepsilon}}$ which we can do with the aid of Table B.1 where values from the literature are quoted for the above mentioned materials. Note that (T) and (S) refer to low frequency (or unclamped) and high frequency (or clamped) values, respectively. They will be briefly discussed in Appendix C.

For our first example we take GaAs and InP, two well-known semiconductors which are isotropic ($\varepsilon_\perp = \varepsilon_\parallel = \varepsilon_r$) in the absence of an electric field. Their only non-zero electro-optic coefficients are

$$r_{231} = r_{41}, \quad r_{132} = r_{52} \quad \text{and} \quad r_{123} = r_{63} \quad (\text{B.15})$$

It is also true that

$$r_{41} = r_{52} = r_{63} = r \quad (\text{B.16})$$

---

[1] Miller indices can be found in any book on the properties of crystals. In the present context it is somewhat confusing to see another set of '1' appearing but its meaning should be clear. [1,0,0], [0,1,0], [0,0,1] mean the planes perpendicular to the X, Y, and Z axes respectively.

Table B.1 Electro-optic coefficients

Substance	Wavelength (µm)	Electro-optic coefficient (pm V^{-1})			
CdTe	1	(T)	$r_{41} = 4.5$		
GaAs	0.9		$r_{41} = 1.1$		
	1.15	(T)	$r_{41} = 1.43$		
InP	1.06	(S)	$r_{41} = 1.34$		
	1.50	(S)	$r_{41} = 1.68$		
Bi$_{12}$SiO$_{20}$ (BSO)	0.63	(T)	$r_{41} = 5$		
	0.51	(T)	$r_{41} = 4.4$		
		(S)	$r_{41} = 3.7$		
Bi$_{12}$GeO$_{20}$ (BGO)	0.63	(T)	$r_{41} = 3.4$		
	0.666	(T)	$r_{41} = 3.22$		
Bi$_{12}$TiO$_{20}$ (BTO)	0.51	(T)	$r_{41} = 5.17$		
LiNbO$_3$	0.633	(T)	$r_{13} = 9.6$;	(S)	$r_{13} = 8.6$
		(T)	$r_{22} = 6.8$;	(S)	$r_{22} = 3.4$
		(T)	$r_{33} = 30.9$;	(S)	$r_{33} = 30.8$
		(T)	$r_{51} = 32.6$;	(S)	$r_{51} = 28$
LiTaO$_3$	0.633	(T)	$r_{13} = 8.4$;	(S)	$r_{13} = 7.5$
		(T)	$r_{33} = 30.5$;	(S)	$r_{33} = 33$
		(T)	$r_{22} = -0.2$;	(S)	$r_{51} = 20$
				(S)	$r_{22} = 1$
BaTiO$_3$	0.546	(T)	$r_{51} = 1640$;	(S)	$r_{51} = 820$
		(T)	$r_{33} = 28$		
SBN	0.633	(T)	$r_{51} = 42$		
		(T)	$r_{13} = 67$		
		(T)	$r_{33} = 1340$		

with which $\underline{\underline{\Delta\varepsilon}}$ takes the simple form

$$\underline{\underline{\Delta\varepsilon}} = -\varepsilon_0 \varepsilon_r^2 r \begin{pmatrix} 0 & E_3 & E_2 \\ E_3 & 0 & E_1 \\ E_2 & E_1 & 0 \end{pmatrix} \tag{B.17}$$

Experimental and theoretical investigations (Yeh 1987b, 1988; Cheng and Yeh 1988; T. Y. Chang et al. 1988; Roy and Singh 1990a,b) have indeed been carried out on wave interaction in a GaAs crystal cut in this way leading to cross-polarization coupling in the reflection geometry due to the off-diagonal components. To amplify a particular polarization, the faces are usually cut perpendicular to the [1,1,0], [1,1,0], and [0,0,1] directions. The new cut is shown in Fig. B.1 with both coordinate systems: X, Y, Z in which the crystal parameters are given, and x, y, z which is perpendicular to the new faces. Let us see how the expression for $\underline{\underline{\Delta\varepsilon}}$ modifies in this case. Note that our new x and y axes may be obtained by rotating them around the $Z = z$ axis (which remains unchanged) by 45°. Using the well-

420   *The electro-optic effect, including the piezoelectric contribution*

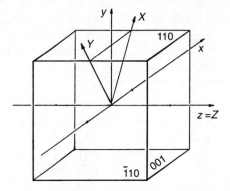

Fig. B.1 Crystal configuration for two-wave mixing experiments.

known laws of coordinate transformations, the new coordinates of the electric field (or of any other vector) are related to the old ones by the transformation

$$\mathbf{E}_{new} = \underline{\underline{T}} \mathbf{E}_{old} \tag{B.18}$$

where

$$\underline{\underline{T}} = \begin{pmatrix} \cos 45° & -\sin 45° & 0 \\ \sin 45° & \cos 45° & 0 \\ 0 & 0 & 0 \end{pmatrix} \tag{B.19}$$

and the new form of the dielectric constant tensor is related to the old one by

$$\underline{\underline{\varepsilon}}_{new} = \underline{\underline{T}} \, \underline{\underline{\varepsilon}}_{old} \underline{\underline{T}}^{-1} = \varepsilon_0 \varepsilon_r \begin{pmatrix} 1 + \varepsilon_r r E_z & 0 & 0 \\ 0 & 1 - \varepsilon_r r E_z & -\varepsilon_r r E_y \\ 0 & -\varepsilon_r r E_y & 0 \end{pmatrix} \tag{B.20}$$

where E_y and E_z are now the components of the electric field in the new (45°-rotated) coordinate system.

Let us now consider the case when $E_y = 0$. This is an often used configuration. The two incident beams are in the x–z plane so that the grating vector and the space charge field are in the z direction. Note that the terms due to the electro-optic tensor appear only in the ε_{xx} and ε_{yy} terms. If the optical waves propagate in (or very close to) the direction of the x axis, then the change in ε_{xx} is irrelevant, but the change in ε_{yy} is of course significant. It tells us that an electric field (static or slowly varying) in the z direction will affect the y component of the electric field associated with the optical wave. Note that the z component of the optical electric field is not affected. If the incident polarization of a linearly polarized optical wave has both y and z components, then the two components will propagate with different velocities. As a consequence, the linearly polarized wave will turn into an elliptically polarized wave at the end of the crystal. This is one of the methods for the measurement of the electro-optic coefficient to be discussed in more detail in Appendix C concerned with the measurement of material parameters.

The electro-optic effect, including the piezoelectric contribution

What happens when $E_z = 0$ and the space charge field is in the y direction? This is also a possible configuration used in practice. The incident beams are then in the x–y plane. The grating vector and the applied field (if present) are in the y direction. Note that for $E_z = 0$ and $E_y = 0$ the dielectric matrix (B.20) is not diagonal. We need to diagonalize it if we wish to find the principal axes. It turns out that we need to rotate the y and z axes around the x axis by $45°$. In the new x', y', z' coordinate system the diagonalized matrix takes the form

$$\underline{\underline{\varepsilon}}_{\text{new}} = \varepsilon_0 \varepsilon_r \begin{pmatrix} 1 & 0 & 0 \\ 0 & 1 - \varepsilon_r r E_y & 0 \\ 0 & 0 & 1 + \varepsilon_r r E_y \end{pmatrix} \tag{B.21a}$$

Note that now both the $\varepsilon_{y'y'}$ and the $\varepsilon_{z'z'}$ terms are affected and the change in the dielectric constant is opposite in these two directions. If an optical wave incident upon the $(1,1,0)$ face is linearly polarized in the $+45°$ or $-45°$ directions, then it can traverse the crystal unchanged, but for any other incident linear polarization the output will be elliptically polarized.

It is of importance to find the principal axes of the dielectric tensor because it improves our physical understanding of the anisotropic behaviour of the waves. When it comes to numerical calculations then it is not necessarily a disadvantage to use matrices with off-diagonal elements. This is what we do in Section 6.4. There the applied field and the grating vector are always taken to be in the z direction. Hence the change in the dielectric tensor when obtained from eqn (B.20) would be written as

$$\underline{\underline{\varepsilon}}_{\text{new}} = \varepsilon_0 \varepsilon_r r \begin{pmatrix} 0 & 0 & 0 \\ 0 & 0 & -E_z \\ 0 & -E_z & 0 \end{pmatrix} \tag{B.21b}$$

Our next example is the sillénite group, with $Bi_{12}SiO_{20}$ (BSO), $Bi_{12}GeO_{20}$ (BGO), and $Bi_{12}TiO_{20}$ (BTO) as representatives. Their properties are quite close to those of the previously mentioned semiconductors GaAs and InP. The non-zero elements of the electro-optic tensor are again $r_{41} = r_{52} = r_{63}$, and their actual values are quite close too. The major difference between the sillénites and the semiconductors is that those in the former group have optical activity, i.e. they rotate the polarization. The usual crystal cut is again that of Fig. B.1 with the electric field in the $[0,0,1]$ direction. The presence of optical activity is obviously harmful for wave interaction. When the polarization rotates, it will have directions in which the optical wave does not 'see' the grating; hence the effective gain constant is bound to be lower. This problem is discussed in great detail in Section 6.8 where both experimental and theoretical results are presented.

We shall next investigate $LiNbO_3$, the photorefractive material in which the effect was first discovered (Ashkin et al. 1966). It is a uniaxial crystal. The non-zero components of the electro-optic coefficients are

$$\begin{aligned} r_{112} &= r_{12}, \quad r_{113} = r_{13}, \quad r_{222} = r_{22}, \quad r_{223} = r_{23}, \quad r_{333} = r_{33}, \\ r_{232} &= r_{322} = r_{42}, \quad r_{131} = r_{311} = r_{51}, \quad r_{121} = r_{211} = r_{61} \end{aligned} \tag{B.22}$$

which are not all independent of each other. Their relationship is given by

$$r_{22} = -r_{12}, \quad r_{23} = r_{13}, \quad r_{61} = r_{12} \tag{B.23}$$

The corresponding form of the dielectric constant tensor may be easily obtained from eqn (B.13). The form we shall give below applies to the most often used configuration when the electric field is in the direction of the c axis (z axis):

$$\underline{\underline{\varepsilon}} = \varepsilon_o \begin{pmatrix} \varepsilon_\perp + \varepsilon_\perp^2 r_{13} E_z & 0 & 0 \\ 0 & \varepsilon_\perp + \varepsilon_\perp^2 r_{23} E_z & 0 \\ 0 & 0 & \varepsilon_\| + \varepsilon_\|^2 r_{33} E_z \end{pmatrix} \tag{B.24}$$

It may be seen that in this case only the diagonal terms are affected. As it happens, $r_{33} \gg r_{13}$, consequently a wave polarized in the z direction will 'see' a larger change in the dielectric constant than one polarized in the y direction. Since the crystal is anisotropic, one may rephrase the previous sentence by saying that the effect is larger for an extraordinary than for an ordinary wave.

Our final example is $BaTiO_3$. For recording a grating it was first used in 1970 (Townsend and LaMacchia 1970). Its non-zero electro-optic coefficients are

$$\begin{aligned} r_{113} = r_{13}, \quad r_{223} = r_{23}, \quad r_{333} = r_{33} \quad r_{232} = r_{322} = r_{42}, \\ r_{131} = r_{311} = r_{51} \end{aligned} \tag{B.25}$$

of which

$$r_{23} = r_{13} \quad \text{and} \quad r_{51} = r_{42} \tag{B.26}$$

If the electric field were only in the z direction, then the dielectric tensor would be of exactly the same form as that of eqn (B.24). In practice, however, one likes to make use of the large value of the r_{42} coefficient, which is effective only in the presence of an electric field in the y direction. We shall therefore give below the dielectric tensor when there is field in both the y and z directions. It is of the form

$$\underline{\underline{\varepsilon}} = \varepsilon_o \begin{pmatrix} \varepsilon_\perp - \varepsilon_\perp^2 r_{13} E_z & 0 & 0 \\ 0 & \varepsilon_\perp - \varepsilon_\perp^2 r_{23} E_z & -\varepsilon_\perp \varepsilon_\| r_{42} E_z \\ 0 & -\varepsilon_\perp \varepsilon_\| r_{42} E_z & \varepsilon_\| - \varepsilon_\|^2 r_{33} E_z \end{pmatrix} \tag{B.27}$$

When the above dielectric tensor is substituted into the wave equation and the coupled wave differential equations are derived, it is possible to define an effective electro-optic coefficient. In Chapter 6 we do a detailed derivation of the differential equations for BSO. Here we shall indicate the main steps, relevant for the determination of r_{eff}, for $BaTiO_3$. The physical configuration is two-wave mixing as shown in Fig. B.2. Waves 1 and 2 are incident at angles α and β with respect to the c axis. The input waves are assumed to be extraordinary waves with polarization vectors e_1 and e_2.

The aim is to turn the vector differential equation into a scalar one. The technique is not a trivial one, as pointed out by Ringhofer et al. (1991). The method usually employed is quite a simple one. It consists of multiplying the $\underline{\underline{\Delta\varepsilon}}$ tensor by e_1 from the left and by e_2 from the right, which serves to replace

The electro-optic effect, including the piezoelectric contribution

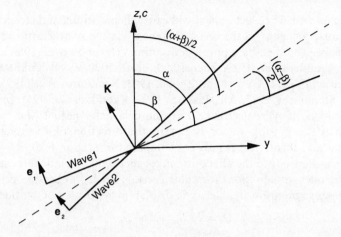

Fig. B.2 Two-wave mixing geometry in a BaTiO$_3$ crystal. The crystal surface is in the x–y plane. Both waves are polarized in the y–z plane.

the two field vectors by their amplitudes. Thus the operation, done in two dimensions in the input plane of the waves, is

$$\begin{pmatrix} -\cos\alpha \\ \sin\alpha \end{pmatrix} \begin{pmatrix} -\varepsilon_\perp^2 r_{23} E_z & \varepsilon_\perp \varepsilon_\| r_{42} E_y \\ -\varepsilon_\perp \varepsilon_\| r_{42} E_y & -\varepsilon_\perp^2 r_{23} E_z \end{pmatrix} \begin{pmatrix} -\cos\beta \\ \sin\beta \end{pmatrix} \tag{B.28}$$

yielding

$$-E_s \sin\left(\frac{\alpha+\beta}{2}\right) \left[\varepsilon_\|^2 r_{23} \cos\alpha\cos\beta + 2\varepsilon_\perp \varepsilon_\| r_{42} \cos^2\left(\frac{\alpha+\beta}{2}\right) + \varepsilon_\|^2 r_{33} \sin\alpha\sin\beta\right] \tag{B.29}$$

where we have used the relationships

$$E_z = E_s \sin\left(\frac{\alpha+\beta}{2}\right), \quad E_y = -E_s \cos\left(\frac{\alpha+\beta}{2}\right) \tag{B.30}$$

for the components of the space charge field, and

$$\mathbf{e}_1 = -\cos\alpha\,\mathbf{i}_y + \sin\alpha\,\mathbf{i}_z, \quad \mathbf{e}_2 = -\cos\beta\,\mathbf{i}_y + \sin\beta\,\mathbf{i}_z \tag{B.31}$$

for the unit vectors.

If both waves are of ordinary polarization, the problem is a scalar one leading to

$$r_{\text{eff}} = r_{13} \sin\left(\frac{\alpha+\beta}{2}\right) \tag{B.32}$$

So far we have taken into account the effect of an electric field upon the dielectric tensor via the electro-optic tensor. This is however not the only way an electric field may influence the dielectric tensor. There is also another possibility, considering that the crystal symmetry is such that all electro-optic crystals are piezoelectric as well. The electric field will cause then a strain, and strains are well known to cause a change in the dielectric tensor via the photoelastic effect. This problem has been studied by a number of authors (Izvanov et al. 1986; Stepanov et al. 1987; Mandel et al. 1989; Shepelevich et al. 1990, 1994; Kukhtarev et al. 1991; Pauliat et al. 1991; Günter and Zgonik 1991; Shandarov et al. 1991; Volkov et al. 1991; Shandarov 1992; Anastassakis 1993; Zgonik et al. 1994). Indeed, the combined effects of piezoelectricity and photoelasticity need to be taken into account when the grating vector is not in the direction of the electric field. Experimental and theoretical results are compared in Section 6.10.

Here we shall introduce the effect first in a general mathematical form and will then give the relevant equations for cubic crystals needed in Section 6.10.

In the presence of strain the dielectric displacement **D** may be written as

$$D_i = e_{ijk} S_{jk} + \varepsilon_o \varepsilon_{ij} E_j \tag{B.33}$$

where again the summation convention over repeated subscripts is used, S_{jk} are elements of the second-rank strain tensor, and e_{ijk} are elements of the third-rank piezoelectric tensor. For the symmetric case the relationship between strain and mechanical displacement, **U**, is given by

$$S_{jk} = \frac{1}{2} \left[\frac{\partial U_j}{\partial x_k} + \frac{\partial U_k}{\partial x_j} \right] \tag{B.34}$$

The other side of the coin is that the stress is no longer related to the strain alone but will also be influenced by the electric field. The relationship is given then by

$$T_{ij} = c_{ijkl} S_{kl} - e_{kij} E_k \tag{B.35}$$

where c_{ijkl} is the stiffness tensor. Stress and mechanical displacement are of course related to each other by the well known equation of acoustics:

$$\rho_o \frac{\partial^2 U_i}{\partial t^2} = \frac{\partial}{\partial x_j} T_{ij} \tag{B.36}$$

where ρ_o is the density of the material.

Let us remember now that we want the solution for an electric field that varies periodically with a spatial period of Λ. Acoustic waves would now come into the picture if the characteristic time constant (the grating formation time, τ_r) were comparable with the temporal acoustic period, τ_a. For the normally used incident intensities however it is true that

$$\tau_a = \Lambda/v_a \ll \tau_r \tag{B.37}$$

where v_a is the acoustic velocity, i.e. whatever happens takes place over many acoustic periods. Hence time variation may be disregarded (temporal derivatives neglected on the left-hand side of eqn (B.36)), which is equivalent to using the equilibrium condition of elasto-statics:

$$\frac{\partial}{\partial x_j} T_{ij} = 0 \tag{B.38}$$

The electro-optic effect, including the piezoelectric contribution

Owing to the condition curl $\mathbf{E} = 0$, the variation of the electric field must be coincident with the direction of the grating vector, $\mathbf{K}$. It follows then that the electric field varies only in the direction of the grating vector, reducing the problem to a one-dimensional one. Consequently,

$$\frac{\partial}{\partial x_j} = n_j \frac{\partial}{\partial \eta} \tag{B.39}$$

where $\mathbf{n}$ is a unit vector in the direction of $\mathbf{K}$ and η is the coordinate in the same direction. Substituting eqn (B.35) into eqn (B.38) we obtain

$$\frac{\partial}{\partial x_j} [c_{ijkl} S_{kl} - e_{kij} E_k] = 0 \tag{B.40}$$

But, using eqns (B.34) and (B.39)

$$\begin{aligned}\frac{\partial S_{kl}}{\partial x_j} &= \frac{1}{2} \frac{\partial}{\partial x_j} \left[\frac{\partial U_k}{\partial x_l} + \frac{\partial U_l}{\partial x_k} \right] \\ &= \frac{1}{2} n_j \left[n_l \frac{\partial^2 U_k}{\partial \eta^2} + n_k \frac{\partial^2 U_l}{\partial \eta^2} \right]\end{aligned} \tag{B.41}$$

and

$$\frac{\partial E_k}{\partial x_j} = n_j \frac{\partial E_k}{\partial \eta} \tag{B.42}$$

Integrating now by η and taking the integration constant equal to zero, eqn (B.40) takes the form

$$\frac{1}{2} c_{ijkl} n_j \left(n_l \frac{\partial U_k}{\partial \eta} + n_k \frac{\partial U_l}{\partial \eta} \right) = e_{kij} n_j E_k \tag{B.43}$$

Defining now Christoffel's tensor

$$\Gamma_{ik} = c_{ijkl} n_j n_l \tag{B.44}$$

and its reciprocal tensor $\gamma = \underline{\underline{\Gamma}}^{-1}$, and taking into account the symmetry properties of the stiffness tensor, eqn (B.43) may be rewritten as

$$\frac{\partial U_k}{\partial \eta} = \gamma_{ki} e_{nij} n_j E_n \tag{B.45}$$

Substituting now eqn (B.45) into eqn (B.34) we may write eqn (B.33) in the form

$$D_m = (\varepsilon_{mn} + e_{mkl} n_l \gamma_{ki} e_{nij} n_j) E_n \tag{B.46}$$

showing clearly the piezoelectric contribution to the change in the dielectric tensor.

An alternative method is to start with the definition

$$\underline{\underline{\Delta \eta^s}} = \underline{\underline{p}}\, \underline{\underline{S}} \tag{B.47}$$

i.e. the change in the impermeability tensor depends linearly on strain, where p is called the elasto-optic or the photoelastic tensor. Noting further that strain depends linearly on the electrical field (eqn B.40), we could express eqn (B.47) in the form

$$\underline{\underline{\Delta \eta^s}} = \underline{\underline{b}}\,\underline{E} \tag{B.48}$$

The two contributions to the change in the impermeability tensor can clearly be added; thus the total contribution is

$$\underline{\underline{\Delta \eta^s}} = (\underline{\underline{r}} + \underline{\underline{b}})\underline{E} \tag{B.49}$$

i.e. our previously discussed electro-optic tensor should be complemented by the new tensor b. A calculation similar to that from eqn (B.41) to eqn (B.46) would yield

$$b_{mnj} = p_{mnkl} n_l \gamma_{ki} e_{pij} n_p \tag{B.50}$$

The expressions we have been wrestling with so far in this appendix have been complicated enough, so we did not wish to add to the difficulties by distinguishing between the strain-free and stress-free coefficients. When it comes to photorefractive two-wave mixing it turns out (Günter and Zgonik 1991) that one needs to add the strain-free electro-optic contribution to the elasto-optic contribution. A correction is also needed owing to the fact that the antisymmetric part of the strain tensor (disregarded in the previous analysis) causes a rotation of the principal axes of the dielectric tensor. When optical activity is present as well, it may be expected to depend on the crystal's piezoelectric properties, but at the moment there is not sufficient information available about the crystal parameters to pursue this matter further. Obviously, further experiments are needed to test the theoretical predictions.

A set of two-wave mixing experimental results is presented in Chapter 6. For comparison with theory we use there the strain-free electro-optic coefficients and add the piezoelectric contribution in the form of eqn (B.50), yielding reasonable agreement.

APPENDIX C
DETERMINATION OF PHOTOREFRACTIVE PARAMETERS

C.1 Introduction

In the main text of this book we saw how the photorefractive effect evolved from a simple, easily understood concept into a complex physical description involving the use of a multitude of material and optical parameters. By happily quoting various parameter values we were subsequently able to demonstrate how well our models represented reality. But how were these values obtained? In this appendix we will take a closer look at how some of the more important values may be determined, and why large discrepancies may sometimes be found between values quoted in the literature.

There are three major reasons for being interested in the determination of material parameters. Firstly, it is extremely important for the understanding and correct modelling of the photorefractive effect in a particular material that we know the true parameter values. Secondly, characterization of materials is invaluable when developing/creating new materials. Finally, knowledge of the material constants is necessary when attempting to predict the performances of devices based on the photorefractive effect.

However, several problems have in the past been a hindrance to the accurate determination of material parameters. The main challenge has been the growth of good quality crystals, by which we mean large crystals (tens of cubic millimetres) of good optical quality, of well-defined doping, without evidence of strain or stress, and, where applicable, completely poled (e.g. no 90° and 180° domains in $BaTiO_3$) and of one phase. In the past it was not uncommon for the photorefractive properties of crystals to vary as much from laboratory to laboratory as from where in a boule a crystal was cut. The upshot was that every crystal was unique and had to be characterized separately. In particular, the carrier mobility μ, the recombination time τ, the effective trap density N_{eff}, and the role of electrons and holes could vary significantly from sample to sample.

C.2 Mobility, lifetime, electron–hole competition factor, and effective trap density

The question is now, having acquired a material to be used in a photorefractive experiment or application, how do we obtain the necessary material parameters? The obvious and by far the most common way is to carry out a simple two-wave mixing experiment and to determine the relevant material parameters from a best fit to theory. This will typically involve using tabulated values for such parameters as refractive index, dielectric constants, and electro-optic coefficients (provided they exist), and then determining the remaining free parameters simultaneously. This clearly involves a large degree of freedom and is strictly speaking scientifically unacceptable, especially when used to support the validity of a new model. Furthermore, the dilemma arises as to which photorefractive model to use. Will the simple band transport model suffice, or must multiple trap levels and electron–hole competition be considered? Unfortunately, this cannot be answered before a series of experiments has been carried out. Typically, this will include the detailed study of two-beam coupling gain as a function of grating spacing (i.e. inter-beam angle), total intensity, applied electric field, wavelength, detuning frequency, beam ratio, and temperature.

We shall here illustrate a two-beam coupling characterization method which is in wide use. We start by choosing a model – the single-species electron–hole competition model. With no external electric field or frequency detuning and for constant intensity we find that the gain coefficient for small grating period (where x is independent of grating spacing) may be written as

$$\Gamma = A\xi \frac{E_D E_q}{E_D + E_q} \quad \text{where} \quad A = \frac{2\pi}{\lambda \cos\theta} n^3 r_{\text{eff}} \tag{C.1}$$

By plotting K/Γ versus K^2, as shown in Fig. C.1 for measurements on CdTe:V (Zielinger *et al.* 1993), we can calculate the electron–hole competition factor, ξ, and the effective trap density, N_{eff}, (Moisan *et al.* 1994):

$$\xi = \frac{e}{Y_o k_B T A} \quad \text{and} \quad N_{\text{eff}} = \frac{\varepsilon}{P A \xi e} \tag{C.2}$$

where Y_o is the ordinate crossing and P is the slope of the best linear fit. Furthermore, we saw (text after eqn 5.14) that for the two-beam coupling to be effective, the photoexcitation (photoconductivity) needed to exceed the thermal excitation (dark conductivity). Thus the gain had the functional form

$$\Gamma = \frac{\Gamma_o}{1 + I_d/I} \tag{C.3}$$

where Γ_o is the saturated (high intensity) gain coefficient, and I_d is the incident intensity at which the photoconductivity equals the dark conductivity. Thus, we notice that by determining I_d from a measurement of the intensity dependence of the normalized gain, as shown in Fig. C.2 for CdTe:V (Partovi *et al.* 1990b), and equating the photoconductivity (assuming unit quantum efficiency):

$$\sigma_p = \frac{\hbar\omega}{e\mu\tau\alpha I} \tag{C.4}$$

at I_d with a measured value of the dark conductivity, we may obtain the mobility–lifetime product. Here α is the optical absorption at the frequency ω.

C.2 Mobility, lifetime, electron–hole competition and effective trap density

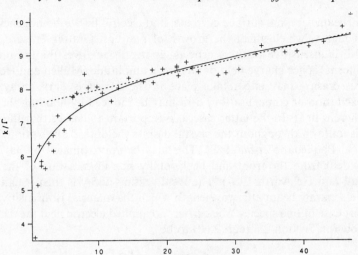

Fig. C.1 Experimental K/Γ as a function of K^2 as measured for a crystal of CdTe:V. The electron–hole competition factor and the effective trap density may be deduced from the dotted line representing the best linear fit for large K (after Zielinger et al. 1987).

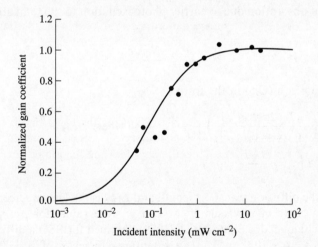

Fig. C.2 Normalized gain (Γ/Γ_o) as a function of incident intensity as measured in CdTe:V (after Partovi et al. 1990b).

The preferred technique for direct photorefractive characterization, however, is the study of grating decay. When a grating is formed during a two-wave mixing experiment, it involves the complex interaction of light with the photorefractive material. The grating formation is therefore very sensitive to external disturbances which cause variations in the phases and intensities of the interacting beams.

Moreover, beam fanning, surface currents, and electric field inhomogeneities are important factors which affect the growth of gratings. Grating erasure, on the other hand, is insensitive to these aspects, as it only involves the erasure of an existing internal space charge field of unknown amplitude (Mullen and Hellwarth 1985). This erasure may either take place in the dark (i.e. dark decay due to thermal excitation or compensation) or it may be induced by flooding the crystal with incoherent light. In the latter case, it is important to ensure that the erasure intensity is uniform throughout the crystal despite the optical absorption (Pauliat et al. 1986a; Baquedano et al. 1989). This may be approximated by having two beams incident from the front and back, or by side illumination of the crystal (Strohkendl and Hellwarth 1987). The local grating decay is then usually monitored by a separate beam at a wavelength which the material is insensitive to. In the simplest case of one species, one carrier, no applied electric field the decay rate τ_r was shown in Section 2.3 (eqn 2.60) to be

$$\tau_g = \frac{\tau_d}{\text{Re}(p)} = \tau_d \frac{1 + \frac{E_D}{E_M}}{1 + \frac{E_D}{E_q}} \tag{C.5}$$

By inserting in eqn (C.5) the expressions for the dielectric relaxation time, τ_d, and the characteristic fields, E_D, E_q, and E_M, and defining the quantum efficiency, η_q, as the ratio of absorption due to carrier photoexcitation to the total absorption α, i.e.

$$\eta_q = \frac{sN_D\hbar\omega}{\alpha} \tag{C.6}$$

we may rewrite eqn (C.5) in the form:

$$\tau_r = \left(\frac{k_B T}{e} \frac{2\pi\varepsilon\hbar\omega}{e\alpha\omega\eta_q I_o} \frac{1}{L_D^2}\right) \frac{1 + K^2 L_D^2}{1 + K^2 L_{\text{Deb}}^2} \quad \text{where} \quad L_D^2 = \frac{k_B T \mu \tau}{e}$$
$$\text{and} \quad L_{\text{Deb}}^2 = \frac{\varepsilon k_B T}{e^2 N_A} \tag{C.7}$$

where L_D is the diffusion transport length and L_{Deb} is the Debye screening length. Thus, from a curve of τ_r versus K^2 we can procure values for $\mu\tau$, η_q, and N_A. Using this technique, dos Santos et al. (1989) obtained for BSO a diffusion length of 0.1 μm ($\mu\tau = 4 \times 10^{-13}$ m^2 V^{-1}), a quantum efficiency of $\eta_q = 0.89$, and a Debye screening length of 0.036 μm ($N_A = 6.3 \times 10^{22}$ m^{-3}). A possible control is to check whether sensitivity, defined as $\partial E/\partial t|_{t=0}$, is maximum for $\Lambda = L_D$, as it should be (Moharam et al. 1979).

A study of grating decay will also fairly quickly give an indication of which photorefractive model to use. If multiple gratings are present in the crystal, this will become apparent during decay measurement, as the grating relaxation times will differ (Valley 1983b). Thus, whereas single-species (one trap level) models yield one time constant, two-species models (see Section 5.2) yield two (Bashaw et al. 1992; Attard et al. 1986; Tayebati 1991). Furthermore, the decay rate of absorption or photochromic gratings is expected to be independent of read or write intensities (Bylsma et al. 1988).

In the following we will briefly recount some of the many other characterization methods which may be used, some of which offer the benefit of being less sensitive to or independent of the choice of photorefractive model. Most of the following methods are concerned with measuring carrier mobility, μ. In fact, it is not quite untrue to say that there are just as many different values for μ as there are measurement methods. This discrepancy deserves some attention before we continue. By all accounts most photorefractive materials do not possess only one deep trap level but also have an appreciable concentration of shallow traps, the corollary being that the overall mobility of carriers is reduced by repeated trapping and thermal excitation from shallow traps. In other words, the mobility becomes 'trap-limited':

$$\mu_t = \mu_c \frac{\beta_s}{\rho_s} \tag{C.8}$$

where μ_t is the trap-limited mobility, μ_c is the mobility in the conduction band, $1/\rho_s$ is the average time the carrier spends in the conduction band between trapping events, and $1/\beta_s$ is the dwell time in a shallow trap (Nouchi et al. 1992; Le Saux and Brun 1987). Measured values of mobility will therefore be sensitive to whether the shallow traps are filled or not (Kostyuk et al. 1980; Tayag et al. 1994). In most cases, however, we may ignore the shallow traps, provided we account for them phenomenologically by using the trap-limited mobility.

The transient photocurrent method. This method consists of briefly flooding a crystal with light while measuring the photocurrent response. The mobility may then be deduced from the maximum value of the photocurrent current, i_o:

$$i_o = sI_o N_D T_p \mu e E_o S \tag{C.9}$$

where T_p is the duration of the optical pulse of intensity I_o, and S is the electrode surface area (Partanen et al. 1991; Le Saux and Brun 1987; Nouchi et al. 1992; Ennouri et al. 1993). Figure C.3 shows a typical result of the transient photocurrent following an optical pulse excitation in BSO (Le Saux and Brun 1987). A study of the photocurrent decay will subsequently not only yield the recombination time, but will reveal the presence and significance of multiple traps as well (i.e. multiple components of decay). Using this technique, Le Saux and Brun found they needed to use a model consisting of one deep-trap and one shallow-trap level, and from this deduced a conduction band mobility of $\mu_c = 3.24 \times 10^{-4} \, \text{m}^2 \, \text{V}^{-1} \, \text{s}^{-1}$, an electron deep-level recombination time of $\tau_e = 10 \, \text{ms}$, and an effective 'trap-limited' electron diffusion length of $L_D = 3.7 \, \mu\text{m}$.

Another related method is based on having a thin sample with semi-transparent electrodes through which light may excite carriers near one electrode. The idea is then that the mobility can be calculated from the time it takes the charge package

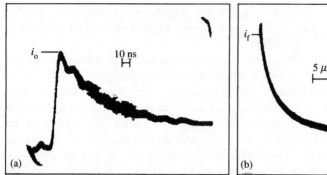

Fig. C.3 Transient photocurrent resulting from optical pulse excitation in BSO shown (a) on a short time scale revealing a decay time constant of 50 ns, and (b) on a long time scale showing a decay with a time constant of 10 μs. The vertical scale in (a) is 6 times that in (b) (after Le Saux and Brun 1987).

to traverse the width, d, of the crystal (Hou *et al.* 1973; Kostyuk *et al.* 1980; Tayag *et al.* 1994):

$$\mu_t = \frac{d}{E_o \tau_t} \tag{C.10}$$

where τ_t is the transit time.

Holographic time-of-flight method. When two pulsed plane waves interfere in a photorefractive crystal, the photoexcitation of carriers leads to the creation of two superimposed gratings: an electronic and a trap grating. At first however, there will be no resultant space charge field, as the two gratings are in phase, compensating each other. The time-of-flight method capitalizes on the fact that the electronic grating is mobile while the trap grating is not. Hence, in an electric field the electronic grating will drift with a velocity μE_o and a space charge grating will appear. The evolution of this grating may subsequently be monitored with the help of an on-Bragg probe beam. Figure C.4 shows typical experimental results in BSO (Nouchi *et al.* 1993). The mobility may now be deduced from the temporal position of the first maximum, as this will coincide with the average time, T, it takes the electrons to drift half a grating spacing (Pauliat *et al.* 1990; Partanen *et al.* 1990):

$$T = \frac{\Lambda}{2\mu E_o} \tag{C.11}$$

We notice from Fig. C.4 that the oscillations in diffraction efficiency diminish with time. This is because diffusion eventually leads to the 'washing out' of the carrier grating, leaving only the trap grating. For the same reason the best

C.2 Mobility, lifetime, electron–hole competition and effective trap density

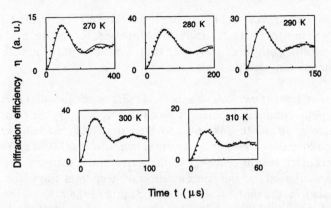

Fig. C.4 Typical results of the holographic time-of-flight method showing the temporal evolution of diffraction efficiency at various temperatures for a crystal of BSO and an applied voltage of 500 V (after Nouchi et al. 1993).

experimental results are obtained when the gratings are written with a small interbeam angle to reduce the effects of diffusion.

The high frequency method. In Section 5.4 we saw that when the AC field or moving grating techniques were used to enhance energy transfer in two-wave mixing, high frequency resonances would appear. Apart from the experimental parameters of frequency, grating spacing, and electric field, the position of these resonances depended only on the 'trap-limited' mobility (Pauliat et al. 1990; Grunnet-Jepsen et al. 1995a). For the moving grating technique we found with good approximation that the high frequency maximum was to be found at

$$f = -\frac{\mu E_o}{\Lambda} \tag{C.12}$$

Thus, the resonance arose when the velocity of the interference pattern matched the average velocity of the electrons (or holes). For the AC enhancement technique the mathematics was not quite as straightforward, but it was none the less possible to empirically obtain simple relationships for the positions of the resonance peaks, $f_{\sin}(n)$ and $f_{sq}(n)$, for sinusoidal and square wave fields, respectively:

$$f_{sq}(n) = \frac{f_o}{n}, \quad f_{\sin}(n) = \frac{f_o}{1.55n - 0.355}, \quad f_o = -\frac{\mu E_{oo}}{2\Lambda} \tag{C.13}$$

where n is the number of the resonance. Figure 5.24(a) shows a typical result obtained by Pauliat et al. for a crystal of BGO:Fe. From this measurement with a sinusoidal field of amplitude $E_o = 3.1\,\text{kV}\,\text{cm}^{-1}$ and a grating spacing of $\Lambda = 29\,\mu\text{m}$ one may estimate a 'trap-limited' mobility of $\mu_t = 2 \times 10^{-6}\,\text{m}^2\,\text{V}^{-1}\,\text{s}^{-1}$.[1]

---

[1] Note that the original qualitative arguments for the resonance peaks proposed by Pauliat et al. (1990) led them to underestimate the mobility by a factor of about 1.52.

The advantages of these techniques are their relatively easy implementation and insensitivity to optical intensity variations. Moreover, for the AC field method the frequency is sufficiently high to reduce field screening and eliminate the uncertainty of the internal electric field.

Induced photocurrent method. This method relies on the fact that currents can be induced in photoconductive materials by either imposing on the crystal an oscillating (Petrov *et al.* 1990; Sokolov and Stepanov 1993; Sochava *et al.* 1993a; Sochava and Stepanov 1994) or a moving (Davidson *et al.* 1994,a,b) interference pattern. Through a best fit of theory and experiment it is subsequently possible to obtain information about the concentration of deep-level traps, the sign of the dominant carriers, the mobility–lifetime product, and the ratio of transverse and longitudinal mobilities. Figure C.5 illustrates a typical result of the photocurrent generated when a crystal of InP:Fe was exposed to a moving interference pattern (Davidson *et al.* 1994). The DC photocurrent is seen to be an odd function of the optical frequency difference, the symmetry of which depends on the sign of the dominant carriers. In this example, the change of symmetry with grating spacing clearly illustrates the change in the polarity of the dominant carrier. The main benefit of the photocurrent method is that it does not rely on measurements of energy coupling between the optical beams and is therefore only weakly influenced by the electro-optic effect.

The screening charge method. This optical method for studying the transport properties of photogenerated electrons in photoconducting materials is based on studying the dynamics of screening charge build-up. When an electric field is applied to a photoconducting material, illuminating a small cross section of the

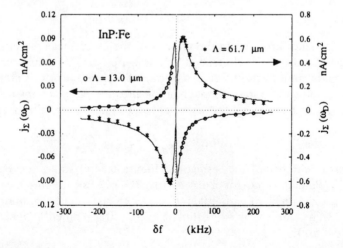

Fig. C.5 Induced DC photocurrents in InP:Fe as a function of frequency detuning for $\Lambda = 61.7\,\mu\text{m}$ (∗) and $\Lambda = 13\,\mu\text{m}$ (o) (after Davidson *et al.* 1994a).

crystal will effect a redistribution of charges near the dark–light interfaces, effectively screening the field in the lit region. By sandwiching the crystal between two crossed polarizers we may study the temporal evolution of the internal field and subsequently obtain values for the quantum efficiency and the mobility–lifetime product (Grousson et al. 1984b).

The optical pulse method. The use of pulsed laser light in photorefractive experiments can be a very valuable tool when it comes to characterizing carrier transport. The main merit is that processes that occur on different time scales, such as photo-excitation and transport of carriers, may be separated. Thus, the carriers are excited during the interference of the two short light pulses, whereas the grating builds up in the dark (Jonathan et al. 1988; Pauliat and Roosen 1990). Moreover, the role of the bulk photovoltaic effect is effectively eliminated, as this can only play a role during illumination. Under these conditions the temporal evolution of the space charge field is given by eqn (5.83) (Ewart et al. 1994; Biaggio et al. 1992). By measuring the evolution of the diffraction efficiency for different grating spacings we may obtain values for the mobility and electron lifetime. One advantage this method has over the time-of-flight method is that no external field is needed.

C.3 Sign of carriers

For material characterization purposes it is often important to know whether the carriers primarily responsible for the photorefractive properties of a material are electrons or holes. This can be unambiguously determined by noting the direction of energy transfer in a two-wave mixing experiment. The direction of energy transfer depends on the relative signs of the carriers and the electro-optic coefficient. The latter can be found from phase retardation or ellipsometric measurements by noting the sign of the birefringence induced by applying an electric field to the crystal (Glass et al. 1985; Pauliat et al. 1987; Partovi and Garmire 1991). Thus, if the field direction shown in Fig. C.6 increases the refractive index, then an

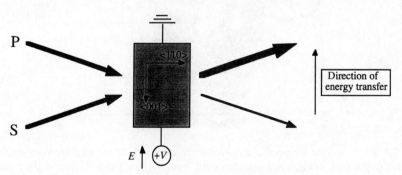

Fig. C.6 Two-wave mixing experiment to determine sign of majority charge carriers.

energy transfer from P to S would imply that electrons are the primary charge carriers.

Another way of settling the sign of the dominant carriers is to carry out a two-wave mixing experiment using the moving grating technique described in Section 5.4.2. Here, if the maximum energy transfer or diffraction efficiency occurs when the interference pattern is moving in the same direction as the field, then the dominant carriers are electrons (Stepanov *et al.* 1982; Strohkendl and Hellwarth 1987). Thus, if in Fig. C.6 beam S has a slightly higher frequency than beam P, then electrons dominate.

Finally, as we mentioned earlier, the sign of the carriers can also be deduced from measurements of induced photocurrents. If the photocurrent is in the same direction as the moving grating, then the current is electron dominated.

C.4 Electro-optic coefficient

Several different methods for measuring the electro-optic coefficients have been proposed and implemented in the past. The simplest of these is the half-wave voltage method, illustrated in Fig. C.7, which is based on using the electric field-induced birefringence to vary the transmission of light through a crystal sandwiched between two crossed polarizers (see e.g. Yariv 1991). The effective electro-optic coefficient can then be read directly from the voltage at which the transmission has its first minimum. However, the concept of half-wave voltage is not always applicable, in particular not for crystals which exhibit optical activity

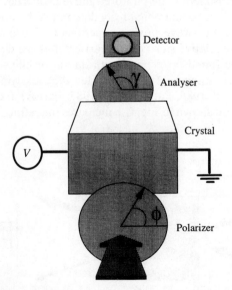

Fig. C.7 Experimental set-up for measuring the electric field-induced birefringence from which the product of the electro-optic coefficient and the internal electric field may be deduced.

C.4 Electro-optic coefficient", 437

cal activity (Pellat-Finet 1984) or in which the half-wave voltage exceeds the crystal break-down voltage. In this case, it is necessary to use either an eigenstate determination technique (Henry et al. 1986; Tanguay 1977), an interferometric technique (Günter 1976; Knöpfle et al. 1994; Ducharme et al. 1987), or a transmission technique (Wilde et al. 1990; Henry et al. 1986; Raymond, 1992). The first of these, the eigenstate determination technique, is based on finding the polarization state which transmits unaltered though the crystal, i.e. the eigenstate. For instance, for a crystal of BSO the eigenstate changes from circular to elliptical polarization when a field is applied. The new eigenstate is a unique function of the electro-optic effect. The second method, the interferometric technique, is for its part conceptually even simpler. The electric field-induced change in refractive index can be measured directly by placing the crystal in one of the two arms of a Michelson interferometer. The variation of output intensity from the interferometer is consequently a measure of the induced change in optical path length and thereby the size of the electro-optic coefficient. The last method is the transmission technique, which is very similar to the aforementioned half-wave voltage method. However, instead of relating the electro-optic coefficient to an ill-defined 'half-wave voltage', the coefficient is now calculated from the exact expression for the optical transmission. This method will be covered in more detail later in this Appendix.

Before continuing, however, we should mention that despite the obvious need for an accurate determination of the electro-optic coefficients and the numerous papers on the subject, there is still a large discrepancy in reported values, even for photorefractive materials which have been the subject of intense study for years. For instance, for BSO the value of the electro-optic coefficient r_{41} varies in the literature from 3.4 (Marrakchi et al. 1981) to 5 pm V^{-1} (Vachss and Hesselink 1987b). The main reason for the discrepancy is by all accounts an intrinsic uncertainty in the true electric field present inside the material. Even though this may be caused by contact losses at the metal–crystal interfaces and internal non-uniformities in conductivity, the dominant factor is generally believed to be the build-up of space charge near the electrodes which acts to screen the field in the bulk of the crystal (Ziari and Steier 1993; Ziari et al. 1992; M. B. Klein et al. 1988; Henry et al. 1986; Tanguay 1977; Steier et al. 1988; Jean et al. 1994; Lemaire and Georges 1992). Thus, owing to the photoconductive property of the materials, screening charges may build up due to non-uniformity in illumination or due to blocking contacts (Astratov et al. 1984; Bryksin et al. 1987), the outcome being that the internal field may vary significantly from the usually assumed value of V_o/d, where V_o is the voltage applied across the crystal width d.

In an attempt to circumvent the problem of electrical contacting, Vachss and Hesselink (1987b) used a photorefractive technique whereby they based their measurement of r_{eff} on the known relationship between the internally generated electric space charge field and the holographic diffraction efficiency, doing away with electric contacts altogether. Other remedies that have been suggested include using low illumination levels, thereby increasing the space charge field build-up time (Henry et al. 1986; Tanguay, 1977), or using side illumination of UV light to effect a redistribution of any space charge (Bayvel et al. 1988). Usually a longitudinal configuration in which the light in the crystal propagates parallel to the applied field direction is preferred (Henry et al. 1986; Tanguay 1977). Here the light beam will experience an integral of the field along its path, and since this integral is equal to the applied voltage, the influence of any field inhomogeneities will be reduced. This is in contrast to the transverse configuration in which the direction of the light and field are orthogonal. Here, the electric field experienced by the light beam is constant but will depend on the position of the beam. Lemaire and Georges (1992) suggested that the electro-optic coefficient could be reliably measured in this configuration by integrating the phase shift along the transverse profile. Another approach, suggested recently, is to eliminate the uncertainty of the internal electric field by applying an AC square wave field to the crystal instead of the usual DC field (Grunnet-Jepsen et al. 1995b). The benefit of this method is that if the frequency of the AC field is high enough, the carriers do not have time to accumulate at the contacts, and as a consequence no screening occurs.

We will here briefly describe the principles of the aforementioned 'transmission' technique. The idea is that any local internal electric field will induce a birefringence which may be observed by placing the crystal between two crossed polarizers, as in Fig. C.7. Thus, by measuring the transmission of a linearly polarized plane wave through the set-up in Fig. C.7, we may visualize the internal electric field distribution. The transmission, T, is given by (Wilde et al. 1990; Henry et al. 1986; Raymond 1992)

$$T = \left[\frac{q_2}{q_3} \sin \mu_3 \cos \mu_2 - \cos \mu_3 \sin \mu_2\right]^2 + \left[\frac{q_1}{q_3} \sin \mu_3 \sin(\mu_2 + 2\phi)\right]^2 \quad \text{(C.14)}$$

where

$$q_1 = -\frac{1}{2} n^2 r_{\text{eff}} E_o, \quad q_2 = \frac{\rho \lambda}{\pi n}, \quad \mu_1 = -\frac{1}{2} \frac{\pi}{\lambda} n^3 r_{\text{eff}} E_o L, \quad \mu_2 = \rho L \quad \text{(C.15)}$$

and E_o is the internal electric field, λ is the wavelength of the light, n is the refractive index, L is the optical path length inside the crystal, and ρ is the optical rotatory power. We have used the term 'crossed polarizers' to signify the condition of zero tranmission for no applied field. For an optically active crystal such as BSO, this is obtained for $\gamma - \phi = \mu_2 + 90°$. We note that without optical activity eqn (C.14) simplifies to the well-known

$$T = (\sin \mu_1 \sin 2\phi)^2 \quad \text{(C.16)}$$

The product of the electric field and the effective electro-optic coefficient can now easily be determined by solving the transcendental equation (C.14) numerically using for example a simple secant method.

Before continuing, it should be recalled that the electro-optic effect in photorefractive materials is a tensor of rank 3 consisting of direct and indirect (i.e. piezoelectric and piezoelastic effect) contributions, as described in Appendix B (Nye 1985; D. F. Nelson 1979; Günter and Zgonik 1991; Pauliat et al. 1991;

Kaminow 1974; Shandarov 1992). The electro-optic coefficient we measure in an experiment will therefore depend on crystal orientation, field orientation, and whether the crystal is clamped or not. A crystal is understood to be clamped when the piezoelectric and photoelastic effects do not contribute to the electro-optic effect. The clamped or 'true' value of a particular electro-optic coefficient can be obtained in one of two ways. It can either be calculated by measuring the unclamped value and subtracting the calculated indirect contributions, or it can be measured directly. The latter may be accomplished by applying an electric field of sufficiently high frequency so that the crystal has no time to deform, i.e. well above the piezoelectric resonance $f_{\text{piez}} = v_s/2d$, where d is the distance between the electrodes and v_s is the velocity of sound in the crystal. In a crystal of BSO with $d = 1$ cm, this frequency is near 100 kHz. Figure C.8 illustrates a typical result obtained for a crystal of $NH_4H_2PO_4$ (Carpenter 1950). It is here easy to distinguish the clamped (strain-free) from the unclamped (stress-free) value. Alternatively, one can apply a step-function voltage and monitor the electro-optic response of the crystal. Immediately after the application of the voltage the crystal has not had time to deform, and we will consequently measure the clamped value (A. R. Johnston 1965). Figure C.9 depicts experimental results for $BaTiO_3$ using this technique (Zgonik et al. 1994). The unclamped value is obtained when the crystal stops 'ringing' and settles to a steady-state value.

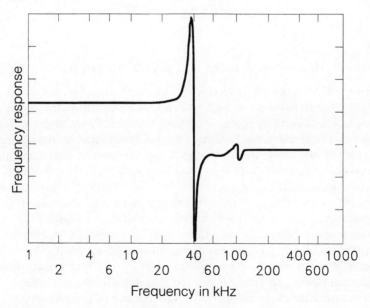

Fig. C.8 High frequency response of electro-optic effect in $NH_4H_2PO_4$ crystal measured by applying a sinusoidal electric field (after Carpenter 1950).

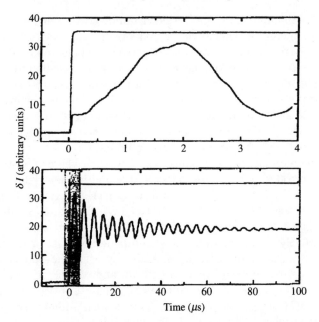

Fig. C.9 The electro-optic response of BaTiO$_3$ following a step change of the applied electric field. The upper curve shows measurements in the first few microseconds. Here the first step corresponds to the electro-optic response of a clamped crystal. The lower curve shows the 'ringing' (the damped mechanical resonance) which settles down to the unclamped state of the crystal (after Zgonik *et al.* 1994).

C.5 The elastic, elasto-optic, dielectric, and piezoelectric properties

In order to calculate the effective electro-optic and dielectric coefficients to be used in photorefractive wave-mixing, it is necessary to know the values of the piezoelectric, elasto-optic (photoelastic), and elastic stiffness tensors. We shall here give only a brief introduction. The interested reader should refer to Zgonik *et al.* (1993, 1994) for an in-depth look at the full characterization of the electro- and acousto-optic properties of photorefractive BaTiO$_3$ and KNbO$_3$.

The elastic constants, c_{ijkl}^E (constant field strength) and c_{ijkl}^D (constant electric displacement), not surprisingly are a measure of the elastic stiffness of a crystal. These constants can be obtained by measuring the sound velocity for different polarizations and propagation directions using a pulse-echo technique. A full determination of all the tensor components thus requires that different crystals be prepared which permit sound propagation along different crystal directions. The elastic constants can subsequently be calculated from the product of the square of the velocity of the ultrasound and the density of the material.

The elasto-optic coefficients, p_{ijkl}, indicate to which extent crystal deformations change the dielectric tensor. It is through this effect that sound and light may be

C.5 The elastic, elasto-optic, dielectric, and piezoelectric properties

made to interact in a crystal. The coefficients may essentially be obtained by measuring the amount of light which is scattered off an ultrasonic pulse which is launched into the crystal. A reliable method for accomplishing this, proposed by Dixon and Cohen (1966), is based on sending a pulse of sound into a reference crystal (e.g. quartz) which is bonded to the crystal to be studied, as illustrated in Fig. C.10. The elasto-optic coefficient (or rather, a particular component) is subsequently related to the ratios of the beams diffracted by the pulse in the reference material and in the crystal of unknown properties. The results are independent of acoustic loss in the material or the quality of the bond.

The coefficients of the piezoelectric strain tensor, d_{ilm}, are a measure of how an electric field deforms a crystal. This effect is used in an untold number of

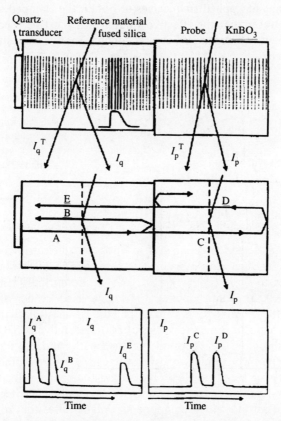

Fig. C.10 Determination of the elasto-optic properties by the Dixon–Cohen method. The transducer generates a short ultrasonic pulse which travels through the reference material and is partially reflected upon reaching the 'probe' material. The elasto-optic properties may be deduced by monitoring the ratio of light diffracted off the gratings in both materials for both the forward and backward propagating pulses (after Zgonik et al. 1993).

applications, from instruments requiring the fine control of displacements such as piezoelectric mirrors, to sensors for measuring stress such as piezoelectric balances. This effect is especially important in photorefractive materials, because in conjunction with the elasto-optic effect it contributes indirectly to the electro-optic effect. These coefficients can be obtained by measuring the change in crystal length which is induced by an electric field. Again we can use an interferometric technique whereby a change in path length is measured by placing the crystal in one arm of a Michelson interferometer (Ducharme et al. 1987). However, instead of transmitting the light through the crystal, the beam is reflected off the front and back surfaces of the crystal, as shown in Fig. C.11. This neatly eliminates any sensitivity to translation of the crystal while ensuring that we measure only the change in crystal length and not the electro-optically induced change in phase.

Finally, the dielectric constants, ε_{ij}, can be obtained by measuring the capacitance over a range of applied frequencies to obtain the clamped and unclamped values. Note however that the effective value which must be used in a photorefractive experiment will depend on crystal configuration, elastic stiffness, and the piezoelectric effect.

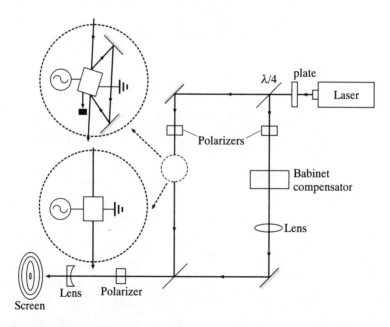

Fig. C.11 Electro-optic interferometers. The Mach–Zehnder interferometer used to measure the electro-optic coefficients r_{31} and r_{33} (lower inset) and the piezoelectric coefficient d_{13} (upper inset). The crystal is placed in one arm and the Babinet–Soleil compensator in the other arm. The crystal c axis and the compensator slow axis are aligned in the plane of the figure (after Ducharme et al. 1987).

C.5 The elastic, elasto-optic, dielectric, and piezoelectric properties

In concluding this appendix, we need to point out that our survey of the characterization methods is by no means a complete one. Among the other methods which are in common use, we should mention that the measurement of absorption, photoluminescence, and photoconductivity are all helpful in indicating at which wavelengths the photorefractive material may operate and the energy levels of the traps. Electron paramagnetic resonance measurements are also a common tool when trying to identify which trap levels participate actively in the photorefractive process. For a recent review of a broad range of characterization methods see Mullen (1989). For detailed comments and for an excellent coverage of the literature on the physical and holographic properties of a number of photorefractive crystals see the Appendix of Petrov *et al.* (1991).

BIBLIOGRAPHY

Abbe, E. (1873). Beiträge zur Theorie des Mikroskopes und der mikroskopischen Wahrnehmung. *Arch. Mikrosk. Anat.*, **9**, 413–68.
Agranat, A., Leyva, V., and Yariv, A. (1989). Voltage-controlled photorefractive effect in paraelectric $KTa_{1-x}Nb_xO_3$: Cu,V. *Opt. Lett.*, **14**, 1017–19.
Albanese, G., Kumar, J., and Steier, W. H. (1986). Investigation of the photorefractive behaviour of chrome-doped GaAs by using two-beam coupling. *Opt. Lett.*, **11**, 650–2.
Alimpiev, V. N. and Gural'nik, I. R. (1984). Space charge waves under inhomogeneous optical generation conditions. *Sov. Phys. Semicond.*, **18**, 978–80.
Alimpiev, V. N. and Gural'nik, I. R. (1986). Parametric instability in a photosensitive semiconductor due to a travelling illumination intensity grating. *Sov. Phys. Semicond.*, **20**, 512–14.
Alphonse, G. A., Alig, R. C., Staebler, D. L., and Phillips, W. (1975). Time-dependent characteristics of photo-induced space charge field and phase holograms in lithium niobate and other photorefractive media. *RCA Rev.*, **36**, 213–29.
Amodei, J. J. (1971). Electron diffusion effects during hologram recording in crystals. *Appl. Phys. Lett.*, **18**, 22–4.
Amodei, J. J. and Staebler, D. L. (1971). Holographic pattern mixing in electro-optic crystals. *Appl. Phys. Lett.*, **18**, 540–2.
Amodei, J. J. and Staebler, D. L. (1972). Holographic recording in lithium niobate. *RCA Rev.*, **33**, 71–93.
Amodei, J. J., Staebler, D. L., and Stephens, A. W. (1971). Holographic storage in doped barium sodium niobate ($Ba_2NaNb_5O_{15}$). *Appl. Phys. Lett.*, **18**, 507–9.
Amodei, J. J., Phillips, W., and Staebler, D. L. (1972). Improved electrooptic materials and fixing techniques for holographic recording. *Appl. Opt.*, **11**, 390–6.
Anastassakis, E. (1993). Photorefractive effects in cubic-crystals – explicit treatment of the piezoelectric contribution. *IEEE J. Quantum Electron.*, **29**, 2239–44.
Andersen, P. E., Petersen, P. M., and Buchhave, P. (1994). Crosstalk in dynamic optical interconnects in photorefractive crystals. *Appl. Phys. Lett.*, **65**, 271–3.
Anderson, D. Z. (1986). Coherent optical eigenstate memory. *Opt. Lett.*, **11**, 56–8.
Anderson, D. Z. and Feinberg, J. (1989). Optical novelty filters. *IEEE J. Quantum Electron.*, **25**, 635–47.
Anderson, D. Z. and Lininger, D. M. (1987). Dynamic optical interconnects – volume holograms as optical 2-port operators. *Appl. Opt.*, **26**, 5031–8.
Anderson, D. Z., Lininger, D. M., and Feinberg, J. (1987). Optical tracking novelty filter. *Opt. Lett.*, **12**, 123–5.
Anderson, D. Z., Benkert, C., Chorbajian, D., and Hermanns, A. (1991). Photorefractive flip-flop. *Opt. Lett.*, **16**, 250–2.
Apostolidis, A. G., Mallick, S., Rouède, D., Herriau, J. P., and Huignard, J. P. (1985). Polarization properties of phase gratings recorded in a $Bi_{12}SiO_{20}$ crystal. *Opt. Commun.*, **56**, 73–8.
Arizmendi, L. (1989). Thermal fixing of holographic gratings in $Bi_{12}SiO_{20}$. *J. Appl. Phys.*, **65**, 423–7.

Ashkin, A., Boyd, G. D., Dziedzic, J. M., Smith, R. G., Ballman, A. A., Levinstein, J. J., and Nassau, K. (1966). Optically-induced refractive index inhomogeneities in $LiNbO_3$, and $LiTaO_3$. *Appl. Phys. Lett.*, **9**, 72–4.

Astratov, V. N., Il'inskii, A. V., and Kiselev, V. A. (1984). Stratification of the space-charge in the case of a screening of a field in crystals. *Sov. Phys. Solid State*, **26**, 1720–5.

Attard, A. E. and Brown, T. X. (1986). Experimental observation of trapping levels in BSO. *Appl. Opt.*, **25**, 3253–9.

Au, L. B. and Solymar, L. (1988a). Space charge field in photorefractive materials at large modulation. *Opt. Lett.*, **13**, 660–2.

Au, L. B. and Solymar, L. (1988b). Higher diffraction orders in photorefractive materials. *IEEE J. Quantum Electron.*, **24**, 162–8.

Au, L. B. and Solymar, L. (1988c). Amplification in photorefractive materials via a higher order wave. *Appl. Phys. B*, **45**, 125–8.

Au, L. B. and Solymar, L. (1989). Transients in photorefractive two-wave mixing: a numerical study. *Appl. Phys. B*, **49**, 339–42.

Au, L. B. and Solymar, L. (1990). Higher harmonic gratings in photorefractive materials at large modulation with moving fringes. *J. Opt. Soc. Am. A*, **7**, 1554–61.

Au, L. B., Solymar, L., and Ringhofer, K. H. (1990). Subharmonics in BSO. In Proceedings of the Topical Conference on Photorefractive Materials, Effects and Devices II, Aussois, France, pp. 87–91.

Aubrecht, I., Ellin, H. C., Grunnet-Jepsen, A., and Solymar, L. (1995a). Space charge fields in photorefractive materials enhanced by moving fringes: comparison of electron-hole transport models. *J. Opt. Soc. Am. B* **12**, 1918–23.

Aubrecht, I., Grunnet-Jepsen, A., and Solymar, L. (1995b). Can trap density limitations be overcome in photorefractive two-beam coupling? *Opt. Commun.*, **117**, 303–5.

AuYeung, J. and Yariv, A. (1979). Phase conjugate optics. *Opt. News*, 1979, (Spring).

Avakyan, E. M., Alaverdyan, S. A., Belabaev, K. G., Sarkisov, V. Kh., and Tumanyan, K. M. (1978). Characteristics of the induced optical inhomogeneity of $LiNbO_3$, crystals doped with iron ions. *Sov. Phys. Solid State*, **20**, 1401–3.

Avakyan, E. M., Belabaev, K. G., and Odoulov, S. G. (1983). Polarization-anisotropic light-induced scattering in $LiNbO_3$, Fe crystals. *Sov. Phys. Solid State*, **25**, 1887–90.

Banerjee, P. P., Yu, H.-L., Gregory, D. A., Kukhtarev, N., and Caulfield, H. J. (1995). Self-organization of scattering in photorefractive $KNbO_3$, into a reconfigurable hexagonal spot array. *Opt. Lett.*, **20**, 10–12.

Baquedano, J. A., Contreras, L., Diéguez, E., and Cabrera, J. M. (1989). Spectral dependence on photorefractive erasure in $Bi_{12}GeO_{20}$ and $Bi_{12}SiO_{20}$. *J. Appl. Phys.*, **66**, 5146–50.

Barry, N. and Damzen, M. J. (1992). Two-beam coupling and response-time measurements in barium titanate using high-intensity laser pulses. *J. Opt. Soc. Am. B*, **9**, 1488–92.

Barry, N., Duffault, L., Troth, R., Ramos-Garcia, R., and Damzen, M. J. (1994). Comparison between continuous-wave and pulsed photorefraction in barium titanate. *J. Opt. Soc. Am. B*, **11**, 1758–63.

Bashaw, M. C., Ma, T. P., Barker, R. C., Mroczkowski, S., and Dube, R. R. (1990a). Introduction, revelation, and evolution of complementary gratings in photorefractive bismuth silicon oxide. *Phys. Rev. B*, **42**, 5641–8.

Bashaw, M. C., Ma, T.-P., Barker, R. C., Mroczkowski, S., and Dube, R. R. (1990b). Theory of complementary holograms arising from electron-hole transport in photorefractive media. *J. Opt. Soc. Am. B*, **7**, 2329–38.

Bashaw, M. C., Ma, T.-P., and Barker, R. C. (1992). Comparison of single- and two-species models of electron-hole transport in photorefractive media. *J. Opt. Soc. Am. B*, **9**, 1666–72.

Bashaw, M. C., Jeganathan, M., and Hesselink, L. (1994). Theory of two-center transport in photorefractive media for low-intensity, continuous-wave illumination in the quasi-steady-state limit. *J. Opt. Soc. Am. B*, **11**, 1743–57.

Baumert, J. C., Hoffnagle, J., and Günter, P. (1985). Nonlinear optical effects in $KNbO_3$ crytals at $Al_xGa_{1-x}As$, dye, ruby and NdYAG laser wavelengths. *Proc. Soc. Photo-opt. Instrum. Eng.*, **492**, 374–85.

Bayvel, P., McCall, M., and Wright, R. V. (1988). Continuous method for measuring the electro-optic coefficient in $Bi_{12}SiO_{20}$ and $Bi_{12}GeO_{20}$. *Opt. Lett.*, **13**, 27–9.

Beck, A. H. W. (1958). *Space charge waves and slow electromagnetic waves*. Pergamon Press, Oxford.

Belabaev, K. G., Kiseleva, I. N., Obukhovskii, V. V., Odoulov, S. G., and Taratuta, R. A. (1986). New parametric holographic-type scattering of light in lithium tantalate crystals. *Sov. Phys. Solid State*, **28**, 321–2.

Belaud, Y., Delaye, P., Launay, J.-C., and Roosen, G. (1994). Photorefractive response of CdTe-V under ac electric-field from 1 to 1.5 μm. *Opt. Commun.*, **105**, 204–8.

Bélic, M. R., Timotijevic, D., and Krolikowski, W. (1991). Multigrating phase conjugation: chaotic results. *J. Opt. Soc. Am. B*, **8**, 1723–31.

Bélic, M. R., Leonardy, J., Timotijevic, D., and Kaiser, F. (1994). Transverse effects in double phase conjugation. *Opt. Commun.*, **111**, 99–104.

Belinicher, V. I. (1978). Space-oscillating photocurrent in crystals without symmetry center. *Phys. Let. A.*, **66**, 213–14.

Belinicher, V. I. and Sturman, B. I. (1980). The photogalvanic effect in media lacking a center of symmetry. *Sov. Phys. Usp.*, **23**, 199–223.

Belinicher, V. I., Malinovsky, V. K., and Sturman, B. I. (1977). Photogalvanic effect in crystals with a polar axis. *Sov. Phys. JETP*, **73**, 692–9.

Benkert, C. and Anderson, D. Z. (1991). Controlled competitive dynamics in a photorefractive ring oscillator: "Winner-takes-all" and the "voting-paradox" dynamics. *Phys. Rev. A*, **44**, 4633–8.

Bernardo, L. M., Lopez, J. C., and Soares, O. C. (1990). Hole electron competition with fast and slow grating in $Bi_{12}SiO_{20}$ crystals. *Appl. Opt.*, **29**, 12–14.

Besson, C., Jonathan, J. M. C., Villing, A., Pauliat, G., and Roosen, G. (1989). Influence of alternating field frequency on enhanced photorefractive gain in two-beam coupling. *Opt. Lett.*, **14**, 1359–61.

Biaggio, I., Zgonik, M., and Günter, P. (1992). Photorefractive effects induced by picosecond light pulses in reduced $KNbO_3$. *J. Opt. Soc. Am. B*, **9**, 1480–7.

Bledowski, A., Otten, J., and Ringhofer, K. H. (1991). Photorefractive hologram writing with modulation 1. *Opt. Lett.*, **16**, 672–4.

Bløtekjaer, K. (1979). Limitations on holographic storage capacity of photochromic and photorefractive media. *Appl. Opt.*, **18**, 57–67.

Bløtekjaer, K. and Quate, C. F. (1964). The coupled modes of acoustic waves and drifting carriers in piezoelectric crystals. *Proc. IEEE*, **52**, 360–77.

Boggess, T. F., White, J. O., and Valley, G. C. (1990). Two-photon absorption and anisotropic transient energy transfer in $BaTiO_3$ with 1-psec excitation. *J. Opt. Soc. Am. B*, **7**, 2255–9.

Boj, S., Pauliat, G., and Roosen, G. (1992). Dynamic holographic memory showing readout, refreshing and updating capabilities. *Opt. Lett.*, **17**, 438–40.

Bosomworth, D. R. and Gerritsen, H. J. (1968). Thick holograms in photochromic materials. *Appl. Opt.*, **7**, 95–8.

Boyd, R. W. (1992). In *Nonlinear Optics*, Ch. 1. Academic Press, San Diego.

Brady, D. and Psaltis, D. (1991). Holographic interconnections in photorefractive waveguides. *Appl. Opt.*, **30**, 2324–33.

Brady, D., Hsu, K., and Psaltis, D. (1990). Periodically refreshed multiply exposed photorefractive holograms. *Opt. Lett.*, **15**, 817–19.

Breugnot, S., Defour, M., Rajbenbach, H., and Huignard, J. P. (1993). Heterodyne detection of weak optical wavefronts buried in a photorefractive fanning noise. *Opt. Commun.*, **104**, 118–22.

Breugnot, S., Dolfi, D., Rajbenbach, H., and Huignard, J. P. (1994). Enhancement of the signal-to-background ratio in photorefractive two-wave mixing by mutually incoherent two-beam coupling. *Opt. Lett.*, **19**, 1070–2.

Brignon, A. and Wagner, K. H. (1993). Polarization state evolution and eigenmode switching in photorefractive BSO. *Opt. Commun.*, **101**, 239–46.

Brignon, A., Breugnot, S., and Huignard, J. P. (1995). High-gain two-wave mixing in $BaTiO_3$ with a self-bended pump beam. *Opt. Lett.*, **20**, 1689–91.

Brost, G. A. (1992). Photorefractive grating formations at large modulation with alternating electric fields. *J. Opt. Soc. Am. B*, **9**, 1454–60.

Brost, G. A. and Motes, R. A. (1990). Photoinduced absorption in photorefractive barium titanate. *Opt. Lett.*, **15**, 538–40.

Brost, G. A., Motes, R. A., and Rotge, J. R. (1988). Intensity-dependent absorption and photorefractive effects in barium titanate. *J. Opt. Soc. Am. B*, **5**, 1879–85.

Brost, G. A., Magde, K. M., Larkin, J. J., and Harris, M. T. (1994). Modulation dependence of the photorefractive response with moving gratings: numerical analysis and experiment. *J. Opt. Soc. Am. B*, **11**, 1764–72.

Brown, W. P. and Valley, G. C. (1993). Kinky beam paths inside photorefractive crystals. *J. Opt. Soc. Am. B*, **10**, 1901–6.

Brubaker, R. M., Wang, Q. N., Nolte, D. D., Harmon, E. S., and Melloch, M. R. (1994). Steady-state four-wave mixing in photorefractive quantum wells with femtosecond pulses. *J. Opt. Soc. Am. B*, **11**, 1038–44.

Bryksin, V. V., Korovin, L. I., and Kuz'min, Yu. I. (1987). Role of injection currents in the evolution of a photoinduced charge in photorefractive crystal. *Sov. Phys. Solid State*, **29**, 757–61.

Bryskin, V. V., Dorogovtsev, S. N., and Korovin, L. I. (1993). Initialization of the holographic current constant component by recording pattern oscillation in photorefractive crystals. *Opt. Lett.*, **18**, 1760–2.

Burke, W. J. and Sheng, P. (1977). Crosstalk noise from multiple thick-phase holograms. *J. Appl. Phys.*, **48**, 681–5.

Buse, K. (1993). Thermal gratings and pyroelectrically produced charge redistribution in $BaTiO_3$ and $KNbO_3$. *J. Opt. Soc. Am. B*, **10**, 1266–75.

Buse, K., Holtmann, L., and Krätzig, E. (1991). Activation of $BaTiO_3$, for infrared holographic recording. *Opt. Commun.*, **85**, 183–6.

Buse, K., Pankrath, R., and Krätzig, E. (1994). Pyroelectrically induced photorefractive effect in $Sr_{0.61}Ba_{0.39}Nb_2O_6$:Ce. *Opt. Lett.*, **19**, 260–2.

Bylsma, R. B., Olsen, D. H., and Glass, A. M. (1988). Photochromic gratings in photorefractive materials. *Opt. Lett.*, **13**, 853–5.

Bylsma, R. B., Glass, A. M., Olson, D. H., and Cronin-Golomb, M. (1989). Self-pumped phase conjugator in InP:Fe. *Appl. Phys. Lett.*, **54**, 1968–70.

Carpenter, R. O'B. (1950). The electro-optic effect in uniaxial crystals of the dihydrogen phosphase type. III. Measurement of coefficients. *J. Opt. Soc. Am.*, **40**, 225–9.

Carrascosa, M. and Agulló-López, F. (1988). Erasure of holographic gratings in photorefractive materials with two active species. *Appl. Opt.*, **14**, 2851–7.

Carrascosa, M. and Agulló-López, F. (1990). Theoretical modeling of the fixing and developing of holographic gratings in $LiNbO_3$. *J. Opt. Soc. Am. B*, **7**, 2317–22.

Carrascosa, M., Cabrera, J. M., and Agulló-López, F. (1988). Role of photovoltaic drift on the initial writing and erasure rates of holographic gratings: some implications. *Opt. Commun.*, **69**, 83–6.

Carrascosa, M., Agullo-Rueda, F., and Agulló-López, F. (1992). Steady holographic gratings in semiconductor multiple quantum wells. *Appl. Phys. A*, **55**, 25–9.

Carson, W. J. (1974). Holographic page synthesis for sequential input of data. *Appl. Opt.*, **13**, 896–903.

Casasent, D. and Psaltis, D. (1976). Position, rotation and scale invariant optical correlation. *Appl. Opt.*, **15**, 1795–9.

Caulfield, H. J. (1982). Role of the Horner efficiency in the optimization of spatial filters for optical pattern recognition. *Appl. Opt.*, **21**, 4391–2.

Caulfield, H. J. and Weinberg, M. H. (1982). Computer recognition of 2-D patterns using generalized matched filters. *Appl. Opt.*, **21**, 1699–1704.

Caulfield, H. J., Shamir, J., and He, Q. (1987). Flexible 2-way optical interconnections in layered computers. *Appl. Opt.*, **26**, 2291–2.

Chang, C. C. and Selviah, D. R. (1995). Mutually pumped phase-conjugate mirror: fishhead configuration. *Opt. Lett.*, **20**, 677–9.

Chang, T. Y. and Hellwarth, R. W. (1985). Optical phase conjugation by backscattering in barium titanate. *Opt. Lett.*, **10**, 408–10.

Chang, T. Y. and Yeh, P. (1987). Dark rings from photorefractive conical diffraction in a $BaTiO_3$ crystal. *Proc. SPIE*, **739**, 109–16.

Chang, T. Y., Chiou, A. E. T., and Yeh, P. (1988). Cross-polarization photorefractive two-beam coupling in GaAs. *J. Opt. Soc. Am. B*, **5**, 1724–9.

Chanussot, G. and Glass, A. M. (1976). A bulk photovoltaic effect due to electron-phonon coupling in polar crystals. *Phys. Lett. A*, **59**, 405–7.

Chen, C.-T., Kim, D. M., and von der Linde, D. (1980). Efficient pulsed photorefractive process in $LiNbO_3$:Fe for optical storage and deflection. *IEEE J. Quantum Electron.* **16**, 126–9.

Chen, F. S. (1967). A laser-induced inhomogeneity of refractive indices in KTN. *J. Appl. Phys.*, **38**, 3418–20.

Chen, F. S. (1969). Optically induced change of refractive indices in $LiNbO_3$ and $LiTaO_3$. *J. Appl. Phys.*, **40**, 3389–96.

Chen, F. S., Denton, R. T., Nassau, K., and Ballman, A. A. (1968a). Optical memory planes using $LiNbO_3$ and $LiTaO_3$. *Proc. IEEE*, **56**, 782–3.

Chen, F. S., LaMacchia, J. T., and Fraser, D. B. (1968b). Holographic storage in lithium niobate. *Appl. Phys. Lett.*, **13**, 223–5.

Chen, W. H., Wang, P. J., San, P. C., and Yeh, P. (1987). Phase conjugate interferometry. *Proc. Soc. Photo-opt. Instrum. Eng.*, **739**, 105–8.

Cheng, L. J. and Yeh, P. (1988). Cross-polarization beam coupling in photorefractive GaAs crystals. *Opt. Lett.*, **13**, 50–2.

Cheng, L.-J. and Partovi, A. (1986). Temperature and intensity dependence of photorefractive effect in GaAs. *Appl. Phys. Lett.*, **49**, 1456–8.

Cheng, L. J., Liu, D. T. H., and Luke, K. L. (1991). Photorefractive image processing using compound semiconductors. *Int. J. Opt. Comput.*, **2**, 111–42.

Chiou, A. E. and Yeh, P. (1986). Parallel image subtraction using a phase-conjugate Michelson interferometer. *Opt. Lett.*, **11**, 306–8.

Chiou, A. E. and Yeh, P. (1992). 2×8 photorefractive reconfigurable interconnect with laser diodes. *Appl. Opt.*, **31**, 5536–41.

Chomsky, D., Sternklar, S., Zigler, A., and Jackel, S. (1992). Laser frequency bandwidth narrowing by photorefractive two-beam coupling. *Opt. Lett.*, **17**, 481–3.

Collicott, S. H. (1991). Numerical modeling of photorefractive response to short-pulsed illumination. *Opt. Lett.*, **16**, 1829–31.

Connors, L., Foote, P., Hall, T. J., Jaura, R., Laycock, L. C., McCall, M. W., and Petts, C. R. (1984). Fidelity of real-time correlation by four-wave mixing. *Proc. SPIE*, **492**, 361–9.

Conwell, E. M. (1958). Properties of silicon and germanium, Part II. *Phys. Rev.*, **156**, 860–

Cooper, I. R., Nicholson, M. G., and Petts, C. R. (1986). Dynamic frequency plane correlator. *IEEE Proc. J*, **133**, 70–6.

Cornish, W. D. and Young, L. (1975). Influence of multiple internal reflections and thermal expansion on the effective diffraction efficiency of holograms stored in lithium niobate. *J. Appl. Phys.*, **46**, 1252–4.

Courant, R. and Hilbert, D. (1953). *Methods of mathematical physics*, Vol. 2. Wiley, New York.

Cronin, Golomb, M. (1987). Analytic solution for photorefractive two beam coupling with time varying signal *Digest of the topical meeting on photorefractive materials, effects and devices*. Optical Society of America, Washington, DC, 142.

Cronin-Golomb, M. (1992). Whole beam method for photorefractive nonlinear optics. *Opt. Commun.*, **89**, 276–82.

Cronin-Golomb, M. and Brandle, C. D. (1989). Ring self-pumped phase conjugator using total internal reflection in photorefractive strontium barium niobate. *Opt. Lett.*, **14**, 462–4.

Cronin-Golomb, M. and Lau, K. Y. (1985). Infrared photorefractive phase conjugation with $BaTiO_3$: demonstrations with GaAlAs and 1.09-μm Ar^+ lasers. *Appl. Phys. Let.*, **47**, 567–9.

Cronin-Golomb, M. and Yariv, A. (1986). Plane-wave theory of nondegenerate oscillation in the linear photorefractive passive phase-conjugate mirror. *Opt. Lett.*, **11**, 242–4.

Cronin-Golomb, M., White, J. O., Fischer, B., and Yariv, A. (1982a). Exact solution of a nonlinear model of four-wave mixing and phase conjugation. *Opt. Lett.*, **7**, 313–15.

Cronin-Golomb, M., Fischer, B., Nilsen, J., White, J. O., and Yariv, A. (1982b). Laser with dynamic holographic intracavity distortion correction capability. *Appl. Phys. Lett.*, **41**, 219–20.

Cronin-Golomb, M., Fischer, B., White, J. O., and Yariv, A. (1982c). Passive (self-pumped) phase conjugate mirror: theoretical and experimental investigations. *Appl. Phys. Lett.*, **41**, 689–91.

Cronin-Golomb, M., Fischer, B., White, J. O., and Yariv, A. (1983). A passive conjugate mirror based on self-induced oscillation in an optical ring cavity. *Appl. Phys. Lett.*, **42**, 919–21.

Cronin-Golomb, M., Fischer, B., White, J. O., and Yariv, A. (1984). Theory and applications of four-wave mixing in photorefractive media. *IEEE J. Quantum Electron.*, **20**, 12–29.

Cronin-Golomb, M., Yariv, A., and Ury, I. (1986). Coherent coupling of diode by phase conjugation. *Appl. Phys. Lett.*, **48**, 1240–2.

Cronin-Golomb, M., Biernacki, A. M., Lin, C., and Kong, H. (1987). Photorefractive time differentiation of coherent optical images. *Opt. Lett.*, **12**, 1029–31.

Crosignani, B., Segev, M., Engin, D., Di Porto, P., Yariv, A., and Salamo, G. (1993). Self-trapping of optical beams in photorefractive media. *J. Opt. Soc. Am. B*, **10**, 446–53.

Cudney, R., Pierce, R. M., and Feinberg, J. (1988). The transient detection microscope. *Nature*, **332**, 424–6.

Cudney, R. S., Pierce, R. M., Bacher, G. D., and Feinberg, J. (1991). Absorption gratings in photorefractive crystals with multiple levels. *J. Opt. Soc. Am. B*, **8**, 1326–32.

Cutrona, L. J. et al. (1960). Optical data processing and filtering systems. *IRE Trans. Inform. Theory*, **6**, 386.

Cutrona, L. J. et al. (1966). On the application of coherent optical processing techniques to synthetic-aperture radar. *Proc. IEEE*, **54**, 1026.

D'Alessandro, G. and Firth, W. J. (1991). Spontaneous hexagon formation in a nonlinear optical medium with feedback mirror. *Phys. Rev. Lett.*, **66**, 2597–600.

Damzen, M. J. and Barry, N. (1993). Intensity-dependent hole-electron competition and photocarrier saturation in $BaTiO_3$, when using intense laser pulses. *J. Opt. Soc. Am. B*, **10**, 600–6.

Daniel, O., Stelmach, A., Jonathan, J.-M. C., and Roosen, G. (1995). Whole-beam method analysis of photorefractive effect in correlators. *Opt. Commun.*, **113**, 559–67.

D'Auria, L., Huignard, J. P., and Spitz, E. (1974). Holographic read-write memory enhancement by 3-D storage. *Appl. Opt.*, **13**, 803–7.

Davidson, F., Field, C., and Sun, X. (1992). 50 Mbps optical homodyne communications receiver with a photorefractive optical beam combiner. *IEEE Photonics Technol. Lett.*, **4**, 1295–8.

Davidson, F., Wang, C.-C., and Trivedi, S. (1994a). Electron-hole competition effects on photocurrents generated by moving space charge fields in photoconductive semiconductors with deep level traps. *Opt. Commun.*, **111**, 470–7.

Davidson, F. M., Wang, C. C., Field, C. T., and Trivedi, S. (1994b). Photocurrents in photoconductive semiconductors generated by a moving space-charge field. *Opt. Lett.*, **19**, 478–80.

Deev, V. N. and Pyatakov, P. A. (1986). Photoacoustic effect in photoconductive piezoelectrics. *Zh. Tekh. Fiz.*, **56**, 1909–15.

Deigen, M. F., Odoulov, S. G., Soskin, M. S., and Shanina, B. D. (1975). Holographic phase gratings in nonmetallic crystals. *Sov. Phys. Solid State*, **16**, 1237–41.

Delaye, P., de Montmorillon, L. A., von Bardeleben, H. J., and Roosen, G. (1994). Photorefractive wave mixing in undoped encapsulated Czochralski GaAs at $1.5\mu m$: validation of photorefractive modelling. *Appl. Phys. Lett.*, **64**, 2640–1.

Denz, C., Pauliat, G., Roosen, G., and Tschudi, T. (1991). Volume hologram multiplexing using a deterministic phase encoding method. *Opt. Commun.*, **85**, 171–6.

Denz, C., Pauliat, G., and Roosen, G. (1992). Potentialities and limitations of hologram multiplexing by using the phase-encoding technique. *Appl. Opt.*, **31**, 5700–5.

De Vré, R. and Hesselink, L. (1994). Analysis of photorefractive stratified volume holographic optical elements. *J. Opt. Soc. Am. B*, **11**, 1800–8.

De Vré, R., Jeganathan, M., Wilde, J. P., and Hesselink, L. (1994). Effect of applied fields on the Bragg condition and the diffraction efficiency in photorefractive crystals. *Opt. Lett.*, **19**, 910–12.

De Vré, R., Jeganathan, M., Wilde, J. P., and Hesselink, L. (1995). Effect of applied electric fields on the writing and the read out of photorefractive gratings. *J. Opt. Soc. Am. B*, **12**, 600–14.

Dixon, R. W. and Cohen, M. G. (1966). A new technique for measuring magnitudes of photoelastic tensors and its application to lithium niobate. *Appl. Phys. Lett.*, **8**, 205–7.

Dolfi, D., Forestier, B., Loiseaux, B., Rajbenbach, H., and Huignard, J. P. (1990). Coherent detection of optically carried microwave signals through 2-wave mixing in a $BaTiO_3$ crystal. *Appl. Opt.*, **29**, 5228–31.

Dolfi, D., Delboulbe, A., and Huignard, J. P. (1993). Forward mixing of two mutually incoherent beams in a photorefractive crystal. *Electron. Lett.*, **29**, 450–1.

Donckers, M. C. J. M., Silence, S. M., Walsh, C. A., Hache, F., Burland, D. M., Moerner, W. E., and Twieg, R. J. (1993). Net two-beam-coupling gain in a polymeric photorefractive material. *Opt. Lett.*, **18**, 1044–6.

Doogin, A. V. and Zel'dovich, B. Ya. (1993). Two-wave mixing in photorefractive crystals at asymmetric a.c. field. *Opt. Commun.*, **99**, 221–4.

Doogin, A. V., Ilinykh, P. N., Nestiorkin, O. P., and Zel'dovich, B. Ya. (1992). Phase-locked detection in photorefractive crystal at the multiple frequency difference of light beams. *Opt. Lett.*, **17**, 889–91.

Doogin, A. V., Ilinykh, P. N., Nestiorkin, O. P., Zel'dovich, B. Ya., and Shershakov, Ye. P. (1993). Recording of a stationary photorefractive grating in an external d.c. field by rapidly modulated beams. *J. Opt. Soc. Am. B*, **10**, 1060–8.

Dos Santos, P. A. M., Garcia, P. M., and Frejlich, J. (1989). Transport length, quantum efficiency, and trap density measurement in $Bi_{12}SiO_{20}$. *J. Appl. Phys.*, **68**, 247–51.

Drummond, P. D. and Friberg, A. T. (1983). Specular reflection cancellation in an interferometer with a phase-conjugate mirror. *J. Appl. Phys.*, **54**, 5618–25.

Ducharme, S. (1991). Pyroelectric-optic phase gratings. *Opt. Lett.*, **16**, 1791–3.

Ducharme, S. and Feinberg, J. (1984). Speed of the photorefractive effect in a $BaTiO_3$ single crystal. *J. Appl. Phys.*, **56**, 839–42.

Ducharme, S. and Feinberg, J. (1986). Altering the photorefractive properties of $BaTiO_3$ by reduction and oxidation at 650°C. *J. Opt. Soc. Am. B*, **3**, 283–92.

Ducharme, S., Scott, J. C., Twieg, R. J., and Moerner, W. E. (1991). Observation of the photorefractive effect in a polymer. *Phys. Rev. Lett.*, **66**, 1846–9.

Ducharme, S., Feinberg, J., and Neurgaonkar, R. R. (1987). Electrooptic and piezoelectric measurements in photorefractive barium titanate and strontium barium niobate. *IEEE J. Quantum Electron.*, **23**, 2116–21.

Dunning, G. J., Marom, E., Owechko, Y., and Soffer, B. H. (1987). All-optical associative memory with shift invariance and multiple-image recall. *Opt. Lett.*, **12**, 346–8.

Dunning, G. J., Pepper, D. M., and Klein, M. B. (1990). Control of self-pumped phase conjugate reflectivity using incoherent erasure. *Opt. Lett.*, **15**, 99–101.

Duree, G., Shultz, J. L., Salamo, G., Segev, M., Yariv, A., Crosignani, B., di Porto, P., Sharp, E. J., and Neurgaonkar, R. R. (1993). Observation of self-trapping of an optical beam due to the photorefractive effect. *Phys. Rev. Lett.*, **71**, 533–6.

Eason, R. W. and Smout, A. M. C. (1987). Bistability and noncommutative behaviour of multiple-beam self-pulsing and self-pumping in $BaTiO_3$. *Opt. Lett.*, **12**, 51–3.

Eichler, H. J., Günter, P., and Pohl, D. W. (1986). *Laser-induced dynamic gratings*. Springer-Verlag, Berlin.

Eichler, H. J., Glotz, M., Kummrov, A., Richter, K., and Yang, X. (1987) Picosecond pulse amplification by coherent wave mixing in silicon. *Phys. Rev. A*, **35**, 4673–8.

Eichler, H. J., Ding, Y., and Smandek, B. (1992). Two-wave mixing in InP:Fe at 1064 nm by linear and quadratic photorefractive effect. *Opt. Commun.*, **94**, 127–32.

El Guibaly (1983). Hologram writing in electro-optic materials with gaussian illumination. *Ferroelectrics*, **47**, 159–67.

El Guibaly, F. and Young, L. (1980). Influence of the envelope field on hologram storage in $LiNbO_3$. *Ferroelectrics*, **27**, 251–3.

El Guibaly, F. and Young, L. (1983). Optically induced light scattering and beam distortion in iron doped lithium niobate. *Ferroelectrics*, **46**, 201–8.

Ellin, H. C. (1994). Aspects of wave interactions in photorefractive materials. Unpublished D. Phil. thesis. University of Oxford.

Ellin, H. C. and Solymar, L. (1994). Scatter rings in bismuth silicate illuminated by a single beam. *Appl. Opt.*, **33**, 4125–7.

Ellin, H. C., Grunnet-Jepsen, A., Solymar, L., and Takacs, J. (1994). Effects of piezoelectricity on the photorefractive gain in a bismuth silicate crystal. *Proc. Soc. Photo-opt. Instrum. Eng.*, **2321**, 107–10.

Engin, D., Segev, M., Orlov, S., and Yariv, A. (1994). Double phase conjugation. *J. Opt. Soc. Am. B*, **11**, 1708–17.

Ennouri, A., Tapiero, M., Vola, J. P., Zielinger, J. P., Moisan, J. Y., and Launay, J. C. (1993). Determination of the mobility and transport properties of photocarriers in $Bi_{12}GeO_{20}$ by time-of-flight technique. *J. Appl. Phys.*, **74**, 2180–91.

Erbschloe, D. R. and Solymar, L. (1988a). Unidirectional ring resonator in photorefractive bismuth silicon oxide with two pump beams. *Appl. Phys. Lett.*, **53**, 1135–7.

Erbschloe, D. R. and Solymar, L. (1988b). Linear resonator in photorefractive BSO with two pump beams. *Electron. Lett.*, **24**, 683–4.

Erbschloe, D. R. and Wilson, T. (1989). A simple analytical solution for transient two-wave mixing in photorefractive materials. *Opt. Commun.*, **72**, 135–8.

Erbschloe, D. R., Solymar, L., Takacs, J., and Wilson, T. (1989). Higher diffracted orders in a BSO crystal: an experimental study of transients. *Appl. Phys. B*, **49**, 431–3.

Erdmann, A., Hesse, G., Kowarschik, R., and Wenke, L. (1987). Optimized phase conjugating mirror for interferometric arrangements. *Exper. Tech. Phys.*, **35**, 225–33.

Everard, J. K. A., Powell, K., Page-Jones, M., and Hall, T. (1992). Selfrouting optical interconnects. *Electron. Lett.*, **28**, 556–8.

Ewart, M., Biaggio, I., Zgonik, M., and Günter, P. (1994). Pulsed-photoexcitation studies in photorefractive $KNbO_3$. *Phys. Rev. B*, **49**, 5263–73.

Ewbank, M. D. (1987). Incoherent beam sharing in photorefractive holograms. In *Topical meeting on photorefractive materials, effects, and devices*, Technical Digest 17, pp. 179–82. Optical Society of America, Washington, DC.

Ewbank, M. D. (1988). Mechanism for photorefractive phase conjugation using incoherent beams. *Opt. Lett.*, **13**, 47–9.

Ewbank, M. D., Khoshnevisan, M., and Yeh, P. (1984). Phase conjugate interferometry. *SPIE Proc.*, **464**, 2–6.

Ewbank, M. D., Yeh, P., and Feinberg, J. (1986). Photorefractive conical diffraction in $BaTiO_3$. *Opt. Commun.*, **59**, 423–8.

Ewbank, M. D., Vazquez, R. A., Neurgaonkar, R. R., and Feinberg, J. (1990). Mutually pumped phase conjugation in photorefractive strontium barium niobate: theory and experiment. *J. Opt. Soc. Am. B*, **7**, 2306–16.

Fabre, J. C., Brotons, E., Halter, P. U., and Roosen, G. (1989). Photorefractive nonlinear optics in semiconductors. *Int. J. Optoelectron.*, **4**, 459–77.

Fainman, Y. and Lee, S. H. (1988). Advances in applying nonlinear optical crystals to optical signal processing. In *Signal processing Handbook*, (ed. C. H. Chen), pp. 349–77. Dekker, New York.

Fainman, Y., Klancnik, E., and Lee, S. H. (1986). Optical coherent image amplification by two-wave coupling in photorefractive $BaTiO_3$. *Opt. Eng.*, **25**, 228–34.

Faria, S. G., Tagliaferri, A. A., and dos Santos, P. A. M. (1991). Photorefractive optical holographic correlation using a $Bi_{12}TiO_{20}$ crystal at $\lambda = 0.633 \mu m$. *Opt. Commun.*, **86**, 29–33.

Feinberg, J. (1980). Real-time edge enhancement using the photorefractive effect. *Opt. Lett.*, **5**, 330–2.

Feinberg, J. (1982a). Asymmetric self-defocusing of an optical beam from the photorefractive effect. *J. Opt. Soc. Am.*, **72**, 46–51.

Feinberg, J. (1982b). Self-pumped, continuous-wave phase conjugator using internal reflection. *Opt. Lett.*, **7**, 486–8.

Feinberg, J. (1983). Interferometer with a self-pumped phase-conjugating mirror. *Opt. Lett.*, **8**, 569–71.

Feinberg, J. (1988). Photorefractive nonlinear optics. *Physics Today*, **41**(10), 46–52.

Feinberg, J. and Hellwarth, R. W. (1980). Phase conjugating mirror with continuous wave gain. *Opt. Lett.*, **5**, 519–21.
Feinberg, J. and MacDonald, K. R. (1989). Phase-conjugate mirrors and resonators with photorefractive materials. In *Photorefractive materials and their applications II*, Topics in Applied Physics 62, (ed. P. Günter and J.-P. Huignard), pp. 151–203, Springer-Verlag, Heidelberg.
Feinberg, J., Heiman, D., Tanguay, A. R. Jr., and Hellwarth, R. W. (1980). Photorefractive effects and light-induced charge migration in barium titanate. *J. Appl. Phys.*, **51**, 1297–305.
Ferrier, J.-L., Gazengel, J., Phu, X. N., and Rivoire, G. (1986). Picosecond holography and 4-wave mixing in BSO. *Opt. Commun.*, **58**, 343–8.
Firth, W. J. (1990). Spatial instabilities in a Kerr medium with single feedback mirror. *J. Mod. Opt.*, **37**, 151–3.
Fischer, B. and Sternklar, S. (1987). Self Bragg matched beam steering using the double color pumped photorefractive oscillator. *Appl. Phys. Lett.*, **51**, 74–5.
Fischer, B. and Weiss, S. (1988). Solvable optimized four-wave mixing configuration with cubic photorefractive crystals. *Appl. Phys. Lett.*, **53**, 257–9
Fischer, B., Cronin-Golomb, M., White, J. O., and Yariv, A. (1981). Amplified, reflection, transmission and oscillation in real-time holography. *Opt. Lett.*, **6**, 519–21.
Fischer, B., Sternklar, S., and Weiss, S. (1989). Photorefractive oscillators. *IEEE J. Quantum Electron.*, **25**, 550–69.
Fish, D. A., Powell, A. K., Hall, T. J., Jeffrey, P. M., and Eason, R. W. (1993). Theoretical analysis of mechanism of photorefractive enhancement of photochromic gratings in BSO. *Opt. Commun.*, **98**, 349–56.
Fisher, R. A. (ed.) (1983). *Optical phase conjugation*. Academic Press, New York.
Flannery, D. L. and Horner, J. L. (1989). Fourier optical signal processors. *Proc. IEEE*, **77**, 1511–27.
Flytzanis, C. (1975). In *Quantum electronics*, (ed. H. Rabin and C. L. Tang), p. 9. Academic Press, New York.
Földvári, I., Martin, J. J., Hunt, C. A., Powell, R. C., Reeves, R. J., and Peter, A. (1993a). Temperature dependence of the photorefractive effect in undoped $Bi_{12}GeO_{20}$. *J. Appl. Phys.*, **74**, 783–9.
Földvári, I., Taheri, B., Reeves, R. J., and Powell, R. C. (1993b). Nonlinear absorption of laser light in Bi_2TeO_5 single crystal. *Opt. Commun.*, **102**, 245–50.
Foote, P. and Hall, T. J. (1986). Influence of optical activity on two beam coupling constants in photorefractive $Bi_{12}SiO_{20}$. *Opt. Commun.*, **57**, 201–6.
Ford, J. E., Fainman, Y., and Lee, S. H. (1988). Time-integrating interferometry using photorefractive fanout. *Opt. Lett.*, **13**, 856–8.
Ford, J. E., Fainman, Y., and Lee, S. H. (1990). Array interconnection by phase-coded optical correlation. *Opt. Lett.*, **15**, 1088–90.
Ford, J. E., Fainman, Y., and Lee, S. H. (1994). Reconfigurable array interconnection by photorefractive correlation. *Appl. Opt.*, **33**, 5363–77.
Forshaw, M. R. B. (1974). Explanation of the two-ring diffraction phenomenon observed by Moran and Kaminow. *Appl. Opt.*, **13**, 2.
Franz, V. W. (1958). Influence of an electric field on an optical absorption edge. *Z. Naturforsch. A*, **13**, 484–9.
Fridkin, V. M. and Popov, B. N. (1978). *Ferroelectrics*, **21**, 611.
Fuchs, B. I., Kagan, M. S., and Suris, R. A. (1977). Capacitance oscillations in compensated germanium. *Phys. Status Solidi A*, **40**, K61–3.
Furman, A. S. (1987). Spontaneous growth of trap charging waves in uniformly illuminated noncentrosymmetric crystals. *Sov. Phys. Solid State*, **29**, 617–22.

Furman, A. S. (1988a). Photovoltaic instabilities. *Ferroelectrics*, **83**, 41–53.
Furman, A. S. (1988b). Theory of stimulated light scattering by trap charging waves. *Sov. Phys. JETP*, **67**, 1034–8.
Gabor, D. (1949). Microscopy by reconstructed wavefronts. *Proc. Roy. Soc. A*, **197**, 454–87.
Gan, X., Ye, S., and Sun, Y. (1988). Alternating electric field enhancement of the two-wave mixing gain in photorefractive BSO. *Opt. Commun.*, **66**, 155–60.
Garrett, M. H., Chang, J. Y., Jenssen, H. P., and Warde, C. (1992). High beam-coupling gain and deep- and shallow-trap effects in cobalt-doped titanate. $BaTiO_3$:Co. *J. Opt. Soc. Am. B*, **9**, 1407–15.
Gauthier, D. J., Narum, P., and Boyd, R. W. (1987). Observation of deterministic chaos in a phase-conjugate mirror. *Phys. Rev. Lett.*, **58**, 1640–3.
Gaylord, T. K., Rabson, T. A., and Tittel, F. K. (1972). Optically erasable and rewritable solid-state holograms. *Appl. Phys. Lett.*, **20**, 47–9.
Gaylord, T. K., Rabson, T. A., Tittel, F. K., and Quick, C. R. (1973). Self-enhancement of $LiNbO_3$ holograms. *J. Appl. Phys.*, **44**, 896–7.
Gheen, G. and Cheng, L. J. (1988). Optical correlators with fast updating speed using photorefractive semiconductor materials. *Appl. Opt.*, **27**, 2756–61.
Glass, A. M. (1978). The photorefractive effect. *Opt. Eng.*, **17**, 470–9.
Glass, A. M., von der Linde, D., and Negran, T. J. (1974). High-voltage bulk photovoltaic effect and the photorefractive process in $LiNbO_3$. *Appl. Phys. Lett.*, **25**, 233.
Glass, A. M., Johnson, A. M., Olson, D. H. Simpson, W., and Ballman, A. A. (1984). Four-wave mixing in semi-insulating InP and GaAs using the photorefractive effect. *Appl. Phys. Lett.*, **44**, 948–50.
Glass, A. M., Klein, M. B., and Valley, G. C. (1985). Photorefractive determination of the sign of photocarriers in InP and GaAs. *Electron. Lett.*, **21**, 220–1.
Glass, A. M., Klein, M. B., and Valley, G. C. (1987). Fundamental limit of the speed of photorefractive effect and impact on device applications and materials research: comment. *Appl. Opt.*, **26**, 3189–91.
Glass, A. M., Nolte, D. D., Olson, D. H., Doran, G. E., Chemla, D. S., and Knox, W. H. (1990). Resonant photodiffractive four-wave mixing in semi-insulating GaAs/AlGaAs quantum wells. *Opt. Lett.*, **15**, 264–6.
Goff, J. R. (1995). Polarization properties of transmission and diffraction in BSO – a unified analysis. *J. Opt. Soc. Am. B*, **12**, 99–116.
Goltz, J. and Tschudi, T. (1988). Angular selectivity of volume holograms recorded in photorefractive crystals: an analytic treatment. *Opt. Commun.*, **67**, 414–16.
Goodman, J. W. (1968). *Introduction to Fourier optics*. McGraw Hill, San Francisco.
Goulkov, M. Yu., Odoulov, S. G., and Sturman, B. I. (1993). Polarization-anisotropic scattering lines in $LiNbO_3$. *Appl. Phys. B*, **56**, 223–8.
Gower, M. C. and Hribek, P. (1988). Mechanisms for internally self-pumped phase-conjugate emission from $BaTiO_3$ crystals. *J. Opt. Soc. Am. B*, **5**, 1750–7.
Gravey, P., Picoli, G., and Labandibar, J. Y. (1989a). Stabilization of photorefractive two-wave coupling in InP:Fe under high dc fields by temperature control. *Opt. Commun.*, **70**, 190–3.
Gravey, P., Picoli, G., Ozkul, C., and Wolffer, N. (1989b). High two-wave mixing gain (11.4 cm^{-1}) in photorefractive InP:Fe by using dc field. *Proc. SPIE*, **1127**, 237–45.
Grousson, R. and Mallick, S. (1980). White-light image processing with $LiNbO_3$. *Appl. Opt.*, **19**, 1762–7.
Grousson, R., Mallick, S., and Odoulov, S. (1984a). Amplified backward scattering in $LiNbO_3$:Fe. *Opt. Commun.*, **51**, 342–6.

Grousson, R., Henry, M., and Mallick, S. (1984b). Transport properties of photoelectrons in $Bi_{12}SiO_{20}$. *J. Appl. Phys.*, **56**, 224–9.
Grunnet-Jepsen, A., Elston, S. J., Richter, I., Takacs, J., and Solymar, L. (1993). Subharmonic domains in a bismuth germanate crystal. *Opt. Lett.*, **18**, 2147–9.
Grunnet-Jepsen, A., Kwak, C. H., Richter, I., and Solymar, L. (1994a). Fundamental space-charge fields for applied alternating electric fields in photorefractive materials. *J. Opt. Soc. Am. B*, **11**, 124–31.
Grunnet-Jepsen, A., Richter, I., Shamonin, M., and Solymar, L. (1994b). Subharmonic instabilities in photorefractive crystals for an applied alternating electric field: theoretical analysis. *J. Opt. Soc. Am. B*, **11**, 132–5.
Grunnet-Jepsen, A., Kwak, C. H., and Solymar, L. (1994c). The effect of subharmonics on two-wave gain in $Bi_{12}SiO_{20}$ under an ac-electric field. *Opt. Lett.*, **19**, 1299–301.
Grunnet-Jepsen, A., Aubrecht, I., and Solymar, L. (1995). Investigation of the internal field in photorefractive materials and measurement of the effective electrooptic coefficient. *J. Opt. Soc. Am. B*, **12**, 921–9.
Grunnet-Jepsen, A., Aubrecht, I., and Solymar, L. (1995a). High frequency resonances in photorefractive materials. *Opt. Lett.*, **20**, 819–21.
Grynberg, G. (1988). Microwave oscillations of current in III–V semiconductors. *Opt. Commun.*, **66**, 321–4.
Grynberg, G., Le Bihan, E., Verkerk, P., Simoneau, P., Leite, J. R. R., Bloch, D., Le Boiteux, S., and Ducloy, M. (1988). Observation of instabilities due to mirrorless four-wave mixing oscillation in sodium. *Opt. Commun.*, **67**, 363–6.
Gu, C., Hong, J., and Campbell, S. (1992). 2-D shift-invariant volume holographic correlator. *Opt. Commun.*, **88**, 309–14.
Gunn, J. B. (1963). Microwave oscillation of current in III–V semiconductors. *Solid State Commun.*, **1**, 88.
Günter, P. (1982). Holography, coherent light amplification and optical phase conjugation with photorefractive materials. *Phys. Rep.*, **93**, 201–99.
Günter, P. (1976). Electro-optical properties of $KNbO_3$. In *Electro-optics/Laser Proceedings International '76 UK*, (ed. H. G. Jerrard), pp. 121–30. IPC Science and Technology Press, London.
Günter, P. and Huignard, J. P. (eds) (1988). *Photorefractive materials and their applications*, I, pp. 1–295. Springer, Berlin.
Günter, P. and Huignard, J. P. (eds) (1989). *Photorefractive materials and their applications*, II, pp. 1–367. Springer, Berlin.
Günter, P. and Micheron, F. (1978). Photorefractive effects and photocurrents in $KNbO_3$:Fe. *Ferroelectrics*, **18**, 27–38.
Günter, P. and Zgonik, M. (1991). Clamped-unclamped electro-optic coefficient dilemma in photorefractive phenomena. *Opt. Lett.*, **16**, 1826–8.
Günter, P., Fluckiger, U., Huignard, J. P., and Micheron, F. (1976). Optically induced refractive index changes in $KNbO_3$:Fe. *Ferroelectrics*, **13**, 297–9.
Günter, P., Voit, E., and Zha, M. Z. (1985). Self-pulsation and optical chaos in self-pumped photorefractive $BaTiO_3$. *Opt. Commun.*, **55**, 210–14.
Hagedorn, P. (1981). *Nonlinear oscillations*. Clarendon Press, Oxford.
Hall, T. J. and Powell, A. K. (1990).Phase conjugation. In *Nonlinear optics in solids*, (ed. O. Keller), pp. 235–64. Springer-Verlag, Heidelberg.
Hall, T. J., Fiddy, M. A., and Ner, M. S. (1980). Detector for an optical fiber acoustic sensor using dynamic holographic interfermetry. *Opt. Lett.*, **5**, 485–7.
Hall, T. J., Jaura, R., Connors, L. M., and Foote, P. D. (1985). The photorefractive effect – a review. *Prog. Quantum Electron.*, **10**, 77–146.

Hamel de Montchenault, G. and Huignard, J. P. (1988). Two-wave mixing with time modulated signal in $Bi_{12}SiO_{20}$: theory and application to homodyne wave-front detection. *J. Appl. Phys.*, **63**, 624–7.
Hamel de Montchenault, G., Loiseaux, B., and Huignard, J. P. (1986). Moving grating during erasure in photorefractive $Bi_{12}SiO_{20}$. *Electron. Lett.*, **22**, 1030–2.
Hamel de Montchenault, G., Loiseaux, B., and Huignard, J. P. (1987). Amplification of high bandwidth signals through two-wave mixing in photorefractive $Bi_{12}SiO_{20}$ crystals. *Appl. Phys. Lett.*, **50**, 1794–6.
He, Q. B. and Yeh, P. (1994). Fanning noise reduction in photorefractive amplifiers using incoherent erasures. *Appl. Opt.*, **33**, 283–7.
He, Q. B., Yeh, P., Gu, C., and Neurgaonkar, R. R. (1992). Multigrating competition effects in photorefractive mutually pumped phase conjugation. *J. Opt. Soc. Am. B.*, **9**, 114–20.
He, Q. B. *et al.* (1993). Shift-invariant photorefractive joint-transform correlator using Fe:$LiNbO_3$ crystal plates. *Appl. Opt.*, **32**, 3113–15.
Heanue, J. F., Bashaw, M. C., and Hesselink, L. (1994). Volume holographic storage and retrieval of data. *Science*, **265**, 749–52.
Heaton, J. M. and Solymar, L. (1985). Transient energy transfer during hologram formation in photorefractive crystals. *Optica Acta*, **32**, 397–408.
Heaton, J. M., Mills, P. A., Paige, E. G. S., Solymar, L., and Wilson, T. (1984). Diffraction efficiency and angular selectivity of volume phase holograms recorded in photorefractive materials. *Optica Acta*, **31**, 885–901.
Heer, C. V. and Griffin, N. G. (1979). Generation of a phase-conjugate wave in the forward direction with thin Na-vapour cells. *Opt. Lett.*, **4**, 239–41.
Hellwarth, R. W. (1977). Generation of time-reversed wave fronts by non-linear refraction. *J. Opt. Soc. Am.*, **67**, 1.
Henry, M., Mallick, S., Rouéde, D., Celaya, L. E., and Garcia Weidner, A. (1986). Propagation of light in an optically active electro-optic crystal of $Bi_{12}SiO_{20}$. *J. Appl. Phys.*, **67**, 2245–52.
Henshaw, P. D. and Todtenkopf, A. B. (1986). *Proc. Soc. Photo-opt. Instrum. Eng.*, **634**, 422–38.
Hermann, J. P., Herriau, J. P., and Huignard, J. P. (1981). Nanosecond four-wave mixing and holography in BSO crystals. *Appl. Opt.*, **20**, 2173–5.
Herriau, J. P. and Huignard, J. P. (1986). Hologram fixing process at room temperature in photorefractive $Bi_{12}SiO_{20}$ crystals. *Appl. Phys. Lett.*, **49**, 1140–2.
Herriau, J. P., Huignard, J. P., and Auborg, P. (1978). Some polarization properties of volume holograms in $Bi_{12}SiO_{20}$ crystals and applications. *Appl. Opt.*, **17**, 1851–2.
Herriau, J. P., Rojas, D., Huignard, J. P., Bassat, J. M., and Launay, J. C. (1987). Highly efficient diffraction in photorefractive BSO-BGO crystals at large applied fields. *Ferroelectrics*, **75**, 271–9.
Hertel, P., Ringhofer, K. H., and Sommerfeldt, R. (1987). Theory of thermal hologram fixing and application to $LiNbO_3$:Cu. *Phys. Status Solidi A*, **104**, 855–62.
Hesselink, L. and Bashaw, M. C. (1993). Optical memories implemented with photorefractive media. *Opt. Quantum Electron.*, **25**, S611–S661.
Hesselink, L. and Redfield, S. (1988). Photorefractive holographic recording in strontium barium niobate fibers. *Opt. Lett.*, **13**, 877–9.
Hester, C. F. and Casasent, D. (1980). Multivariant technique for multiclass pattern recognition. *Appl. Opt.*, **19**, 1758–61.
Hofmeister, R. and Yariv, A. (1992). Vibration detection using dynamic photorefractive gratings in KTN/KLTN crystals. *Appl. Phys. Lett.*, **61**, 2395–7.

Holtmann, L. (1989). A model for the nonlinear photoconductivity of $BaTiO_3$. *Phys. Status Solidi*, **113**, K89–K93.

Honda, T. (1991). Hexagonal pattern formation due to counterpropagation in $KNbO_3$. *Opt. Lett.*, **18**, 598–600.

Honda, T. and Matsumoto, H. (1995a). Controlling motion and size of hexagonal pattern spontaneously formed in $KNbO_3$.

Honda, T. and Matsumoto, H. (1995b). Buildup of spontaneous hexagonal patterns in photorefractive $BaTiO_3$ with a feedback mirror, *Opt. Lett.*, **20**, 1755–7.

Hong, J. (1993). Applications of photorefractive crystals for optical neural networks. *Opt. Quantum Electron.*, **25**, S551–S568.

Hong, J. H., Chiou, A. E., and Yeh, P. (1990). Image amplification by two-wave mixing in photorefractive crystals. *Appl. Opt.*, **29**, 3026–9.

Horner, J. L. (1982). Light utilization in optical correlators. *Appl. Opt.*, **21**, 4511–14.

Horner, J. L. and Gianino, P. D. (1984). Phase-only matched filtering. *Appl. Opt.*, **23**, 812–16.

Horowitz, M. and Fischer, B. (1992). Parametric scattering with constructive and destructive light patterns induced by two mutually incoherent beams in photorefractive crystals. *Opt. Lett.*, **17**, 1082–4.

Horowitz, M., Kligler, D., and Fischer, B. (1991). Time-dependent behaviour of photorefractive two- and four-wave mixing. *J. Opt. Soc. Am. B*, **8**, 2204–16.

Horowitz, M., Bekker, A., and Fischer, B. (1993). Image and hologram fixed method with $Sr_xBa_{1-x}Nb_2O_6$ crystals. *Opt. Lett.*, **18**, 1964–6.

Hou, S. L., Lauer, R. B., and Aldrich, R. E. (1973). Transport processes of photoinduced carriers in $Bi_{12}SiO_{20}$. *J. Appl. Phys.*, **44**, 2652–8.

Hu, G., Zhang, Z., Jiang, Y., and Ye, P. (1989). A new kind of anisotropic conical scattering generated via wave mixing in $BaTiO_3$. *Opt. Commun.*, **71**, 202–8.

Huignard, J. P. and Herriau, J. P. (1977). Real-time double-exposure interferometry with $Bi_{12}SiO_{20}$ crystals in transverse electro-optic configuration. *Appl. Opt.*, **16**, 1807–9.

Huignard, J. P. and Herriau, J. P. (1978). Real time coherent object edge reconstruction with $Bi_{12}SiO_{20}$ crystals. *Appl. Opt.*, **17**, 2671–2.

Huignard, J. P. and Marrakchi, A. (1981a). Coherent signal beam amplification in two-wave mixing experiments with photorefractive $Bi_{12}SiO_{20}$ crystals. *Opt. Commun.*, **38**, 249–54.

Huignard, J. P. and Marrakchi, A. (1981b). Two-wave mixing and energy transfer in $Bi_{12}SiO_{20}$ crystals: application to image amplification and vibration analysis. *Opt. Lett.*, **6**, 622–4.

Huignard, J. P. and Micheron, F. (1976). High-sensitivity read-write volume holographic storage in $Bi_{12}SiO_{20}$ and $Bi_{12}GeO_{20}$ crystals. *Appl. Phys. Lett.*, **29**, 591–3.

Huignard, J. P. and Rajbenbach, H. (1993). Noise reduction in photorefractive image amplifiers and applications. *J. Phys. III (France)*, **3**, 1357–67.

Huignard, J.-P., Herriau, J.-P., and Micheron, F. (1975a). Coherent selective erasure of superimposed volume holograms in $LiNbO_3$. *Appl. Phys. Lett.*, **26**, 256–8.

Huignard, J.-P., Herriau, J.-P., and Micheron, F. (1975b). Optical storage in $LiNbO_3$:Fe with selective erasure capability. *Rev. Phys. Appl.*, **10**, 417–23.

Huignard, J. P., Herriau, J. P., and Micheron, F. (1976a). Selective erasure and processing in volume holograms superimposed in photosensitive ferroelectrics. *Ferroelectrics*, **11**, 393–6.

Huignard, J.-P., Herriau, J.-P., and Micheron, F. (1976b). Les materiaux electro-optiques photosensibles pour le stockage holographique d'informations. *Rev. Tech. Thomson-CSF*, **8**, 671–97.

Huignard, J. P., Herriau, J. P., and Valentin, T. (1977). Time average holographic interferometry with photoconductive elctrooptic $Bi_{12}SiO_{20}$ crystals. *Appl. Opt.*, **16**, 2796–9.
Huignard, J. P., Herriau, J. P., Aubourg, P., and Spitz, E. (1979). Phase-conjugate wavefront generation via real-time holography in $Bi_{12}SiO_{20}$ crystals. *Opt. Lett.*, **4**, 21–3.
Huignard, J. P., Herriau, J. P., Pichon, L., and Marrakchi, A. (1980a). Speckle-free imaging in four-wave mixing experiments with $Bi_{12}SiO_{20}$ crystals. *Opt. Lett.*, **5**, 436–8.
Huignard, J. P., Herriau, J. P., Rivet, G., and Günter, P. (1980b). Phase-conjugation and spatial-frequency dependence of wave-front reflectivity in $Bi_{12}SiO_{20}$ crystals. *Opt. Lett.*, **5**, 102–4.
Huignard, J. P., Rajbenbach, H., Refregier, P., and Solymar, L. (1985). Wave mixing in photorefractive bismuth silicon oxide crystals and its applications. *Opt. Eng.*, **24**, 586–92.
Hussain, G. and Eason, R. W. (1991). Velocity filtering using complementary gratings in photorefractive BSO. *Opt. Commun.*, **86**, 106–12.
Hussain, G., James, S. W., and Eason, R. W. (1990). Observation and modeling of dynamic instabilities in the mutually pumped bird-wing phase conjugator in $BaTiO_3$. *J. Opt. Soc. Am. B*, **7**, 2294–8.
Hutson, A. R., McFee, J. H., and White, D. L. (1961). Ultrasonic amplification in CdS. *Phys. Rev. Lett.*, **7**, 237–9.
Hyde, S. C. W. et al. (1995). Depth-resolved holographic imaging through scattering media by photorefraction. *Opt. Lett.*, **20**, 1331–3.
Ilinykh, P. N., Nestiorkin, O. P., and Zel'dovich, B. Ya. (1989). Degenerate two-wave mixing in the photorefractive crystal $Bi_{12}TiO_{20}$. *Sov. Tech. Phys. Lett.*, **15**, 820–2.
Ilinykh, P. N., Nestiorkin, O. P., and Zel'dovich, B. Ya. (1991a). Recording a static hologram with laser beams of different frequencies in photorefractive crystals. *J. Opt. Soc. Am. B*, **8**, 1042–6.
Ilinykh, P. N., Nestiorkin, O. P., and Zel'dovich, B. Ya. (1991b). Phase shift in nondegenerate coupling of waves in the photorefractive crystal. *Opt. Commun.*, **80**, 249–52.
Ilinykh, P. N., Nestiorkin, O. P., and Zel'dovich, B. Ya. (1991c). Nondegenerate two-wave mixing in a photorefractive crystal in an external detecting field. *Opt. Lett.*, **16**, 414–16.
Ilinykh, P. N., Nestiorkin, O. P., and Zel'dovich, B. Ya. (1991d). Nondegenerate photorefractive four-wave mixing without feedback in $Bi_{12}TiO_{20}$. *Opt. Commun.*, **86**, 75–80.
Ilyenkov, A. V., Odoulov, S. G., Soskin, M. S., and Vasnetsov, M. V. (1992). Phase-matched light induced scattering, mirrorless self-oscillation, and generation of nearly retropropagating waves in $LiNbO_3$:Fe in 'forbidden' interaction geometry. *Appl. Phys.*, **B55**, 509–12.
Imbert, B. (1986). Etude de la sensibilité des cristaux BSO à $\lambda = 633$nm et $\lambda = 840$ nm. Rapport de stage, Université de Paris XI, DEA d'Optique et Photonique, pp. 1–34.
Imbert, B., Rajbenbach, H., Mallick, S., Herriau, J. P., and Huignard, J. P. (1988). High photorefractive gain in two-beam coupling with moving fringes in GaAs:Cr crystals. *Opt. Lett.*, **13**, 327–9.
Ing, R. K. and Monchalin, J.-P. (1991). Broadband optical detection of ultrasound by two-wave mixing in a photorefractive crystal. *Appl. Phys. Lett.*, **59**, 3233–5.
Ingold, M., Keller, J., Pauliat, G., and Günter, P. (1989). Optical bistability of a nematic liquid crystal with photorefractively amplified feedback. *Ferroelectrics*, **92**, 269–76.
Ingold, M., Duelli, M., Günter, P., and Schadt, M. (1992). All-optical associative memory based on a nonresonant cavity with image-bearing beams. *J. Opt. Soc. Am. B*, **9**, 1327–37.
Ishida, A., Mikami, O., Miyazawa, S., and Sumi, M. (1972). Rh-doped $LiNbO_3$ as an improved new material for reversible holographic storage. *Appl. Phys. Lett.*, **21**, 192–3.

Ito, F. and Kitayama, K. (1992). Real-time holographic storage of a temporal bit sequence by using angular multiple recording of spectral components. *Opt. Lett.*, **17**, 1152–4.
Ito, F., Kitayama, K., and Oguri, H. (1992). Holographic storage in LiNbO$_3$ fibers with compensation for intrasignal photorefractive coupling. *J. Opt. Soc. Am. B*, **9**, 1432–9.
Itoh, M., Kuroda, K., Shimura, T., and Ogura, I. (1990). Generation of phase-conjugate wave in a photorefractive GaP at 633 nm. *Jap. J. Appl. Phys. Lett.*, **29**, L1542–L1543.
Iturbe-Castillo, M. D., Marquez-Aguilar, P. A., Sanchez-Mondragon, J. J., Stepanov, S., and Vysloukh, V. (1994). Spatial solitons in photorefractive Bi$_{12}$TiO$_{20}$ with drift mechanism of nonlinearity. *Appl. Phys. Lett.*, **64**, 408–10.
Izvanov, A. A., Mandel, A. E., Khat'kov, N. D., and Shandarov, S. M. (1986). Influence of the piezoelectric effect on hologram writing and reconstruction in photorefractive crystals. *Optoelectron. Instrum. Data Process.*, (2), 79–84.
Ja, Y. H. (1980). Real-time double-exposure holographic interferometry in four-wave mixing with photorefractive Bi$_{12}$GeO$_{20}$ crystals. *Appl. Opt.*, **21**, 3230–1.
Jariego, F. and Agulló-Lopéz, F. (1991). Holographic writing and erasure in unipolar photorefractive materials with multiple active centers: theoretical analysis. *Appl. Opt.*, **33**, 4615–21.
Jaros, M. (1982). *Deep levels in semiconductors*. Hilger, Bristol.
Javidi, B. (1989). Nonlinear joint power spectrum based optical correlation. *Appl. Opt.*, **28**, 2358–67.
Jean, B., Couturier, G., and Joffre, P. (1994). Electro-optic measurements in a semiconductor: the case of CdIn$_2$Te$_4$. *J. Appl. Phys.*, **75**, 3579–85.
Jeffrey, P. M. *et al.* (1993). Mechanism of photorefractive enhancement of photochromic gratings in BSO – experimental results and phenomenological modelling. *Opt. Commun.*, **98**, 357–65.
Jeganathan, M. and Hesselink, L. (1994). Diffraction from thermally fixed gratings in a photorefractive medium: steady-state and transient analysis. *J. Opt. Soc. Am. B*, **11**, 1791–9.
Jeganathan, M., Bashaw, M. C., and Hesselink, L. (1994). Trapping the grating envelope in bulk photorefractive media. *Opt. Lett.*, **19**, 1415–17.
Jeganathan, M., Bashaw, M. C., and Hesselink, L. (1995). Evolution and propagation of grating envelopes during erasure in bulk photorefractive media. *J. Opt. Soc. Am.*, **12**, 1370–83.
Jermann, F. and Otten, J. (1993). Light-induced charge transport in LiNbO$_3$:Fe at high light intensities. *J. Opt. Soc. Am. B*, **10**, 2085–92.
Johnson, R. V. and Tanguay, A. R. (1988). Stratified volume holographic optical elements. *Opt. Lett.*, **13**, 189–91.
Johnston, A. R. (1965). The strain-free electro-optic effect in single-crystal barium titanate. *Appl. Phys. Lett.*, **7**, 195–8.
Johnston, W. D., Jr. (1970). Optical index damage in LiNbO$_3$ and other pyroelectric insulators. *J. Appl. Phys.*, **41**, 3279–85.
Jonathan, J. M. C., Roosen, G., and Roussignol, P. (1988). Time-resolved buildup of a photorefractive grating induced in Bi$_{12}$SiO$_{20}$ by picosecond light pulses. *Opt. Lett.*, **13**, 224–6.
Jones, B. E. *et al.* (1994). Photoconductivity and grating response time of a photorefractive polymer. *J. Opt. Soc. Am. B*, **11**, 1064–72.
Jones, D. C. and Solymar, L. (1989). Competition between subharmonic and resonating beams for photorefractive gain in bismuth silicon oxide. *Opt. Lett.*, **14**, 743–4.
Jones, D. C. and Solymar, L. (1991a). Transient-response of photorefractive bismuth silicon-oxide in pulsed regime. *Opt. Commun.*, **85**, 372–80.

Jones, D. C. and Solymar, L. (1991b). Comparison of 2-wave and 3-wave forward mixing in bismuth silicon oxide – theory and experiment. *IEEE J. Quantum Electron.*, **27**, 121–7.

Jones, D. C., Lyuksyutov, S. F., and Solymar, L. (1990). 3-Wave and 4-wave forward phase-conjugate imaging in photorefractive bismuth silicon-oxide. *Opt. Lett.*, **15**, 935–7.

Jones, D. C., Lyuksyutov, S. F., and Solymar, L. (1991). Competition between subharmonic and signal beams for photorefractive gain in BSO with two pump beams. *Appl. Phys. B*, **52**, 173–5.

Joseph, J., Pillai, P. K. C., and Singh, K. (1990). A novel way of noise reduction in image amplification by two-beam coupling in photorefractive $BaTiO_3$ crystal. *Opt. Commun.*, **80**, 84–8.

Joseph, J., Pillai, P. K. C., and Singh, K. (1991a). High-gain, low-noise signal beam amplification in photorefractive $BaTiO_3$. *Appl. Opt.*, **30**, 3315–18.

Joseph, J., Singh, K., and Pillai, P. K. C. (1991b). Crystal orientation dependence of the SNR for signal beam amplification in photorefractive $BaTiO_3$. *Opt. Laser Technol.*, **23**, 237–40.

Joseph, J., Singh, K., and Pillai, P. K. C. (1991c). Spatial amplification via photorefractive two-beam coupling: real time image processing using controllable erasure of Fourier spectrum. *Opt. Commun.*, **85**, 389–92.

Joseph, J., Kamra, K., Singh, K., and Pillai, P. K. C. (1992). Real-time image processing using selective erasure in photorefractive two-wave mixing. *Appl. Opt.*, **31**, 4769–72.

Kaczmarek, M., Solymar, L., and Pun, P. (1994a). Multi-colour-pumped oscillator in photorefractive $BaTiO_3$. *Opt. Commun.*, **108**, 176–84.

Kaezmarek, M., Richter, I., and Solymar, L. (1994b). Threshold conditions and intensity relations in double-color-pumped oscillations. *J. Opt. Soc. Am. B*, **11**, 136–42.

Kalinin, V. A. and Solymar, L. (1988a). Numerical solution of four-wave mixing in passive phase conjugate mirrors. *IEEE J. Quantum Electron.*, **24**, 2070–5.

Kalinin, V. and Solymar, L. (1988b). Transient effects in four-wave mixing in photorefractive passive phase conjugate mirrors. *Appl. Phys. B*, **45**, 129–45.

Kaminow, I. P. (1974). *An introduction to electrooptic devices*. Academic Press, New York.

Kamshilin, A. A. and Mokrushina, E. V. (1986). Possible use of photorefractive crystals in holographic vibrometry. *Sov. Tech. Phys. Lett.*, **12**, 149–51.

Kamshilin, A. A. and Petrov, M. P. (1980). Holographic image conversion in a $Bi_{12}SiO_{20}$ crystal. *Sov. Tech. Phys. Lett.*, **6**, 144–5.

Kandidova, V. V., Lemanov, V. V., Sukharev, B. V., and Furman, A. S. (1988). Motion of space charge waves in $LiNbO_3$:Fe. *JETP Lett.*, **46**, 552–6.

Kawata, Y. and Kawata, S. (1991). Gain enhancement by signal beam chopping for two-wave coupling with a BSO crystal. *Appl. Opt.*, **30**, 2453–7.

Kawata, Y., Kawata, S., and Minami, S. (1992). Gain dependence on external electric field in two-wave coupling with a BSO crystal. *Optik*, **90**, 27–31.

Kazarinov, R. F., Suris, R. A., and Fuks, B. I. (1972). 'Thermal current' instability in compensated semiconductors. *Sov. Phys. Semicond.*, **6**, 500–2.

Keldysh, L. V. (1958). Effect of a strong electric field on the optical properties of insulating crystals. *Sov. Phys. JETP*, **34**, 788–90.

Khizhnyak, A., Kondilenko, V., Kremenitski, V., Odoulov, S., and Soskin, M. (1979). Degenerate four-wave mixing in nonlinear media with local response: free carrier and space charge gratings, strong coupling. *Proc. SPIE*, **213**, 18–25.

Khizhnyak, A., Kondilenko, V., Kucherov, Yu., Lesnik, S., Odoulov, S., and Soskin, M. S. (1984). Phase conjugation by degenerate forward four-wave mixing. *J. Opt. Soc. Am. A*, **1**, 169–75.

Khoo, I. C. and Liu, T.-H. (1987). Probe beam amplification via 2-wave and 4-wave mixings in a nematic liquid-crystal film. *IEEE J. Quantum Electron.*, **23**, 171–3.

Khoo, I. C., Yan, P. Y., Finn, G. M., Liu, T.-H., and Michael, R. R. (1988). Lower power (10.6μm) laser-beam amplification by thermal-grating-mediated degenerate 4-wave mixing in a nematic liquid-crystal film. *J. Opt. Soc. Am. B*, **5**, 202–6.

Khoury, J. A., Hussain, G., and Eason, R. W. (1989). Optical tracking and motion detection using photorefractive $Bi_{12}SiO_{20}$. *Opt. Commun.*, **71**, 138–44.

Khoury, J., Woods, C. L., and Cronin-Golomb, M. (1991a). Photorefractive holographic interference novelty filter. *Opt. Commun.*, **82**, 533–8.

Khoury, J., Ryan, V., Woods, C. L., and Cronin-Golomb, M. (1991b). Photorefractive time correlation motion detection. *Opt. Commun.*, **85**, 5–9.

Khoury, J., Cronin-Golomb, M., Gianino, P., and Woods, C. (1994a). Photorefractive two-beam-coupling nonlinear joint-transform correlator. *J. Opt. Soc. Am. B*, **11**, 2167–74.

Khoury, J., Kane, J. S., Asimellis, G., Cronin-Golomb, M., and Woods, C. (1994b). All-optical joint Fourier transform correlator. *Appl. Opt.*, **33**, 8216–25.

Khromov, A. L., Kamshilin, A. A., and Petrov, M. P. (1990). Photochromic and photorefractive gratings induced by pulsed excitation in BSO crystals. *Opt. Commun.*, **77**, 139–43.

Kim, D. M., Shah, R. R., Rabson, T. A., and Tittel, F. K. (1976). Nonlinear dynamic theory for photorefractive phase hologram formation. *Appl. Phys. Lett.*, **28**, 338–40.

Kim, Y. P. and Hutchinson, M. R. (1989). Intensity-induced nonlinear effects in uv window materials. *Appl. Phys. B*, **49**, 469–78.

Kippelen, B., Tamura, K., Peyghambarian, N., Padias, A. B., and Hall, H. K., Jr. (1993). Photorefractivity in a functional side-chain polymer. *Phys. Rev. B*, **48**, 10 711–18.

Kiseleva, I. N., Obukhovskii, V. V., and Odoulov, S. G. (1987). Parametric scattering of the holographic type in class 3m crystals. *Sov. Phys. Solid State*, **28**, 1673–6.

Klein, M. B. (1984). Beam coupling in undoped GaAs at 1.06μm using the photorefractive effect. *Opt. Lett.*, **9**, 350–2.

Klein, M. B. and Valley, G. C. (1985). Beam couplings in BaTiO at 442 nm. *J. Appl. Phys.*, **57**, 4901–5.

Klein, M. B., Dunning, G. J., Valley, G. C., Lind, R. C., and O'Meara, T. R. (1986). Imaging threshold detector using a phase-conjugate resonator in $BaTiO_3$. *Opt. Lett.*, **11**, 575–7.

Klein, M. B., McCahon, S. W., Boggess, T. F., and Valley, G. C. (1988). High-accuracy, high-reflectivity phase conjugation at 1.06μm by four-wave mixing in photorefractive gallium arsenide. *J. Opt. Soc. Am. B*, **5**, 2467–72.

Klein, W. R. and Cook, B. D. (1967). Unified approach to ultrasonic light diffraction. *IEEE Trans. Sonics Ultrasonics*, **14**, 123–34.

Knöpfle, G., Bosshar, C., Schlesser, R., and Günter, P. (1994). Optical, nonlinear optical, and electrooptical properties of 4'-nitrobenzylidene-3-acetamino-4methaoxyaniline (MNBA) crystals. *IEEE J. Quantum Electron.*, **30**, 1303–11.

Knyaz'kov, A. V. and Lobanov, M. N. (1985). Influence of the amplitude component on the properties of dynamic holograms in ferroelectrics. *Sov. Tech. Phys. Lett.*, **11**, 365–7.

Kobialka, T., Herden, A., Tschudi, T. (1989). Spatial chaos in a phase-conjugate ring-resonator during self-oscillation. *Ferroelectrics*, **92**, 189–97.

Kogelnik, H. (1965). Holographic image projection through inhomogeneous media. *Bell Syst. Tech. J.*, **44**, 2451.

Kogelnik, H. (1969). Coupled wave theory for thick hologram gratings. *Bell Syst. Tech. J.*, **48**, 2909–47.

Kondilenko, V. P., Odoulov, S. G., Oleinik, O. I., and Soskin, M. S. (1982). Holographic technique for photovoltaic constant determination. *Ferroelectrics*, **45**, 13–18.

Korneev, N. A. and Sochava, S. L. (1995). Double phase-conjugate mirror: oscillation and amplification properties. *Opt. Commun.*, **115**, 539–44.

Korneev, N., Mansurova, S., and Stepanov, S. (1995). Non-steady-state photoelectromotive force in bipolar photoconductors with arbitrary level structure. *J. Opt. Soc. Am. B*, **12**, 615–20.

Kostyuk, B. Kh., Kudzin, A. Yu., and Sokolyanskii, G. Kh. (1980). Phototransport in $Bi_{12}SiO_{20}$ and $Bi_{12}GeO_{20}$ single crystals. *Sov. Phys. Solid State*, **22**, 1429–32.

Kovasnay, L. S. G. and Arman, A. (1957). Optical autocorrelation measurement of two-dimensional random patterns. *Rev. Sci. Instrum.*, **28**, 793.

Krainak, M. A. and Davidson, F. M. (1989). Two-wave mixing gain in $Bi_{12}SiO_{20}$ with applied alternating electric fields: self-diffraction and optical activity effects. *J. Opt. Soc. Am. B*, **6**, 634–8.

Krainak, M. A., Cohen, J. D., and Attard, A. E. (1988). Photorefractive adaptive beam combiner: use in a crossed acoustooptic cell correlator. *Appl. Opt.*, **27**, 747–51.

Krile, T. F., Hagler, M. O., Redus, W. D., and Walkup, J. F. (1979). Multiplex holography with chirp-modulated binary phase-coded reference-beam masks. *Appl. Opt.*, **18**, 52–6.

Królikowski, W., Bélic, M. R., Cronin-Golomb, M., and Bledowski, A. (1990). Chaos in photorefractive four-wave mixing with a single grating and a single interaction region. *J. Opt. Soc. Am. B*, **7**, 1204–9.

Kukhtarev, N. V. (1976). Kinetics of hologram recording and erasure in electro-optic crystals. *Sov. Tech. Phys. Lett.*, **2**, 438–40.

Kukhtarev, N. V. (1978). Effect on transient self-enhancement during volume hologram read-out. *Ukr. Fiz. Zh.*, **23**, 1947–53.

Kukhtarev, N. V. and Odoulov, S. G. (1979a). Wave front convolution in 4-wave interaction in media with nonlocal nonlinearity. *Sov. Phys. JETP Lett.*, **30**, 6–11.

Kukhtarev, N. V. and Odoulov, S. G. (1979b). Wavefront conjugation via degenerate four-wave mixing in electro-optic crystals. *Proc. SPIE*, **213**, 2–9.

Kukhtarev, N. V. and Odoulov, S. G. (1980). Wavefront inversion in anisotropic diffraction of laser beams. *Sov. Tech. Phys. Lett.*, **6**, 503–4.

Kukhtarev, N., Markov, V., and Odoulov, S. (1977). Transient energy transfer during hologram formation in $LiNbO_3$ in external electric field. *Opt. Commun.*, **23**, 338–43.

Kukhtarev, N. V., Markov, V. B., Odoulov, S. G., Soskin, M. S., and Vinetskii, V. L. (1979a). Holographic storage in electrooptic crystals. I. Steady state. *Ferroelectrics*, **22**, 949–60.

Kukhtarev, N. V., Markov, V. B., Odoulov, S. G., Soskin, M. S., and Vinetskii, V. L. (1979b). Holographic storage in electrooptic crystals. II. Beam coupling – light amplification. *Ferroelectrics*, **22**, 961–4.

Kukhtarev, N. V., Dovgalenko, G. E., and Starkov, V. N. (1984a). Influence of the optical activity on hologram formation in photorefractive crystals. *Appl. Phys. A*, **33**, 227–30.

Kukhtarev, N. V., Krätzig, E., Külich, M. C., Rupp, R. A., and Albers, J. (1984b). Anisotropic self-diffraction in $BaTiO_3$. *Appl. Phys. B*, **35**, 17–21.

Kukhtarev, N. V., Krätzig, E., Külich, H. C., and Rupp, R. A. (1985). Anisotropic self-diffraction in $BaTiO_3$. *Appl. Phys. B*, **35**, 17–21.

Kukhtarev, N. V., Semenets, T. I., and Hribek, P. (1991). The influence of photoelasticity on the self-diffraction of light in cubic photorefractive crystals. *Ferroelectrics Lett.*, **13**, 29–35.

Külich, H.-C. (1991). Reconstructing volume holograms without image field losses. *Appl. Opt.*, **30**, 2850–7.

Kulikov, V. V. and Stepanov, S. I. (1979). Mechanisms of holographic recording and thermal fixing in photorefractive LiNbO$_3$:Fe. *Appl. Phys. Lett.*, **26**, 1849–51.

Kumar, J., Albanese, G., and Steier, W. H. (1987a). Photorefractive two-beam coupling with applied radio-frequency fields: theory and experiment. *J. Opt. Soc. Am. B*, **4**, 1079–82.

Kumar, J., Albanese, G., Steier, W. H., and Ziari, M. (1987b). Enhanced two-beam mixing gain in photorefractive GaAs using alternating electric fields. *Opt. Lett.*, **12**, 120–2.

Kumar, J., Albanese, G., and Steier, W. H. (1987c). Measurement of two-wave mixing gain in GaAs with a moving grating. *Opt. Commun.*, **63**, 191–3.

Kuroda, K., Okazaki, Y., Shimura, T., Okamara, H., Chihara, M., and Itoh, M. (1990). Photorefractive effect in GaP. *Opt. Lett.*, **15**, 1197–9.

Kwak, C. H., Takacs, J., and Solymar, L. (1992). Spatial subharmonic instability in photorefractive Bi$_{12}$SiO$_{20}$ crystal. *Electron. Lett.*, **28**, 530–1.

Kwak, C. H., Takacs, J., and Solymar, L. (1993a). Spatial subharmonic instabilities. *Opt. Commun.*, **96**, 278–82.

Kwak, C. H., Shamonin, M., Takacs, J., and Solymar, L. (1993b). Spatial subharmonics in photorefractive Bi$_{12}$SiO$_{20}$ with a square wave applied field. *Appl. Phys. Lett.*, **62**, 328–30.

Kwong, N. S.-K., Tamita, Y., and Yariv, A. (1988). Optical tracking filter using transient energy coupling. *J. Opt. Soc. Am. B*, **5**, 1788–91.

Kwong, S. K. and Yariv, A. (1986). One-way, real time wave front converters. *Appl. Phys. Lett.*, **48**, 564–6.

Kwong, S. K., Cronin-Golomb, M., and Yariv, A. (1984). Optical bistability and hysteresis with a photorefractive self-pumped phase conjugate mirror. *Appl. Phys. Lett.*, **45**, 1016–18.

Kwong, S. K., Rakuljic, G. A., and Yariv, A. (1986). Real time image subtraction and "exclusive or" operation using a self-pumped phase conjugate mirror. *Appl. Phys. Lett.*, **48**, 201–3.

Laeri, F., Tschudi, T., and Albers, J. (1983). Coherent cw image amplifier and oscillator using two-wave interaction in a BaTiO$_3$ crystal. *Opt. Commun.*, **47**, 387–90.

Lam, J. F. (1983). Spectral response of nearly degenerate four-wave mixing in photorefractive materials. *Appl. Phys. Lett.*, **42**, 155–7.

Lam, L. K., Chang, T. Y., Feinberg, J., and Hellwarth, R. W. (1981). Photorefractive-index gratings formed by nanosecond optical pulses in BaTiO$_3$. *Opt. Lett.*, **6**, 475–7.

LaMacchia, J. T. and White, D. L. (1968). Coded multiple exposure holograms. *Appl. Opt.*, **7**, 91–4.

Landau, L. D. and Lifshitz, E. M. (1984). *Electrodynamics of continuous media*, (2nd edn). Pergamon Press, Oxford.

Landolt–Börnstein (1979). *Numerical data and functional relationships in science and technology*, Vol. 11. Springer-Verlag, Berlin.

Larkin, J., Harris, M., Cormier, J. E., and Armington, A. (1993). Hydrothermal growth of bismuth silicate (BSO). *J. Cryst. Growth*, **128**, 871–5.

Lee, H., Gu, X., Psaltis, D. (1989). Volume holographic interconnections with maximal capacity and minimal cross-talk. *J. Appl. Phys.*, **65**, 2191–4.

Lee, K.-Y., Kin, Y.-J., and Park, H.-K. (1993). Residue lookup table processor using an optical phase conjugation correlator. *Appl. Opt.*, **32**, 3684–9.

Lee, T. C., Rebholz, J., Tamura, P. N., and Lindquist, J. (1980). Ambiguity processing by joint Fourier transform holography. *Appl. Opt.*, **19**, 895–9.

Lee, Y. H. and Hellwarth, R. W. (1992). Spatial harmonics of photorefractive gratings in a barium titanate crystal. *J. Appl. Phys.*, **72**, 916–23.

Leger, J. R. and Lee, S. H. (1982). Hybrid optical processor for pattern recognition and classification using a generalized set of pattern functions. *Appl. Opt.*, **21**, 274–87.

Lemaire, Ph. and Georges, M. (1992). Electro-optic coefficient measurement: correction of the electric-field inhomogeneities in the transverse configuration. *Opt. Lett.*, **17**, 1411–13.

Lemeshko, V. V. and Obukhovskii, V. V. (1985). Autowaves of photoinduced light scattering. *Sov. Tech. Phys. Lett.*, **11**, 573–4.

Le Saux, G. and Brun, A. (1987). Photorefractive material response to short pulse illumination. *IEEE J. Quantum Electron.*, **23**, 1680–8.

Le Saux, G., Roosen, G., and Brun, A. (1986). Nanosecond light energy transfer in $Bi_{12}SiO_{20}$ at 532nm. *Opt. Commun.*, **58**, 238–40.

Levanyuk, A. P. and Osipov, V. V. (1977). Mechanisms for the photorefractive effect. *Bull. Acad. Sci. USSR, Phys. Ser.*, **41**(4), 83–98.

Levenson, M. D., Johnson, K. M., Hanchett, V. C., and Chiang, K. (1981). Projection photolithography by wave-front conjugation. *J. Opt. Soc. Am.*, **71**, 737–43.

Li, H.-Y. S., Qiao, Y., and Psaltis, D. (1993). Optical network for real-time pattern recognition. *Appl. Opt.*, **32**, 5026–35.

Liberman, V. S. and Zel'dovich, B. Ya. (1993). Spatio-temporal resonances in generation of subharmonics, sum and difference frequencies in photorefractive crystals. *Opt. Quantum Electron.*, **25**, 231–9.

Lin, L. H. (1968). Holographic measurements of optically induced refractive index inhomogeneities in bismuth titanate. *Proc. IEEE*, **57**, 252–3.

Lindsay, I. and Dainty, J. C. (1986). Partial cancellation of specular reflection in the presence of a phase-conjugate mirror. *Opt. Commun.*, **59**, 405–10.

Litvinenko, A. and Odoulov, S. (1984). Copper-vapor laser with self-starting $LiNbO_3$ nonlinear mirror. *Opt. Lett.*, **9**, 68–70.

Litvinov, R. and Shandarov, S. (1994). Influence of piezoelectric and photoelastic effects on pulse hologram recording in photorefractive crystals. *J. Opt. Soc. Am. B*, **11**, 1204–10.

Liu, D. T. H. and Cheng, L.-J. (1992). Real-time VanderLugt optical correlator that uses photorefractive GaAs. *Appl. Opt.*, **31**, 5675–80.

Liu, H. K. (1993). Bifurcating optical pattern recognition in photorefractive crystals. *Opt. Lett.*, **18**, 60–2.

Liu, H. M., Powell, R. C., and Boatner, L. A. (1991a). Origin of picosecond-pulse-induced degenerate 4-wave-mixing signals in $KTa_{1-x}Nb_xO_3$. *J. Appl. Phys.*, **70**, 20–8.

Liu, H. M., Powell, R. C., and Boatner, L. A. (1991b). Effect of niobium doping on the properties of picosecond laser-induced transient gratings in $KTa_{1-x}Nb_xO_3$. *Phys. Rev. B*, **44**, 2461–9.

Liu, S. R. and Indebetouw, G. (1992). Periodic and chaotic spatiotemporal states in a phase-conjugate resonator using a photorefractive $BaTiO_3$ phase-conjugate mirror. *J. Opt. Soc. Am. B*, **9**, 1507–20.

Ma, J., Liu, L., Wu, S., Wang, Z., and Xu, L. (1989). Grating-encoded multichannel photorefractive incoherent-to-coherent optical conversion. *Opt. Lett.*, **14**, 572–4.

Ma, J., Ford, J. E., Taketomi, Y., and Lee, S. H. (1991a). Moving grating for enhanced holographic recording in cerium doped $Sr_{0.6}Ba_{0.4}Nb_2O_6$. *Opt. Lett.*, **16**, 270–2.

Ma, J., Taketomi, Y., Fainman, Y., Ford, J. E., and Lee, S. H. (1991b). Moving grating and dc external field in photorefractive GaP at 633nm. *Opt. Lett.*, **16**, 1080–2.

Ma, J., Catanzaro, B., Ford, J. E., Fainman, Y., and Lee, S. H. (1994). Photorefractive holographic lenses and applications for dynamic focusing and dynamic image shifting. *J. Opt. Soc. Am. A*, **11**, 2471–80.

MacCormack, S., Feinberg, J., and Garrett, M. H. (1994). Injection locking a laser-diode array with a phase-conjugate beam. *Opt. Lett.*, **19**, 120–2.

MacDonald, K. R. and Feinberg, J. (1983). Theory of a self-pumped phase conjugator with two coupled interaction regions. *J. Opt. Soc. Am.*, **73**, 548–53.

Macdonald, R. and Eichler, H. J. (1992). Spontaneous optical pattern formation in a nematic liquid crystal feedback mirror. *Opt. Commun.*, **89**, 289–95.

Magana, L. F., Agullo-Lopez, F., and Carrascosa, M. (1994). Role of physical parameters on the photorefractive performance of semiconductor multiple quantum wells. *J. Opt. Soc. Am. B*, **11**, 1651–4.

Magnusson, R. and Gaylord, T. K. (1974). Laser scattering induced holograms in lithium niobate. *Appl. Opt.*, **13**, 1545–8.

Magnusson, R. and Gaylord, T. K. (1976). Use of dynamic theory to describe experimental results from volume holography. *J. Appl. Phys.*, **47**, 190–9.

Magnusson, R., Mitchell, J. H., III, Black, T. D., and Wilson, D. R. (1987). Holographic interferometry using iron-doped lithium niobate. *Appl. Phys. Lett.*, **51**, 81–2.

Magnusson, R., Wang, X., Black, T. D., and Tello, L. N. (1992). Diffraction efficiency dependence of holographic interferometry in Fe:LiNbO$_3$. *Appl. Opt.*, **31**, 3350–63.

Mahgerefteh, D. and Feinberg, J. (1990). Explanation of the apparent sublinear photoconductivity of barium titanate. *Phys. Rev. Lett.*, **64**, 2195–8.

Mainguet, B. (1988). Characterization of the photorefractive effect in InP:Fe by using two-wave mixing under electric fields. *Opt. Lett.*, **13**, 657–9.

Mainguet, B. (1991). Caractérisation par la technique de mélange a deux ondes, de l'effet photoréfractif dans le phosphure d'indium dope au fer. Unpublished doctoral thesis. Université de Bretagne Occidentale.

Mainguet, B., Le Guiner, F., and Picoli, G. (1990). Moving grating and intrinsic electron-hole resonance in two-wave mixing in photorefractive InP:Fe. *Opt. Lett.*, **15**, 938–40.

Mallick, S., Rouéde, D., and Apostolidis, A. G. (1987). Efficiency and polarization characteristics of photorefractive diffraction in a Bi$_{12}$SiO$_{20}$ crystal. *J. Opt. Soc. Am. B*, **4**, 1247–59.

Mallick, S., Imbert, B., Ducollet, H., Herriau, J. P., and Huignard, J. P. (1988). Generation of spatial subharmonics by two-wave mixing in a nonlinear photorefractive medium. *J. Appl. Phys.*, **63**, 5660–3.

Mandel, A. E., Shandarov, S. M., and Shepelevich, V. V. (1989). Influence of piezoelectric effect and gyrotropy on light diffraction in cubic photorefractive crystals. *Opt. Spectrosec.*, **67**, 481–4.

Maniloff, E. S. and Johnson, K. M. (1991). Maximized photorefractive holographic storage. *J. Appl. Phys.*, **70**, 4702–7.

Maniloff, E. S. and Johnson, K. M. (1992). Incremental recording for photorefractive hologram multiplexing: comment. *Opt. Lett.*, **17**, 961.

Maniloff, E. S. and Johnson, K. M. (1993). Effects of scattering on the dynamics of holographic recording and erasure in photorefractive lithium niobate. *J. Appl. Phys.*, **73**, 541–7.

Marotz, J., Ringhofer, K. H., Rupp, R. A., and Treichel, S. (1986). Light-induced scattering in photorefractive crystals. *IEEE J. Quantum Electron.*, **22**, 1376–83.

Marrakchi, A. (1988a). Photorefractive spatial light modulation based on enhanced self-diffraction in sillenite crystals. *Opt. Lett.*, **13**, 654–6.

Marrakchi, A. (1988b). Two-beam coupling photorefractive spatial light modulation with reversible contrast. *Appl. Phys. Lett.*, **53**, 634–6.

Marrakchi, A. (1989). Continuous coherent erasure of dynamic holographic interconnects in photorefractive crystals. *Opt. Lett.*, **14**, 326–8.

Marrakchi, A., Huignard, J. P., and Herriau, J. P. (1980). Application of phase conjugation in $Bi_{12}SiO_{20}$ crystals to mode pattern visualisation of diffuse vibrating structures. *Opt. Commun.*, **34**, 15–18.

Marrakchi, A., Huignard, J. P., and Günter, P. (1981). Diffraction efficiency and energy transfer in two-wave mixing experiments with $Bi_{12}SiO_{20}$. *Appl. Phys.*, **24**, 131–8.

Marrakchi, A., Tanguay, A. R., Jr, Yu, J., and Psaltis, D. (1984). Photorefractive incoherent-to-coherent optical converter: physical and materials considerations. *Proc. Soc. Photo-opt. Instrum. Eng.*, **465**, 82–96.

Marrakchi, A., Johnson, R. V., and Tanguay, A. R., Jr (1986a). Polarization properties of enhanced self-diffraction in sillenite crystals. *IEEE J. Quantum Electron.*, **23**, 2142–51.

Marrakchi, A., Johnson, R. V., and Tanguay, A. R., Jr (1986b). Polarization properties of photorefractive diffraction in electrooptic and optically active sillenite crystals (Bragg regime). *J. Opt. Soc. Am. B*, **3**, 321–36.

Marrakchi, A., Hubbard, W. M., Habiby, S. F., and Patel, J. S. (1990). Dynamic holographic interconnects with analog weights in photorefractive crystals. *Opt. Eng.*, **29**, 215–24.

Martin, J. J., Földvári, I., and Hunt, C. A. (1991). The low-temperature photochromic response of bismuth germanium oxide. *J. Appl. Phys.*, **70**, 7554–9.

Mathey, P., Pauliat, G., Launay, J. C., and Roosen, G. (1991). Overcoming the trap density limitation in photorefractive two-beam coupling by applying pulsed electric fields. *Opt. Commun.*, **82**, 101–6.

Mathieu, E. (1868). Mémoire sur le mouvement vibratoire d'une membrane de forme elliptique. *J. Math. Pures Appl.*, **13**, 137–203.

McCahon, S. W., Rytz, D., Valley, G. C., Klein, M. B., and Wechsler, B. A. (1989). Hologram fixing in $Bi_{12}TiO_{20}$ using heating and an a.c. electric field. *Appl. Opt.*, **28**, 1967–9.

McCall, M. W. and Petts, C. R. (1985). Grating modification in degenerate four wave mixing. *Opt. Commun.*, **53**, 7–12.

McClelland, T. E., Webb, D. J., Sturman, B. I., and Ringhofer, K. H. (1994). Generation of spatial subharmonic gratings in the absence of photorefractive beam coupling. *Phys. Rev. Lett.*, **73**, 3082–4.

McClelland, T. E., Webb, D. J., Sturman, B. I., Shamonina, E., and Ringhofer, K. H. (1995). Low frequency peculiarities of the photorefractive response in sillenites. *Opt. Commun.*, **113**, 371–7.

McMichael, I. and Yeh, P. (1987). Phase shifts of photorefractive gratings and phase-conjugate waves. *Opt. Lett.*, **12**, 48–50.

McRuer, R., Wilde, J., Hesselink, L., and Goodman, J. (1989). Two-wavelength photorefractive dynamic optical interconnect. *Opt. Lett.*, **14**, 1174–6.

Medrano, C., Voit, E., Amrhein, P., and Günter, P. (1988). Optimization of the photorefractive properties of $KNbO_3$ crystals. *J. Appl. Phys.*, **64**, 4668–73.

Medrano, C., Zgonik, M., Berents, S., Bernasconi, P., and Günter, P. (1994). Self-pumped and incoherent phase conjugation in Fe-doped $KNbO_3$. *J. Opt. Soc. Am. B*, **11**, 1718–26.

Meerholz, K., Volodin, B. L., Sandalphon, M., Kippelen, B., and Peyghambarian, N. (1994). A photorefractive polymer with high optical gain and diffraction efficiency near 100%. *Nature*, **371**, 497–500.

Meigs, A. D. and Saleh, B. E. A. (1994a). Spatial fidelity of photorefractive image correlators. *IEEE J. Quantum Electron.*, **30**, 3025–32.

Meigs, A. D. and Saleh, B. E. A. (1994b). Spatial and temporal fidelity of photorefractive image correlators. *J. Opt. Soc. Am. B*, **11**, 1848–57.

Merkle, F. and Lörch, T. (1984). Hybrid optical-digital pattern recognition. *Appl. Opt.*, **23**, 1509–16.
Meyer, W., Wurfel, P., Munser, R., and Muller-Vogt, G. (1979). Kinetics of fixation of phase holograms in $LiNbO_3$. *Phys. Stat. Sol.*,(a) **53**, 171–80.
Micheron, F. and Bismuth, G. (1972). Electrical control of fixation and erasure of holographic patterns in ferroelectric materials. *Appl. Phys. Lett.*, **20**, 79–81.
Micheron, F. and Bismuth, G. (1973). Field and time thresholds for the electrical fixation of holograms recorded in $(Sr_{0.75}Ba_{0.25})Nb_2O_6$ crystals. *Appl. Phys. Lett.*, **23**, 71–2.
Micheron, F., Mayeux, C., and Trotier, J. C. (1974). Electrical control in photorefractive materials for optical storage. *Appl. Opt.*, **13**, 784–7.
Millerd, J. E., Koehler, S. D., Garmire, E. M., Partovi, A., Glass, A. M., and Klein, M. B. (1990). Photorefractive gain enhancement in InP:Fe using band-edge resonance and temperature stabilization. *Appl. Phys. Lett.*, **57**, 2776–8.
Millerd, J. E., Garmire, E. M., and Klein, M. B. (1992a). Self-pumped phase conjugation in InP:Fe using band-edge resonance and temperature stabilization: theory and experiments. *Opt. Lett.*, **17**, 100–2.
Millerd, J. E., Garmire, E. M., and Klein, M. B. (1992b). Investigation of photorefractive self-pumped phase-conjugate mirrors in the presence of loss and high modulation depth. *J. Opt. Soc. Am. B*, **9**, 1499–506.
Mills, P. and Paige, E. (1985). Holographically formed, highly selective, infra-red filter in iron-doped lithium niobate. *Electron. Lett.*, **21**, 885–6.
Miridonov, S. V., Petrov, M. P., and Stepanov, S. I. (1978). Light diffraction by volume holograms in optically active photorefractive crystals. *Sov. Tech. Phys. Lett.*, **4**, 393–4.
Miteva, M. and Nikolova, L. (1982). Polarization characteristics of volume holograms in $Bi_{12}SiO_{20}$. *Opt. Commun.*, **42**, 307–9.
Miteva, M. and Nikolova, L. (1988). Oscillating behaviour of diffracted light on uniform illumination of holograms in photorefractive $Bi_{12}TiO_{20}$ crystals. *Opt. Commun.*, **67**, 192–4.
Moerner, W. E. (1994). Polymers scale new heights. *Nature*, **371**, 475–7.
Moerner, W. E. and Peyghambarian, N. (1995). Advances in photorefractive polymers: plastics for holography and optical processing. *Opt. and Phot. News*, March 1995, 24–9.
Moerner, W. E. and Silence, S. M. (1994). Polymeric photorefractive materials. *Chem. Rev.*, **94**, 127–55.
Moerner, W. E., Silence, S. M., Hache, F., and Bjorklund, G. C. (1994). Orientationally enhanced photorefractive effect in polymers. *J. Opt. Soc. Am.*, B **11**.
Moharam, M. G. and Young, L. (1976). Hologram writing by the photorefractive effect with Gaussian beams at constant applied voltage. *J. Appl. Phys.*, **47**, 4048–51.
Moharam, M. G., Gaylord, T. K., Magnusson, R., and Young, L. (1979). Holographic grating formation in photorefractive crystals with arbitrary electron transport lengths. *J. Appl. Phys.*, **50**, 5642–51.
Moisan, J.-Y., Wolffer, N., Moine, O., Gravey, P., Martel, G., Aoudia, A., Repka, E., Marfaing, Y., and Triboulet, R. (1994). Characterization of photorefractive CdTe:V: high two-wave mixing gain with an optimum low-frequency periodic external field. *J. Opt. Soc. Am. B*, **11**, 1655–67.
Mok, F. H. (1993). Angle-multiplexed storage of 5000 holograms in $LiNbO_3$. *Opt. Lett.*, **18**, 915–17.
Mok, F. H., Tackitt, M. C., and Stoll, H. M. (1991). Storage of 500 high-resolution holograms in a $LiNbO_3$ crystal. *Opt. Lett.*, **16**, 605–7.
Montemezzani, G., Zgonik, M., and Günter, P. (1993a). Photorefractive charge compensation at elevated temperatures and application to $KNbO_3$. *J. Opt. Soc. Am. B*, **10**, 171–85.

Montemezzani, G., Rogin, P., Zgonik, M., and Günter, P. (1993b). Interband photorefractive effects in $KNbO_3$ induced by ultraviolet illumination. *Opt. Lett.*, **18**, 1144–6.

Montemezzani, G., Rogin, P., Zgonik, M., and Günter, P. (1994). Interband photorefractive effects: theory and experiments in $KNbO_3$. *Phys. Rev. B*, **49**, 2484–503.

Montemezzani, G., Zozulya, A. A., Czaia, L., Anderson, D. Z., Zgonik, M. and Günter, P. (1995). Origin of the lobe structure in photorefractive beam fannings, *Phys. Rev. A*, **52**, 1791–4.

Moran, J. M. and Kaminow, I. P. (1973). Properties of holographic gratings photoinduced in polymethyl methacrylate. *Appl. Opt.*, **12**, 1964–70.

Motes, A. and Kim, J. J. (1987). Intensity-dependent absoption coefficient in photorefractive $BaTiO_3$ crystals. *J. Opt. Soc. Am. B*, **4**, 1379–81.

Mu, X., Shao, Z., Yue, X., Chen, J., Guan, Q., and Wang, J. (1995). High reflectivity self-pumped conjugation in an unusually cut Fe-doped $KTa_{1-x}Nb_xO_3$ crystal. *Appl. Phys. Lett.*, **66**, 1047–9.

Mullen, R. A. (1989). Photorefractive measurements of physical parameters. In *Photorefractive materials and their applications*, ed. P. Günter and J. P. Huignard, Ch. 6. Springer-Verlag, Berlin.

Mullen, R. A. and Hellwarth, R. W. (1985). Optical measurement of the photorefractive parameters of $Bi_{12}SiO_{20}$. *J. Appl. Phys.*, **58**, 40–4.

Müller, R., Santos, M. T., Arizmendi, L., and Cabrera, J. M. (1994a). A narrow-band interference filter with photorefractive $LiNbO_3$, *J. Phys. D.*, **27**, 241–6.

Müller, R., Alvàrez-Bravo, J.V., Arizmendi, L., and Cabrera, J. M. (1994b). Tuning of photorefractive interference filters in $LiNbO_3$, *J. Phys. D.* **27**, 1628–32.

Mullin, T. (ed.) (1994). *The nature of chaos*. Clarendon Press, Oxford.

Neifeld, M. A. and Psaltis, D. (1993). Programmable image associative memory using an optical disk and a photorefractive crystal. *Appl. Opt.*, **32**, 4398–409.

Nelson, D. F. (1979). *Electric, optic, and acoustic interactions in dielectrics*. Wiley, New York.

Nelson, K. A. and Fayer, M. D. (1980). Laser-induced phonons: a probe of intermolecular interactions in molecular solids. *J. Chem. Phys.*, **72**, 5202–18.

Nestiorkin, O. P. (1991). Instability of spatial subharmonics under hologram recording in a photorefractive crystal. *Opt. Commun.*, **81**, 315–20.

Nestiorkin, O. P. and Shershakov, Y. P. (1993). Parametric generation of a spatial subharmonic grating in photorefractive crystals: theory. *J. Opt. Soc. Am. B*, **10**, 1907–18.

Nestiorkin, O. P. and Zel'dovich, B. Ya (1992). Nonlinear theory of nondegenerate beam coupling through phase-locked detection in an a.c. field. *Opt. Lett.*, **17**, 16–18.

Nicholson, M. G., Gibbons, G. G., Laycock, L. C., and Petts, C. R. (1986). Image correlation via pulsed dynamic holography. *Electron. Lett.*, **22**, 1200–2.

Ninomiya, Y. (1973). Recording characteristics of volume holograms. *J. Opt. Soc. Am.*, **63**, 1124–30.

Nolte, D. D., Olson, D. H., and Glass, A. M. (1989). Nonequilibrium screening of the photorefractive effect. *Phys. Rev. Lett.*, **63**, 891–4.

Nolte, D. D., Olson, D. H., Doran, G. E., Knox, W. H., and Glass, A. M. (1990). Resonant photodiffractive effect in semi-insulating multiple quantum wells. *J. Opt. Soc. Am. B*, **7**, 2217–25.

Nolte, D. D., Wang, Q., and Melloch, M. R. (1991). Robust infrared gratings in photorefractive quantum wells generated by an above-bandgap laser. *Appl. Phys. Lett.*, **58**, 2067–9.

Nordin, G. P., Johnson, R. V., and Tanguay, A. R. (1992). Diffraction properties of stratified volume holographic optical elements. *J. Opt. Soc. Am. A*, **9**, 2206–17.

Notni, G., Kowarschik, R., and Rehn, H. (1992). Incoherent-to-coherent conversion via photorefractive fan-out and two-wave mixing. *J. Mod. Opt.*, **39**, 871–9.

Nouchi, P., Partanen, J. P., and Hellwarth, R. W. (1992). Temperature dependence of the elctron mobility in photorefractive $Bi_{12}SiO_{20}$. *J. Opt. Soc. Am. B*, **9**, 1428–31.

Nouchi, P., Partanen, J. P., and Hellwarth, R. W. (1993). Simple transient solutions for photoconduction and the space-charge field in a photorefractive material with shallow traps. *Phys. Rev. B*, **47**, 15 581–7.

Novikov, A. D., Obukhovskii, V. V., Odoulov, S. G., and Sturman, B. I. (1986a). "Explosive instability" and optical generation in photorefractive crystals. *JETP Lett.*, **44**, 538–42.

Novikov, A., Odoulov, S., Oleinik, O., and Sturman, B. (1986b). Beam-coupling, four-wave mixing and optical oscillation due to spatially oscillating photovoltaic currents in lithium niobate crystals. *Ferroelectrics*, **66**, 295–315.

Novikov, A., Obukhovskii, V., Odoulov, S., and Sturman, B. (1988). Mirrorless coherent oscillation due to six-beam vectorial mixing in photorefractive crystals. *Opt. Lett.*, **13**, 1017–19.

Novikov, A., Odoulov, S., Jungen, R., and Tschudi, T. (1992). Spatial subharmonic generation in $BaTiO_3$. *J. Opt. Soc. Am. B*, **9**, 1654–60.

Nye, J. F. (1957a). *Physical properties of crystals*, Ch. XIV. Oxford University Press.

Nye, J. F. (1957b). *Physical properties of crystals*. Ch. XIII. Oxford University Press.

Nye, J. F. (1985). *Physical properties of crystals*. Oxford University Press.

Obukhovskii, V. V. (1989). The nature of photoinduced light scattering in ferroelectric crystals. *Ferroelectrics*, **89**, 231–4.

Obukhovsky, V. V. and Stoyanov, A. V. (1982). A model of photogalvanic centers in crystals. *Ferroelectrics*, **43**, 137–41.

Obukhovskii, V. V. and Stoyanov, A. V. (1985). Photoinduced light scattering in crystals with a nonlocal response. *Sov. J. Quantum Electron.*, **15**, 367–71.

Obukhovskii, V. V., Stoyanov, A. V., and Lemeshko, V. V. (1987). Photoinduced scattering of light by fluctuations of photoelectric parameters of a medium. *Sov. J. Quantum Electron.*, **17**, 64–8.

Obukhovskii, V. V., Odoulov, S. G., and Karabekian, S. I. (1993). Backward conical photorefractive scattering in $LiNbO_3$. *Opt. Commun.*, **104**, 123–8.

Ochoa, E., Vachss, F., and Hesselink, L. (1986). Higher-order analysis of the photorefractive effect for large modulation depths. *J. Opt. Soc. Am. A*, **3**, 181–7.

Odintsov, V. I. and Rogacheva, L. F. (1982). Efficient phase conjugation under parametric-feedback condition. *JETP Lett.*, **36**, 344–7.

Odoulov, S. G. (1982). Spatially oscillating photovoltaic current in iron-doped lithium niobate crystals. *JETP Lett.*, **35**, 10–13.

Odoulov, S. G. (1984). Self-excitation of lasing in lithium niobate during recording of dynamic phase gratings by circular photogalvanic currents. *Sov. J. Quantum Electron.*, **14**, 360–4.

Odoulov, S. G. and Soskin, M. (1983). Lithium niobate laser with frequency-degenerate pumping. *JETP Lett.*, **37**, 289–93.

Odoulov, S. G. and Soskin, M. (1989). Amplification, oscillation and light-induced scattering in photorefractive crystals. In *Photorefractive materials and their applications*, (ed. P. Günter and J. P. Huignard), Ch. 2. Springer-Verlag, Berlin.

Odoulov, S. G. and Sturman, B. I. (1992). Parametric polarization scattering of light in photorefractive $BaTiO_3$. *Sov. Phys. JETP*, **75**, 241–9.

Odoulov, S. G. and Sukhoverkhova, L. G. (1984). Steady-state characteristics of a laser utilizing transmission gratings in crystals with a diffusion nonlinearity. *Sov. J. Quantum Electron.*, **14**, 390–3.

Odoulov, S., Belabaev, K., and Kiseleva, I. (1985). Degenerate stimulated parametric scattering in LiTaO$_3$. *Opt. Lett.*, **10**, 31–3.
Odoulov, S. G., Soskin, M. S., and Khyzhniak, A. I. (1990). *Optical oscillators with OFWM (dynamic lasers)*. Harwood Academic, London.
Odoulov, S., Sturman, B., Holtmann, L., and Kratzig, E. (1991). Nonlinear scattering in BaTiO$_3$ induced by two orthogonally polarized waves. *Appl. Phys. B*, **52**, 317–22.
Odoulov, S., Sturman, B., Holtmann, L., and Kratzig, E. (1992). Parametric conical scattering of two orthogonally polarized waves in BaTiO$_3$. *J. Opt. Soc. Am. B*, **9**, 1648–53.
Okamato, K., Sawada, T., and Ujihara, K. (1993). Transient response of a photorefractive grating produced in a BSO crystal by short light pulse. *Opt. Commun.*, **99**, 82–8.
Okamura, H., Kuroda, K., Yajima, H., Iizuka, K., and Itoh, M. (1993). Measurement of the envelope function of the index grating in photorefractive crystals. *Opt. Lett.*, **18**, 1305–7.
Oppenheim, A. V. and Lim, J. S. (1981). The importance of phase in signals. *Proc. IEEE*, **69**, 529–41.
Orlov, S., Segev, M., Yariv, A., and Neurgaonkar, R. R. (1994). Light-induced absorption in photorefractive strontium barium niobate. *Opt. Lett.*, **19**, 1293–5.
Orlowski, R. and Krätzig, E. (1978). Holographic method for the determination of photo-induced electron and hole transport in electro-optic crystal. *Solid State Commun.*, **27**, 1351–4.
Owechko, Y. (1989). Nonlinear holographic associative memories. *IEEE J. Quantum Electron.*, **25**, 619–34.
Owechko, Y. and Soffer, B. H. (1991). Optical interconnection method for neural networks using self-pumped phase-conjugate mirrors. *Opt. Lett.*, **16**, 675–7.
Owechko, Y., Dunning, G. J., Marom, E., and Soffer, B. H. (1987). Holographic associative memory with nonlinearities in the correlation domain. *Appl. Opt.*, **26**, 1900–10.
Ozkül, C., Picoli, G., Gravey, P., and Wolffer, N. (1990). High gain coherent amplification in thermally stabilized InP:Fe crystals under dc fields. *Appl. Opt.*, **18**, 2711–17.
Ozkül, C., Picoli, G., Gravey, P., and Le Rouzic, J. (1991). Energy transfer enhancement in photorefractive InP:Fe crystals using an auxiliary incoherent beam and a negative thermal gradient. *Opt. Eng.*, **30**, 397–402.
Ozkül, C., Jamet, S., Gravey, P., Turki, K., and Bremond, G. (1994). Photorefractive effect in InP:Fe dominated by holes at room temperature: influence of the indirect transitions. *J. Opt. Soc. Am. B*, **11**, 1668–73.
Pampaloni, E., Residori, S., and Arecchi, F. T. (1993). Roll-hexagon transition in a Kerr-like experiment. *Europhys. Lett.*, **24**, 647–52.
Pankove, J. I. (1971). *Optical processes in semiconductors*. Dover, New York.
Partanen, J. P., Jonathan, J. M. C., and Hellwarth, R. W. (1990). Direct determination of electron mobility in photorefractive Bi$_{12}$SiO$_{20}$ by a holographic time-of-flight technique. *Appl. Phys. Lett.*, **57**, 2404–6.
Partanen, J. P., Nouchi, P., Jonathan, J. M. C., and Hellwarth, R. W. (1991). Comparisons between holographic and transient-photocurrent measurements of electron mobility in photorefractive Bi$_{12}$SiO$_{20}$. *Phys. Rev. B*, **44**, 1487–91.
Partovi, A. and Garmire, E. M. (1991). Band-edge photorefractivity in semiconductors: theory and experiment. *J. Appl. Phys.*, **69**, 6885–98.
Partovi, A., Glass, A. M., Olson, D. H., Zydzik, G. J., Short, K. T., Feldman, R. D., and Austin, R. F. (1991). High sensitivity optical image processing device based on CdZnTe/ZnTe multiple quantum well structures. *Appl. Phys. Lett*, **59**, 1832–4.
Partovi, A., Glass, A. M., Olson, D. H., Zydzik, G. J., Short, K. T., Feldman, R. D., and Austin, R. F. (1992). High-speed photodiffractive effect in a semi-insulating CdZnTe/ZnTe multiple quantum wells. *Opt. Lett.*, **17**, 655–7.

Partovi, A., Glass, A. M., Olson, D. H., Zydzik, G. J., O'Bryan, H. M., Chiou, T. H., and Knox, W. H. (1993). Cr-doped GaAs/AlGaAs semi-insulating multiple quantum well photorefractive devices. *Appl. Phys. Lett.*, **62**, 464–6.

Partovi, A., Kost, A., Garmire, E. M., Valley, G. C., and Klein, M. B. (1990a). Band-edge photorefractive effect in semiconductors. *Appl. Phys. Lett.*, **56**, 1089–91.

Partovi, A., Millerd, J., Garmire, E. M., Ziari, M., Steier, W. H., Trivedi, S. B., and Klein, M. B. (1990b). Photorefractivity at 1.5μm in CdTe:V. *Appl. Phys. Lett.*, **57**, 846–8.

Pauliat, G. and Roosen, G. (1987). Theoretical and experimental study of diffraction in optically active and linearly birefringent sillenite crystals. *Ferroelectrics*, **75**, 281–94.

Pauliat, G. and Roosen, G. (1990). Photorefractive effect generated in sillenite crystals by picosecond pulses and comparison with the quasi-continuous regime. *J. Opt. Soc. Am. B*, **7**, 2259–68.

Pauliat, G. and Roosen, G. (1991). New advances in photorefractive holographic memories. *Int. J. Opt. Comput.*, **2**, 271–92.

Pauliat, G., Cohen-Jonathan, J. M., Allain, M., Launay, J. C., and Roosen, G. (1986a). Determination of the photorefractive parameters of $Bi_{12}GeO_{20}$ crystals using transient grating analysis. *Opt. Commun.*, **59**, 266–71.

Pauliat, G., Herriau, J. P., Delboulbé, A., Roosen, G., and Huignard, J. P. (1986b). Dynamic beam deflection using photorefractive gratings in $Bi_{12}SiO_{20}$ crystals. *J. Opt. Soc. Am. B*, **3**, 306–11.

Pauliat, G., Allain, M., Launay, J.-C., and Roosen, G. (1987). Optical evidence of a photorefractive effect due to holes in $Bi_{12}GeO_{20}$ crystals. *Opt. Commun.*, **61**, 321–4.

Pauliat, G., Besson, C., and Roosen, G. (1989). Polarization properties of two-wave mixing under an alternating electric field in BSO crystals. *IEEE J. Quantum Electron.*, **25**, 1736–40.

Pauliat, G., Villing, A., Launay, J. C., and Roosen, G. (1990). Optical measurements of charge-carrier mobilities in photorefractive sillenite crystals. *J. Opt. Soc. Am. B*, **7**, 1481–6.

Pauliat, G., Mathey, P., and Roosen, G. (1991). Influence of piezoelectricity on the photorefractive effect. *J. Opt. Soc. Am. B*, **8**, 1942–6.

Pedersen, H. C. and Johansen, P. M. (1994). Observation of angularly tilted subharmonic gratings in photorefractive BSO. *Opt. Lett.*, **19**, 1418–20.

Pedersen, H. C. and Johansen P. M. (1995). Parametric oscillation in photorefractive media. *J. Opt. Soc. Am. B*, **12**, 1065–73.

Pellat-Finet, P. (1984). Measurement of the electro-optic coefficient of BSO crystals. *Opt. Commun.*, **50**, 275–80.

Pencheva, T. G., Petrov, M. P., and Stepanov, S. I. (1982). Selective properties of volume phase holograms in photorefractive crystals. *Opt. Commun.*, **40**, 175–8.

Pender, J. and Hesselink, L. (1989). Conical emissions and phase conjugation in atomic sodium vapor. *IEEE J. Quantum Electron.*, **25**, 395–402.

Pender, J. and Hesselink, L. (1990). Degenerate conical emissions in atomic sodium vapor. *J. Opt. Soc. Am. B*, **7**, 1361–73.

Pepper, D. M. (1986). Hybrid phase conjugate modulators using self-pumped 0-degrees-cut and 45-degrees-cut $BaTiO_3$ crystals. *Appl. Phys. Lett.*, **49**, 1001–3.

Pepper, D. M., Feinberg, J., and Kukhtarev, N. V. (1990). The photorefractive effect. *Sci. Am.*, **263**(4), 34–40.

Petersen, P. M. (1991). Theory of one-grating nondegenerate four-wave mixing and its application to a linear photorefractive oscillator. *J. Opt. Soc. Am. B*, **8**, 1716–22.

Petersen, P. M. (1994). Optical phase conjugation and optical signal processing in photorefractive materials.

Petersen, P.M., Johansen, P. M., and Skettrup, T. (1991). Laser induced interference filters in photorefractive media from "Photorefractive materials, effects and devices". *Techn. Digest Series,* **14**, *Opt. Soc. Am.*, 195–8.

Petersen, P. M., Edvold, B., Buchhave, P., Andersen, P. E., and Marrakchi, A. (1992). Photorefractive particle image velocimetry: performance enhancement with bismuth silicon oxide crystals. *Opt. Lett.*, **17**, 619–21.

Petrov, M. P., Stepanov, S. I., and Kamshilin, A. A. (1979a). Holographic storage of information and peculiarities of light diffraction in birefringent electro-optic crystals. *Opt. Laser Technol.*, 149–51.

Petrov, M. P., Stepanov, S. I., and Kamshilin, A. A. (1979b). Light diffraction from the volume holograms in electrooptic birefringent crystals. *Opt. Commun.*, **29**, 44–8.

Petrov, M. P., Miridonov, S. V., Stepanov, S. I., and Kulikov, V. V. (1979c). Light diffraction and nonlinear image processing in electro-optic $Bi_{12}SiO_{20}$ crystals. *Opt. Commun.*, **31**, 301–5.

Petrov, M. P., Pencheva, T. G., and Stepanov, S. I. (1981). Light diffraction from volume phase holograms in electro-optic photorefractive crystals. *J. Opt. (Paris)*, **12**, 287–92.

Petrov, M. P., Stepanov, S. I., and Khomenko, A. V. (1983). *Photosensitive electro-optic media in holography and optical information processing.* Nauka, Leningrad [in Russian].

Petrov, M. P., Stepanov, S. I., and Trofimov, G. S. (1986). Time-varying emf in a non-uniformly illuminated photoconductor. *Sov. Tech. Phys. Lett.*, **12**, 379–81.

Petrov, M. P., Sochava, S. L., and Stepanov, S. I. (1989). Double phase-conjugation using a photorefractive $Bi_{12}SiO_{20}$ crystal. *Opt. Lett.*, **14**, 284–6.

Petrov, M. P., Sokolov, I. A., and Trofimov, G. S. (1990). Non-steady-state photoelectromotive-force induced by dynamic gratings in partially compensated photoconductors. *J. Appl. Phys.*, **68**, 2216–25.

Petrov, M. P., Stepanov, S. I., and Khomenko, A. V. (1991). *Photorefractive crystals in coherent optical systems*, Springer Series in Optical Sciences, Vol. 59. Springer-Verlag, Berlin.

Petrovic, M. S., Suchocki, A., Powell, R. C., Valley, G. C., and Cantwell, G. (1991). Picosecond two-beam coupling and polarization rotation by scalar gratings in undoped cadmium telluride at $1.064 \mu m$. *Phys. Rev. B*, **43**, 2228–33.

Phariseau, P. (1956). On the diffraction of light by progressive supersonic waves. *Proc. Indian Acad. Sci. A*, **44**, 165–70.

Phillips, W. and Staebler, D. L. (1974). Control of the $Fe2+$ concentration in iron-doped lithium niobate. *J. Electron. Mater.*, **3**, 601–17.

Phillips, W., Amodei, J. J., and Staebler, D. L. (1972). Optical and holographic storage properties of transition metal doped $LiNbO_3$. *RCA Rev.*, **33**, 94–109.

Pichon, L. and Huignard, J. P. (1981). Dynamic joint Fourier-transform correlator by Bragg diffraction in photorefractive $Bi_{12}SiO_{20}$ crystals. *Opt. Commun.*, **36**, 277–80.

Picoli, G., Gravey, P., and Ozkul, C. (1989a). Model for resonant intensity dependence of photorefractive two-wave mixing in InP:Fe. *Opt. Lett.*, **14**, 1362–4.

Picoli, G., Gravey, P., Ozkul, C., and Vieux, V. (1989b). Theory of two-wave mixing gain enhancement in photorefractive InP:Fe: a new mechanism of resonance. *J. Appl. Phys.*, **66**, 3798–813.

Pierce, R. M., Cudney, R. S., Bacher, G. D., and Feinberg, J. (1990). Measuring the photorefractive trap density without the electro-optic effect. *Opt. Lett.*, **15**, 414–16.

Pilipetski, N. F., Sukhov, A. V., Tabiryan, N. V., and Zel'dovich, B. Ya. (1981). The orientational mechanism of nonlinearity and self-focusing of He-Ne laser radiation in nematic liquid crystal mesophase (theory and experiment). *Opt. Commun.*, **37**, 280–4.

Pipes, L. A. (1965). *Operational methods in nonlinear mechanics.* Dover, New York.

Porter, A. B. (1906). On the diffraction theory of microscope vision. *Lond., Edinb. Dublin Phil. Mag. J. Sci.*, **11**, 154–66.
Press, W. H., Teutolsky, S. A., Vetterling, W. T., and Flannery, B. P. (1992). *Numerical recipes*, (2nd edn). Cambridge University Press.
Psaltis, D., Yu, J., and Hong, J. (1985). Bias-free time integrating optical correlator using a photorefractive crystal. *Appl. Opt.*, **24**, 3860–5.
Psaltis, D., Brady, D., and Wagner, K. (1988). Adaptive optical networks using photorefractive crystals. *Appl. Opt.*, **27**, 1752–9.
Osaltis, D., Neifeld, M. A., Yamamura, A., and Kobayashi, S. (1990). Optical memory disks in optical information processing. *Appl. Opt.*, **29**, 2038–57.
Psaltis, D., Mok, F., and Li, H.-Y. S. (1994). Nonvolatile storage in photorefractive crystals. *Opt. Lett.*, **19**, 210–12.
Qiao, Y. and Psaltis, D. (1992). Sampled dynamic holographic memory. *Opt. Lett.*, **17**, 1376–8.
Qiao, Y., Psaltis, D., Gu, C., Hong, J., Yeh, P., and Neurgaonkar, R. R. (1991). Phase-locked sustainment of photorefractive holograms using phase conjugation. *J. Appl. Phys.*, **70**, 4646–8.
Rabinovich, W. S. and Feldman, B. J. (1991). Photorefractive two beam coupling with white light. *Opt. Lett.*, **16**, 708–10.
Rabinovich, W. S., Feldman, B. J., and Gilbreath, C. G. (1991a). Suppression of photorefractive beam fanning using achromatic gratings. *Opt. Lett.*, **16**, 1147–9.
Rabinovich, W. S., Gilbreath, C. G., and Feldman, B. J. (1991b). Achromatic multibeam coupling in $KNbO_3$:Rb. *Opt. Commun.*, **80**, 317–??.
Rai, S. (1986). Studies in static and dynamic volume holography. Unpublished M.Sc. thesis, University of Oxford.
Rajbenbach, H. and Huignard, J. P. (1985). Self-induced coherent oscillations with photorefractive $Bi_{12}SiO_{20}$ amplifier. *Opt. Lett.*, **10**, 137–9.
Rajbenbach, H., Huignard, J. P., and Loiseaux, B. (1983). Spatial frequency dependence of the energy transfer in two-wave mixing experiments. *Opt. Commun.*, **48**, 247–52.
Rajbenbach, H., Fainman, Y., and Lee, S. H. (1987). Optical implementation of an iterative algorithm for matrix inversion. *Appl. Opt.*, **26**, 1024–31.
Rajbenbach, H., Delboulbe, A., and Huignard, J. P. (1989a). Noise suppression in photorefractive image amplifiers. *Opt. Lett.*, **14**, 1275–7.
Rajbenbach, H., Imbert, B., Huignard, J. P., and Mallick, S. (1989b). Near-infrared four-wave mixing with gain and self-starting oscillators with photorefractive GaAs. *Opt. Lett.*, **14**, 78–80.
Rajbenbach, H., Huignard, J. P., and Günter, P. (1990). Optical processing with nonlinear photorefractive crystals. In *Nonlinear photonics*, (ed. H. M. Gibbs, G. Khitrova and N. Peyghambrian). Springer-Verlag, Berlin.
Rajbenbach, H., Delboulbe, A., and Huignard, J. P. (1991). Low-noise amplification of ultraweak optical wave fronts in photorefractive $Bi_{12}SiO_{20}$. *Opt. Lett.*, **16**, 1481–3.
Rajbenbach, H., Bann, S., and Huignard, J.-P. (1992a). Long term readout of photorefractive memories by using a storage/amplification two-crystal configuration. *Opt. Lett.*, **17**, 1712–14.
Rajbenbach, H., Bann, S., Refregier, P., Joffre, P., Huignard, J. P., Buchkremer, H.-S., Jensen, A. S., Rasmussen, E., Brenner, K.-H., and Lohman, G. (1992b). Compact photorefractive correlator for robotic applications. *Appl. Opt.*, **31**, 5666–74.
Rak, D., Ledoux, I., and Huignard, J. P. (1984). Two-wave mixing and energy transfer in $BaTiO_3$: application to laser beamsteering. *Opt. Commun.*, **49**, 302–6.
Rakuljic, G. A. and Leyva, V. (1993). Volume holographic narrow band optical filter. *Opt. Lett.*, **18**, 459–61, 753.

Rakuljic, G. A., Leyva, V., and Yariv, A. (1992). Optical data storage by using orthogonal wavelength-multiplexed volume holograms. *Opt. Lett.*, **17**, 1471–3.

Raman, C. V. and Nath, N. S. N. (1935). The diffraction of light by high frequency sound waves. *Proc. Indian Acad. Sci. A*, **2**, 406–20.

Ramazza, P. L. and Zhao, M. (1993). Experimental study of two-wave mixing amplification in Cu-doped KNSBN. *Opt. Commun.*, **102**, 93–9.

Rana, R. S., Nolte, D. D., Steldt, R., and Monberg, E. M. (1992). Temperature dependence of the photorefractive effect in InP:Fe: role of multiple defects. *J. Opt. Soc. Am. B*, **9**, 1614–25.

Rana, R. S., Eunsoon, O., Chua, K., Ramdas, A. K., and Nolte, D. D. (1994). Voigt-photorefractive two-wave mixing in $Cd_{0.9}Mn_{0.1}Te$. *J. Luminesc.*, **60&61**, 56–9.

Rastani, K. and Hubbard, W. M. (1992). Large interconnects in photorefractives: grating erasure problem and a proposed solution. *Appl. Opt.*, **31**, 598–605.

Rayleigh, Lord, (1888). On maintained vibrations. *Phil. Mag.*, **15**, 229.

Raymond, C. J. (1992). Electric fields in photorefractive materials. Unpublished M.Sc. thesis. University of Oxford.

Reeves, R. J., Jani, M. J., Jassemnejad, B., Powell, R. C., Mizell, G. J., and Fay, W. (1991). Photorefractive properties of $KNbO_3$. *Phys. Rev. B*, **43**, 71–82.

Refregier, P., Solymar, L., Rajbenbach, H., and Huignard, J. P. (1984). Large signal effects in an optical BSO amplifier. *Electron. Lett.*, **20**, 656–7.

Refregier, P., Solymar, L., Rajbenbach, H., and Huignard, J. P. (1985). Two-beam coupling in photorefractive $Bi_{12}SiO_{20}$ crystals with moving grating: theory and experiments. *J. Appl. Phys.*, **58**, 45–57.

Reiner, G., Belic, M. R., and Meystre, P. (1988). Optical turbulence in phase-conjugate resonator. *J. Opt. Soc. Am. B*, **5**, 1193–210.

Reinfelde, M. J., Ozols, A. O., and Shvarts, K. K. (1984). *Izv. Akad. Nauk LatvSSR, Ser. Fiz. Tekh. Nauk*, (6), 88.

Richter, I., Grunnet-Jepsen, A., Takacs, J., and Solymar, L. (1994). An experimental and theoretical study of spatial subharmonics in photorefractive $Bi_{12}GeO_{20}$ crystal induced by DC-field and moving grating technique. *IEEE J. Quantum Electron.*, **30**, 1645–50.

Ridley, B. K. and Watkins, T. B. (1961). The possibility of negative resistance in semiconductors. *Proc. Phys. Soc.*, **78**, 293–304.

Ringhofer, K. H. and Solymar, L. (1988a). New gain mechanism for wave amplification in photorefractive materials. *Appl. Phys. Lett*, **53**, 1039–40.

Ringhofer, K. H. and Solymar, L. (1988b). 3-wave and 4-wave forward mixing in photorefractive materials. *Appl. Phys.*, **48**, 395–400.

Ringhofer, K. H., Tao, S., Takacs, J., and Solymar, L. (1991). The role of the longitudinal component of the electric field vector in two-wave mixing in photorefractive $BaTiO_3$. *Appl. Phys. B*, **52**, 259–61.

Rittner, E. S. (1956). Electron processes in photoconductors. In *Photoconductivity Conference*, (ed. R. G. Breckenbridge, B. R. Russell and E. E. Hahn) pp. 215–68. Wiley.

Roblin, M. L., Gires, F., Grousson, R., and Lavallard, P. (1987). Enregistrement par holographie de volume d'une loi de phase spectrale: application à la compression d'impulsion picoseconde. *Opt. Commun.*, **62**, 209–14.

Ross, G. W. and Eason, R. W. (1992). Highly efficient self-pumped phase conjugation at near-infrared wavelengths by using nominally undoped $BaTiO_3$. *Opt. Lett.*, **17**, 1104–6.

Ross, G. W., Hribek, P., Eason, R. W., Garrett, M. H., and Rytz, D. (1993). Impurity enhanced self-pumped phase conjugating in the near infrared in 'blue' $BaTiO_3$. *Opt. Commun.*, **101**, 60–4.

Rossomakhin, I. and Stepanov, S. (1991). Linear adaptive interferometers via diffusion recording in cubic photorefractive crystals. *Opt. Commun.*, **86**, 199–204.

Rouède, D., Kukhtarev, N., Khitrova, G., Wang, L., and Gibbs, H. M. (1989). Photorefractive energy exchange requiring optical activity and an electric field. *Opt. Lett.*, **14**, 740–2.

Roy, A. and Singh, K. (1990a). Combined effects of cross and parallel coupling in two-wave mixing in photorefractive crystals: reflection geometry. *Opt. Commun.*, **75**, 51–6.

Roy, A. and Singh, K. (1990b). Combined effects of cross and parallel coupling in two-wave mixing in photorefractive crystals: transmission geometry. *Optik*, **84**, 47–50.

Rupp, R. A. and Drees, F. W. (1986). Light-induced scattering in photorefractive crystals. *Appl. Phys. B*, **39**, 223–9.

Rupp, R. A., Marotz, J., Ringhofer, K. H., Treichel, S., Feng, S., and Krätzig, E. (1987). Four-wave interaction phenomena contributing to holographic scattering in $LiNbO_3$ and $LiTaO_3$. *IEEE J. Quantum Electron.*, **23**, 2136–41.

Russell, J. S. (1844). Report on waves. In *Rep. 14th Meeting of the Brit. Assoc. Adv. Sci.*, pp. 311–90.

Rustamov, F. A. (1993). Two-level model of recording and reading of holographic gratings in photorefractive crystals: nonstationary case. *Opt. Quantum Electron.*, **25**, 351–8.

Rytz, D. and Shen, D. Z. (1989). Self-pumped phase conjugator in potassium niobate ($KNbO_3$). *Appl. Phys. Lett.*, **54**, 2625–7.

Rytz, D., Wechsler, B. A., Garrett, M. H., Nelson, C. C., and Schwartz, R. N. (1990). Photorefractive properties of $BaTiO_3$:Co. *J. Opt. Soc. Am. B*, **7**, 2245–54.

Saffman, M., Benkert, C., and Anderson, D. Z. (1991). Self-organizing photorefractive frequency demultiplexer. *Opt. Lett.*, **16**, 1993–5.

Salamo, G., Miller, M. J., Clark, W. W., Wood, G. L., and Sharp, E. J. (1986). Strontium barium niobate as a self-pumped phase-conjugator. *Opt. Commun.*, **59**, 417–22.

Salamo, G. J., Miller, M. J., Clark, W. W., III, Wood, G. L., Sharp, E. J., and Neurgaonkar, R. (1988). Photorefractive rainbows. *Appl. Opt.*, **27**, 4356–8.

Sanchez, F., Kayoun, P. H., and Huignard, J. P. (1988). Two-wave mixing with gain in liquid crystals at $10.6\mu m$ wavelength. *J. Appl. Phys.*, **64**, 26–31.

Sasaki, H., Ma, J., Fainman, Y., Ford, J. E., Taketomi, Y., and Lee, S. H. (1991). Dynamic photorefractive optical memory. *Opt. Lett.*, **16**, 1874–6.

Sasaki, H., Fainman, Y., Ford, J. E., Lee, S. H., and Taketomi, Y. (1992). Fast update of dynamic photorefractive optical memory. *Opt. Lett.*, **17**, 1468–70.

Sasaki, H., Fainman, Y., and Lee, S. H. (1993). Gray-scale fidelity in volume-multiplexed photorefractive memory. *Opt. Lett.*, **18**, 1358–60.

Saxena, R. and Chang, T. Y. (1992). Perturbative analysis of higher-order photorefractive gratings. *J. Opt. Soc. Am. B*, **9**, 1467–72.

Saxena, R., Gu, C., and Yeh, P. (1991). Properties of photorefractive gratings with complex coupling constants. *J. Opt. Soc. Am. B*, **8**, 1047–52.

Sayano, K., Rakuljic, G. A., and Yariv, A. (1988). Thresholding semilinear phase conjugate mirror. *Opt. Lett.*, **13**, 143–5.

Schroeder, W. A., Stark, T. S., Dawson, M. D., Boggess, T. F., Smirl, A. L., and Valley, G. C. (1991a). Picosecond separation and measurement of coexisting photorefractive, bound-electronic, and free-carrier grating dynamics in GaAs. *Opt. Lett.*, **16**, 159–61.

Schroeder, W. A., Stark, T. S., Smirl, A. L., and Valley, G. C. (1991b). Picosecond enhancement of photorefractive beam coupling in CdTe:V at 960 nm. *Opt. Commun.*, **84**, 369–73.

Schroeder, W. A., Stark, T. S., Boggess, T. F., Smirl, A. L., and Valley, G. C. (1991c). Photorefractive nonlinearities caused by Dember space-charge field in undoped CdTe. *Opt. Lett.*, **16**, 799–801.

Schroeder, W. A., Stark, T. S., and Smirl, A. L. (1991d). Hot-carrier enhancement of photorefractive space charge fields in zinc-blende semiconductors. *Opt. Lett.*, **16**, 989–91.

Schuster, H. G. (1988). *Deterministic chaos.* Physikverlag, Weinheim.

Segev, M., Crosignani, B., Yariv, A., and Fischer, B. (1992). Spatial solitons in photorefractive media. *Phys. Rev. Lett.*, **68**, 923–6.

Segev, M., Yariv, A., Salamo, G., Duree, G., Shultz, J., Crosignani, B., Porto, P., and Sharp, E. (1993). Photorefractive spatial solitons. *Opt. Photon. News*, **4**(Dec), 9.

Segev, M., Crosignani, B., di Porto, P., Yariv, A., Duree, G., Salamo, G., and Sharp, E. (1994a). Stability of photorefractive spatial solitons. *Opt. Lett.*, **19**, 1296–8.

Segev, M., Valley, G. C., Crosignani, B., di Porto, P., and Yariv, A. (1994b). Steady-state spatial screening solitons in photorefractive materials with external applied field. *Phys. Rev. Lett.*, **73**, 3211–14.

Serrano, E., Lopez, V., Carrascosa, M., and Agullo-Lopez, F. (1994a). Recording and erasure of kinetics in photorefractive materials at large modulation depths. *J. Opt. Soc. Am. B*, **11**, 670–75.

Serrano, E., Lopez, V., Carrascosa, M., and Agullo-Lopez, F. (1994b). Steady-state photorefractive gratings in $LiNbO_3$ for strong light modulation depths. *IEEE J. Quantum Electron.*, **30**, 875–80.

Serrano, E., Carrascosa, M., Agullo-Lopez, F., and Solymar, L. (1994c). Subharmonic instability taking into account higher harmonics. *Appl. Phys. Lett.*, **64**, 658–60.

Shamonin, M. (1993a). Theory of spatial subharmonic instability under hologram recording in a photorefractive crystal with applied ac field. *Appl. Phys. A*, **57**, 153–6.

Shamonin, M. (1993b). Soliton-shaped nonlinear waves of space charge in photorefractive materials. *Appl. Phys. A*, **56**, 467–8.

Shandarov, S. (1992). The influence of piezoelectric effect on photorefractive gratings in electro-optic crystals. *Appl. Phys. A*, **55**, 91–6.

Shandarov, S. M., Shepelevich, V. V., and Khatkov, N. D. (1991). Variation of the permittivity tensor in cubic photorefractive piezoelectric crystals under the influence of the electric field of a holographic grating. *Opt. Spectrosc.*, **70**, 627–30.

Sharp, E. J. *et al.* (1990). Double phase conjugation in tungsten bronze crystals. *Appl. Opt.*, **29**, 743–49.

Sharp, E. J., Wood, G. I., Clark, W. W., III, Salamo, G. J., and Neurgaonkar, R. R. (1992). Incoherent-to-coherent conversion using a photorefractive self-pumped phase conjugator. *Opt. Lett.*, **17**, 207–9.

Sheng, Z.-M., Cui, Y., Wang, F., and Wei, F. (1995). Spatial subharmonic instabilities in photorefractive crystals. *Opt. Commun.*, **115**, 545–50.

Shepelevich, V. V. and Khramovich, E. M. (1991). Simultaneous diffraction of two light waves on holographic gratings in cubic gyrotropic photorefractive crystals. *Opt. Spectrosc.*, **70**, 618–21.

Shepelevich, V. V., Shandarov, S. M., and Mandel, E. A. (1990). Ligh diffraction by holographic gratings in optically active photorefractive piezocrystals. *Ferroelectrics*, **110**, 235–49.

Shepelevich, V. V., Egorov, N. N., and Shepelevich, V. (1994). Orientation and polarization effects of two-beam coupling in a cubic optically active photorefractive piezoelectric BSO crystal. *J. Opt. Soc. Am. B*, **11**, 1394–402.

Sheppard, C. J. R. (1976). The application of the dynamical theory of X-ray diffraction to holography. *Int. J. Electron.*, **41**, 365–73.

Sheridan, J. T. P. W. (1990). Diffraction by volume gratings. Unpublished D. Phil. thesis. University of Oxford.

Shershakov, Ye. P. and Nestiorkin, O. P. (1993). Nondegenerate spatial subharmonic grating generation in photorefractive crystal. *Opt. Commun.*, **96**, 271–7.

Shi, Y., Psaltis, D., Marrakchi, A., and Tanguay, A. R., Jr (1983). Photorefractive incoherent-to-coherent optical converter. *Appl. Opt.*, **22**, 3665–7.

Shih, M., Segev, M., Valley, G. C., Salamo, G., Crosignani, B., and di Porto, P. (1995). Observation of two-dimensional steady-state photorefractive screening solitons. *Electron. Lett.* (to be published).

Shkunov, V. V. and Zel'dovich, B. Ya. (1985). Optical phase conjugation. *Sci. Am.*, **253**(6), 400–5.

Shkunov, V. V. and Zolotarev, M. V. (1995). Theory of the photorefractive effect for multiple-quantum-well structures in perpendicular field reflection grating geometry. *J. Opt. Soc. Am. B*, **12**, 913–20.

Shvarts, K., Ozols, A., Augustov, P., and Reinfelde, M. (1987). Photorefraction and self-enhancement of holograms in $LiNbO_3$ and $LiTaO_3$ crystals. *Ferroelectrics*, **75**, 231–49.

Silence, S. M., Walsh, C. A., Scott, J. C., Matray, T. J., Twieg, R. J., Hache, F., Bjorklund, G. C., and Moerner, W. E. (1992a). Subsecond grating growth in a photorefractive polymer. *Opt. Lett.*, **17**, 1107–9.

Silence, S. M., Walsh, C. A., Scott, J. C., and Moerner, W. E. (1992b). C_{60} sensitization of a photorefractive polymer. *Appl. Phys. Lett.*, **61**, 2967–9.

Silence, S. M., Donckers, M. C. J. M., Walsh, C. A., Burland, D. M., Twieg, R. J., and Moerner, W. E. (1993). Optical properties of poly(N-vinylcarbazole) based guest-host photorefractive polymer systems. *Appl. Opt.*, **33**, 2218–22.

Silence, S. M., Twieg, R. J., Bjorklund, G. C., and Moerner, W. E. (1994a). Quasinondestructive readout in a photorefractive polymer. *Phys. Rev. Lett.*, **73**, 2047–50.

Silence, S. M., Bjorklund, G. C., and Moerner, W. E. (1994b). Optical trap activation in a photorefractive polymer. *Opt. Lett.*, **19**, 1822–4.

Smirl, A. L., Valley, G. C., Mullen, R. A., Bohnert, K., Mire, C. D., and Boggess, T. F. (1987). Picosecond photorefractive effect in $BaTiO_3$. *Opt. Lett.*, **12**, 501–3.

Smirl, A. L., Valley, G. C., Bohnert, K. M., and Boggess, T. F. (1988). Picosecond photorefractive and free-carrier transient energy transfer in GaAs at $1\mu m$. *IEEE J. Quantum Electron.*, **24**, 289–303.

Smirl, A. L., Dubard, J., Gui, A. G., Boggess, T. F., and Valley, G. C. (1989a). Polarization-rotation switch using picosecond pulses in GaAs. *Opt. Lett.*, **14**, 242–4.

Smirl, A. L., Bohnert, K., Valley, G. C., Mullen, R. A., and Boggess, T. F. (1989b). Formation, decay, and erasure of photorefractive gratings in barium-titanate by picosecond pulses. *J. Opt. Soc. Am. B*, **6**, 606–15.

Smout, A. M. C. and Eason, R. W. (1987). Analysis of mutually incoherent beam coupling in $BaTiO_3$. *Opt. Lett.*, **12**, 498.

Sochava, S. L. and Stepanov, S. I. (1994). Optical excitation of Hall current in a GaAs crystal. *J. Appl. Phys.*, **75**, 2941–4.

Sochava, S. L., Troth, R. C., and Stepanov, S. I. (1992). Holographic interferometry using -1 order diffraction in photorefractive $Bi_{12}SiO_{20}$ and $Bi_{12}TiO_{20}$ crystals. *J. Opt. Soc. Am. B*, **9**, 1521–7.

Sochava, S., Buse, K., and Kratzig, E. (1993a). Non-steady-state photocurrent technique for the characterization of photorefractive $BaTiO_3$. *Opt. Commun.*, **98**, 265–8.

Sochava, S. L., Mokrushina, E. V., Prokof'ev, V. V., and Stepanov, S. I. (1993b). Experimental comparison of the a.c. field and the moving grating holographic recording technique for BSO and BTO photorefractive crystals. *J. Opt. Soc. Am. B*, **10**, 1600–4.

Soffer, B. H., Dunning, G. J., Owechko, Y., and Marom, E. (1986). Associative holographic memory with feedback using phase-conjugate mirrors. *Opt. Lett.*, **11**, 118–20.

Sokolov, I. A. and Stepanov, S. I. (1990). Non-steady-state photovoltage in crystals with long photoconductivity relaxation times. *Electron. Lett.*, **26**, 1275–7.
Sokolov, I. A. and Stepanov, S. I. (1993). Non-steady-state photoelectromotive force in crystals with long photocarrier lifetimes. *J. Opt. Soc. Am. B*, **10**, 1483–8.
Sokolov, I. A., Stepanov, S. I., and Trofimov, G. S. (1989). Unsteady photo-EMF under two-frequency nonlinear excitation. *Sov. Phys. Tech. Phys.*, **34**, 1165–7.
Solymar, L. (1987a). An equivalent circuit of spatial variations in photorefractive materials. *Opt. Commun.*, **63**, 413–14.
Solymar, L. (1987b). Theory of volume hologram formation in photorefractive crystals. In *Electro-optic and photorefractive materials*, Proceedings in Physics 18, (ed. P. Günter). Springer-Verlag, Berlin.
Solymar, L. and Ash, E. A. (1966). Some travelling wave interactions in semiconductors: theory and design considerations. *Int. J. Electron.*, **20**, 127–48.
Solymar, L. and Cooke, D. J. (1981). *Volume holography and volume gratings*. Academic Press, New York.
Solymar, L. and Heaton, J. M. (1984). Transient energy transfer in photorefractive materials: an analytical solution. *Opt. Commun.*, **51**, 76–8.
Solymar, L. and Riddy, D. G. D. (1990). Noise gratings for single and double exposures in silver halide emulsions. *J. Opt. Soc. Am. A*, **7**, 1554–61.
Solymar, L. and Ringhofer, K. H. (1988). Equivalent circuit representation of spatial frequency dependence for wave interactions in photorefractive materials. *Opt. Commun.*, **66**, 31–4.
Solymar, L. and Shamonin, M. S. (1992). Space charge waves in photorefractive materials. Technical Digest, OSA 1992 Annual Meeting, p. 1.
Solymar, L., Wilson, T., and Heaton, J. M. (1984). Space charge fields in photorefractive materials. *Int. J. Electron.*, **57**, 125–7.
Solymar, L., Webb, D. J., and Grunnet-Jepsen, A. (1994). Forward wave interactions in photorefractive materials. *Prog. Quantum Electron.*, **18**, 377–450.
Soutar, C., Cartwright, C. M., Gillespie, W. A., and Wang, Z. Q. (1991). Tracking novelty filter using transient enhancement of gratings in photorefractive BSO. *Opt. Commun.*, **86**, 255–9.
Soutar, C., Wang, Z. Q., Cartwright, C. M., and Gillespie, W. A. (1992). Real-time optical intensity correlator using photorefractive BSO and a liquid crystal television. *J. Mod. Opt.*, **39**, 761–9.
Stace, C., Powell, A. K., Walsh, K., and Hall, T. J. (1989). Coupling modulation in photorefractive materials by applying electric fields. *Opt. Commun.*, **70**, 509–14.
Staebler, D. L. (1977). Ferroelectric crystals. In *Holographic recording materials*, (ed. H. M. Smith), Ch. 4. Springer-Verlag, Berlin.
Staebler, D. L. and Amodei, J. J. (1972a). Coupled wave analysis of holographic storage in $LiNbO_3$. *J. Appl. Phys.*, **43**, 1042–9.
Staebler, D. L. and Amodei, J. J. (1972b). Thermally fixed holograms in $LiNbO_3$. *Ferroelectrics*, **3**, 107–13.
Staebler, D. L. and Phillips, W. (1974). Fe-doped $LiNbO_3$ for read-write applications. *Appl. Opt.*, **13**, 788–94.
Staebler, D. L., Burke, W. J., Phillips, W., and Amodei, J. J. (1975). Multiple storage and erasure of fixed holograms in Fe-doped $LiNbO_3$. *Appl. Phys. Lett.*, **26**, 182–4.
Stankus, J. J., Silence, S. M., Moerner, W. E., and Bjorklund, G. C. (1994). Electric-field switchable stratified volume holograms in photorefractive polymers. *Opt. Lett.*, **19**, 1480–2.
Steier, W. H., Kumar, J., and Ziari, M. (1988). Infrared power limiting and self-switching in CdTe. *Appl. Phys. Lett.*, **53**, 840–1.

Stepanov, S. I. (1982). Light refraction in crystals with bipolar photoconductivity. *Sov. Phys. Tech. Phys.*, **27**, 1300–1.

Stepanov, S. I. (1989). *Optical holography with recording in three dimensional media.* Nauka, Leningrad. (In Russian.)

Stepanov, S. I. and Petrov, M. P. (1984a). Photorefractive crystals of the $Bi_{12}SiO_{20}$ type for interferometry, wavefront conjugation and processing of non-stationary images. *Optica Acta*, **31**, 1335–43.

Stepanov, S. I. and Petrov, M. P. (1984b). Efficient phase conjugation in the photorefractive crystal $Bi_{12}TiO_{20}$. *Sov. Tech. Phys. Lett.*, **10**, 572–3.

Stepanov, S. I. and Petrov, M. P. (1985). Efficient unstationary holographic recording in photorefractive crystals under external alternating electric field. *Opt. Commun.*, **53**, 292–5.

Stepanov, S. I. and Sochava, S. L. (1987). Effective energy transfer in a two wave interaction in $Bi_{12}TiO_{20}$. *Sov. Phys. Tech. Phys.*, **32**, 1054–6.

Stepanov, S. I. and Trofimov, G. S. (1989). Transient emf in crystals having ambipolar photoconductivity. *Sov. Phys. Solid State*, **31**, 49–50.

Stepanov, S. I., Petrov, M. P., and Kamshilin, A. A. (1977). Diffraction of light with rotation of the plane of polarization in volume holograms in electro-optic crystals. *Sov. Tech. Phys. Lett.*, **3**, 345–6.

Stepanov, S. I., Kulikov, V. V., and Petrov, M. P. (1982). "Running" holograms in photorefractive $Bi_{12}SiO_{20}$ crystals. *Opt. Commun.*, **44**, 19–23.

Stepanov, S. I., Shandarov, S. M., and Khatkov, N. D. (1987). Photoelastic contribution to the photorefractive effect in cubic crystals. *Sov. Phys. Solid State*, **29**, 1754–6.

Stepanov, S. I., Sokolov, I. A., Trofimov, G. S., Vlad, V. I., Popa, D., and Apostol, I. (1990). Measuring vibration amplitudes in the picometer range using moving light gratings in photoconductive GaAs:Cr. *Opt. Lett.*, **15**, 1239–41.

Sternklar, S. and Fischer, B. (1987). Double-color-pumped photorefractive oscillator and image color conversion. *Opt. Lett.*, **12**, 711–13.

Sternklar, S., Weiss, S., Segev, M., and Fischer, B. (1986). Beam coupling and locking of lasers using photorefractive four-wave mixing. *Opt. Lett.*, **11**, 528–30.

Strait, J., Reed, J. D., and Kukhtarev, N. V. (1990). Orientational dependence of photorefractive two-beam coupling in InP:Fe. *Opt. Lett.*, **15**, 209–11.

Strohkendl, F. P. and Hellwarth, R. W. (1987). Contribution of holes to the photorefractive effect in n-type $Bi_{12}SiO_{20}$. *J. Appl. Phys.*, **62**, 2450–5.

Strohkendl, F. P., Jonathan, J.-M. C., and Hellwarth, R. W. (1986). Hole-electron competition in photorefractive gratings. *Opt. Lett.*, **11**, 312–14.

Strohkendl, F. P., Tayebati, P., and Hellwarth (1989). Comparative study of photorefractive $Bi_{12}SiO_{20}$ crystals. *J. Appl. Phys.*, **66**, 6024–9.

Sturman, B. I., Bledowski, A., Otten, J., and Ringhofer, K. H. (1992a). Spatial subharmonics in photorefractive crystals. *J. Opt. Soc. Am. B*, **9**, 672–81.

Sturman, B. I., Mann, M., and Ringhofer, K. H. (1992b). Instability of moving gratings in photorefractive crystals. *Appl. Phys. A*, **55**, 235–41.

Sturman, B. I., Mann, M., Otten, J., Ringhofer, K. H., and Bledowski, A. (1992c). Subharmonic generation in photorefractive crystals: application of theory to experiment. *Appl., Phys. A*, **55**, 55–60.

Sturman, B. I., Mann, M., and Ringhofer, K. H. (1992d). Instability of spatial grating induced by a.c. fields in photorefractive crystals. *Opt. Lett.*, **17**, 1620.

Sturman, B., Odoulov, S., Holtmann, L., and van Olfen, U. (1992e). Dynamics of parametric conical scattering of orthogonally polarized waves in $BaTiO_3$. *Appl. Phys. A*, **55**, 65–72.

Sturman, B. I., Mann, M., Otten, J., and Ringhofer, K. H. (1993a). Space-charge waves in photorefractive crystals and their parametric excitation. *J. Opt. Soc. Am. B*, **10**, 1919–32.

Sturman, B. I., Mann, M., and Ringhofer, K. H. (1993b). Instability of the resonance enhancement of moving photorefractive gratings. *Opt. Lett.*, **18**, 702–4.

Sturman, B., Goulkov, M., and Odoulov, S. (1993c). Polarization-degenerate parametric light scattering in photorefractive crystals. *Appl. Phys. B*, **56**, 193–9.

Sturman, B. I., Webb, D. J., Kowarschik, R., Shamonina, E., and Ringhofer, K. H. (1994). Exact solution of the Bragg-diffraction problem in sillenites. *J. Opt. Soc. Am. B*, ??

Sturman, B. I., McClelland, T. E., Webb, D. J., Shamonina, E., and Ringhofer, K. H. (1995a). Investigation of photorefractive subharmonics in the absence of wavemixing. *J. Opt. Soc. Am. B*, **12**, 1621–7.

Sturman, B. I., Shamonina, E., Mann, M., and Ringhofer, K. H. (1995b). Space charge waves in photorefractive ferroelectrics. *J. Opt. Soc. Am. B*, **12**, 1642–50.

Sugg, B., Kahmann, F., Rupp, R. A., Delaye, P., and Roosen, G. (1993). Diffraction and two-beam-coupling in GaAs along [111]-direction. *Opt. Commun.*, **102**, 6–12

Suris, R. A. and Fuks, B. I. (1975). Influence of the excitation of spatial trap-charging waves on the impedance of a compensated semiconductor. *Sov. Phys. Semicond.*, **9**, 1130–5.

Suzuki,T. and Sata, T. (1992). Novelty imaging system with a desired long-time scale using $BaTiO_3$ and a controlled shutter sequence. *Appl. Opt.*, **31**, 606–12.

Swinburne, G. A., Hall, T. J., and Powell, A. K. (1989). Large modulation effects in photorefractive crystals. In *2nd International Conference on Holographic Systems, Components and Applications*, p. 116.

Sylla, M., Rouède, D., Chevalier, R., Phu, X. N., and Rivoire, G. (1992). Picosecond nonlinear absorption and phase conjugation in BSO and BGO crystals. *Opt. Commun.*, **90**, 391–8.

Syms, R. R. A. (1990). *Practical volume holography*. Clarendon Press, Oxford.

Syms, R. R. A. and Solymar, L. (1983). Planar volume phase holograms formed in bleached photographic emulsions. *Appl. Opt.*, **22**, 1479–96.

Sze, S. M. (1985). *Semiconductor devices: physics and technology*. Wiley, New York.

Sze, S. M. and Irvin, J. C. (1968). Resistivity, mobility, and impurity levels in GaAs, Ge, and Si at 300 K. *Solid State Electron.*, **11**, 599.

Takacs, J. and Solymar, L. (1992). Subharmonics in $Bi_{12}SiO_{20}$ with an applied ac electric field. *Opt. Lett.*, **17**, 247–8.

Takacs, J., Schaub, M., and Solymar, L. (1992a). Subharmonics in photorefractive $Bi_{12}TiO_{20}$ crystal. *Opt. Commun.*, **91**, 252–4.

Takacs, J., Ellin, H. C., and Solymar, L. (1992b). Multiple forward phase conjugation in photorefractive bismuth silicate crystal. *Opt. Commun.*, **93**, 223–6.

Takahashi, H., Zaleta, D., Ma, J., Ford, J. E., Fainman, Y., and Lee, S. H. (1994). Packaged optical interconnection system based on photorefractive correlation. *Appl. Opt.*, **33**, 2991–7.

Taketomi, Y., Ford, J. E., Sasaki, H., Ma, J., Fainman, Y., and Lee, S. H. (1991). Incremental recording for photorefractive hologram multiplexing. *Opt. Lett.*, **16**, 1774–6.

Taketomi, Y., Ford, J. E., Sasaki, H., Fainman, Y., and Lee, S. H. (1992). Incremental recording of photorefractive hologram multiplexing: reply to comment. *Opt. Lett.*, **17**, 962.

Tamburrini, M., Bonavita, M., Wabnitz, S., and Santamato, E. (1993). Hexagonally patterned beam filamentation in a thin liquid-crystal film with a single feedback mirror. *Opt. Lett.*, **18**, 855–7.

Tanguay, A. R. (1977). The Czochralski growth and optical properties of bismuth silicon oxide. Unpublished Ph. D. thesis. Yale University.
Tanguay, A. R. and Johnson, R. V. (1986). Stratified volume holographic optical-elements. *J. Opt. Soc. Am. A*, **3**, P 53.
Tao, S., Selviah, D. R., and Midwinter, J. E. (1993). Spatioangular multiplexed storage of 750 holograms in an Fe:LiNbO$_3$ crystal. *Opt. Lett.*, **18**, 912–14.
Tayag, T., Batchman, T. E., and Sluss, J. J., Jr (1994). Electric-field dependence of the photocarrier hopping mobility in bismuth silicon oxide. *J. Appl. Phys.*, **76**, 967–73.
Tayebati, P. (1991). The effect of shallow traps on the dark storage of photorefractive gratings in Bi$_{12}$SiO$_{20}$. *J. Appl. Phys.*, **70**, 4082–94.
Tayebati, P. (1992). Effect of shallow traps on electron-hole competition in semi-insulating photorefractive materials. *J. Opt. Soc. Am. B*, **9**, 415–19.
Tayebati, P. and Mahgerefteh, D. (1991). Theory of photorefractive effect for Bi$_{12}$SiO$_{20}$ and BaTiO$_3$ with shallow traps. *J. Opt. Soc. Am. B*, **8**, 1053–64.
Temple, D. A. and Ward, C. (1986). Anisotropic scattering in photorefractive crystals. *J. Opt. Soc. Am. B*, **3**, 337–41.
Temple, D. A. and Ward, C. (1988). High order anisotropic diffraction in photorefractive crystals. *J. Opt. Soc. Am. B*, **5**, 1800–5.
Thaxter, J. B. (1969). Electrical control of holographic storage in strontium-barium-niobate. *Appl. Phys. Lett.*, **15**, 210–12.
Thaxter, J. B. and Kestigian, M. (1974). Unique properties of SBN and their use in a layered optical memory. *Appl. Opt.*, **13**, 913–24.
Tikhonchuk, V. T. and Zozulya, A. A. (1991). Structure of light-beams in self-pumped 4-wave-mixing geometries for phase conjugation and mutual conjugation. *Prog. Quantum Electron.*, 15.
Tomita, Y. and Ishii, H. (1994). Dynamics of photoexcited carriers and space-charge field by picosecond two-beam excitation in photorefractive semiconductors. *Jap. J. Appl. Phys.*, **33**, 1892–8.
Tomita, Y., Yahalom, R., and Yariv, A. (1988). Real-time image subtraction with the use of wave polarization and phase conjugation. *Appl. Phys. Lett.*, **52**, 425–7.
Townsend, R. L. and LaMacchia, J. T. (1970). Optically induced refractive index changes in BaTiO$_3$. *J. Appl. Phys.*, **41**, 5188–92.
Trofimov, G. S. and Stepanov, S. I. (1986). Time-dependent holographic currents in photorefractive crystals. *Sov. Phys. Solid State*, **28**, 1559–62.
Trofimov, G. S. and Stepanov, S. I. (1988). Steady-state holographic currents in Bi$_{12}$SiO$_{20}$. *Sov. Phys. Solid State*, **30**, 534–5.
Trofimov, G. S., Stepanov, S. I., Petrov, M. P., and Krasin'kova, M. V. (1987). Non-stationary photo-emf under the spatially heterogeneous surface excitation of GaAs–Cr. *Sov. Tech. Phys. Lett.*, **13**, 108.
Troth, R. C., Sochava, S. L., and Stepanov, S. I. (1991). Noise and sensitivity characteristics of Bi$_{12}$SiO$_{20}$ crystals for optimization of a real-time self-diffraction holographic interferometer. *Appl. Opt.*, **30**, 3756–61.
Tschudi, T., Herden, A., Goltz, J., Klumb, H., Laeri, F., and Albers, J. (1986). Image amplification by 2-wave and 4-wave mixing in BaTiO$_3$ photorefractive crystals. *IEEE J. Quantum Electron.*, **22**, 1493–502.
Turin, G. L. (1960). An introduction to matched filters. *IRE Trans. Inform. Theory*, **6**, 311–29.
Thuring, B., Neubecker, R., and Tschudi, T. (1993). Transverse pattern formation in liquid crystal light valve feedback system. *Opt. Commun.*, **102**, 111–15.

Turki, K. (1993). Contribution a l'ètude des performances et des limites photoréfractives de InP:Fe. Application à la commutation optique. Unpublished Ph. D thesis. Université Joseph Fourier, Grenoble.

Vachss, F. (1994). Frequency-dependent photorefractive response in the presence of applied d.c. electric fields. *J. Opt. Soc. Am. B*, **11**, 1045–8.

Vachss, F. and Hesselink, L. (1987a). Holographic beam coupling in anisotropic photorefractive media. *J. Opt. Soc. Am. A*, **4**, 325–39.

Vachss, F. and Hesselink, L. (1987b). Measurement of the electrogyratory and electro-optic effects in BSO and BGO. *Opt. Commun.*, **62**, 159–65.

Vachss, F. and Hesselink, L. (1988a). Selective enhancement of spatial harmonics of a photorefractive grating. *J. Opt. Soc. Am. B*, **5**, 1814–21.

Vachss, F. and Hesselink, L. (1988b). Nonlinear photorefractive response at high modulation depths. *J. Opt. Soc. Am. A*, **5**, 690–701.

Vachss, F. and Hesselink, L. (1988c). Synthesis of a holographic image velocity filter using the nonlinear photorefractive effect. *Appl. Opt.*, **27**, 2887–94.

Vachss, F. and Yeh, P. (1989). Image-degradation mechanisms in photorefractive amplifiers. *J. Opt. Soc. Am. B*, **6**, 1834–44.

Vachss, F., Hong, J., and Yeh, P. (1991). Photorefractive square law converter. *Opt. Lett.*, **16**, 1204–6.

Vachss, F., Hong, J., Campbell, S., and Yeh, P. (1992). Stable photorefractive square-law conversion using moving grating techniques. *Appl. Opt.*, **31**, 1783–6.

Vahey, D. W. (1975). A nonlinear coupled-wave theory of holographic storage in ferroelectric materials. *J. Appl. Phys.*, **46**, 3510–15.

Vainos, N. A. and Gower, M. C. (1991). High-fidelity image amplification and phase conjugation in photorefractive $Bi_{12}SiO_{20}$ crystals. *Opt. Lett.*, **16**, 363–5.

Vainos, N. A., Clapham, S. L., and Eason, R. W. (1989). Multiplexed permanent and real time holographic recording in photorefractive BSO. *Appl. Opt.*, **28**, 4381–5.

Valley, G. C. (1983a). Short-pulse grating formation in photorefractive materials. *IEEE J. Quantum Electron.*, **19**, 1637–45.

Valley, G. C. (1983b). Erase rates in photorefractive materials with two photoactive species. *Appl. Opt.*, **22**, 3160–4.

Valley, G. C. (1984). Two-wave mixing with an applied field and a moving grating. *J. Opt. Soc. Am. B*, **1**, 868–73.

Valley, G. C. (1986). Simultaneous electron/hole transport in photorefractive materials. *J. Appl. Phys.*, **59**, 3363–6.

Valley, G. C. (1987). Competition between forward- and backward-stimulated photorefractive scattering in $BaTiO_3$. *J. Opt. Soc. Am. B*, **4**, 14–19.

Valley, G. C. and Dunning, G. (1984). Observation of optical chaos in phase-conjugate resonator. *Opt. Lett.*, **9**, 513–15.

Valley, G. C. and Klein, M. B. (1983). Optical properties of photorefractive materials for optical data processing. *Opt. Eng.*, **22**, 704–11.

Valley, G. C., Klein, M. B., Mullen, R. A., Rytz, D., and Wechsler, B. (1988). Photorefractive crystals. *Ann. Rev. Mater. Sci.*, **18**, 165–88.

Valley, G. C., Boggess, T. F., Dubard, J., and Smirl, A. L. (1989). Picosecond pump-probe technique to measure deep-level, free-carrier, and two photon cross sections in GaAs. *J. Appl. Phys.*, **66**, 2407–13.

Valley, G. C., Dubard, J., and Smirl, A. L. (1990). Theory of high gain transient energy transfer in GaAs and Si. *IEEE J. Quantum Electron.*, **26**, 1058–66.

Valley, G. C., Segev, M., Crosignani, B., Yariv, A., Fejer, M., and Bashaw, M. C. (1995). Dark and bright photovoltaic spatial solitons. *Phys. Rev. A*, **50**, R4457–60.

Vander Lugt, A. B. (1964). Signal detection by complex spatial filtering. *IEEE Trans. Inform. Theory*, **10**, 139–??.
Van Heerden, P. J. (1963). Theory of optical information storage in solids. *Appl. Opt.*, **2**, 393–400.
Van Olfen, U., Hesse, H., Jakel, G., and Krätzig, E. (1992). Anisotropic grating recording in photorefractive $KNbO_3$:Fe. *Opt. Commun.*, **93**, 219–22.
Verdiell, J. M., Rajbenbach, H., and Huignard, J. P. (1990). Efficient diffraction-limited beam combining of semiconductor laser diode arrays using photorefractive $BaTiO_3$. *IEEE Photonics Technol. Lett.*, **2**, 568–70.
Vinetskii, V. L. and Itskovskii, M. A. (1978). Pyroelectric mechanism of holographic grating recording. *Ferroelectrics*, **18**, 81.
Vinetskii, V. L. and Kukhtarev, N. V. (1975). Theory of the conductivity induced by recording holographic gratings in nonmetallic crystals. *Sov. Phys. Solid State*, **16**, 2414–15.
Vinetskii, V. L. and Kukhtarev, N. V. (1976). Energy coupling between optical beams by dynamic hologram grating. *Sov. Tech. Phys. Lett.*, **2**, 364–5.
Vinetskii, V. L., Kukhtarev, N. V., Markov, V. B., Odoulov, S. G., and Soskin, M. S. (1977). Amplification of coherent light beams by dynamic holograms in ferroelectric crystals. *Bull. Acad. Sci. USSR, Phys. Ser.*, **41**, 135–43.
Voit, E. (1987). Anisotropic Bragg diffraction in photorefractive crystals. In *Electro-optic and photorefractive materials*. Springer Proceedings in Physics, Vol. 18, (ed. P. Günter), pp. 246–65. Springer-Verlag, Berlin.
Voit, E. and Günter, P. (1987). Photorefractive spatial light modulation by anisotropic self-diffraction in $KNbO_3$ crystals. *Opt. Lett.*, **12**, 769–71.
Voit, E., Zaldo, C., and Günter, P. (1986). Optically induced variable light deflection by anisotropic Bragg diffraction in photorefractive $KNbO_3$. *Opt. Lett.*, **11**, 309–11.
Volkov, V. I. et al. (1991). Influence of photoelasticity on self-diffraction of light in electro-optic crystals. *Kvant. Elektron.*, **18**, 1237–40.
Von der Linde, D. and Glass, A. M. (1975). Photorefractive effects for reversible holographic storage of information. *Appl. Phys.*, **8**, 85–100.
Von der Linde, D., Glass, A. M., and Rodgers, K. F. (1974). Multiphoton photorefractive processes for optical storage in $LiNbO_3$. *Appl. Phys. Lett.*, **25**, 155–7.
Von der Linde, D., Glass, A. M., and Rodgers, K. F. (1975). High-sensitivity optical recording in KTN by two-photon absorption. *Appl. Phys. Lett.*, **26**, 22–4.
Von der Linde, D., Glass, A. M., and Rodgers, K. F. (1976). Optical storage using refractive index changes induced by two-step excitation. *J. Appl. Phys.*, **47**, 217–20.
Von der Linde, D., Schirmer, O. F., and Kurz, H. (1978). Intrinsic photorefractive effect in $LiNbO_3$. *Appl. Phys.*, **15**, 153–6.
Vormann, H. and Krätzig, E. (1984). 2 step excitation in $LiTaO_3$-Fe for optical-data storage. *Solid State Commun.*, **49**, 843–7.
Vormann, H., Weber, G., Kapphan, S., and Krätzig, E. (1981). Hydrogen as origin of thermal fixing in $LiNbO_3$:Fe. *Solid State Commun.*, **40**, 543–5.
Voronov, V. V., Dorosh, I. R., Kuz'minov, Yu. S., and Tkachenko, N. V. (1980). Photoinduced light scattering in cerium-doped barium strontium niobate crystals. *Sov. J. Quantum Electron.*, **10**, 1346–9.
Wagner, K. and Psaltis, D. (1987). Multilayer optical learning networks. *Appl. Opt.*, **26**, 5061–76.
Walsh, C. A. and Moerner, W. E. (1992). Two-beam coupling measurements of grating phase in a photorefractive polymer. *J. Opt. Soc. Am. B*, **8**, 1642–7.
Walsh, K. and Hall, T. J. (1988). Photorefractive two-wave mixing in GaAs using a diode-pumped Nd:YLF laser at $1.31\mu m$. *Electron. Lett.*, **24**, 477–8.

Walsh, K., Hall, T. J., and Burge, R. E. (1987). Influence of polarization state and absorption gratings on photorefractive two-wave mixing in GaAs. *Opt. Lett.*, **12**, 1026–8.

Walsh, K., Powell, A. K., Stace, C., and Hall, T. J. (1990). Techniques for enhancement of space charge fields in photorefractive materials. *J. Opt. Soc. Am. B*, **7**, 288–303.

Wang, D., Zhang, Z., Zhu, Y., Zhang, S., and Ye, P. (1989). Observations on the coupling channel of two mutually incoherent beams without internal reflections in $BaTiO_3$. *Opt. Commun.*, **73**, 495–500.

Wang, J. et al. (1992). Photorefractive properties and self-pumped phase conjugation of tetragonal Fe-doped $KTa_{1-x}Nb_xO_3$ crystal. *Appl. Phys. Lett.*, **61**, 2761–3.

Wang, Q. N., Nolte, D. D., and Melloch, M. R. (1991). Spatial-harmonic gratings at high modulation depths in photorefractive quantum wells. *Opt. Lett.*, **16**, 1944–6.

Wang, Q. N., Brubaker, R. M., Nolte, D. D., and Melloch, M. R. (1992). Photorefractive quantum wells: transverse Franz-Keldysh geometry. *J. Opt. Soc. Am. B*, **9**, 1626–41.

Wang, Q. N., Brubaker, R. M., and Nolte, D. D. (1994a). Photorefractive phase shift induced by nonlinear electronic transport. *Opt. Lett.*, **19**, 822–4.

Wang, Q. N., Brubaker, R. M., and Nolte, D. D. (1994b). Photorefractive phase shift induced by hot-electron transport: multiple-quantum-well structures. *J. Opt. Soc. Am. B*, **11**, 1773–9.

Wang, Z. Q., Gillespie, W. A., and Cartwright, C. M. (1994). Holographic-recording improvement in a bismuth silicon oxide crystal by the moving-grating technique. *Appl. Opt.*, **33**, 7627–33.

Wardzynski, W., Lukasiewicz, T., and Zmija, J. (1979). Reversible photo-chromic effects in doped single crystals and bismuth germanium ($Bi_{12}GeO_{20}$) and bismuth silicon ($Bi_{12}SiO_{20}$) oxide. *Opt. Commun.*, **30**, 203–5.

Weaver, C. S. and Goodman, J. W. (1966). A technique for optically convolving two functions. *Appl. Opt.*, **5**, 1248–9.

Webb, D. J. and Solymar, L. (1990a). The amplification of an amplitude modulated signal beam via two-wave mixing in photorefractive media. *Digest of the topical meeting on photorefractive materials, effects and devices.* JP11-1. Optical Society of America, Washington, DC.

Webb, D. J. and Solymar, L. (1990b). Amplification of temporally modulated signal beams by two-wave mixing in $Bi_{12}SiO_{20}$. *J. Opt. Soc. Am. B*, **7**, 2369–73.

Webb, D. J. and Solymar, L. (1990c). Observation of spatial subharmonics arising during two-wave mixing in BSO. *Opt. Commun.*, **74**, 386–8.

Webb, D. J. and Solymar, L. (1991a). The effects of optical activity and absorption on two wave mixing in $Bi_{12}SiO_{20}$. *Opt. Commun.*, **83**, 287–94.

Webb, D. J. and Solymar, L. (1991b). Phase dependent forward 3-wave amplification in photorefractive BSO. *Electron. Lett* **27**, 889–90.

Webb, D. J. and Solymar, L. (1991c). Comparison of three wave mixing in photorefractive and nematic liquid crystals. In *IEE Conference Publication No. 342*, pp. 108–12. Institution of Electrical Engineers, London.

Webb, D. J., Au, L. B., Jones, D. C., and Solymar, L. (1990). Onset of subharmonics generated by forward wave interactions in $Bi_{12}SiO_{20}$. *Appl. Phys. Lett.*, **57**, 1602–4.

Webb, D. J., Kiessling, A., Sturman, B. I., Shamonina, E., and Ringhofer, K. H. (1994). Verification of the standard model of the photorefractive nonlinearity. *Opt. Commun.*, **108**, 31–6.

Wei, J., Guan, Q., Wang, J., Yue, X., Shao, Z., and Liu, Y. (1994). Self-pumped phase conjugation of KLTN crystal. *Opt. Commun.*, **107**, 129–32.

Weiss, S., Sternklar, S., and Fischer, B. (1987). Double phase-conjugate mirror: analysis, demonstration, and applications. *Opt. Lett.*, **12**, 114–16.

Weiss, S., Segev, M., Sternklar, S., and Fischer, B. (1988). Photorefractive dynamic optical interconnects. *Appl. Opt.*, **27**, 3422–8.

Wemple, S. H., di Domenico, M., and Camlibel, I. (1968). Relationship between linear and quadratic electro-optic coefficients in $LiNbO_3$ and oxygen-octahedro ferroelectrics based on direct measurement of spontaneous polarization. *Appl. Phys. Lett.*, **12**, 209–11.

White, J. O. and Yariv, A. (1980). Real-time image processing via four-wave mixing in a photorefractive medium. *Appl. Phys. Lett.*, **37**, 5–7.

White, J. O. and Yariv, A. (1982). Spatial information processing and distortion correction via four-wave mixing. *Opt. Eng.*, **21**, 224–30.

White, J. O., Cronin-Golomb, M., Fischer, B., and Yariv, A. (1982). Coherent oscillation by self-induced gratings in the photorefractive crystal $BaTiO_3$. *Appl. Phys. Lett.*, **40**, 450–2.

White, J. O., Kwong, S.-K., Cronin-Golomb, M., Fisher, B., and Yariv, A. (1989). Wave propagation in photorefractive media. In *Photorefractive materials and their applications II*, Topics in Applied Physics 62, (ed. P. Günter and J.-P. Huignard), pp. 119–150. Springer-Verlag, Heidelberg.

Wilde, J., McRuer, R., Hesselink, L., and Goodman, J. (1987). Dynamic holographic interconnections using photorefractive crystals. *Proc. Soc. Photo-opt. Instrum. Eng.*, **752**, 200–8.

Wilde, J. P., Hesselink, L., McCahon, S. W., Klein, M. B., Rytz, D., and Wechsler, B. A. (1990). Measurement of electro-optic and electrogyratory effects in $Bi_{12}TiO_{20}$. *J. Appl. Phys.*, **67**, 2245–52.

Wolffer, N. and Gravey, P. (1991). Two-wave mixing in photorefractive InP:Fe with an external alternative field. *Ann. Phys. (France)*, Colloque No. 1, Supplement au No. 1, Vol. 16, pp. 143–51.

Wolffer, N. and Gravey, P. (1994). High quality phase conjugation in a double phase conjugate mirror using InP:Fe at 1.3 μm. *Opt. Commun.*, **107**, 115–19.

Wu, S., Song, Q., Mayers, A., Gregory, D. A., and Yu, F. T. S. (1990). Reconfigurable interconnections using photorefractive holograms. *Appl. Opt.*, **29**, 1118–25.

Wu, W., Yang, C., Campbell, S., and Yeh, P. (1995). Photorefractive fuzzy-logic processor based on grating degeneracy. *Opt. Lett.*, **20**, 922–4.

Wunsch, D. C., II, Morris, D. J., McGann, R. L., and Caudell, T. P. (1993). Photorefractive adaptive resonance neural network. *Appl. Opt.*, **32**, 1399–1407.

Xie, C., Itoh, M., Kuroda, K., and Ogura, I. (1991). Vibration analysis using photorefractive two-wave mixing. *Opt. Commun.*, **82**, 544–8.

Xu, H., Yuan, Y., Yu, Y., Xu, K., and Xu, Y. (1990). Performance of real time associative memory using a photorefractive crystal and liquid crystal electrooptic switches. *Appl. Opt.*, **29**, 3375–9.

Xu, H., Yuan, Y., Xu, K., and Xu, Y. (1992). Real-time parallel optical logic operation using photorefractive two-wave mixing and fringe-shifting techniques. *Appl. Opt.*, **31**, 1769–73.

Xu, J., Zhang, G., Liu, S., Liu, J., and Men, L. (1994). Noise suppression for photorefractive image amplification in the $LiNbO_3$:Fe crystal sheet. *Appl. Phys. Lett.*, **64**, 2332–4.

Xu, K., Xu, H., Yuan, Y., Hong, J., and Xu, Y. (1990). Real time holographic optical storage. *Proc. Soc. Photo-opt. Instrum. Eng.*, **1078**, 331–5.

Yakimovich, A. P. (1980). Multilayer, three-dimensional holographic gratings. *Opt. Spectrosc.*, **49**, 85–8.

Yang, G., Siahmakoun, A., and Khorana, B. M. (1991). Tunable self-reference phase conjugate interferometer. *Appl. Opt.*, **30**, 2714–17.

Yao, X. S., Dominic, V., and Feinberg, J. (1990). Theory of beam coupling and pulse shaping of mode-locked laser-pulses in a photorefractive crystal. *J. Opt. Soc. Am. B*, **7**, 2347–55.
Yariv, A. (1978). Four wave nonlinear optical mixing as real time holography. *Opt. Commun.*, **25**, 23–5.
Yariv, A. (1991). *Optical electronics*, (4th edn). Saunders College Publishing, New York.
Yariv, A. and Kwong, S. K. (1986). Associative memories based on message-bearing optical modes in phase-conjugate resonators. *Opt. Lett.*, **11**, 186–8.
Yariv, A. and Pepper, D. M. (1977). Amplified reflection, phase conjugation, and oscillation in degenerate four wave mixing. *Opt. Lett.*, **1**, 16–18.
Yariv, A. and Yeh, P. (1984a). *Optical waves in crystals*, Section 4.9. Wiley, New York.
Yariv, A. and Yeh, P. (1984b). *Optical waves in crystals*, Section 3.4. Wiley, New York.
Yariv, A., Kwong, S. K. and Kyuma, K. (1986). Demonstration of an all-optical associative holographic memory. *Appl. Phys. Lett.*, **48**, 1114–16.
Yeh, P. (1987a). Fundamental limit of the speed of photorefractive effect and its impact on device applications and materials research. *Appl. Opt.*, **26**, 602–4.
Yeh, P. (1987b). Photorefractive two-beam coupling in cubic crystals. *J. Opt. Soc. Am. B*, **4**, 1382–6.
Yeh, P. (1988). Photorefractive two-beam coupling in cubic crystals. II. General case ($\phi \neq \pi/2$). *J. Opt. Soc. Am. B*, **5**, 1811–13.
Yeh, P. (1989). Photorefractive nonlinear optics and optical computing. *Opt. Eng.*, **28**, 328–43.
Yeh, P. (1993). *Introduction to photorefractive nonlinear optics*. Wiley, New York.
Yeh, P. and Gu, C. (ed.) (1993). *Photorefractive materials, effects and applications*. CR48. SPIE Optical Engineering Press, Bellingham, WA.
Yeh, P., Chiou, A. E. T., and Hong, J. (1988). Optical interconnection using photorefractive dynamic holograms. *Appl. Opt.*, **27**, 2093–6.
Yeh, P., Chiou, A. E., Hong, J., Beckwith, P., Chang, T., and Koshnevishan, M. (1989). Photorefractive nonlinear optics and optical computing. *Opt. Eng.*, **28**, 328–43.
Ye, P., Blouin, A., Demers, C., Roberge, M.-M. D., and Wu, X. (1991). Picosecond photoinduced absorption in photorefractive $BaTiO_3$. *Opt. Lett.*, **16**, 980–2.
Young, L., Wong, W. K. Y., Thewalt, M. L. W., and Cornish, W. D. (1974). Theory of formation of phase holograms in lithium niobate. *Appl. Phys. Lett.*, **24**, 264–5.
Yu, F. T. S. and Yin, S. (1991). Applications of photorefractive crystals to signal processing. *Int. J. Opt. Comput.*, **2**, 143–64.
Yu, F. T. S., Shundong, W., Rajan, S., and Gregory, D. A. (1992). Compact joint transform correlator with a thick photorefractive crystal. *Appl. Opt.*, **31**, 2416–18.
Yu, F. T. S., Yin, S., and Wang, C.-M. (1994). A content addressable polychromatic neural net using a (Ce:Fe)-doped $LiNbO_3$ photorefractive crystal. *Opt. Commun.*, **107**, 300–8.
Yue, X., Shao, Z., Lui, X., Song, Y., and Chen, H. (1989). *Opt. Commun.*, **89**, 59.
Zabusky, N. J. and Kruskal, M. D. (1965). Interaction of 'solitons' in a collisionless plasma and the recurrence of initial states. *Phys. Rev. Lett.*, **15**, 240–3.
Zel'dovich, B. Ya. (1981). The orientational mechanism of nonlinearity and self-focusing of He-Ne laser radiation in nematic liquid crystal mesophase (theory and experiment). *Opt. Commun.*, **37**, 280–4.
Zel'dovich, B. Ya. and Nestiorkin, O. P. (1991). Comparative analysis of time-dependent mechanisms for recording holograms in photorefractive crystals. *J. Moscow Phys. Soc.*, **1**, 231–42.
Zel'dovich, B. Ya. and Yakovleva, T. V. (1984). Theory of a two-layer hologram. *Sov. J. Quantum Electron.*, **14**, 323–8.

Zel'dovich, B. Ya., Popovichev, V. I., Raguel'sky, V. V., and Faizullov, F. S. (1972). Connection between the wavefronts of the reflected and exciting light in stimulated Mandel'shtam–Brillouin scattering. *Sov. Phys. JETP*, **15**, 109–13.
Zel'dovich, B. Ya., Mirovitskii, D. I., Rostovtseva, N. V., and Serov, O. B. (1984). Characteristics of two-layer phase holograms. *Sov. J. Quantum Electron.*, **14**, 364–9.
Zel'dovich, B. Ya., Pilipetsky, N. F., and Shkunov, V. V. (1985). *Principles of phase conjugation.* Springer-Verlag, Berlin.
Zelenskaya, T. E. and Shandarov, S. M. (1986). Photogeneration of acoustic waves on holographic grating in photorefractive crystals. *Dokl. Akad. Nauk SSSR*, **289**, 600–3.
Zernicke, F. (1935). Das Phasenkontrastverfahren bei der mikroskopischen Beobachtung. *Z. Tech. Phys.*, **16**, 454.
Zgonik, M., Biaggio, I., Amrhein, P., and Günter, P. (1990). Time resolved investigation of transient photorefractive gratings in $KnBO_3$. *Ferroelectrics*, **107**, 15–20.
Zgonik, M., Biaggio, I., Bertele, U., and Günter, P. (1991). Degenerate four-wave mixing in $KNbO_3$: picosecond and photorefractive nanosecond response. *Opt. Lett.*, **16**, 977–9.
Zgonik, M., Schlesser, R., Biaggio, I., Voit, E., Tscherry, J., and Günter, P. (1993). Materials constants of $KNbO_3$ relevant for electro- and acousto-optics. *J. Appl. Phys.*, **74**, 1287–97.
Zgonik, M. *et al.* (1994). Dielectric, elastic, piezoelectric, electro-optic, and elasto-optic tensors of $BaTiO_3$ crystals. *Phys. Rev. B*, **50**, 5941–9.
Zhang, H. J., Tang, Z. S., Moore, T., and Boyd, R. W. (1993). High-order diffraction in photorefractive SBN:Ce due to non-sinusoidal gratings formed by beams of comparable intensity. *Int. J. Nonlinear Opt. Phys.*, **2**, 221–7.
Zhang, H. Y., He, X. H., Chen, E., and Liu, Y. (1990). High-reflectivity self-pumped phase conjugator using total internal reflection in $KNbO_3$:Fe. *Appl. Phys. Lett.*, **57**, 1298–300.
Zhang, Y., Cui, Y., and Prasad, P. N. (1992). Observation of photorefractivity in a fullerene doped polymer composite. *Phys. Rev. B*, **46**, 9900–2.
Zhang, Z., Hu, G., Wu, X., Zhang, S., and Ye, P. (1988). Isotropic conical scattering in $BaTiO_3$. *Opt. Commun.*, **69**, 66–70.
Zhdanova, N. G., Kagan, M. S., Suris, R. A., and Fuks, B. I. (1978). Trap exchange waves in compensated germanium. *Sov. Phys. JETP*, **47**, 189–92.
Zheng, Y., Sasaki, A., Gao, X., and Aoyama, H. (1995). Origin and elimination of dynamic instability in a self-pumped phase-conjugate mirror. *Opt. Lett.*, **20**, 267–9.
Zhivkova, S. (1992). Quasinondestructive readout of holograms stored in photorefractive sillenites. *J. Appl. Phys.*, **71**, 581–5.
Zhivkova, S. and Miteva, M. (1990). Holographic recording in photorefractive crystals with simultaneous electron-hole transport and two active centers. *J. Appl. Phys.*, **68**, 3099–103.
Zhivkova, S. and Miteva, M. (1991). Investigations of the characteristics of fixed holograms in $Ba_{12}TiO_{20}$ photorefractive crystals. *Opt. Commun.*, **86**, 449–53.
Zhou, G. and Anderson, D. Z. (1993). Photorefractive delay line for the visualization and processing of time-dependent signals. *Opt. Lett.*, **18**, 167–9.
Zhou, S., He, Q. B., and Yeh, P. (1993). Spatial fidelity in photorefractive image amplification. *Opt. Commun.*, **99**, 18–24.
Ziari, M. and Steier, W. H. (1993). Optical switching in cadmium telluride using a light-induced electrode non-linearity. *Appl. Opt.*, **32**, 5711–23.
Ziari, M., Steier, W. H., Ranon, P. M., Klein, M. B., and Trivedi, S. (1992). Enhancement of photorefractive gain at 1.3–1.5μm in CdTe using alternating electric fields. *J. Opt. Soc. Am. B*, **9**, 1461–6.

Zielinger, J. P., Tapiero, M., Guellil, Z., Roosen, G., Delaye, P., Launay, J. C., and Mazoyer, V. (1993). Optical, photoelectrical, deep level and photorefractive characterization of CdTe:V. *Mater. Sci. Eng. B*, **16**, 273–8.

Zozulya, A. A. (1991). Double phase-conjugate mirror is not an oscillator. *Opt. Lett.*, **16**, 545–7.

Zozulya, A. A. and Tikhonchuk, V. T. (1989). Investigation of stability of four-wave mixing in photorefractive media. *Phys. Lett. A*, **135**, 447–51.

Zozulya, A. A., Saffman, M., and Anderson, D. Z. (1994). Propagation of light-beams in photorefractive media: fanning, self-bending, and formation of self-pumped four-wave mixing phase-conjugation geometries. *Phys. Rev. Lett.*, **73**, 818–21.

Zurita, G. R., Kowarschik, R., and Erdmann, A. (1991). Heterodyne interferometer with photorefractive BGO crystal. *J. Mod. Opt.*, **38**, 2203–19.

INDEX

aberration correction 255
absorption
  constant 22, 190
  grating 212
adaptive interferometry 403–5
AlGaAs 147
amplification (AC signals) 90
angular multiplexing 386–90, 398
anisotropic
  diffraction 218–223
  media 409–15
  scattering 274–8
  self-diffraction 220
applied AC field 52, 58, 144, 161–169, 205, 206, 215, 234, 235, 285, 289, 290, 292, 378, 433, 438
associative memory 365–71

$Ba_2NaNb_5O_{15}$ 72–4
band edge photorefractivity 305–6
band transport model 16, 107
$BaTiO_3$
  anisotropic self-diffraction 220
  anisotropic scattering 273–4
  associative memory 367
  beam fanning 321–2, 338
  doping, 122
  electrical fixing 380
  electro-optic coefficient 418, 419, 422, 423
  electro-optic response 440
  electron-hole competition 63, 127–9
  figures of merit 43–7
  gain 198–200
  holes 120
  image amplification 319
  intensity dependence 143, 173–7
  microwave amplification 232
  noise reduction 280
  novelty filter 332–3
  parameters 31, 45, 172, 419
  passive phase conjugate mirror 263
  phase angle 205
  phase encoding 393
  photoinduced absorption 137
  pulsed illumination 170, 181
  pyroelectric effect 180
  rise time 36–8
  scattering 272–3
  self-organizing circuit 406
  self-phase conjugation 96, 97, 296
  space charge field 33–36
  spatial light modulation 338
  thermo-optic effect 180
beam clean-up 406
beam coupling 68–72
beam deflection 406–7
beam distortion 269
beam fanning 320–2, 332–8
beam ratio 73, 78, 81
BGO 109, 122, 123, 141, 164–7, 182, 217, 282, 285, 286, 291, 339, 378, 418, 419, 421
$Bi_{12}GeO_{20}$, see BGO
bipolar transport 127–35
birefringence 275, 369, 410
$Bi_{12}SiO_{20}$, see BSO
bistable flip-flop 405
$Bi_4Ti_3O_{12}$ 105
$Bi_{12}TiO_{20}$, see BTO
Bragg condition 8, 12, 20, 81, 85
Bragg degeneracy 388, 389, 402
BSKNN 273
BSO
  anisotropic diffraction 222
  applied AC field 205
  coasting 139
  correlator 345, 359
  defects 122
  diffusion length 430
  double exposure interferometry 339
  electro-optic coefficient 418, 419, 437
  external current 169
  forward phase conjugation 252
  gain 72, 91, 198, 211–12, 325, 327
  higher diffraction resonance 160–1
  high frequency resonance 160–1
  intensity modulation 91
  lifetime 18, 172
  losses 196
  mobility–lifetime product 31
  noise reduction 280
  nonlinear absorption 182
  optical activity 219
  parameters 31
  piezoelectricity 227–30, 439
  phase dependent amplification 244
  pulsed illumination 170, 431, 432
  rise time 48
  scattering 270
  space charge field 50
  space charge wave 154–5
  spatial light modulation 337
  state of polarization 224
  subharmonics 282
  thermal fixing 376

BSO (cont'd)
  three-wave mixing 242
  trapping 122, 132
BTO 59, 132, 169, 205, 206, 260, 282, 285, 286, 310, 349, 378, 418, 419, 421

'cat' phase conjugator 263
CdTe 122, 144, 206, 207, 283, 285, 286, 293, 418, 419, 428, 429
CdMnTe 212–3
chaos 287, 295, 296
characteristic
  fields 27, 41
  lengths 41–3
  times 43
chromophore 303, 304
circular photovaltaic current 299, 301
coasting 139
Co doping 122
complementary grating 132, 134, 136, 140, 145
compression 322
conduction current 23
conical scattering, 1 beam 275–6
conical scattering, 2 beam 277–8
content addressable memory 365–71
continuity equation 23
contrast reduction 324
convolution 342–4
coupling
  inter-mode 220–1, 274
  intra-mode 219–20, 222
  phase 73–4, 404
correlation 342–4
correlator 345–9, 353–5, 407
coupled wave equation
  vectorial 192
  scalar 70, 77, 195–6
  paraxial 193, 195
Cr doping 122
cross-polarization coupling 199
current 23, 60–62
Curie temperature 379
cutoff wavelength 122

damped Mathieu equation 286
dark conductivity 29
data page 388
Debye screening length 34
Dember effect 181, 298
detuning 49–52, 57–60
deep traps 123
dielectric constant 7
dielectric relaxation time 27, 29, 33, 34, 37
  holes 65
diffraction efficiency 8, 11, 14, 80–84, 215–26
diffusion 16, 18
  constant 30

current 23
field 27
length 30
regime 30, 34, 41, 42, 48, 56
time 43, 171
displacement current 23
domains 290
donors and acceptors 117
doping 122
double exposure interferometry 337
double phase conjugate mirror 97–98, 264–7, 403
drift 16, 18
drift length 41
drift time 43
dynamic range 324

edge enhancement 329–32
efficiency
  diffraction 8, 11, 14, 80–4, 215–26
  Horner 362
  quantum 22
elastic constant 40
electrical fixing 378–81
electron density 23
electron lifetime 18
electron–hole competition 62–65, 129, 132, 428, 429
electro-optic coefficient 17, 43, 67, 76, 193, 197, 416–26, 436–40
electro-optic effect 18, 66
electro-optic response 440
electro-optic tensor 416
electrorefraction 306
electrostriction 180
EL2 defect 122
E_M 27, 65
enhancement 48–55, 76, 110
E_q 27, 28
Ewald diagram 14, 219, 245, 249, 271, 272, 281, 388, 391
excitons 147, 307
explosive instability 279
external current 23, 60–2, 169
extraordinary wave 411

fanning 320–2
figures of merit 43–7
filtering
  amplitude-only 362–4
  matched 351, 352, 362–4
  phase-only 363–4
  spatial 349
fixing 121, 144, 374–81
  electrical 378
  refreshed memories 384–5
  storage amplification 385
  thermal 373–8

Index

two photon 381–2
two wavelength 382–3
forward four-wave mixing 247–51
forward phase conjugation 251–3
forward phase conjugate imaging 252–3
forward three-wave mixing 236–47
Franz–Keldysh effect 305–7
free-carrier grating 180
fringe bending 75, 81

GaAs
  applied AC field 164, 203, 205
  band-edge photorefractivity 306
  correlator 349
  Cr doping 122
  detuning 203
  electro-optic coefficient 45, 418, 419, 421
  EL2 defect 122
  external current 169
  four-wave mixing 260
  free-carrier grating 181
  gain 203, 212
  hot electron effect 147
  modulation transfer function 325
  resonator 93
  parameters 45
  piezoelectricity 180
  pulsed illumination 171
  quantum wells 307
  subharmonics 282
GaAs/AlGaAs 147, 307–8
GaP 203
gain 71
generation rate 22
glass transition temperature 303, 304
grating
  decay 28, 29, 373
  diffraction 7, 8
  dynamic 7
  free-carrier 180
  ionic 374
  reflection 15, 76–80
  spacing 7
  switched polarization 380, 381
  static 7–16
  thick 7, 8, 10, 85
  thin 7, 10, 85
  vector 8
  volume 7, 8, 10

high frequency resonance 160–1
high modulation 55–60, 110, 157
higher diffraction orders 10, 69, 84–7
higher order beam 245–6
holes 62–65, 110, 117, 127–30
hologram fixing
  electrical 378

refreshed memories 384–5
storage amplification 385
thermal 373–8
two photon 381–2
two wavelength 382–3
hologram multiplexing
  spatial 386
  angular 386–90
  wavelength 390–1
  phase encoding 391–4
holographic
  imaging 407
  interferometry 339–41
  lens 407
Horner efficiency 362
hot electron effect 147, 308
hot electron transport 147

image amplification 111, 319–28
image processing 328–41
impermeability tensor 190
incoherent erasure 280
incoherent-to-coherent conversion 336–8
incremental recording 396–9
induced photocurrent 434
InP:Fe 109, 136, 141–3, 149, 169, 264, 282, 285, 293, 306, 418, 421
instability
  explosive 279
  space charge wave 152
  spatio-temporal 293–6
  subharmonic 164, 281–293
intensity dependent gain 138
intensity modulation 91
intensity resonance 141, 159, 306
intermode coupling 220, 221, 274
interconnects 400–3
interference filter 400, 401
interferometry 339–41, 403–5, 406
intramode coupling 219–20, 222
ionic grating 374
ionized acceptors 16, 117
ionized donors 16, 117
ions 121, 144–5

joint transform correlator 352
Jones vector 194
journals 113–116

KLTN 264
$KNbO_3$ 127, 129, 148, 170, 172, 174–8, 180, 181, 220, 221, 264, 275, 294, 295, 300, 338, 418, 440
Kramers–Kronig relations 306–307

Index

large modulation 55–60, 202–3, 206
lifetime 18, 172
LiNbO$_3$
  angular multiplexing 389
  anisotropic self-diffraction 220
  correlator 369
  diffraction efficiency 216–221
  doping 122
  double exposure interferometry 339
  electro-optic coefficient 418, 419, 421
  electron-hole competition 63, 127
  image amplification 111, 319, 322
  intermode diffraction 221
  interference filter 400
  lifetime 18
  parameters 45, 172, 395, 413, 419
  photoinduced absorption 174
  photovoltaic effect 38, 41, 297, 298
  piezoelectricity 226
  pulsed illumination 170
  reflection grating 400
  scattering 271, 274, 275, 277, 278, 281
  solitons 310
  storage capacity 399
  thermal fixing 374–8
  two-photon process 145–6, 381
  wavelength multiplexing 390
linearization 24–28
LiTaO$_3$ 105, 145, 274, 275, 372, 418, 419, 421
local response 245
logic operation 406
longitudinal geometry 193–4, 197
low frequency resonance 159

matched filter 351–3, 362–4
Matthieu equation, damped 286–7
microwave amplification 232
mobility 23, 24, 427–35
mobility–lifetime product 31
model
  bipolar two species 133
  monopolar one species 128
  monopolar shallow and deep trap 137
  monopolar two species 137
modulation
  interference pattern 27
  transfer function 325–8
moving grating 50, 282, 285, 288, 306, 325, 328, 338, 348, 433, 436
multiple quantum well 147
multiplexing
  angular 386–90
  spatial 386
  wavelength 390–1, 403, 407

noise grating 270–2
noise reduction 279–81

nonlinear absorption 182
nonlinear optical chromophore 303
nonlinear susceptibility 318
novelty filters 332–6

off-Bragg diffraction 245, 247
off-Bragg parameter 12, 14–15, 85, 236, 240, 245, 248, 250
optical activity 190, 208–11, 213, 215, 219, 224, 225, 260, 270, 414–15
optical axis 410
optical damage 105, 269
ordinary wave 411

passive phase conjugate mirror 260–64
permittivity tensor 189, 190
phase conjugate mirror 93, 260–4
phase conjugation 93–8, 109, 251–67, 327, 328
phase conjugator
  bird wing 265
  bridge 265
  fish head 265
  frog-legs 265
  self-pumped 260–4, 295
  mutually pumped 264–7
phase coupling 73–4, 404
phase-dependent amplification 244
phase encoding 391–4
photoconductivity 18
photodiffractive effect 307
photoelasticity 180, 226, 229, 424–6, 440–3
photoelectric effect 226–7, 229
photoexcitation 121
photoinduced absorption 137, 140
photoinduced transparency 137
photoionization 22, 215, 355
photon-assisted tunnelling 305, 306
photorefractive fibre 390
photorefractive polymers 150, 302–5
photovoltaic
  effect 38–41, 124, 150, 277, 278, 298–302
  field 40
  regime 41, 42
  tensor 299
  transport length 41
piezoelectric effect 180, 187, 226–30, 424–6, 440–3
piezoelectric resonance 43
Poisson's equation 24, 28, 47, 107, 125, 137, 138, 144, 153
polarization eigenstates 219
polarization optic coefficient 44
polymers, photorefractive 302–5
power conservation 72
power transfer 74–6, 85, 86
pulsed illumination 170, 431, 432
pump depletion 322–5

pyroelectric effect 180

quadratic electro-optic effect 305
quantum efficiency 22
quantum wells 146–8, 306–9

recombination 18, 19, 22, 117
recombination time 22
reflection grating 15, 76–80, 400
refreshed memory 384–5
reconfigurable interconnection 400–3
recording schedules
   incremental 397–9
   scheduled 394–9
   sequential 394–7
resonance 50, 143, 159–61
resonator 92–93, 237, 238, 279, 405, 406, 407
rise time 36–8
rotation invariance 359–62

saturation field 28
SBN 104, 105, 174, 175, 180, 198, 204, 205, 270, 310, 317, 321, 325, 338, 360, 379, 380, 381, 382, 418, 419
scale invariance 359–62
scattered light 320–2
scattering 92–93, 269–81
   anisotropic 274–8
scattering and beam fanning 320–2
selective erasure 394
self-organizing optical circuits 405–6
self-phase conjugation 96–97, 296
self-trapping 310
sensitivity 44
sequential recording, 396–9
shallow traps 123, 135–44, 173–7, 431
shift invariance 359–62
short-time limit 30–34, 42
silicon 122, 181
slowly varying envelope approximation 10, 69, 192
solitons 89, 90, 309–11
space charge 18
space charge field 19, 27–38
space charge field enhancement 48–55, 57
space charge wave 52, 92, 151–7, 165, 167
   dispersion 153–6
   subharmonic 288
spatial
   filtering 349–51
   frequency 7
   light modulation 336–8
   multiplexing 386

period, *see* grating spacing
spatio-temporal instabilities 293–6
spurious beams 268
state of polarization 224–5
stationary solution 34–6, 42
statistics 113–6
stiffness 424, 440
storage 111, 372–99
storage capacity 387–9, 399
strain tensor 424
stratified holographic optical elements 309
stress tensor 424
subharmonics 24, 111, 164, 281–3
   competitions effects 291–3
   domains 290–1
superposition 13
switched polarization grating 380, 381

temporal modulation 89–91
thermal excitation 22, 29, 121, 123, 124, 126, 135–143, 373
thermal fixing 144–5
thermoelasticity 180
thermo-optic effect 180
three wave mixing 236–247
thresholding 328–9
time-of-flight method 432–3
time reversal 254
total current 23
transients 38, 86–9, 126, 230–5
transverse geometry 193–4
trap exchange wave, *see* space charge waves
trap intercommunication 149
trapless photorefractivity 148–9
traps
   deep 123
   shallow 122, 123, 135–43, 174–5
two photon processes 121, 145–6, 381–2
two-wavelength storage 382–3
two wave mixing 17, 195–7

undepleted pump approximation 198
uniaxial crystal 410

V doping 122
Vander Lugt filter 352

wave equation 8, 67
wavelength multiplexing 390–1, 403, 407
weighted correlation 356–9